17-11-76

Physical Geology

Physical Geology

JOHN E. SANDERS
BARNARD COLLEGE

ALAN H. ANDERSON JR.

ROBERT CAROLA

HARPER'S COLLEGE PRESS

A DEPARTMENT OF HARPER & ROW PUBLISHERS
NEW YORK HAGERSTOWN SAN FRANCISCO LONDON

Copyright © 1976 by Harper & Row, Publishers, Inc.
Pangaea flip drawings, pages 457-535, copyright © 1976 by Robert Carola.

All rights reserved. Printed in the United States of America.
No part of this book may be used or reproduced in any manner whatsoever without written permission except in the case of brief quotations embodied in critical articles and reviews. For information address Harper & Row, Publishers, Inc., 10 East 53 Street, New York, N.Y. 10022

Library of Congress Cataloging in Publication Data

Sanders, John Essington, 1926-
   Physical geology.

   Includes bibliographies.
   1. Physical geology.   I. Anderson, Alan H., Jr., joint author.   II. Carola, Robert, joint author.
QE28.2.S26   550   75-26831
ISBN 0-06-163403-4

A Robert Carola Book

Photo Editor   Rhoda Galyn
Designer   Robert Carola
Compositor   The Clarinda Company
Printer and Binder   Halliday Lithograph Corporation

The illustrations were designed by V. Lorenzo Porcelli and drawn by Vantage Art, Inc.
The handlettering for the cover and title pages was done by Tom Carnase.

The cover photograph of Earth from space was provided by the National Aeronautics and Space Administration (NASA).

The quotation on the back cover originally appeared in *The New York Times*.

Preface

The object of a college course in physical geology should be to enable the students to acquire an appreciation of science by showing them how to look at the world through new eyes, the eyes of a geologist. In this way, students inevitably gain a greater understanding and appreciation of how the remarkable Planet Earth works.

In this book we have tried to make physical geology as understandable and enjoyable to the student as possible. We have included the most up-to-date information wherever feasible (in Chapter 13 for example, Earthquakes and Seismology), but have also organized the chapters and sections in a rather traditional fashion, one section and chapter logically leading to the next. The Introduction has been designed to give the student a comprehensive view of geology, but like the rest of the book, is nontechnical and related to human interests.

Several pedagogical features have been employed in an effort to help the teacher and student, among them the several special pictorial features such as the simple explanation of forms of energy on pages 60-61, and other double-page spreads illustrating kinds of mountains, plate movements, downslope movements, and so forth. Also, we have included Field Trips, which were chosen to give the student a first-hand look at geology; the sources of the Field Trips vary from Thoreau to the San Francisco Call-Chronicle-Examiner the day after the 1906 earthquake.

The Pangaea flip drawings on pages 457-535 were designed to present the idea of drifting continents as vividly as possible. To achieve the proper effect the reader should flip the corners of the pages with the right hand, from front to back.

Acknowledgments
The authors wish to give thanks to Paul Panes (Queensborough Community College) for his suggestions toward adopting a suitable reading level; to John Conron (Middlebury College) for his help in selecting

Preface

Field Trips—Professor Conron's book, *The American Landscape: A Critical Anthology of Prose and Poetry* (Oxford University Press, 1974) is recommended for those who care about our landscape; to Mark R. Chartrand III (Chairman, The American Museum—Hayden Planetarium) for assisting in the preparation of Appendix A, "The Origin of the Earth"; to Claire Thompson, Kendra Crossen, and Leslie Carola for invaluable editorial assistance; to Bruce Caplan, G.M. Friedman, Robert LaFleur, B. Charlotte Schreiber, R. J. Cheng, J. C. Crowell, Charles Haberman, Joanne Bourgeois, and American Airlines for allowing us to use their photographs; to the staff of the Picture Collection of the New York Public Library, main branch, for their generous and expert assistance in acquiring pictorial material; to Alexis N. Moiseyev (California State University, Hayward), Jack H. Hyde (Tacoma Community College), Walter S. Skinner (Duquesne University), Tom Walker (Olympic College), Richard G. Shorc (Santa Rosa Junior College), and Robert E. Boyer (The University of Texas at Austin) for reviewing the complete outline and portions of the manuscript; to Howard R. Level (Ventura College), W. Hilton Johnson (University of Illinois), and Alex R. Ross (Oklahoma State University) for reviewing the entire outline and manuscript, sometimes in more than one draft—as is the custom, the authors accept full responsibility for the final text; to the staff of Harper's College Press, especially Janice Kuta, Marketing Manager, and our editors, Caroline Eastman and Raleigh Wilson.

The authors are grateful to Archibald MacLeish and the American astronauts whose unique vision helped us to see the Earth with new eyes also.

Contents

Introduction
The Study of Geology

The Nature of Geology 4. The Beginnings of Geology: Catastrophism and Uniformitarianism, 5. The Structure of the Earth, 7. **Rocks and the Earth's Crust** 7. The Three Great Groups of Rocks, 9. *Igneous rocks. Sedimentary rocks. Metamorphic rocks.* The Rock Cycle, 10. *An idealized rock cycle: magma to magma.* **The Concept of Geologic Time** 11. Geologic Change, 12. The Geologic Time Scale. 12. Sediments and Time, 12. Radioactivity and Time, 15. **Recent Trends** 15. Magnetism, 15. Oceanography, 15. *The mid-oceanic ridge. Island arcs and trenches.* Seismology, 17. Plate Tectonics, 17. *Sea-floor spreading and continental drift. Unanswered questions.* **Geology and You** 20. Concerns for the Future, 21. Our View of Geology Today, 23. **Chapter Review** 24. Questions 25. Suggested Readings 25.

Chapter One
Riders on the Earth Together

The Geologic Cycle As It Affects Life on Earth 28. The Lithosphere, 29. The Hydrosphere, 29. The Atmosphere, 30. The Biosphere, 31. **Our Dependence on the Geologic Cycle** 33. The History of Our Dependence on Geologic Processes, 33. *The fossil-fuel revolution.* **Human Beings As a Geologic Force** 36. Our Impact on the Lithosphere, 37. *Mining. Erosion.* Our Impact on the Hydrosphere, 42. Our Impact on the Atmosphere, 43. *Jet-age pollution. Dust. Fluorocarbon gases ("Freon").* Our Impact on the Biosphere, 45. **Resources and Human Populations** 45. The Great "Escape from Nature," 46. **Physical Geology in Perspective** 47. **Chapter Review** 48. Questions 49. Suggested Readings 49.

Field Trip Man and Nature *George P. Marsh* 40.

Contents

Chapter Two
Matter, Energy, and Minerals

Matter 52. Atomic Structure of Matter, 52. *Protons, electrons, and neutrons. Electric charge. Atomic size, mass, number. Elements. Isotopes. Radioactivity. Molecules and compounds. States of matter. Elements in the Earth's crust.* Ions and Bonding, 57. *Energy-level shells, Ionic bonding. Covalent bonding. Metallic bonding.* **Energy and Forces of the Earth** 59. Kinds of Energy, 60. *Nuclear energy. Radiant energy. Heat energy. Chemical energy. Electrical energy.* Mass and Energy, 62. Earth Forces, 63. *Gravity. Magnetism.* **Minerals of the Earth's Crust** 67. Composition and Structure of Minerals, 67. *Composition of minerals. Structure of minerals.* Properties of Minerals, 70. *Color. Streak. Luster. Crystal form. Cleavage. Hardness. Specific gravity. Other properties of minerals.* Noncrystalline Solids, 74. **Chemical Groups of Minerals** 74. Silicate Minerals, 74. *Kinds of silicates. Silicate structure. Variations in silicate structure.* Oxide Minerals, 75. Sulfide Minerals, 77. Carbonate and Sulfate Minerals, 77. Minor Groups of Minerals, 78. Why Study Minerals?, 78. **Chapter Review** 79. Questions 80. Suggested Readings 81.

Field Trip Mount Katahdin *Henry David Thoreau* 64.

Chapter Three
Igneous Activity and Volcanoes

Why Study Igneous Activity? 84. Igneous Activity and the Geologic Cycle, 84. *Igneous activity is continuous.* The Causes of Igneous Activity, 84. *Heat, pressure, and generation of magma.* **Volcanoes** 86. Products of Volcanoes, 86. *Liquids. Solid-gaseous mixtures. Solids. Gases.* Volcanic Landforms, 90. *Domes. Cones. Craters and calderas. Volcanic shields. Lava plains. Composite cones.* Kinds of Eruptions, 97. *Effusive eruptions. Explosive eruptions. Fumarolic eruptions.* Distribution of Volcanoes, 99. *Geographic distribution. Geologic distribution.* Some Classic Volcanic Eruptions, 102. *Vesuvius. Krakatoa. Mt. Pelée. Other volcano types. Volcanism in the continental United States.* **Underwater Volcanoes** 109. Differences from Subaerial Volcanoes, 112. *Hydrostatic pressure. Vertical movement and rebuilding.* New and Disappearing Islands, 112. *The formation of volcanic islands. Surtsey.* Underwater Eruptions and Lava, 113. *Pillow structure. Broken-glass rocks. Subglacial eruptions.* **Fumaroles, Hot Springs, Geysers** 116. Causes of Underground Activity, 117. *Causes of underground activity. Locations of underground activity. Old Faithful and Big Geysir.* **Igneous Activity and Seismic Activity** 119. Seismic Activity and Volcanoes, 119. *Examples of seismic activity.* **Prediction and Use of Igneous Activity** 120. Forecasting Igneous Activity, 121. *Seismic, magnetic, heat indicators. Tiltmeters and satellites. Center for Short-Lived Phenomena.* Living with Igneous Activity, 122. *The bright side of volcanoes. Reducing volcanic damage. Fighting Helgafell.* **Geothermal Energy** 123. Importance of Geothermal Energy, 123. World Locations of Geothermal Energy, 124. Sources of Geothermal Energy, 125. **Chapter Review** 125. Questions 126. Suggested Readings 127.

Field Trip The Discovery of Yellowstone Park *Nathaniel Pitt Langford* 110.

Chapter Four
Igneous Rocks

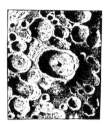

Plutons 130. Tabular Plutons, 130. *Dikes. Sills.* Lens-shaped Plutons, 134. *Laccoliths.* Massive Plutons, 135. *Stocks. Batholiths.* Geologic Dating of Plutons, 137. **Textures of Igneous Rocks** 138. Kinds of Textures, 138. *Equigranular texture. Porphyritic texture.* **Mineral Composition of Igneous Rocks** 140. Felsic Rocks, 141. *Granite. Rhyolite. Obsidian.* Intermediate Rocks, 143. *Diorite. Andesites.* Mafic Rocks, 144. *Gabbro. Dolerite. Basalt. Basalt glass.* Ultramafic Rocks, 145. *Dunites. Anorthosites.* **Magma and the Formation of Igneous Rocks** 148. The Relation of Igneous Rocks to Magma, 149. *Kinds of lava. Varieties of igneous rocks. Laboratory experiments. Bowen's Reaction Principle.* Origins of Ultramafic Rocks, 151. **The Igneous Rocks of the Moon** 152. Basic Composition of the Moon, 153. *Results of Apollo landings.* Moon Mysteries, 154. *Volcanism on the Moon. Origin of the Moon.* **Chapter Review** 155. Questions 156. Suggested Readings 156.

Field Trip Fuji-No-Yama *Lafcadio Hearn* 146.

Chapter Five
Weathering and Soils

Weathering and the Geologic Cycle; Everyday Importance of Weathering 160. **The Environment of Weathering** 161. Structure of Water Molecules, 162. Chemical Composition of Rainwater, 163. Effects of Climate, 163. *Humid tropics. Humid temperate zones. Warm-arid regions. Cold-arid regions.* Slope Situation, 166. Pore Spaces in Bedrock and Regolith, 167. Organisms and Organic Materials, 167. Visible Evidence of Weathering, 169. **Processes of Weathering** 170. Physical Weathering: Disintegration, 171. *Granular disintegration. Near-surface fractures. Thermal effects, frost wedging, heating and cooling of rocks in deserts.* Chemical Weathering: Decomposition, 173. *Oxidation. Dissolution. Hydrolysis. Carbonation.* Complex Weathering, 175. *Crystallization of salts. Combined effects of chemical and physical weathering. Effects of organisms.* Weathering of **Some Common Rocks**, 177. *Weathering of granite. Weathering of mafic igneous rocks. Weathering of limestone. Weathering of sandstone.* **Rates of Weathering** 179. Selected Examples of Weathering Rates, 179. Effects of Air Pollution, 180. **Soils** 182. Definition of Soil, 182. Relationship of Soil to Regolith: Soil Zones and Soil Profiles, 183. *Kinds of regolith. Soil zones and soil profiles. Humus, bacteria, and organic acids.* Examples of Modern Soils, 186. *Latosols. Pedalfers. Pedocals. Other soils.* Management of Soils, 189. Ancient Soils, 190. **Chapter Review** 190. Questions 192. Suggested Readings, 192.

Field Trip Realms of the Soil *Rachel Carson* 184.

Chapter Six
Sedimentary Rocks and Metamorphic Rocks

Sediments and Their Conversion to Sedimentary Rocks 196. Sedimentary Particles, 196. *Relationship of sedimentary particles to bedrock.* Description of Sedimentary Particles, 197. *Detritus. Particles derived from ions carried in solution. Organic material. Volcanic particles.* Lithification, 200. *Cementation. Recrystallization and replacement.* Sedimentary Strata, 201. *Origin of strata. Surface features of strata.* **How Sedimentary Rocks Are Classified** 204. Basis for Classification, 204. *Mineral composition. Textures of sedimentary rocks.* Sedimentary Rocks Formed by Cementation of Sediments, 206. *Detrital sedimentary rocks. Sedimentary rocks composed of volcanic material. Carbonate rocks.* Sedimentary Rocks Not Formed by Cementation of Sediments, 213. *"Instant" limestones. Evaporites. Cherts. Dolomites.* **Metamorphic Rocks** 214. The Concept of Metamorphism, 214. *Parent rocks. Metamorphic change.* The Basis of Classification of Metamorphic Rocks, 215. *Mineral composition. Texture. Presence or absence of foliation in small specimens.* Metamorphic Rocks Having Crystalline Textures, 217. *Foliates. Nonfoliates.* Cataclastic Metamorphic Rocks, 220. **Surface and Subsurface Environments** 220. Conditions of Subsurface Environments, 220. *Pressure. Temperature. Salinity.* Subsurface Changes in Organic Materials, 222. *Peat to coal to graphite. Organic matter to petroleum to graphite.* Subsurface Changes in Rock-forming Silicate Minerals, 223. *Changes in clay minerals.* Concept of Metamorphic Facies, 224. *Contact-metamorphic aureoles. Contact-metamorphic minerals.* The Metamorphic Granite Facies, 226. **Chapter Review** 228. Questions 229. Suggested Readings 229.

Field Trip Rock Formations: The Missouri *Prince Maxmilian* 208.

Chapter Seven
Stability of Slopes

Introduction to Landscapes and Downslope Movement 232. Why Is Mass-wasting Important?, 232. *Some questions about slopes.* Components of Gravity on a Slope, 234. Reduction of Resisting Force, 235. *Water and fluid pressure. Other factors.* Increase in Pulling Force, 237. **Kinds of Downslope Movements** 238. Rapid Movements, 239. *Rock fall. Rock slide. Rock avalanche. Slump. Slumping in sand. Debris flow.* Slow Movements, 249. *Creep. Solifluction. Earth flow. Talus creep and rock-glacier creep.* **Mass-wasting and Land Use** 254. Engineering Aspects of Mass-wasting, 255. *Damages and costs. Engineering solutions.* Prediction and Prevention of Slope Failure, 256. *Early warnings. Prevention of slope failure. Prediction of slope failure.* **Chapter Review** 257. Questions 258. Suggested Readings 258.

Field Trip Broken Country *George W. Kendall* 250.

Chapter Eight
Running Water As a Geologic Agent

The Hydrologic Cycle 262. **Water and the Biosphere** 264. Consumption of Water, 264. *Household uses of water. Industrial uses of water. Agricultural uses of water.* Water Problems: Uneven Distribution, 265. Rivers and Civilization, 266. *Uses of river water. Floods and flood control. Environmental impact upon rivers.* **The Mechanics of Streams** 270. Energy Relationships, 270. Discharge, 270. Speed, 272. Slope, Profile, and Base Level, 273. Channels, 273. Load, 273. Concept of a Graded Stream, 274. Laminar and Turbulent Flow, 274. The Geologic Work of Running Water, 275. *Erosion of bedrock. Dissolved load. Sediment load.* **Erosion by Running Water** 279. Playfair's Law, 279. Headward Erosion, 280. Factors Shaping Valley-side Slopes, 281. *Raindrops. Sheet flows. Rills, gullies, and badlands.* Drainage Patterns, 286. *Dendritic patterns. Trellis-rectangular patterns. Radial patterns. Annular patterns.* Rates of Erosion, 286. *How erosion is measured. Factors affecting erosion.* **The Concept of the Erosion Cycle** 288. The Three Stages of the Erosion Cycle, 289. *Youthful stage. Mature stage. Old age.* **Streams and Stream Sediments** 290. Features of Flood-plain Rivers, 291. *Channels, natural levees, flood-plain basins. Meanders and point bars. Effects of channel migration.* Deltas, 294. Braided Streams, 296. **Stream Response to Major Geologic Changes** 298. Stream Terraces, 299. Valley-Fill Sediments, 299. **Chapter Review** 300. Questions 302. Suggested Readings 303.

Field Trip The Grand Canyon *John Wesley Powell* 284.

Chapter Nine
Groundwater

The Origin and Storage of Groundwater 306. Where Does Groundwater Come From, 306. How Is Groundwater Stored?, 306. *Zone of aeration. Zone of saturation. Water table. Replenishing the supply of groundwater.* **The Mechanics of Groundwater Flow** 308. Characteristics of the Underground Pores, 308. *Porosity. Permeability. Factors affecting porosity and permeability.* Aquifers, 310. *Nonconfined aquifers. Confined aquifers. Size of aquifers. Rate of movement. Darcy's law.* **Groundwater Consumption and Supply** 312. Groundwater Consumption, 313. *Increasing demands.* Groundwater Supply, 313. *Springs. Wells. Problems of extracting groundwater. Artificial recharge.* Problems of Groundwater Use, 316. *Seeking new sources. Groundwater pollution.* **The Geologic Effects of Groundwater** 318. Groundwater and the Landscape, 318. *Caves. Sinkholes. Deposits around hot springs and geysers.* Groundwater in the Rock Cycle, 324. *Dissolution. Cementation of sediment. Replacement. Concretions and geodes.* **Chapter Review** 326. Questions 327. Suggested Readings 327.

Field Trip The Mammoth Cave of Kentucky *Frank Blackwell Mayer* 320.

Contents

Chapter Ten
Glaciers and Glaciation

Historical Background, 331. **Modern Glaciers** 332. Snowfall and the Origin of Glacial Ice, 332. *Snowfields. Firn. Glacier ice.* Geologic Significance of Glaciers, 334. *Glaciers as historical archives. Grinding bedrock into sediment. Creating distinctive landforms.* Kinds of Glaciers, 335. *Valley or Alpine glaciers. Piedmont glaciers. Ice sheets.* World Distribution of Glaciers, 338. *Antarctic Ice Sheet. Greenland Ice Sheet. Other major areas.* Glacial Movements, 339. *Measurement and rates. Mechanics. The zones of fracture and flow.* **Results of Glaciation and Related Processes** 343. Erosion, 343. *Quarrying and plucking. Abrasion. Striae and grooves.* Glacial Landforms Eroded in Bedrock, 346. *Cirques. Glaciated and hanging valleys; fjords. Arêtes, cols, horns. Roches moutonnées.* Deposition, 351. *Sediments. Erratics. Till. Other glacier-controlled sediments. Outwash. Varved lake deposits.* Depositional Landforms Made by Glaciers or Related to Glaciers, 357. *Moraine ridges. Drumlins. Eskers. Kames and kame terraces. Kettles.* **Former Glaciers** 361. Pleistocene Glaciations, 361. *Evidence. Extent.* Older Glaciers, 362. *Evidence. Age and extent.* **Theories About Climate Change** 363. Theories Based on External Factors, 363. Theories Based on Factors on Earth, 364. *Variations in the Earth's magnetic field. Atmospheric conditions; variations in volcanic dust. Circulation of the oceans. Altitude of continents. Shifts of crust with respect to the Earth's interior.* Theories of Quaternary Climatic Oscillations, 365. **The Great Lakes and Their Geologic History** 366. The Great Lakes, 366. *What are they? What is their importance?* The Ancestral Great Lakes, 367. *The Wisconsin glacier. Deglaciation.* The Modern Great Lakes, 368. *How new are they? Human-made perils.* **Chapter Review** 369. Questions 371. Suggested Readings 371.

Field Trip The Glaciers *John Muir* 352.

Chapter Eleven
Deserts and the Wind

Causes and Distribution of Deserts 374. What Is a Desert?, 374. *Precipitation and evaporation. Temperature. Vegetation.* Types and Distribution of Modern Deserts, 377. *Subtropical deserts. Continental-interior deserts. Rainshadow deserts. Coastline deserts.* **The Geologic Cycle in Deserts: Weathering, Sediments, and Landforms** 381. The Effects of Desert Climate on the Geologic Cycle, 381. *Temperature. Distribution of rainfall. Running water. Interior drainage. Wind.* Weathering and Mass-wasting, 384. *Mechanisms of desert weathering. Soils. Mass-wasting in deserts.* Desert Landforms, 385. *Fans. Bolsons and bajadas. Pediments. Playas. Uniqueness of desert landforms.* **Desert Lakes** 389. Description of Desert Lakes, 389. *Water content. Sensitivity to climate. Evidence of past existence.* **The Geologic Work of Wind** 391. Wind Erosion of Loose Sediment Particles, 391. *Deflation and deflation basins.* Mechanics of Sediment Transport by the Wind, 393. *Movement of bed load. Movement of suspended load.* Sediments Deposited by the Wind, 395. *Loess. Sand Dunes, 397. Dune formation. Dune movement and stabilization. Kinds of dunes. Ancient dunes.* Features Formed in Rock by Natural Sandblasting, 401. *Abrasion. Ventifacts, yardangs, and grooves.* **Geologic History of Deserts** 403. **Chapter Review** 403. Questions 404. Suggested Readings 405.

Field Trip Coral Dunes *Joseph Wood Krutch* 398

Chapter Twelve
Oceans and Shorelines

The Oceans As Part of the Biosphere 408. Distribution and Characteristics of Modern Oceans, 408. *Properties of sea water. Depth zones and habitats. Can We "Farm" the Sea?*, 411. **Movements of Sea Water** 412. Movements of Surface Waters, 412. *Relation to climate zones and major wind belts. The Coriolis effect. Major surface currents.* Circulation and Density Currents, 414. *Turbidity currents.* Waves, 415. *Definitions and characteristics of deep-water waves. Wind-generated waves. Tsunami.* Tides, 418. *Astronomic aspects of tides. Kinds of tides. Effect of basin shape. Tidal currents.* **Coastal Features** 421. Estuaries and Tide-dominated Coasts, 421. Wave-dominated Coasts: Beaches, 422. *Definition. Shallow-water waves and oscillation ripples. Very shallow-water waves; wave refraction. Breakers and surf. Nearshore water circulation; rip currents. Beach profiles; barriers; longshore drift. Kinds of beaches.* Bedrock Coasts, 428. Tropical Shorelines and Reefs, 428. Classification and Evolution of Coasts, 429. **Marine Sediments** 429. Coastal Sediments, 430. Shelf Sediments, 430. Sediments of the Continental Margin, 431. Deep-sea Sediments, 431. *Effects of climate. Effects of sea-level changes.* **The Oceans and You** 433. Coastal-zone Management, 433. *Beach problems. Pollution.* Political and Economic Aspects of the Sea, 435. *Ownership of the oceans. Law of the Sea Conference.* **Chapter Review** 436. Questions 437. Suggested Readings 437.

Field Trip The Marginal World *Rachel Carson* 424.

Chapter Thirteen
Earthquakes and Seismology

Earthquakes and Seismic Waves 441. What Is an Earthquake?, 441. **Some Historic Earthquakes** 443. Lisbon, Portugal, 1755, 444. New Madrid, Missouri, 1811-1812, 445. Prince William Sound, Alaska, 1964, 446. Managua, Nicaragua, 1972, 448. **Seismology and Seismic Waves** 449. Types of Seismic Waves, 449. The Scientific Use of Seismic Waves, 450. *Locating earthquake epicenters. Seismic waves in liquids and solids. Speeds of waves and densities of materials.* **The Distribution of Earthquake Epicenters** 452. Mapping Earthquake Epicenters, 452. **Prediction of Earthquakes** 452. The Search for Predictable Behavior, 456. *Changes in speeds of seismic waves. Changes in the tilt of the ground.* **Can We Learn To Live with Earthquakes?** 457 The Effects of Earthquakes, 459. *Why buildings fall down. Building regulations and common sense.* **Chapter Review** 463. Questions 464. Suggested Readings 464.

Field Trip Earthquake and Fire: San Francisco in Ruins 454.

Chapter Fourteen
Plate Tectonics and the Interior of the Earth

The Interior of the Earth 468. The Crust-Mantle Boundary, 468. *The mantle.* The Core of the Earth, 470. **Challenges to the Established Views of the Earth** 470. Alfred Wegener and Continental Drift, 471. *Laurasia and Gondwanaland.* Paleomagnetism and the New Look at Continental Drift, 472. *Paleomagnetism. The sea floor and continental drift.* **Sea-floor Spreading** 474. Harry Hess and the Moving Sea Floor, 474. The "Magnetic Tape Recorder", 476. *Magnetic characteristics of the ocean floor. The Vine-Matthews hypothesis.* Other Implications of Sea-floor Spreading, 477. *Predicted ages of sea-floor crust and sediment. The Glomar Challenger. Fracture zones and transform faults. Present status of the concept of sea-floor spreading.* **Plate Tectonics** 479. Current Concepts of Plate Tectonics: Predictions and Tests, 479. *Plates and plate boundaries. Trenches and subduction. Mantle hot spots and rows of volcanic islands.* Plate Tectonics and Continental Drift, 484. *The break-up of Pangaea. Predictions for future plate movements. Some dissenting opinions about plate tectonics.* Plate Tectonics in Relation to Crustal Deformation and Mountains, 486. **Chapter Review** 486. Questions 488. Suggested Readings 488.

Chapter Fifteen
Mountains, Crustal Deformation, and Recycling of Continents

Orogenic Belts and Their Mountains 492. **The Anatomy of Orogenic Belts** 496. The Rocks of Orogenic Belts, 496. Geologic Structures, 497. *Folds. Thrusts. Other faults.* **Examples of Kinds of Mountains Based on Geologic Structures** 499. Large Domal Uplifts Exposing "Basement" Rocks, 499. Fault-Block Mountains, 501. Fold Mountains, 502. Complex Mountains, 503. **Dynamic History of Orogenic Belts** 503. Geosynclinal Stage, 503. Terminal-Orogeny Stage, 504. Post-Orogenic Stage of Block Faulting and Epeirogenic Movements, 506. **Causes of Epeirogeny: Isostasy** 506. Strength of the Lithosphere, 507. Flow of the Asthenosphere, 509. Changes in the Lithosphere, 509. *Effects of erosion. Thermal Expansion or Contraction. Effects of orogeny.* **Mechanisms of Orogeny** 510. Global Cooling and Shrinking, 510. Gravity Sliding of Strata, 511. Orogenic Belts and Plate Tectonics, 511. Unsolved Problems about Mountains, 515. **Orogenic Belts and Continents** 516. Ancient Orogenic Belts, 516. Can We Identify Active Modern Orogenic Belts?, 517. Orogenic Belts and the Geologic Cycle, 517. **Chapter Review** 518. Questions 519. Suggested Readings 519.

Field Trip A Near View of the High Sierra *John Muir* 512.

Chapter Sixteen
The Earth's Resources

Renewable Energy Sources 522. Solar Energy, 522. Plant Products, 523. Geothermal Energy, 523. **Nonrenewable Energy Sources** 523. Fossil Fuels, 524. *Coal. Petroleum.* Uranium, 528. **Renewable Resources** 530. Water, 530. **Nonrenewable Resources** 531. The Concept of Ore, 531. Resources and Reserves, 532. Metallic Resources, 532. The Formation of Metallic Ores, 533. *Metallic ores formed by igneous and metamorphic activity. Residual ore deposits. Sedimentary ores. Shields and metallic ore deposits.* Nonmetallic Resources, 535. **Resource-use Curves and Human Activities** 537. Geology and Resources, 538. **Future Resources: Will They Come from the Ocean?** 539. **Chapter Review** 540. Questions 541. Suggested Readings 541.

Appendices

Appendix A **The Origin of the Earth** 545. The Origin of the Solar System, 546. *Collision concept. Nebular-cloud ideas. Other factors.* From Nebular Cloud to Solid Planet, 547. Appendix B **Identification of Minerals** 548. Luster, 548. Effects of Breaking (Fracture and Cleavage), 549. Minerals and Properties, 550. Appendix C **Identification of Rocks** 552. Igneous Rocks, 552. Special Features and Ultramafic Varieties, 553. Sedimentary Rocks: Clastic Varieties, 554. Sedimentary Rocks: Nonclastic Varieties, 555. Metamorphic Rocks: Common Foliates and Nonfoliates, 556. Appendix D **Metric Equivalents and Powers of Ten** 557. Metric Conversion Table, 557. Some Useful Metric Conversion Tables, 558. Multiples (Powers of 10) and Prefixes, 559. Appendix E **Understanding Topographic Maps** 560. Map Symbols, 560. Scale and Measurement of Distance, 560. Orientation and the Expression of Direction, 561. *Expression of direction. Expression of location.* Contour Maps, 562. *Constructing topographic profiles.*

Glossary   564

Index   576

Physical Geology

"I see no sign of a beginning or of an end."

Introduction
The Study of Geology

Before the serious study of any subject can begin, fundamental questions need to be asked. In the case of this book and the study of physical geology, we should ask first of all: Why study science? And even more to the point: Why study geology?

Science is a means by which we can analyze the material universe. Of course, geology is not the only science; we will also look at some of the other sciences and their contributions to human knowledge.

Mathematics, the most abstract of the sciences, is the pure basis for analyzing numerical information and abstract concepts of quantitative data. *Physics* is concerned with the mathematical analysis of such things as energy and motion, sub-atomic particles that are even smaller than atoms, and radioactivity. *Chemistry* deals with the properties of matter and the way matter reacts under different conditions. *Biology* is the study of cells, plants, and animals and their relationships to one another, including ecology. *Astronomy* addresses the great distances in the universe, energy from stars, and the possibility of life on other planets. *Geology* is the study of all these things as they apply to the Earth. Geology is generally divided into two parts: (1) the study of processes that shape the Earth, **physical geology,** and (2) the systematic study of the past history of the Earth and life on Earth, **historical geology.**

The Nature of Geology

It is appropriate to start our over-all discussion of geology by emphasizing the great contributions of geology to intellectual understanding.

Geology has had a revolutionary impact on our understanding of **time.** A great "time revolution" started in the early part of the nineteenth century and culminated early in the twentieth century. By that time, radioactivity had been discovered and applied to the universal time scale, and geologists were able to assign numbers such as 4.6 billion years for the approximate age of the Earth. This age of the Earth is now accepted almost universally. The expanse of time with which geologists work is so vast that it is called **geologic time** to separate it from other kinds of time that can be measured with a clock or a calendar.

Some other great contributions of geology include the discovery of marine fossils on the land, which proved that in past times what is now land was on the bottom of the ocean. In fact, the land and the sea have exchanged places—not once, but many times. Also, the idea of the great Ice Ages was a revolutionary concept when it was introduced in Europe in the second half of the nineteenth century.

Another revolution we might consider is the concept of the **geologic cycle** provided by James Hutton—the notion that the activities we see taking place today explain the history of the Earth, and that the Earth's history is recorded in its rocks. We also have a better understanding of the properties of minerals, which has led to the use of tiny mineral crystals as transistors, lasers, and other essential tools of modern electronics.

All of these revolutions have culminated in what we call the revolution of **plate tectonics** and **continental drift,** the latest in a series of theoretical formulations which have expanded our outlook about the Earth. And the plate-tectonics revolution was possible only because of an outburst of new information about the **ocean floor.** As we shall describe, this latest revolution was the culmination of many different discoveries of the 1950s and 1960s—seemingly unrelated but essential to these new theoretical concepts in the earth sciences.

The Beginnings of Geology: Catastrophism and Uniformitarianism

James Hutton, a Scottish physician (1726–1797), is generally regarded as the founder of modern geology (Figure 1). But even before Hutton, many western European scientists had noticed layers of rocks (strata) which implied the notion that much time had been involved in the formation of such strata. Basically, these scientists were religious men, and the implications of great periods of time began to create a conflict between what they saw in nature and what they had been taught to believe by the Bible.

As late as the nineteenth century, many naturalists believed that the Earth's landscape had been shaped by sudden and dramatic events. They thought that canyons had been carved, oceans filled, and mountains raised by catastrophic activities. This viewpoint, called the **doctrine of catastrophism,** was challenged in 1785 when Hutton published his *Theory of the Earth*. Probably because of Hutton's technical writing style his book had little impact upon the general community. But in 1802 John Playfair published Hutton's ideas in the more readable *Illustrations of the Huttonian Theory of the Earth*.

Basically, Hutton had said that the landscape was created not by catastrophes, but by the continuing process of change that had been going on in the past just as it is in the present. Hutton's theory of slow and steady change came to be known as the **doctrine of uniformitarianism,** or as it is sometimes called today, the **uniformity of process.**

In order to harmonize the findings of geologists, which implied a long history of the Earth, and biblical teachings, Baron Georges

1 James Hutton, the observant Scotsman, who first stated the concept of "the great geological cycle" in 1785. (New York Public Library.)

2 Baron Georges Cuvier, whose "great compromise" based on catastrophism brought peaceful coexistence between theologians and geologists during the first third of the nineteenth century. (New York Public Library.)

Cuvier (Figure 2) and others proposed what came to be known as the "great compromise." According to them, the Earth's history should be divided into three parts: (1) the flood of Noah, or diluvian times, as described in the Old Testament, (2) the period from the flood to modern times, or the "post-diluvian" period, and (3) the "ante-diluvian" or pre-flood period.

Cuvier suggested that the period from the flood to modern times was 6000 years long—the age of the Earth and of humans—and that this period was to be translated literally from the scriptures, even to the point of having 24-hour days during the Creation. But according to Cuvier, the flood and the post-diluvian times had nothing to do with "geology." So geology was a period of Earth history that was a supernatural time. It was full of catastrophes, it had all sorts of strange animals in it, but no humans existed. That was geology, and you could do anything you wanted with it, and it would not upset the religious community as long as it remained part of an undated, unknown, supernatural time. And nothing that exists now could help you understand the ante-diluvian period. Cuvier's "great compromise" seemed like a perfect solution to the "time dilemma"; this viewpoint soon attracted a large following.

All was calmly accepted until Charles Lyell (Figure 3) came along in 1830 with his book *Principles of Geology*. Lyell was convinced that Hutton was right, and that the so-called geologic period was not a supernatural time. His book renewed world-wide interest in Hutton's ideas.

Lyell's books (he published the classic *Elements of Geology* in 1838) were written so diplomatically and so tactfully that if it hadn't been for the advent of Charles Darwin's *Origin of Species* in 1859 it would have taken much longer for Lyell's work to take effect. Darwin, with his theory of biological evolution, was the last straw, and the safe world of the catastrophists could survive no longer. Eventually, geology obliterated the great compromise entirely.

It is important to stress that not all Earth changes are caused by the inexorably slow and uniform processes described by Hutton and Lyell. Lyell himself had also written of the importance of sudden, nongradual natural events. We can still see today the dramatic effects of floods, earthquakes, volcanoes, and other dramatic geologic "catastrophes." In this sense, Hutton was describing only the geologic cycle when he said, "In the economy of nature I see no sign of a beginning or of an end." He had concluded that the oldest rocks had been formed by the same processes that had formed the youngest ones. We now know that the oldest rocks are approximately 3.8 billion years old, and that the geologic cycle as envisioned by Hutton had been operating that long. Nothing that has happened in the meantime has changed

3 **Sir Charles Lyell**, lawyer-turned-geologist, whose books and other writings utterly refuted the "great compromise" being popularized by Baron Georges Cuvier. (New York Public Library.)

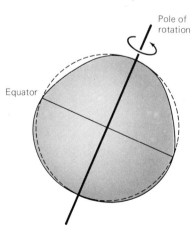

4 **Rapid rotation on polar axis** has caused the Southern Hemisphere to bulge and the polar regions to be correspondingly flattened, as seen in this schematic representation.

his dramatic statement: "I see no sign of a beginning or of an end."

Hutton's notion that there are no rocks on Earth that have not been through the rock cycle has been totally supported. It stands as one of the greatest scientific insights ever made.

The Structure of the Earth

Measurements of selected parts of the Earth show that its shape approximates an *oblate spheriod,* a spherical mass that is slightly flattened at the poles (Figure 4). Recent measurements made on the orbits of satellites have also revealed a slight bulge in the Southern Hemisphere. These measurements suggest that the Earth is slightly pear-shaped and flattened at the top and bottom.

From a human point of view, probably the most interesting aspect of the Earth is its **atmosphere** (from the Greek *atmos,* vapor, and the latin *sphaera,* sphere), a gaseous envelope surrounding the planet. The atmosphere makes human and other life possible by retaining oxygen and water vapor and by screening out harmful rays from the Sun. Equally important is the **hydrosphere** (sphere of water), which includes liquid water in the oceans, rivers, lakes, and underground, and solid water in glaciers.

The Earth's **core** consists of two parts. The inner kernel is a solid sphere with a radius of about 1216 kilometers (km), a bit larger than the Moon (Figure 5). This solid core is surrounded by a transitional layer slightly more than 500 km thick, and this layer, in turn, is surrounded by a liquid outer core 1700 km thick. Outside the core is a **mantle** of solid rock about 2900 km thick. The mantle is covered by a thin, rocky **crust** about 40 km thick beneath the continents and 10 km thick beneath the deep oceans. The crust and the outermost portion of the mantle, to a depth of 100 km, compose the **lithosphere,** or "sphere of rock."

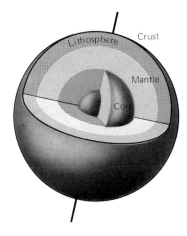

5 Cutaway view of interior of the Earth shows two-part core, thought to be made of metal, surrounded by the mantle, which is composed of rock. The outermost shell of the Earth, also composed of rock, is the lithosphere, of which the top part is named the crust.

The composition and origin of rocks will be discussed in the early chapters of this book, but they deserve some mention here. Rocks not only are the basic historical documents of the Earth, but also are fundamental to many aspects of modern geologic processes. Geologists are concerned with the nature and behavior of the rocks involved in many geologic activities. Whether investigating the eruption of a volcano, the causes of an earthquake, the age of the ocean basins, the

Rocks and the Earth's Crust

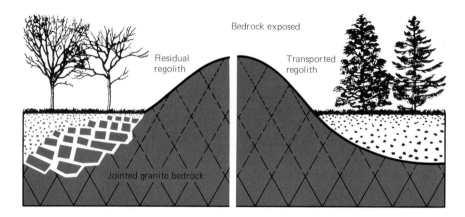

6 Schematic profile and section through the surface of the Earth showing continuous solid rock, the bedrock, overlain by an overburden of noncemented particles known as the regolith. According to one of the fundamental principles of geology, at the Earth's surface various processes break down bedrock and convert it into regolith; but inside the Earth, subsided regolith can be converted into bedrock.

location of a petroleum deposit, or attempting to determine the geologic history of a region, geologists study the related rocks.

The material underlying the surface of the Earth is divided into bedrock and regolith (Figure 6). **Regolith** (from the Greek, for blanket of rock) is loose, noncemented fragments of rocks, such as gravel, sand, and soil. **Bedrock** is continuous, solid rock that everywhere underlies regolith and in some places forms the surface itself.

Most of our knowledge of the Earth is restricted to its crust. The crust is dominated by two major features: the ocean basins and the continental blocks. The average height of the continents above sea level is about 1 km; the maximum height, represented by Mt. Everest, in the Himalayas, is 9.2 km. The average depth of the oceans is 4 km; the maximum depth is 10.8 km, in the Marianas Trench in the Pacific Ocean.

Beneath both the oceans and the continents is a layer of dense, dark-colored rock called **basalt**. The continents themselves are composed largely of a lighter-colored, less-dense rock called **granite**. The difference between the densities of granite (2.6 grams per cubic centimeter) and basalt (3.0 g/cm^3) is about 10 per cent—the same difference that exists between ice (0.9 g/cm^3) and water (1.0 g/cm^3), respectively. Thus, even though the granitic continents are, in some regions, billions of years old, they have remained "floating" in the denser basaltic material below, much as ice floats in water.

The Three Great Groups of Rocks

When we speak of the origin of a rock, we mean the process by which it took its present, solid form. There are three principal ways rocks may form: (1) solidification of molten material, which forms *igneous rocks*, (2) accumulation of layers of particles called sediment, such as sand and clay, and their later cementation, which forms *sedimentary rocks*, and (3) extreme modification (metamorphism) of already formed rocks to make new rocks named *metamorphic rocks* (Figure 7).

Igneous rocks We have said that the Earth's mantle consists of solid rock. Under certain conditions, however, this rock may melt and, in molten form, move toward the surface. This molten matter is called **magma**; if it emerges above the surface, it is **lava.** The cooling and solidification of magma and lava are the processes by which igneous rocks form, that is, become solid. Igneous rocks compose most of the Earth's crust; **granite** and **basalt** are the two most abundant igneous rocks.

Sedimentary rocks Igneous rocks, like other rocks, are gradually broken into smaller particles through various physical and chemical processes known as **weathering.** The products of weathering may be sorted, transported, and deposited as **sediment.** As sediment accumulates, usually at the bottom of lakes, seas, or oceans, the weight of new layers of sediment presses together the particles in the older layers, thus squeezing out their water. Eventually, substances dissolved in the water may cement the particles, forming **sedimentary rocks.** Common sedimentary rocks formed in this way are conglomerate, sandstone, and shale. In some cases, sedimentary rocks, such as limestones, form by the cementing of shells. In other cases, sedimentary rocks such as layers of salt are precipitated chemically out of salty water.

7 Typical arrangement of particles in the three main groups of rocks. (a) Interlocking arrangement formed by solidification of molten material to form igneous rock. (B. M. Shaub.) (b) Broken, worn, and rounded particles that have been cemented together to make one kind of sedimentary rock. (Nat Messik.) (c) Smeared-out interlocking arrangement of particles of metamorphic rock that have been heated and deformed, but not melted. (B. M. Shaub.)

Metamorphic rocks As we have said, the Earth is far more active and dynamic than early geologists supposed it to be. Parts of the crust are being pulled and pushed by other parts, buckling downward as deep trenches or crumpling upward as mountains. Rocks may be crushed, heated, or buried for many millions of years, and may reappear later at the Earth's surface. Rocks that are substantially changed by such heat, pressure, or chemical action are **metamorphic rocks.** Abundant metamorphic rocks are slate, gneiss, schist, and marble.

The Rock Cycle

Rocks may change from one type to another through sedimentation and metamorphism. These changes are part of a larger complex of

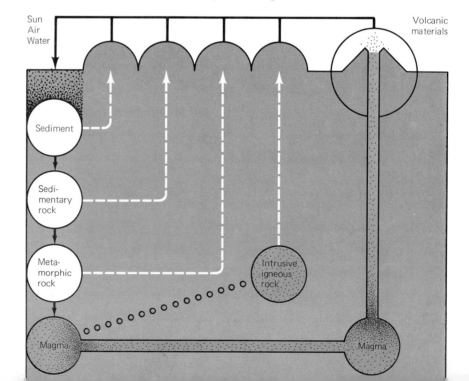

8 In the rock cycle the internal energy of the Earth, the energy received from the Sun, the atmosphere, the hydrosphere, and the rock materials are constantly reacting and changing. Movement of the Earth's outer shell shifts materials upward and downward. Rock materials thus raised to the surface and spewed out of volcanoes are broken down to form sediment. After the sediment has been spread out into layers and these have been buried by still other layers of sediment in an area that is moving downward, the sediments become cemented into sedimentary rocks. Deeper burial plus deformation can convert any kind of material into metamorphic rocks. Still-deeper burial forms magma. Once material melts, its previous history is obliterated, and when the molten material cools and solidifies it begins the cycle all over again as an igneous rock.

processes, the **rock cycle,** which operates continuously on and within the Earth's crust.

An idealized rock cycle: magma to magma Let us follow a parcel of rock material as it might travel through a simple, hypothetical rock cycle (Figure 8). Its journey begins 60 km below the Earth's surface, where it is part of the solid mantle. A gigantic, slow-motion heaving of the mantle creates cracks in the crust and reduces the pressure above our parcel. The material then melts and, as a stiff liquid, moves slowly upward. Let us assume it stops one kilometer below the surface where it solidifies into basalt. After several million years the overlying crust has been removed by rain and by wind erosion. Finally, one day the minerals in the parcel of basalt itself are altered to form clay. Still later the clay is washed into a stream, and in a few weeks is carried downstream, and eventually settles at the bottom of the sea. Other particles are deposited on top of it until, during many thousands of years, it is pressed into shale.

Pressure increases as the mantle forces continue; the shale is forced downward so that pressure and temperature increase. During millions of years the shale is metamorphosed into schist. The schist continues to plunge deeper within the mantle until it melts into magma. This time, however, the composition of the magma is not the same as before. Instead of having the composition of basalt, the magma is granitic. This change from basalt to granite is a direct effect of the reactions that took place at the Earth's surface.

A remarkable thing about the rock cycle is that certain parts of it, particularly those relating to igneous rocks and volcanoes, apply to other planets. We can make generalizations about Earth that help us to understand the Moon, Mars, Jupiter, and Venus. This kind of study is now going forward under the heading of "planetary geology." The principles of planetary geology can be learned from rocks and minerals right here on Earth.

The Concept of Geologic Time

Geologic events occur so slowly that geologists use the phrase "geologic time" to refer to periods of time that are longer than those in any human experience. A billion years, or even a million, has little meaning to us who live only 70 years or so. One way of illustrating the extent of geologic time is to imagine all 4.6 billion years of it (the approximate age of the Earth) compressed into a single year, beginning January 1 and ending December 31.

The oldest rocks (about 3.8 billion years old) would date from mid-March. The first living creatures would appear in May, and land plants and animals in late November. The Rocky Mountains would not arise until December 26, and the first humanlike animals would appear at approximately three o'clock in the afternoon (about 5 million years ago) of December 31. The last glaciers of North America would begin their retreat from the Great Lakes about one minute before midnight. The seeds of modern geology would be sown by James Hutton just one second before the end of the year.

Geologic Change

The importance of geologic time can hardly be overemphasized. Before geologists realized how old the Earth is, they could not account for such huge features as the Grand Canyon. They could see that the water of the Colorado River was carving it out of rock, grain by grain, but there did not seem to be enough time for the job. When they discovered that time is essentially *unlimited*, geologists realized that even subtle processes, such as the falling of raindrops, could eventually reduce entire mountain ranges to flat land.

The Geologic Time Scale

For classifying historical time, historians most often use such units as years, decades, centuries, and even millennia. For geologic time, however, the units must be far larger, stretching to include millions of years. Geologists have worked out a scale known as the **geologic time scale** (opposite page). There are four large units on the scale known as *Eras*—the Precambrian, Paleozoic, Mesozoic, and Cenozoic Eras. The last three Eras are named after the Greek words for ancient life, medieval life, and modern life. Each Era is divided into smaller divisions called *Periods*. The Periods are further divided into shorter *Epochs*.

The Precambrian Era includes 80 to 85 per cent of the entire age of the Earth—everything that occurred from the Earth's creation until 570 million years ago. Yet little is known of Precambrian processes, because subsequent changes of the Earth's surface have erased most of the evidence of what was happening.

Sediments and Time

Of the three major kinds of rocks—igneous, sedimentary, and metamorphic—those laid down as sediment tell us the most about the past. There are three reasons for this: (1) sedimentary rocks account

The Geologic Time Scale

Eons	Eras	Periods	Epochs (of Cenozoic)	Selected events in history of Earth	
PHANEROZOIC	Cenozoic	Quaternary	Holocene	HOMO SAPIENS	
				Many mammals become extinct	Ice age
			Pleistocene		Widespread volcanic activity
			—1.8my—		
		Tertiary	Pliocene	Grazing mammals abundant	Erosion of Grand Canyon begins
			—5.2my—	Grasses spread widely	
			Miocene		
			—24my—		
			Oligocene	Many early mammals become extinct	Lands elevated Deformation in Alps
			—38my—		
			Eocene		
			—54my—	First primates	Climax of Rocky Mountain deformation
			Paleocene		
		—65my—			
	Mesozoic	Cretaceous		Modern plants appear	Coal swamps
		—145my—			Atlantic ocean starts to open
		Jurassic		Time of dinosaurs	
		—190my—		First birds and mammals	Palisades sheet intruded
		Triassic			
		—225my—			
	Paleozoic	Permian		Mammal-like reptiles	Ice age
		—280my—			Climax of Appalachian deformation
		Carboniferous — Pennsylvanian		First reptiles	
		—325my— Mississippian		Oldest fossil footprint (amphibian)	Coal swamps widespread Lands elevated
		—345my—			
		Devonian		Oldest rooted land plants	Ice age
		—395my—			
		Silurian		First fish (oldest vertebrates)	Shallow seas cover continents
		—430my—			
		Ordovician			
		—500my—		First abundant fossils	
		Cambrian			
		—570my—			
CRYPTOZOIC	Precambrian			Widespread ice age	
				Early ice age	
				Oldest one-celled plants	Abundant iron-bearing sediments deposited
	—3500my—				
AZOIC				Oldest rocks	
				Origin of Earth	
—4500my—					

9 Fossils of various kinds of organisms. At top, insect, represented by fossil dragonfly, natural size; Bavaria (U.S. Department of Agriculture). Bottom left, marine invertebrates on surface of ancient limestone represented by bryozoa (rod-like features showing many small pits) and brachiopods, twice natural size; Silurian, Lockport, New York (Smithsonian Institution). Bottom right, dinosaur limb bones, width of view about 3 meters; Utah (U.S. National Park Service, photo by Harry B. Robinson).

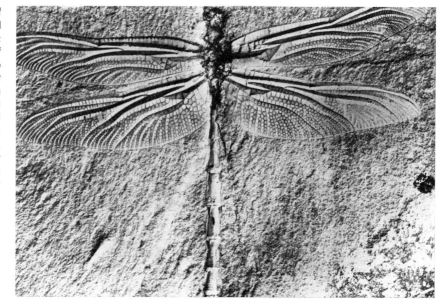

for about 66 per cent of all exposed rocks; (2) they alone are formed at "normal" temperatures and pressures; and (3) only sedimentary rocks contain **fossils,** the remains of ancient organisms (Figure 9).

In terms of measuring geologic events, an essential feature of sedimentary rocks is that they are deposited in **strata,** or layers. Each stratum is a layer of rock that represents an individual episode of sedimentation, or deposition, and may be paper thin or several meters thick, depending upon the duration of that episode. Each stratum is younger than those beneath it, and older than those above.

Radioactivity and Time

A more precise way to date certain ancient rocks is by the decay of minute amounts of radioactive matter they contain. This radioactive material (including certain kinds of carbon, uranium, potassium, and other substances) spontaneously changes at a constant rate into slightly different material. Knowing the rate of change, geologists can measure the amount of "parent" and "daughter" material present in a rock sample and calculate its age. Radioactivity will be described more fully in Chapter 2.

Recent Trends

Although the study of rocks and minerals remains basic to geology, a great many recent trends incorporate techniques from other scientific fields. These trends will be described in detail in the text, but a brief description of some of them is given here to help the student place the subjects of the book in context.

Magnetism

Studies of magnetism (a force, to be described in Chapter 2, which is similar in many ways to gravity) have in recent years taught us new lessons about the Earth. We know, for example, that the Earth, like a toy magnet, has a magnetic field of its own which changes and even reverses by means that are not understood. By using highly sensitive magnetometers towed behind ships and airplanes, geologists have been able to detect and to study evidence in the rocks of magnetic changes that occurred millions of years ago. These changes have provided strong evidence for the notion that the crust of the Earth is in slow but persistent motion, an idea that seemed preposterous only a few decades ago.

Oceanography

Since World War II, oceanography has been one of the fastest-growing fields of science. Much of the impetus for this growth came from the war itself, and its accelerated research in submarine warfare, underwater sound, and navigation techniques. Most of the developed nations, led by the United States, expanded their fleets of oceanographic vessels and set about applying wartime advances in instrumentation to exploring the oceans.

One early result of this intensified effort was the discovery that the sediment covering the ocean floor was not nearly as thick as had been supposed.

The mid-oceanic ridge Coupled with this surprise of thin ocean sediment was the discovery of the mid-oceanic ridge system. Oceanographic ships had discovered unusual "mountain chains" in the mid-Atlantic Ocean as long ago as the nineteenth century (Figure 10). But the flurry of postwar bottom-sounding revealed these features to be far more extensive and regular than had ever been supposed. The mid-Atlantic ridge proved to be part of an enormous, interconnected system that extends for some 60,000 km throughout the oceans of the world.

Island arcs and trenches As detailed mapping of the ocean floor progressed, another puzzling feature became clear: deep, regular trenches bordering certain continents and chains of islands (Figure 11). Why, oceanographers asked themselves, should these oceanic "deeps" be located at ocean margins? It seemed that material eroded from adjacent land areas should have filled such trenches with sediment. Although the trenches contained some sediment, there was not nearly enough, if the trenches had existed as long as the continents. It now seems that the trenches, and the ocean floor of which they are a part, are not nearly as old as the continents. We can see the reasons for the relatively young age of the ocean floor when we discuss sea-floor spreading and the notion that the sea floor is being generated by the rise of igneous material at mid-oceanic ridges.

10 Rough, rocky bottom of the Atlantic Ocean forms a feature which stands above its surroundings as high as any mountain range on land. This profile was made from a moving ship by sending out sound waves every six seconds and by recording the returning sound echoes on a strip chart. The time required for the sound to travel through the water to the bottom, and to return to the ship as echoes, indicates water depth. (B. C. Heezen, M. Tharp, and M. Ewing, 1959.)

Seismology

The study of earthquakes, *seismology*, has expanded almost as rapidly as oceanography. With the aid of newly accurate space-age technology, seismologists have been mapping the sites of earthquakes since the 1950s, noting them as dots on maps of the world (Figure 12). As they filled in the dots, a curious pattern emerged: almost all earthquakes were occurring either along oceanic ridges or beneath oceanic trenches. This pattern, by the time it became clear in the mid-1960s, formed one of the strongest pillars of evidence supporting the idea that the Earth's crust consists of separate sections, or plates, and that these plates are continuously straining and lurching against one another. Most geologists now think that energy released by sudden movements along the boundaries of these plates creates waves that we sometimes feel as earthquakes. However, not all earthquakes start at plate boundaries.

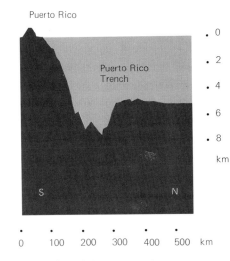

11 Profile of the Puerto Rico Trench shows the deepest part of the Atlantic Ocean is not out in the middle, but near its margin. The same is true of the Pacific Ocean. (B. C. Heezen, M. Tharp, and M. Ewing, 1959.)

Plate Tectonics

Much of the evidence gathered in the study of magnetism, oceanography, and seismology has been united recently to formulate a new geological theory. The theory, called **plate tectonics,** touches nearly every aspect of geology, offers answers to many old and new mysteries, and will be mentioned throughout this book. The exotic-sounding name of the theory has its root in the name Tecton, a carpenter described by Homer in the *Iliad*. The word *tectonics* has come to refer to the art or science of construction, and the term *plate tectonics* implies that the Earth's crust is constructed of a dozen or so moving plates (Figure 13). "Tectonic" may describe features which are the result of such movements, such as *tectonic valleys*, or *tectonic mountains*.

Sea-floor spreading and continental drift When the theory of plate tectonics first began to take form, in the early 1960s, it was dubbed **sea-floor spreading,** a term that describes the movement of the plates away from a "spreading center." According to the theory, igneous rock arises continuously along the mid-oceanic ridges. The rock forms new sea floor, which spreads away from both sides of the

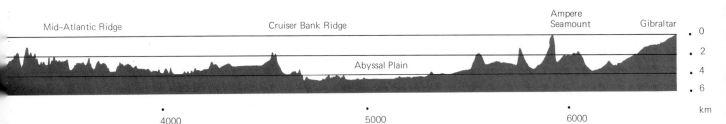

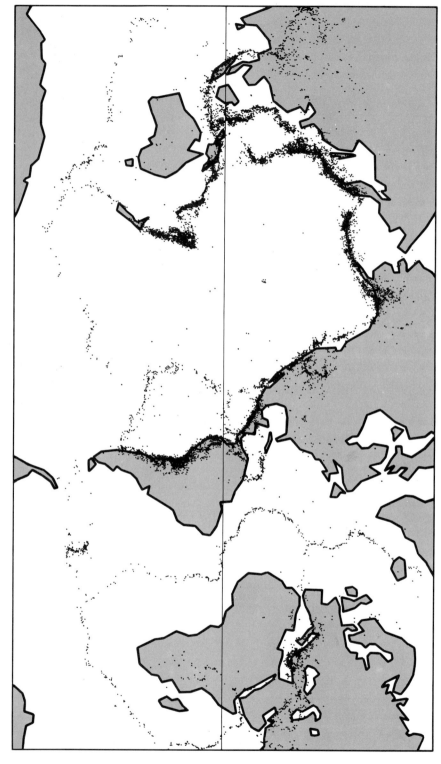

12 World distribution of earthquake epicenters (points on the surface directly above the places where the earthquakes originated). (U.S. Geological Survey)

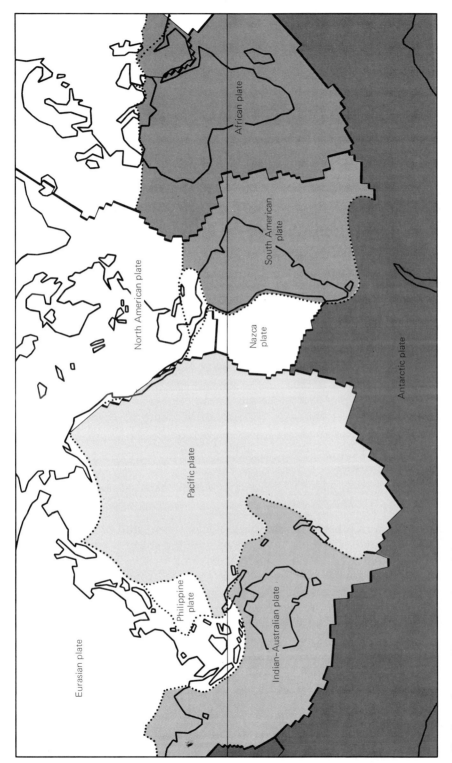

13 One of several interpretations of the large plates of the Earth's lithosphere. The boundaries of the plates have been drawn on the supposition that the only places where earthquakes happen are at the margins of plates.

The Study of Geology

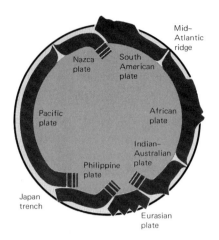

14 Side views of lithosphere plates as seen in a schematic section through the Earth. (Modified from Robert Pelletier, New York Times.)

ridge at rates of a centimeter or so a year. Eventually, hundreds or thousands of kilometers away and millions of years later, the sea floor is thought to plunge downward beneath the oceanic trenches (Figure 14). A single plate, then, is bounded by ridges and/or trenches.

Where two or more plates adjoin, one plate may be forced under or past the other. According to the theory, a plate may carry a load of continental material, much as a river in spring flood carries blocks of ice. The continental material, mostly granite, is lighter than the plate material, mostly basalt, so that the continents have remained "afloat" for hundreds of millions of years. The term **continental drift** is used to describe this alleged movement of land areas.

Unanswered questions The biggest remaining mystery in plate tectonics is what makes the plates move. What force or forces inside the Earth are so powerful and persistent that they can maintain plate movement? Many geologists think that the rock of the mantle is somewhat "plastic" and deformable. Heat from the Earth's interior, they say, can cause motion in mantle rock, forcing it to rise toward the surface in certain regions much as air rises when heated by a radiator. The details of such motion are far from clear, however, and there are serious objections to the entire scheme. As might be expected, the new theory of plate tectonics has raised almost as many questions as it has answered. We will be discussing some of these questions, particularly in Chapter 14.

Geology and You

In addition to broadening our outlook about time and interpreting the past, geology is also closely involved with the study of environmental processes today. Geologists have discovered that human beings have become capable of causing geologic change through their activities. At the same time, other geologists are intent upon the search for raw materials to sustain these activities.

Ultimately, human populations are dependent upon geologic processes for fresh water, food, building materials, and energy resources—in short, for survival. Unfortunately, all of these limited supplies are being used at such a profligate rate that shortages are already occurring. The challenge before modern civilization is to learn to conserve the Earth's resources so that future generations have sufficient supplies. Geologists, with their knowledge of the Earth and its processes, are becoming more and more involved in meeting this challenge.

Geology and You

Concerns for the Future

Perhaps the most pressing problems facing us today are those related to population and pollution. These issues, in turn, are so closely related to natural resources that it is almost impossible to deal with one at a time. And all of them are, to varying degrees, related to geology.

According to several estimates, the world's population will grow from about 4 billion in the mid-1970s to approximately 7 billion by the year 2000 (Figure 15). Such projections are based on the idea that in the future, supplies of food, of energy, and of natural resources can continue to expand at the rate established in the first half of the twentieth century. More people use more resources: more metals, more petroleum, more water. But, because resources are limited, it is obvious that population cannot forever continue to increase as it has in recent years. At present, some 80 million people are being added each year to the world's population. Feeding these additional mouths requires the equivalent of a pile of grain 2 meters deep, 18 meters wide, and 1125 km long—the distance from New Orleans to St. Louis.

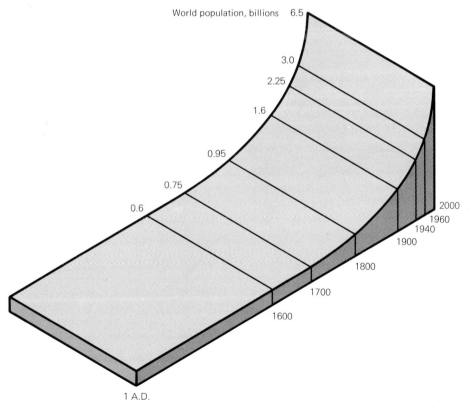

15 The Earth's human populations through time. From the beginning of human time to 1650 world population grew very slowly. During the years 1650 to 1950 the population increased 10 times more rapidly than previously. The most dramatic growth took place during the eighteenth century. Since 1950, population has grown at a rate which indicates that even if the birth rate decreases in the next few years, and if no great wars or famines lead to wholesale human slaughter, in the year 2000 world population will still be more than six billion. (Data courtesy The Population Council.)

16 Characteristic relics of the modern epoch accumulating in a junkyard in Vermont. (United Nations.)

    The end results of the use of natural resources are garbage, smoke, sewage, or other forms of pollution (Figure 16). This pollution is a doubly detrimental phenomenon. It not only fouls the water, air, or land, but it also represents the consumption of raw materials that must be replaced by further mining of the Earth's limited resources. Recycling of industrial and other materials would not only clean up our environment but also preserve our finite supply of resources. Technology, whose processes are now largely wasteful and dirty, could and must become a major force in combating pollution and recycling our raw materials.

    Until recently, the basic concern of most geologists was interpreting the history of the Earth. In other words, geologists wanted to find out what the Earth was like before humans began to affect it. There was a tremendous concern in geology about looking into the ancient past and not really much care about now or the future. One of the big turnarounds in geology has been the impact of the environmental movement. Geologists realize they cannot study only the present and the past, but that they have to see what humans are doing now that will affect the future.

Our View of Geology Today

Now, because of the space program, we are finding out as much about ourselves as we are finding out about other planets. As the Apollo 8 astronauts were returning to Earth on Christmas Day, 1968, after they had circled the Moon for the first time, poet Archibald MacLeish recorded his impressions with these words:

> For the first time in all of time men have seen the earth: seen it not as continents or oceans from the little distance of a hundred miles or two or three, but seen it from the depths of space—seen it whole and round and beautiful and small. . . . To see the earth as it truly is, small and blue and beautiful in that eternal silence where it floats, is to see ourselves as riders on the earth together, brothers on that bright loveliness in the eternal cold—brothers who know now they are truly brothers.

17 **The Earth as seen from a distance** of 160,000 km in December 1968. (NASA.)

This new color-television view of the Earth (Figure 17) has made it possible for us to see the world as geologists have now taught themselves to see it—as an entity, as a system operating within the geologic cycle—and to see what a remarkable place the Earth really is. It is terribly important for us to understand this fragile thing called the Earth and how it works, and what you can do with it, and what you can't do with it.

If you are diligent about your study of geology and you begin to understand how the Earth works, you will be able to go anywhere on the surface of the Earth and not feel like a stranger. You may not know the people, and you may not know the language, but you will find something that is familiar—and you will feel at home because you do understand it. You will probably know more about a place in just a few minutes of looking at it than the people who have lived there all their lives. This is probably the greatest thing you can expect from your study of geology—to acquire this new power.

Chapter Review

1 Geology is generally divided into two parts: *physical geology*, the study of the processes that shape the Earth, and *historical geology*, the systematic study of the past history of the Earth and life on Earth.

2 *Geologic time* is so vast that it is separated from other kinds of time that can be measured with a clock or a calendar.

3 The concept of the *geologic cycle* provides the idea that the activities we see taking place today explain the history of the Earth, and that the Earth's history is recorded in its rocks.

4 Previous geologic revolutions have culminated in what we call the revolution of *plate tectonics* and *continental drift*, which was made possible because of new information about the ocean floor.

5 James Hutton, regarded as the founder of modern geology, set forth a theory of slow and steady Earth change, which came to be known as the *doctrine of uniformitarianism*. Ultimately, this theory upset the *doctrine of catastrophism*, which claimed that the Earth's landscape had been shaped exclusively by sudden and dramatic events.

6 The Earth consists of a two-part *core*, a *mantle* of solid rock, and an outermost thin, rocky *crust*. The crust and the outer part of the mantle compose the *lithosphere*.

7 The surface of the Earth is underlain by *regolith*, loose noncemented fragments of rocks, and *bedrock*, continuous, solid rock that everywhere underlies regolith and in some places forms the surface itself.

8 Molten material solidifies to create *igneous rocks*. Sediment particles, accumulated in layers, become cemented to form sedimentary rocks.

Extreme modification (metamorphism) of already formed rocks creates new kinds named *metamorphic rocks.*

9   As rocks change from one type to another through sedimentation and metamorphism they are taking part in a larger process called the *rock cycle,* which operates continuously on and within the Earth's crust.

10 The *geologic time scale* has been devised by geologists for classifying geologic time. In the attempt to find out more about the Earth's history, the geologist uses such dating devices as sediments (including fossils and strata), and radioactivity.

11 In the study of the Earth, the geologist incorporates techniques from other scientific fields. Examples of such related areas are physics, chemistry, biology, and oceanography.

12 A new geologic theory, *plate tectonics,* touches nearly every aspect of geology and offers answers to many old and new mysteries. The theory claims that the Earth's crust is constructed of a dozen or so moving plates. One of the many unanswered questions about the theory is what makes the plates move.

Questions

1   How did the Hutton-Lyell concept of geology differ from Cuvier's concept?
2   Describe the over-all shape of the solid Earth.
3   List the main features of the interior of the Earth.
4   Viewed broadly, the outer part of the Earth can be thought of as consisting of three concentric spheres. What are the names of these three spheres? What is the chief composition of each sphere?
5   What is the *rock cycle?* Trace the movements of material through this cycle.
6   If the age of the Earth were scaled down to span one calendar year, in what month would we celebrate the "birthday" of the oldest living creatures?

Suggested Readings

Bailey, Sir Edward, *James Hutton — The Founder of Modern Geology.* New York: American Elsevier Publishing Co., Inc., 1967.

Beiser, Arthur, *The Earth.* New York: Time-Life Books, 1964.

Branley, Franklyn M., *The Earth: Planet Number Three.* New York: Thomas Y. Crowell Co., 1966.

Cailleux, A., *Anatomy of the Earth.* New York: World University Library, McGraw-Hill Book Company, 1968.

Eisley, Loren C., "Charles Lyell." *Scientific American,* August, 1959, pp. 98–106. (Offprint No. 846. San Francisco: W. H. Freeman and Company.)

Gamow, George, *A Planet Called Earth.* New York: The Viking Press, 1963.

Melson, William G., *The Forging of Our Continent.* New York: American Heritage Publishing Co., Inc., 1968.

"We have begun to realize how each living thing affects all the others."

Chapter One
Riders on the Earth Together

When the founders of modern geology visualized the geologic cycle it was not obvious that people were factors in that cycle. But about 100 years later, when we began to mobilize energy on a large scale in the late nineteenth century, a new situation was created: human beings had become active agents of geologic change. Before James Hutton introduced the idea that the Earth was continually changing, everyone thought that it had not changed substantially since the Creation. With a largely benevolent climate, seemingly limitless land, and fairly good conditions to grow food, the Earth seemed a true Garden of Eden. The whole concept of a frontier, and unlimited possibilities for growth and movement are typically American ideas; it is not easy for us to imagine that perhaps there are no more frontiers.

As we have come to understand the processes of nature, the picture has grown more complex. Most important, perhaps, has been our realization that the Earth and its resources are indeed limited. We have begun to understand that *Homo sapiens* is only one of many thousands of species of organisms inhabiting this planet, and that a single species cannot continue to multiply indefinitely. We have also begun to realize more fully how each living thing affects all the others, and how we are truly "riders on the earth together."

In this chapter we will attempt to sort out the effects of people on the geologic cycle in order to have a clearer understanding of the future of the cycle. Such an understanding is crucial to insure the survival of human beings.

The Geologic Cycle As It Affects Life on Earth

In broad terms, the geologic cycle includes all of the interactions between the materials of the Earth and its sources of energy. The **materials** of greatest interest to us are those near us, and consist mostly of rock and regolith, liquid water, and gaseous air. The **energy sources** affecting these materials are solar energy and, to a lesser degree, heat from the Earth's interior.

The number of possible interactions among energy, rock, water, and air is nearly infinite. Thus, the geologic cycle is an extremely complex amalgam of interrelated processes. In order to simplify the picture somewhat, we may divide these processes into a number of smaller cycles, all of which are related to human activities. In very general terms, the many cycles composing the geologic cycle can be described by the way they react with the four "spheres": (1) the lithosphere, (2) the hydrosphere, (3) the atmosphere, and (4) the biosphere.

It may appear peculiar to mention the biosphere (the part of the

Earth in which life exists) in a discussion of physical geology. However, cycles of birth, death, and rebirth in the biosphere are sustained by energy from the Sun, and by grand-scale cycles of chemical elements. These cycles, which involve the lithosphere, hydrosphere, and atmosphere, are all affected by the activities of humans—members of the biosphere. Humans, in turn, depend upon these same cycles for their existence. Therefore, human activities, the biosphere, and the geologic cycle are delicately interlinked.

The Lithosphere

The **lithosphere,** or "sphere of rock," includes all the rocky material of the Earth's outer shell, extending from the surface to a depth of about 100 km. The lithosphere, described briefly in the Introduction, is thought by some to consist of a mosaic of about a dozen giant "plates" which are in constant slow motion over the upper mantle. The lithosphere contains most of the bulk of the material involved in the geologic cycle, and its processes occur with extreme slowness. A single parcel of lithosphere, for example, may exist as part of the crust for several hundred million years before it returns to the mantle. In the Introduction the lithosphere was described in terms of the rock cycle—the creation, weathering, transporting, deposition, and alteration processes that produce soils, sediments, and the rocks of the crust.

The Hydrosphere

The **hydrosphere** is the body of liquid water on and near the surface of the Earth. Most of this water is in the oceans, which cover 70.8 per cent of the surface to an average depth of 4 km.

Of the many ways in which the hydrosphere is essential to the geologic cycle and to humans, three stand out. One is the role of water as transporter of regolith and as shaper of the Earth's surface. As we shall discuss in Chapter 8 and elsewhere in the book, water falling as rain erodes mountains, carves river valleys, and transports sediment from one place to another. Without the counteracting mountain-building processes, the actions of water would quickly, in geologic terms, produce a nearly level surface.

Second, the oceans act as a collection reservoir for most soluble materials on Earth, including those dissolved from the rocks and regolith by moving water. Third, water is a crucial ingredient for the basis of life, not only as drinking water for us, but as a life-giving source for all living things.

As the rock cycle is a sub-cycle of the master geologic cycle, so

Riders on the Earth Together

is the **hydrologic cycle,** or water cycle (Figure 1.1). Keeping in mind the over-all idea of the cyclical pattern of nature—constant movement, constant return to the source, constant rebirth—we can describe the hydrologic cycle in the following simple terms: With the help of solar energy, water *evaporates* from the ocean into the atmosphere, and the atmosphere carries it inland as water vapor. At this point, the water vapor may *condense* and fall from clouds (with the help of gravity) as *precipitation*—rain or snow, for example. The precipitated water may fall directly into rivers and be carried to the ocean, or it may return to the ocean via the more indirect route of underground flow. One way or the other, the water returns to the ocean source, ready to repeat the cycle again.

During the course of this cycle, water has exhibited persistent force and energy, and has inexorably altered the Earth's landscape. We will see later how such stupendous natural monuments as the Grand Canyon have been sculpted by the steady work of water, and we will also see how water can work its changes through more drastic means such as floods and rainstorms.

The Atmosphere

Like the water of the hydrosphere, the gases of the **atmosphere** may become part of various cycles of the geologic cycle. Both nitrogen and oxygen are essential to all living things, and may become part of an organism until it dies. The gases are then liberated to return to the atmosphere (Figure 1.2). Alternatively, nitrogen and oxygen may be-

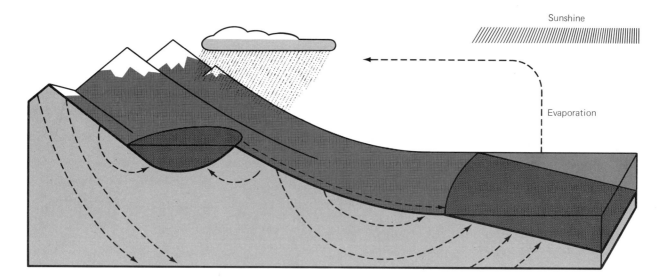

1.1 Major movements of water in the hydrologic cycle, from the ocean by evaporation, through the atmosphere and to the lands as rain or snow, and back to the oceans as flow over the ground and underground.

The Geologic Cycle As It Affects Life on Earth

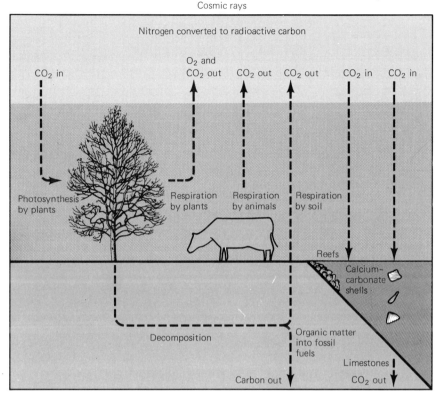

1.2 Some of the ways in which carbon dioxide moves cyclically from the atmosphere into and out of the lithosphere, the biosphere, and the hydrosphere. Volcanic activity and combustion of fossil fuels not shown.

come locked into the solid fabric of rocks, where they may remain for thousands or even millions of years.

Oxygen, in combined form, is the most abundant building block of crustal rocks, accounting for 29.3 per cent of the mass of the entire Earth. Likewise, carbon dioxide, which is needed by plants to make starch, is far more abundant in limestone, coal, petroleum, and other plant-derived materials of the lithosphere than it is in the atmosphere. Carbon dioxide is of great importance as a gas because of its temperature-regulating capacity. In the atmosphere, carbon dioxide tends to "trap" incoming solar heat in the so-called "greenhouse effect." High concentrations of this gas produce higher temperatures, just as a greenhouse does. Hence the complex carbon dioxide cycle is crucial to the habitability of the Earth.

The Biosphere

The **biosphere** is defined as that part of the Earth in which life exists. In general, the biosphere is about 20 km thick, limited to regions pro-

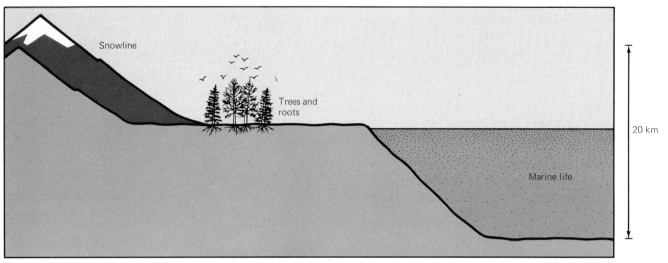

1.3 The biosphere illustrated by a schematic profile of the outer part extending from the top of the highest mountain to the deepest part of the sea. The upper limit of the biosphere for rooted plants is the snowline, above which the ground is covered by snow all year. In lower-lying land areas, the biosphere ranges from the depths of penetration of tree roots up into the atmosphere where birds and insects live. In the sea, most organisms live near the surface. The deep-sea organisms depend on plant food that is transported to the bottom.

viding ample quantities of liquid water and energy from the Sun.

It may at first seem foolish to worry about harming the biosphere, with its vast forests, endless schools of fish, and, of course, human beings themselves. Yet, we see that forests, fish, and humans alike owe their existence to an exceedingly thin and fragile film of plant life spread unevenly over the Earth's surface. The energy source upon which this film of life depends is the Sun. Solar energy enters the biological cycle by way of green plants, which need the Sun's energy to manufacture starch through the process of photosynthesis. Animals, which are unable to manufacture food, depend upon consumption of plants or other animals for survival.

In photosynthesis, the energy of sunlight powers the chemical reactions between carbon dioxide and water to produce oxygen and plant tissue, usually starch. Once again, nature provides a cycle of activity, this time utilizing the Sun's energy to help plants produce the food and energy they need to live. In the bargain, while they are transforming sunlight, plants "exhale" oxygen, which humans and other animals breathe; to pay our debt, we give off carbon dioxide as a waste product, the same carbon dioxide that plants need to start the cycle all over again. Of course, besides this over-simplified version of the exchange of gases, animals also benefit indirectly from the converted solar energy in the plants when they eat the plants.

Each of the grand cycles discussed above is related to one another and, in turn, to smaller cycles (Figure 1.3). The cycling of water is essential to many lithospheric processes, as we shall see in later chapters. Chemicals circulating through the atmosphere, hydrosphere, and lithosphere, such as carbon, oxygen, and nitrogen, are essential to living organisms. Modern civilizations are dependent upon parts of all these cycles. It is essential to our survival to learn more about them—and not to upset them.

Our Dependence on the Geologic Cycle

In increasingly sophisticated ways, humans have learned to tap the resources of the geologic cycle and use them to support ever-larger populations. As we have said, there are basically two kinds of resources associated with the geologic cycle: energy and materials. With development of industrial societies, the demand upon all resources has soared. Modern civilization differs from all past civilizations in the extent to which it has mobilized and harnessed energy resources and used them to refine and alter material resources. To our skill in doing this we owe our great wealth, high standard of living, and many of our worst problems.

The History of Our Dependence on Geologic Processes

Homo sapiens as a species is several million years old. Through almost all of this species' time on Earth, the individuals have been scattered in small, nomadic tribes that used only primitive tools, lived by hunting and fishing, and clothed themselves in animal hides. The materials used by these primitive creatures were products of the biosphere—wood or bone for implements, animal or plant products for food, animal hides for clothing. The only energy primitive people used was the solar energy used by the plants which fed them, and this energy was recycled to the biosphere through decaying implements and animal hides and excremental wastes from food.

Within the last 100,000 years or so, however, humans began to make technological innovations which increased their chances for survival—as well as their need for material and energy resources. Perhaps the earliest example was the use of fire, which provided warmth, extended the range of foods that could be eaten, and frightened large predators. With the use of fire came a demand for increasing amounts of firewood.

Riders on the Earth Together

About 10,000 years ago agriculture began, leading to huge increases in resource demands. For the first time, some members of society did not have to hunt for their food: farmers grew it for them. The non-farming members clustered in cities, which needed greater supplies of stone, wood, and other construction materials, drinking water, and wood or charcoal for heating. People became specialists. Some specialized in growing food, or in large-scale fishing; others quarried stone or dug up ores from which copper, iron, or other metals could be refined; still others organized systems of distribution or functioned as agents of government. Demand upon the Earth steadily increased; wealth and population increased also (Figure 1.4). New technologies were developed, leading to greater usage of raw materials such as stone, wood, clay, fiber, and skin.

As ancient urban centers used greater quantities of resources, they confronted the problem of how to dispose of the wastes and garbage—a problem that continues to the present day. Pigs, dogs, and scavenger birds, such as the kites of modern Calcutta, undoubtedly consumed some of this garbage, but even so, the sights, smells, and

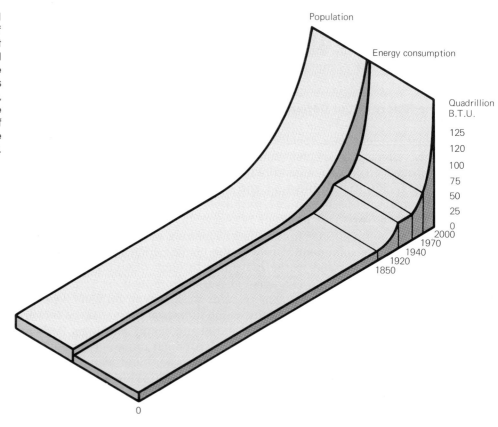

1.4 Graph of total energy used compares closely with graph of population increase. In fact, without the great increase in energy, it would not have been possible for the population to expand as much as it has. B.T.U. means British Thermal Unit, the amount of heat required to raise the temperature of one pound of water by 1 degree Fahrenheit (the standard degree being 39° to 40°F).

1.5 A wood-fired reverberatory furnace used to melt iron ore into steel, an engraving by T. Lamsvelt, made in 1647. (New York Public Library, Astor, Lenox, Tilden Foundation.)

bulk of these wastes must have been oppressive, and the garbage itself a source of diseases.

The melting of metals from rocks greatly increased the need for energy. Copper was the first metal to be used on a large scale. Because of its low melting point, copper required little heat for refining and was therefore very popular. Iron, however, was much more difficult to extract, and furnaces (Figure 1.5) hot enough to do the job were not developed until about 1100 B.C. in the Middle East. Forests throughout Europe were hacked into firewood to fuel iron furnaces, and the trees were never replaced.

The fossil-fuel revolution As England ran out of trees, the iron industry, threatened with extinction, had to find another source of heat. The development of coal for this purpose in the late seventeenth and eighteenth centuries in England launched the Industrial Revolution (Figure 1.6). The linking of coal to iron production was second in importance only to agriculture in changing the course of human history.

The further development of the steam engine was made possible by iron as a structural material and coal as an energy source. The advent of the steam engine rapidly overturned many aspects of life, such as the shipbuilding industry, where wooden sailing ships were quickly replaced by iron steamships. As the nineteenth century ended,

1.6 Manchester, England during Industrial Revolution. Engraving by W. Wyld. (Royal Collection, London, courtesy New York Public Library Astor, Lenox, Tilden Foundation.)

steam combines appeared on farms and quickly took the place of horses and mules—only to be replaced themselves by the internal combustion engine.

The internal combustion engine functions on petroleum products, most often gasoline. Coal, petroleum, and natural gas (the gaseous fractions of petroleum) are called **fossil fuels** because they are formed from the altered remains of organisms, mainly plants. Thus, fossil fuels represent an "accident" of the biological cycle. Instead of being recycled in the biosphere upon the death of the organisms, these fossilized remains are withdrawn from the biological cycle indefinitely. The prodigious rate at which we are now burning these remains is causing their sudden abnormal reintroduction to the atmospheric cycle—abnormal both because of the suddenness and because of the high temperatures at which they are burned. These temperatures produce nitrogen oxides, sulfur oxides, and other gases not normally produced by decomposition of organisms in the biosphere.

Human Beings as a Geologic Force

Since the beginning of the Industrial Revolution, when consumption of both material and energy resources began to rise as never before in history, human activities have rapidly begun to have an impact upon the geologic cycle. Americans use energy at a rate equivalent to burning about 10 tons of coal per year per person. A rough formula for the energy required just to maintain the fabricated steel in use can be cited: for each ton of steel, we must burn about one ton of coal or an equivalent energy source. Another figure often used is that Americans, composing about 6 per cent of the world population, account for about a third of the world's annual energy consumption.

All this activity cannot fail to have a significant impact upon the geologic cycle, and an important lesson to be learned is that activities in one area inevitably affect what happens elsewhere. Let us examine a few examples of specific effects of human activity upon segments of the geologic cycle. We will look at our effect on the lithosphere, the hydrosphere, the atmosphere, and the biosphere.

Our Impact on the Lithosphere

To an ever more obvious degree, the lithosphere has been altered by the extraction of resources, the construction of cities, factories, and residential areas, and other human activities.

Mining After gold was discovered in the Sierra Nevada in 1849, hydraulic mining began in the old stream deposits exposed on the valley sides in the foothills. Streams of water were shot with high pressure against the gravel (Figure 1.7). Gold, gravel, and water

1.7 Hydraulic mining for gold in the foothills of the Sierra Nevada, California.

were washed down, and passed through special troughs, where the gold was recovered. Then the water and the gravel were simply turned back into the streams. This waste material flowed downstream and spread out on the farmland below, ruining the fields and orchards. An inevitable struggle of conflicting economic interests arose, since both the miners and farmers had a high economic stake in the outcome. Ultimately, the farmers were victorious, and hydraulic mining was halted.

More recently, we have seen how the surface can be drastically modified by the machines of a strip-mining operation (Figure 1.8). Thousands of tons of rock and soil are displaced, morphology is altered, stream flow may be displaced, and fresh-water supplies may be contaminated by acids leached from newly exposed rock. Vegetation is destroyed and may not become re-established for many decades. As demand for coal rises in the coming years, strip mining may increase, and unless care is taken, so will undesirable environmental effects.

Erosion Numerous human activities expose regolith to erosion. One is clearing the land for construction, which is described in Chapter 8. Another is agriculture. In order to produce crops, it is necessary to "clear" large parcels of land. This clearing eliminates

1.8 Strip mining and land reclamation in southeastern Ohio west of Caldwell, about halfway between Zanesville and Marietta. (Ohio Power Company, American Electric Power System.)

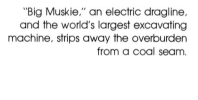

"Big Muskie," an electric dragline, and the world's largest excavating machine, strips away the overburden from a coal seam.

the normal plant cover, whose roots hold soil and whose leaves break the fall of eroding raindrops. Where plant cover is absent, the erosion of topsoil is often severe (Figure 1.9). When farming practices are poor, as they have been in much of the world until the present century, topsoil may be carried away so rapidly that farmers must move every few years, clearing new lands with fires and machetes. This "slash-and-burn" agriculture is still practiced in many parts of South America and Asia.

Overgrazing, too, can destroy the fertility of soil. The introduction of goats in much of the Middle East many centuries ago led to the elimination of virtually all forms of protective vegetation over large regions. Overgrazing by cattle and sheep in the United States during the 1930s, and more recently in Africa, has destroyed grasses and other plants, exposing bare soil to erosion by wind and rain.

Here we see how the alteration of one "sphere" can affect other spheres. Strip mining, which alters the lithosphere, also changes stream flow (hydrosphere) and local vegetation (biosphere). Poor farming practices change the composition of the soil (lithosphere), which severely restricts plant cover (biosphere). Without plants, less water is released from the soil through transpiration (atmosphere) and less water seeps into the subsurface (lithosphere). Thus, changes in one cycle may unbalance others in severe and unpredictable ways.

Park and pond formed after grading and planting of trees in 1961; photographed in May 1972. Last strip mining nearby took place in 1959–60.

Man and Nature George P. Marsh

Man has too long forgotten that the earth was given to him for usufruct alone, not for consumption, still less for profligate waste. Nature has provided against the absolute destruction of any of her elementary matter, the raw material of her works; the thunderbolt and the tornado, the most convulsive throes of even the volcano and the earthquake, being only phenomena of decomposition and recomposition. But she has left it within the power of man irreparably to derange the combinations of inorganic matter and of organic life, which through the night of aeons she had been proportioning and balancing, to prepare the earth for his habitation, when, in the fulness of time, his Creator should call him forth to enter into its possession.

Apart from the hostile influence of man, the organic and the inorganic world are, as I have remarked, bound together by such mutual relations and adaptations as secure, if not the absolute permanence and equilibrium of both, a long continuance of the established conditions of each at any given time and place, or at least, a very slow and gradual succession of changes in those conditions. But man is everywhere a disturbing agent. Wherever he plants his foot, the harmonies of nature are turned to discords. The proportions and accomodations which insured the stability of existing arrangements are overthrown. Indigenous vegetable and animal species are extirpated, and supplanted by others of foreign origin, spontaneous production is forbidden or restricted, and the face of the earth is either laid bare or covered with a new and reluctant growth of vegetable forms, and with alien tribes of animal life. These intentional changes and substitutions constitute, indeed, great revolutions; but vast as is their magnitude and importance, they are, as we shall see, insignificant in comparison with the contingent and unsought results which have flowed from them.

There are, indeed, brute destroyers, beasts and birds and insects of prey—all animal life feeds upon, and, of course, destroys other life,—but this destruction is balanced by compensations. It is, in fact, the very means by which the existence of one tribe of animals or of vegetables is secured against being smothered by the encroachments of another; and the reproductive powers of species, which serve as the food of others, are always proportioned to the demand they are destined to supply. Man pursues his victims with reckless destructiveness; and, while the sacrifice of life by the lower animals is limited by the cravings of appetite, he unsparingly persecutes, even to extirpation, thousands of organic forms which he cannot consume.

The earth was not, in its natural condition, completely adapted to the use of man, but only to the sustenance of wild animals and wild vegetation. These live, multiply their kind in just proportion, and attain their perfect measure of strength and beauty, without producing or requiring any change in the natural arrangements of surface, or in each other's spontaneous tendencies, except such mutual repression of excessive increase as may prevent the extirpation of one species by the encroachments of another. In short, without man, lower animal and spontaneous vegetable life would have been constant in type, distribution, and proportion, and the physical geography of the earth would have remained undisturbed for indefinite periods, and been subject to revolution only from possible, unknown cosmical causes, or from geological action.

But man, the domestic animals that serve him, the field and garden plants the products of which supply him with food and clothing, cannot subsist and rise to the full development of their higher properties, unless brute and unconscious nature be effectually combated, and, in a great degree, vanquished by human art. Hence, a certain measure of transformation of terrestrial surface, of suppression of natural and stimulation of artificially modified productivity becomes necessary. This measure man has unfortunately exceeded. He has felled the forests whose network of fibrous roots bound the mould to the rocky skeleton of the earth; but had he allowed here and there a belt of woodland to reproduce itself by spontaneous propagation, most of the mischiefs which his reckless destruction of the natural protection of the soil has occasioned would have been averted. He has broken up the moun-

From MAN AND NATURE by George P. Marsh, 1864. The Belknap Press of Harvard University Press, 1965. Abridged with permission.

Field Trip

tain reservoirs, the percolation of whose waters through unseen channels supplied the fountains that refreshed his cattle and fertilized his fields; but he has neglected to maintain the cisterns and the canals of irrigation which a wise antiquity had constructed to neutralize the consequences of its own imprudence. While he has torn the thin glebe which confined the light earth of extensive plains, and has destroyed the fringe of semi-aquatic plants which skirted the coast and checked the drifting of the sea sand, he has failed to prevent the spreading of the dunes by clothing them with artificially propagated vegetation. He has ruthlessly warred on all the tribes of animated nature whose spoil he could convert to his own uses, and he has not protected the birds which prey on the insects most destructive to his own harvests.

Purely untutored humanity, it is true, interferes comparatively little with the arrangements of nature, and the destructive agency of man becomes more and more energetic and unsparing as he advances in civilization, until the impoverishment, with which his exhaustion of the natural resources of the soil is threatening him, at last awakens him to the necessity of preserving what is left, if not of restoring what has been wantonly wasted. The wandering savage grows no cultivated vegetable, fells no forest, and extirpates no useful plant, no noxious weed. If his skill in the chase enables him to entrap numbers of the animals on which he feeds, he compensates this loss by destroying also the lion, the tiger, the wolf, the otter, the seal, and the eagle, thus indirectly protecting the feebler quadrupeds and fish and fowls, which would otherwise become the booty of beasts and birds of prey.

It has been maintained by authorities as high as any known to modern science, that the action of man upon nature, though greater in *degree*, does not differ in *kind*, from that of wild animals. It appears to me to differ in essential character, because though it is often followed by unforeseen and undesired results, yet it is nevertheless guided by a self-conscious and intelligent will aiming as often at secondary and remote as at immediate objects. The wild animal, on the other hand, acts instinctively, and, so far as we are able to perceive, always with a view to single and direct purposes. The backwoodsman and the beaver alike fell trees; the man that he may convert the forest into an olive grove that will mature its fruit only for a succeeding generation, the beaver that he may feed upon their bark or use them in the construction of his habitation. Human differs from brute action, too, in its influence upon the material world, because it is not controlled by natural compensations and balances. Natural arrangements, once disturbed by man, are not restored until he retires from the field, and leaves free scope to spontaneous recuperative energies; the wounds he inflicts upon the material creation are not healed until he withdraws the arm that gave the blow. On the other hand, I am not aware of any evidence that wild animals have ever destroyed the smallest forest, extirpated any organic species, or modified its natural character, occasioned any permanent change of terrestrial surface, or produced any disturbance of physical conditions which nature has not, of herself, repaired without the expulsion of the animal that caused it.

The earth is fast becoming an unfit home for its noblest inhabitant, and another era of equal human crime and human improvidence, and of like duration with that through which traces of that crime and that improvidence extend, would reduce it to such a condition of impoverished productiveness, of shattered surface, of climatic excess, as to threaten the depravation, barbarism, and perhaps even extinction of the species. 1864

1.9 **Eroded, gullied land** being rapidly carried away after former cover of vegetation was removed by a fire. (U.S. Department of Agriculture, Soil Conservation Service photo by J. G. James.)

Our Impact on the Hydrosphere

Human effects upon the hydrosphere first became widely publicized during the 1950s, when "water pollution" became a household phrase. Foamy suds formed on streams (Figure 1.10) as the result of the heavy use of detergents for washing dishes and clothes, and industrial and municipal sewage marred many streams, rivers, and lakes.

Since this early alarm, far more subtle effects have been revealed, often far from the source of pollution. Minute amounts of pesticides have been discovered in the bodies of penguins. Because these birds feed upon oceanic fish of the Southern Hemisphere, far from any pesticide sources, we can only conclude that pesticides, once released into the environment, may disperse and travel almost without limit.

Similarly, particles of plastic were recently found in the bodies of fishes along the east coast of North America. Tiny clumps of tar may be collected in drag nets virtually anywhere in the world's oceans, as may patches of floating petroleum which leaks from tankers. Many other chemicals, produced and used by humans, are carried by pipes and rivers to the sea where they may circulate wherever the currents take them.

Other effects on the hydrosphere include withdrawal of underground water. When underground water is used faster than it is replenished by natural cycles the water level is lowered. Such lowering causes the land surface to subside. Extensive use of water for irrigation exposes the water to the Sun and the resultant evaporation increases the salt content of the water, ultimately making it useless for agriculture.

Our Impact on the Atmosphere

Some of us may think that only city dwellers suffer from atmospheric pollution, but the problem is far more general (Figure 1.11). Already, the use of materials and energy by humans has produced measurable and perhaps dangerous changes in the atmosphere. These changes are by no means confined to cities, but extend around the world.

1.10 Detergent suds on a stream near Gainesville, Florida. (Lovett Williams, courtesy National Audubon Society.)

Jet-age pollution As mentioned above, we can trace pollutants to areas far from their sources. This is especially true in the case of high-altitude atmospheric pollutants. For example, we are still receiving evidence of nuclear fallout appearing in isolated localities at unpredictable distances from atmospheric weapons tests. Curiously, sites in between show no sign of fallout.

The fallout pattern from weapons tests was mysterious at first, but finally the evidence became clear that the fallout was inserting itself into the high-altitude atmospheric jet streams. Because of observations made from high-flying jet aircraft, rockets, and high-altitude balloons, we have learned about a kind of stratospheric circulation that James Hutton could never have dreamed about. This is a space-age discovery, but despite its newness we can see that such atmospheric circulation coincides with other aspects of the geologic cycle. Obviously, the interlocking, interlinking facets of the geologic cycle are always in operation.

Dust The total amount of particulate matter (less than five millionths of a meter, or five micrometers, in diameter) in the atmosphere is estimated at 40 million tons. As much as 10 million tons of

1.11 Polluted air from industrial wastes hovers thickly over town in Moselle Valley near Cologne, West Germany. (April 1963; United Nations.)

this amount is derived from human activities, such as farming and burning fossil fuels. The effect of these particles upon temperature is still not clear, because it depends upon the absorption of different fractions of solar energy. But there is certain to be some thermal effect, and also an effect upon cloud formation. Because the dust content of the atmosphere is expected to grow by about 60 per cent by the end of this century, there is urgent need to understand the interaction between particulates and the atmosphere.

But so-called pollutants are not always of human origin. In October 1974 a volcanic eruption in Guatemala shot dust high into the atmosphere. This dust has now traveled around the world, and has caused spectacular multicolored sunsets from Hawaii to France. The sunsets are lovely to observe, but there is still a possibility that the dust may be reducing our source of solar energy, however slightly, by blocking the Sun's rays.

Fluorocarbon gases ("Freon") A rather thin layer of ozone (molecules consisting of three oxygen atoms) surrounds the Earth. Ozone molecules are created by the collision of cosmic rays and other incoming particles with oxygen atoms. Ozone is responsible for blocking the Sun's long-wave ultraviolet radiation. Without the ozone layer, this radiation would be lethal to all forms of life. Some scientists think that the use of aerosol spray cans, which propel such products as insecticides, room sprays, hair sprays, and shaving cream, are changing this layer by releasing molecules of fluorocarbon gases.

Although fluorocarbon gases (most of which are sold under the trade name of "Freon") are used because they are compressible and stable at the Earth's surface, Freon contains chlorine, which can react with ozone. The Freon circulates upward to the ozone layer. Here molecules of ozone interact with the chlorine bound up in the Freon. This interaction destroys both the Freon and the ozone, and thus could threaten the usefulness of the ozone layer to us. What is not known is how the rate at which new ozone forms compares with the rate at which ozone is being destroyed.

Our Impact on the Biosphere

Human effects upon the biosphere are probably greater than upon any other sphere of the geologic cycle. Some of these effects are easy to measure, such as the near-extinction of the buffalo, the blue whale, and other animals, and the alteration of global vegetation patterns by agriculture. Other effects are not so easy to perceive, and these may be the most consequential of all. For example, we all carry in our tissues measureable levels of certain human-made or human-liberated substances, such as DDT, carbon monoxide, and lead. The meat we eat may contain minute amounts of hormones and antibiotics injected into pigs, chickens, and cattle. Fruits and vegetables bear residues of insecticides, herbicides, fungicides, and other chemicals that human bodies have never encountered before. Such chemicals are similarly dispersed through the biosphere, threatening the existence of some species and favoring others. By this means, we may be subtly changing the balance of insects and other members of the biosphere. There is no way to tell whether such changes are beneficial or detrimental to our existence as a species.

It is now clear that humans are extracting some materials from the geologic cycle and introducing others at rates that cannot continue forever. With our new-found ability to tap the energy of coal, petroleum, and even the atomic nucleus, we have learned to support a world population that would have been impossible only a few hundred years ago.

Resources and Human Populations

The Great "Escape from Nature"

Our tremendous use of energy and materials makes it possible to sustain highly mechanized agricultural and industrial systems that produce inexpensive food, clothing, appliances, and electricity. Just 100 years ago, 94 per cent of the power used in industry came from people's muscles. Today, human power accounts for less than 8 per cent. The rest is supplied for us by fossil fuels and other sources of energy. So it may seem to many city dwellers that we have, in many ways, "escaped from nature." Apartments are heated in winter by crude oil or natural gas, and cooled in summer by electrically powered air conditioning (Figure 1.12). Food is brought from farms by gasoline-powered trucks, which are made of extracted iron and aluminum and roll on tires woven from synthetic petroleum products. Food is readily available at huge supermarkets. Some is frozen or cooled by electricity, which in turn requires petroleum, coal, or nuclear

1.12 A completely controlled environment inside the Galleria, Houston, Texas ranges from an ice-skating rink on the lowest level shown to sidewalks for shoppers higher up. (Owner/developer, G. D. Hines Interests.)

energy to produce. Other food is "processed," which requires more energy, and most of it is wrapped in plastic or paper whose production also consumes energy.

In short, "nature" exists in the minds of many people only as an abstraction, something to read about or to visit on weekends (at the cost of many gallons of gasoline), much as we might stare curiously at animals in a zoo. Geology, as part of nature, seems equally remote and unreal. People who inhabit the great urban and suburban regions of the United States and other countries often seem to have broken out of the great cycles we have been discussing.

This escape from nature is, of course, an illusion, one which is quickly shattered during times of geologic crisis. The Arab oil embargo of 1973 produced such a crisis, when supplies of imported oil suddenly dropped by more than 10 per cent. In the space of a few weeks, urban populations rediscovered their links—and debts—to geology and the geologic cycle. During a nightmarish period of gasoline shortages, long waiting lines at fuel pumps, and rapid price rises, people were jolted into an awareness of their total dependence upon petroleum and the geologic cycle. The importance of the word "energy" became universally recognized, and government planners realized that the age of cheap energy and gluttonous consumption was drawing to a close. We can withdraw from natural cycles only at great expense of energy. It has suddenly become apparent just how expensive those sources of energy are.

Physical Geology in Perspective

As it began to seem that we were successful in insulating ourselves from nature some people believed that we had become immune from natural laws and that only the other animals had to worry about survival. Indeed, there may have been some basis for such thinking in the past, but now we realize how dependent we are on our energy sources, and how treacherous the struggle for survival can be.

In order to participate in the struggle more intelligently we must think of the geologic cycle not as an abstract generalization, but as the basic framework for the study of geology. We can best understand the background against which the struggle takes place if we start by looking at the Earth's materials.

Our objective is to understand rocks, but first we must have a basic understanding of silicate minerals, energy, atoms, and matter in general. In the next chapters we will start with atoms, proceed to rocks, and then move on to the larger processes and structures of the

Riders on the Earth Together

Earth. Ultimately, in the spirit of the natural cycles we have been describing, we will see how small the Earth is, how much like an atom it really is. Maybe we will see the Earth "whole and round and beautiful and small." And the smaller it is, the more we need one another.

Chapter Review

1 Since the late nineteenth century human beings have become factors in the geologic cycle, and active agents of geologic change. We have now begun to realize that the resources of the Earth are limited.

2 In broad terms, the *geologic cycle* includes all of the interactions between the materials of the Earth and its sources of energy.

3 The many cycles composing the geologic cycle can be described by the way they react with the four "spheres": (1) the lithosphere, (2) the hydrosphere, (3) the atmosphere, and (4) the biosphere.

4 The *lithosphere* includes all the rocky material of the Earth's outer shell, extending from the surface to a depth of about 100 km. The lithosphere is thought to consist of a mosaic of about a dozen giant "plates" which are in constant slow-motion over the upper mantle.

5 The *hydrosphere* is the body of liquid water on and near the surface of the Earth. Most of this water is in the ocean, which covers 70.8 per cent of the surface to an average depth of 4 km. Water transports regolith and shapes the Earth's surface, the oceans act as a collection reservoir for most soluble materials on Earth, and water is a basic life-giving source for all living things.

6 As the rock cycle is a sub-cycle of the master geologic cycle, so is the *hydrologic cycle*. With the help of solar energy the hydrologic cycle removes water from the ocean and eventually returns it to the ocean source, where the entire cycle is repeated again. During the course of this cycle water has inexorably altered the Earth's landscape.

7 Like the water of the hydrosphere, the gases of the *atmosphere* may become part of the various cycles of the geologic cycle. Circulating gases such as nitrogen, oxygen, and carbon dioxide are crucial to the habitability of the Earth.

8 The *biosphere* is defined as that part of the Earth in which life exists. Human activities, the biosphere, and the geologic cycle are delicately interlinked. Solar energy enters the biological cycle by way of green plants, which need the Sun's energy to manufacture starch. They do this through the process of *photosynthesis*, which is also ultimately essential to sustain animal life.

9 As humans began to make technological innovations the demand upon the Earth steadily increased; wealth and population increased also. Dur-

ing the Industrial Revolution in England the linking of coal to iron production was second in importance only to agriculture in changing the course of human history.

10 To an ever-more obvious degree, the lithosphere has been altered by the extraction of resources, the construction of cities, of factories, and of residential areas, and other human activities.

11 Human effects upon the hydrosphere and atmosphere are most evident in water pollution and air pollution. Such materials as dust and "Freon" are becoming important factors in the over-all pollution of the atmosphere.

12 Human effects upon the biosphere are probably greater than upon any other sphere of the geologic cycle. Some familiar examples of human influence are the near-extinction of some animals, residual effects of insecticides and other chemicals, and the alteration of global vegetation by agriculture.

Questions

1 The rocky material of the Earth's outer shell to a depth of about 100 km is called the _____.
2 That part of the Earth in which life exists is defined as the _____.
3 Why has the biosphere recently become more important in the over-all study of physical geology?
4 Briefly describe the *hydrologic cycle*.
5 How does modern civilization differ from past civilizations in its use of energy sources and material resources?
6 Define *fossil fuels*, and describe the impact of the *fossil-fuel revolution*.
7 Describe some of the human activities that have altered the lithosphere.

Suggested Readings

National Research Council and National Academy of Sciences, *The Earth and Human Affairs.* San Francisco: Canfield Press, 1972.

Murdoch, W. W., editor, *Environment, Resources, Pollution, and Society.* Stamford: Sinauer Associates, Inc., 1971.

A Scientific American Book, *The Biosphere.* San Francisco: W. H. Freeman and Company, 1970.

Thomas, W. L., Jr., editor, *Man's Role in Changing the Face of the Earth.* Chicago: University of Chicago Press, 1956.

"Periodically the North and South Poles exchange places."

Chapter Two
Matter, Energy, and Minerals

52
Matter, Energy, and Minerals

In geology, some knowledge of the fundamental structure of matter is essential in order to understand how minerals and rocks form and change through time.

Matter

The minerals and rocks that geologists study present an astounding array of colors, shapes, and textures. Yet despite this diversity, all minerals are very closely related for a simple reason: all minerals are composed of a few kinds of minute particles called **atoms.** As long ago as 400 B.C., the Greek philosopher Democritus proposed that all matter was composed of tiny indivisible blocks. Not until the twentieth century, however, did scientists discover that each atom is not indivisible at all, but is composed of even smaller **sub-atomic particles.** We define **matter** as anything that has mass and occupies volume.

Atomic Structure of Matter

So many sub-atomic particles have been discovered in recent years that they are sometimes said to be part of the "nuclear zoo." Three particles, however, are by far the best known.

2.1 Proton, electron, and neutron, drawn to scale. Neutral charge of neutron is shown as (±) instead of (0) to indicate that removal of an electron can convert a neutron into a proton. The difference between values of mass of neutron and combined masses of proton and electron represents energy needed to join the particles.

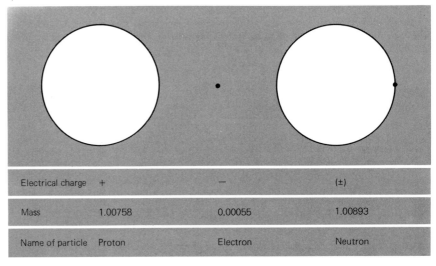

Protons, electrons, and neutrons The three most important sub-atomic particles are *protons, electrons,* and *neutrons* (Figure 2.1). Varying numbers of protons and neutrons make up the **atomic nucleus.** The nucleus is so dense that it holds 99.95 per cent of the **mass,** or amount of matter, of an atom—and sometimes more. (Mass is a measurement of the amount of solid matter in a material. In a gravitational field, such as that of the Earth, mass may be thought of as weight, or the force by which an object is attracted by the Earth.) The nucleus is so small that if the hydrogen atom were the size of the Houston Astrodome, the nucleus would be only the size of a marble placed on second base. The path of the electron of the hydrogen atom would be represented by the curve of the dome itself. A group of 1837 electrons together would have the same mass as a single proton. Thus, because almost all of the mass of an atom is tightly packed into the nucleus, the atoms that compose even the heaviest rocks are mostly empty space.

Electrical charge The force that keeps the whirling particles together is known as *strong interaction force.* Protons, which bear positive **electrical charge,** hold a strong attraction for electrons, which bear opposite, or negative electrical charge. Neutrons have no charge. The charges of protons and electrons effectively cancel one another, so that an entire atom is electrically neutral, or non-charged. A charged atom, which may have gained or lost one or more electrons, is an **ion.** Because of their electrical charge, ions can react with other ions; non-charged atoms cannot enter into chemical reactions.

Atomic size, mass, number An individual atom is so small that if enlarged 100 million times, it would be only the size of a pea—and its sub-atomic particles would still be invisible under a microscope. Atomic mass (or weight), too, is minute; it would take, for example, 270 trillion trillion protons (270 with 24 zeros) to make a mass weighing $\frac{1}{2}$ kg. Instead of using such unwieldy numbers, scientists say a proton has a **mass number** of one; other sub-atomic particles can then be related to this figure. An electron is small enough so that its mass usually can be disregarded. By contrast, a neutron has almost the same mass as that of a proton. The helium nucleus, which contains two protons and two neutrons, has a mass number of 4. This atom bears an **atomic number** of 2 because it has two protons. For example, the element oxygen, of atomic number 16, contains 16 protons and 16 neutrons, giving it a mass number of 32 (Figure 2.2).

2.2 Atoms of helium and oxygen, schematic profiles. The single large dot in outside circle of helium atom (left) represents two electrons. In oxygen atom (right) the heavy dots represent 2, 8, and 6 electrons, respectively (moving outward from the center).

Elements As we have seen, the number and behavior of electrons are determined by the number of protons in the nuclei of atoms. All atoms having the same number of protons are therefore considered the same kind of matter, or **element**. All atoms of the element hydrogen have one proton; all atoms of the element helium have two protons, and so on. In nature, only 92 elements are found (Table 2.1). In the laboratory, man has made more than a dozen additional elements, all more massive than those occurring naturally.

Isotopes Atoms with the same number of protons but a different number of neutrons are sister forms called **isotopes**. The element oxygen, for example, has several isotopes. Oxygen-16 is the most abundant, having 16 protons and 16 neutrons. Oxygen-17 has 16 protons and 17 neutrons; oxygen-18 has 16 protons and 18 neutrons; and so on.

Radioactivity Some isotopes have combinations of protons and neutrons that are stable, and some have combinations that are unstable. An isotope with an unstable nuclear package tends to "decay" spontaneously into a "daughter" isotope. Decay may be accomplished by emitting radiation (X-rays and gamma rays) or by firing out any of a variety of particles, such as electrons or neutrons. The spontaneous decay that gives rise to these emissions is **radioactivity**.

To geologists, the most important aspect of radioactive decay is its constancy. Particles and waves are emitted at a constant rate, and this rate can be used as a long-term timer, or "clock." The rate of decay of a radioactive isotope is defined by its half-life—the time required for one-half of its mass to be converted into a daughter isotope. One-half of a given mass of carbon-14 changes into its daughter product, nitrogen-14, in 5570 years, so that its half-life is 5570 years. Likewise, the half-life of potassium-40 is 1.3 billion years, and of uranium-238, 4.5 billion years.

Radioactive dating has become one of the geologist's most useful tools—and one of geology's greatest contributions to science. With this tool, archeologists can date human artifacts created several thousand years ago; paleontologists have dated the remains of human ancestors who lived several million years ago; and geologists have dated rocks that solidified several billion years ago. After centuries of confusion, radioactive dating has given us the first accurate approximations of the age of oldest rocks on Earth (3.8 billion years) and of the age of the Earth itself—about 4.6 billion years.

Molecules and compounds Ions (but not neutral atoms) of one element, or of more than one element, may combine to form **mole-**

Table 2.1
SELECTED ELEMENTS OF THE PERIODIC CHART OF THE ELEMENTS

Name	Symbol	Atomic Number	Atomic Weight
Hydrogen	H	1	1.0
Helium	He	2	4.0
Lithium	Li	3	6.9
Boron	B	5	10.8
Carbon	C	6	12.0
Nitrogen	N	7	14.0
Oxygen	O	8	16.0
Fluorine	F	9	19.0
Neon	Ne	10	20.2
Sodium	Na	11	22.9
Magnesium	Mg	12	24.3
Aluminum	Al	13	26.9
Silicon	Si	14	28.0
Phosphorus	P	15	30.9
Sulfur	S	16	32.0
Chlorine	Cl	17	35.5
Argon	A	18	39.9
Potassium	K	19	39.1
Calcium	Ca	20	40.0
Titanium	Ti	??	47.9
Manganese	Mn	25	54.9
Iron	Fe	26	55.8
Cobalt	Co	27	58.9
Nickel	Ni	28	58.7
Copper	Cu	29	63.5
Zinc	Zn	30	65.4
Bromine	Br	35	80.0
Krypton	Kr	36	83.8
Rubidium	Rb	37	85.5
Strontium	Sr	38	87.6
Silver	Ag	47	107.8
Tin	Sn	50	118.7
Iodine	I	53	126.9
Xenon	Xe	54	131.3
Platinum	Pt	78	195.2
Gold	Au	79	197.0
Mercury	Hg	80	200.6
Lead	Pb	82	207.2
Radium	Ra	88	226.0
Uranium	U	92	238.0

cules. When two ions of chlorine bond together, the result is a molecule of chlorine gas. Such a molecule is the smallest unit having the properties of chlorine gas. When ions of different elements combine in a liquid, the molecule they form is called a **compound.** When an ion of sodium joins an ion of chlorine, the result is a molecule of the compound sodium chloride. This compound, known to most people as table salt and to geologists as *halite,* is quite unlike either of the elements that compose it—the metal sodium and the gas chlorine. Other molecules of compounds found in rock materials may contain dozens of ions.

States of matter Matter is found in three states: (1) solid, (2) liquid, and (3) gas. In solid form, such as ice, the ions remain rigidly in place like soldiers at attention. In liquid, such as water or petroleum (the only two found in great abundance in nature), the ions are nearly as closely spaced as in solids, but are surrounded by thin envelopes of molecules, hence are free to slip around one another; this freedom of movement causes a liquid to flow. In gases, such as water vapor, ions are even more excited, or hotter, than in liquids. In gases, atoms travel in pairs, or as molecules, moving at higher speed and colliding frequently. When enough energy is added to a liquid it evaporates, becoming a gas. Therefore, evaporation of water is speeded by the addition of heat. Similarly, warm air can hold more water vapor than cool air: the molecules of nitrogen and oxygen composing air move more energetically when the air is warm. Conversely, when air cools, molecules of water vapor lose energy and tend to condense into rain.

Table 2.2
COMPOSITION OF THE EARTH'S CRUST

Element		Proportion (per cent)	
Name	Chemical Symbol	By Weight	By Volume
Oxygen	O	46.6	93.8
Silicon	Si	27.7	0.9
Aluminum	Al	8.1	0.5
Iron	Fe	5.0	0.4
Calcium	Ca	3.6	1.0
Sodium	Na	2.8	1.3
Potassium	K	2.6	1.8
Magnesium	Mg	2.1	0.3
All other elements		1.5	—
		100.0	100.0

Source: Brian Mason, Principles of Geochemistry, Second Edition. New York: John Wiley and Sons, Inc., 1958.

Elements in the Earth's crust Although 92 elements are found in nature, just eight of them make up nearly 99 per cent of the Earth's rocky crust (Table 2.2). Contrary to what might be expected, oxygen is by far the most common element in the crust. This oxygen is not, of course, in a gaseous state, but is locked in solid compounds with other elements. Because oxygen is a relatively light element, it accounts for nearly 94 per cent of the solid rocks in terms of volume. On the average, the rocks of the Earth can be pictured as intricate arrangements of oxygen atoms linked by varying amounts of the elements silicon, aluminum, iron, calcium, sodium, potassium, and magnesium.

Ions and Bonding

Even though atoms are mostly empty space, they are "hard" in one sense. The clouds of tiny electrons are bound to the nucleus by the strongest force known in nature, called the *strong interaction force*. In this way they form "bumpers" for the nucleus, protecting it against collisions with other atoms.

Energy-level shells Electrons do not fly just anywhere around the nucleus. They are organized in regions called **energy-level shells** that may be visualized as similar to the layers of an onion (Figure 2.3). Each shell is capable of holding a fixed number of electrons—and no more. The first shell can hold 2, the second can hold 8, the third 18, and so on. In the largest atoms, as many as seven shells have been identified. The farther from the nucleus a shell is located, the weaker the force binding its electrons to the nucleus. Thus a third-shell electron has lower energy than a first-shell electron, and is easier to knock out of orbit than a first-shell electron. When atoms become ions the changes take place in their outer, not inner, electrons.

Ionic bonding Atoms are most stable when their energy-level shells are full. An atom of neon has 10 protons and 10 electrons. Two of these electrons fill the first shell, and eight fill the second. There are no missing or extra electrons, so neon never loses or gains any electrons. And because neon always remains in its electrically neutral atomic form it does not interact chemically with other atoms.

An atom of sodium has 11 electrons. Two fill the first shell and eight fill the second, but one is left over. This situation is highly unstable; sodium tends to give away this electron. Chlorine has seven electrons in its outer shell, a situation quite different from that of sodium but equally unstable. Whenever possible, chlorine will

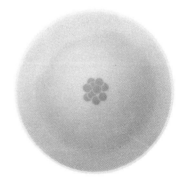

2.3 Neon atom showing energy-level shells. The single, inner shell contains two electrons. The double second shell contains two electrons in the inner shell and six electrons in the outer shell.

Ionic

Covalent

gain one electron. When sodium loses a unit of negative charge, it takes on a positive charge of +1. It is then a *positive ion*. When chlorine has picked up an extra electron, its negative charge increases. It thus becomes a *negative ion* with a charge of −1. Ions are atoms or groups of atoms which have gained or lost electrons and thus have acquired electrical charges. A sodium ion is attached to a chlorine ion because the positive and negative ions attract each other. When the two join they form an **ionic bond** (Figure 2.4).

Covalent bonding Another way for atoms to achieve stability is to form a **covalent bond.** In this case, instead of losing or gaining electrons and becoming ions, an atom *shares* one or more electrons with another atom (Figure 2.4). By *sharing*, it is meant that an electron orbits both nuclei. This contrasts with ionic bonding, where the electrons orbit only one nucleus. An atom of chlorine, for example, may share one of its seven outer electrons with another atom of chlorine in a covalent bond.

Metallic bonding Metallic bonding is a less-formal arrangement than ionic or covalent bonding. Metallic bonding is responsible for the peculiar qualities of metals. In metals, positively charged ions are packed together, while detached electrons wander freely through the structure. Because electricity is caused by electrons in motion, it is easy to understand why metals such as copper are good electrical conductors.

2.4 Ionic and covalent bonding, opposite page. Ionic bonds result from attraction between two ions having opposite charges. The result is comparable to the N and S poles of cylindrical magnets sticking together. In covalent bond one or more electrons orbit two nuclei, instead of only one nucleus.

Energy and Forces of the Earth

It is impossible to discuss matter—or geology—without understanding energy in various forms. **Energy** may be defined as the capacity to do work; to create motion or change (Figure 2.5). In this sense, "work" has a far broader meaning than it does in popular usage. The kinds of events we usually associate with work are everyday human activities: painting a house, digging a ditch. The work associated with energy, however, is often imperceptible: we cannot always see electrical or heat energy.

Yet the work done by energy is no less real—and no less important—than the more familiar kinds of work. Without energy, the matter composing the Earth would be inert and lifeless. There would be no rain, no rivers, no volcanoes, no earthquakes—no change. No new rock would spew out of the Earth's interior; loose regolith would not be re-converted into new rock. Solids could not be melted into liquid, nor liquid evaporated into gas.

2.5 Some Forms of Energy

Potential energy and **kinetic energy** are both shown in the illustration below. When the gymnast is in the motionless, or stopped, position in the center he is demonstrating latent, or potential, energy. As soon as he moves from one position to another he is releasing actual, or kinetic, energy.

Nuclear energy is not unlike the motion of one toy marble striking another, which then strikes yet another in an abbreviated chain reaction. In a nuclear power plant neutrons are fired into uranium-235, which creates the unstable isotope uranium-236, which splits into smaller atoms whose neutrons strike other atoms of uranium-235, and so on, in an enormous release of energy which we hope to harness more and more for peaceful purposes.

Radiant energy may also be called electromagnetic radiation. The Sun is the Earth's most important energy source, emitting waves that travel to us through 155 million kilometers of space. The specific energy of the Sun is also called solar energy. Solar energy is currently being used to heat and cool some homes in the United States, and much related research is underway throughout the world.

The Bettmann Archive

As we shall see throughout this book, the most important source of the Earth's energy is the Sun—beamed to us across 155 million kilometers (km) of interstellar space. The Sun itself is a turbulent ball of gaseous matter that produces almost unimaginably large quantities of energy.

Kinds of Energy

In mechanics, energy is considered in two forms: (1) potential, and (2) kinetic. **Potential energy** is latent or stored energy; a boulder perched at the top of a cliff is said to have potential energy, as is a lump of coal which may later be ignited. **Kinetic energy** is released or actualized energy; a boulder falling down a mountainside now has kinetic energy.

Heat energy
is demonstrated easily with a hot-air balloon. What we call heat is really an effect produced by the constant motion of atoms. As heated, or agitated, atoms rise they carry the balloon aloft with them.

Chemical energy
probably is easiest to understand if we think of it as accompanying matter as matter changes its state. The formation of rock, for example, involves changes of state from liquid to solid, and releases huge quantities of chemical energy. Chemical energy may also be absorbed, as when water evaporates.

Electrical energy
is a flow of electrons through a conducting medium, and is so accessible to us in convenient forms that we are likely to take it for granted. Geologists are helping to find ways to generate electricity from the Earth's heat.

The Bettmann Archive

Nuclear energy Huge quantities of energy are released when the particles of atomic nuclei are split *(fission)* or added to *(fusion)*. This energy is called **nuclear energy** or **atomic energy**. The great energy of the Sun is produced by nuclear reactions within its core. Every second, in a complex series of reactions, 657 million tons of hydrogen are converted to 653 million tons of helium by nuclear fusion. The missing four tons of mass are converted into energy.

Nuclear energy may also become man's most powerful tool on Earth. Nuclear energy can be extremely destructive when released explosively as atomic bombs or hydrogen bombs, but the power of the nucleus can also be released slowly in a controlled "chain reaction" and used for peaceful purposes. This is done in a nuclear power plant by firing neutrons into uranium-235, creating the unstable isotope uranium-236.

The uranium-236 then splits into two smaller atoms, emitting neutrons which strike other atoms of uranium-235, and so on. The energy released in this reaction is so enormous that utility companies hope to produce most of this country's electricity in nuclear power plants by the next century.

Radiant energy The energy produced by the Sun travels to the Earth in the form of **radiant energy,** or electromagnetic radiation. Radiation is a general term for many kinds of energy waves, such as radio waves, visible light waves, and X-rays. All of these waves travel at the same speed—the speed of "light," 297,600 km per second (186,000 miles per second). Radiant energy is essential to *photosynthesis*, the process by which plants convert sunlight into living matter.

Heat energy About a third of the radiant energy from the Sun is absorbed by the Earth's surface and atmosphere and converted to heat. What we call heat is really an effect produced by the *constant motion of atoms;* the greater the motion, the greater the heat. The resulting *heat energy* is directly responsible for producing wind, waves, rain, and many other geologic agents.

Chemical energy When *ions form bonds* with other ions, and when *matter changes its state,* energy is either released or absorbed. When water evaporates, energy is absorbed, and when water falls as rain, energy is released. The formation of rock involves changes of state from liquid to solid, releasing enormous quantities of chemical energy. All of these processes produce or consume chemical energy.

Electrical energy Man has been using electrical energy on a large scale for nearly a century. Electricity, which is a *flow of electrons through a conducting medium,* is easily generated by spinning a magnet inside a copper coil. The magnet is usually spun by the energy of falling water or by steam pushed through turbine blades during the burning of coal or oil. In addition, about nine per cent of the electricity in the United States is generated by steam heated in nuclear power plants. Many geologists today are involved in the search for new ways to generate electricity from the Earth's internal heat.

Mass and Energy

The reason an atomic bomb or a controlled nuclear reaction emits so much energy is one of the most important discoveries of science: *matter can be converted into energy.* This conversion is responsible for the enormous energy output of the Sun, described above. When radioactivity was first discovered, physicists could not account for the tremendous energy coming from a tiny bit of matter. It seemed as though this energy were being "created"—a violation of the traditional Law of Conservation of Energy, which states that energy can neither be created nor destroyed. In 1905, Albert Einstein explained this puzzle with his Theory of Relativity, expressing the equivalence of matter and energy in the now-famous equation, $E = mc^2$. E is energy, m is mass, and c is a constant representing the speed of light. When a neutron changes to a proton, the mass of the particle drops from 1.0089 to 1.0076 (Figure 2.1); the rest is converted to energy. The same thing happens when uranium-235 is split: a tiny fraction of matter is changed into energy. If 1 gram of matter—any matter—were completely transformed into energy, it would release as much energy as burning 24,000 tons of gasoline.

Earth Forces

The forms of energy described above power and participate in many local and regional geologic reactions. Two other "influences" which affect the Earth as a whole are the forces of gravity and magnetism.

Gravity Gravity, a force described by Isaac Newton in 1666 when he was only 24 years old, applies equally throughout the universe. Newton used the concept of **gravity** to define the force exerted between two objects, when the objects are solids that can be considered as points.

We use units of *weight* to measure the pull of gravity upon an object by a large mass, such as a planet. For a rock that weighs one kilogram on Earth, this force could be measured as 2½ kg on Jupiter; 26 kg on the Sun; and less than ⅙ kg on the Moon. The weights vary because the masses of Earth, Jupiter, Sun, and Moon are not the same.

In geology, the importance of gravity is that the Earth attracts all objects toward its center. The intensity of this force varies with distance from the center, and with the density of material beneath the object. As we shall see, these variations serve as important clues in locating structures of interest, such as bodies of low-density salt which may be associated with pools of petroleum, or bodies of dense rock which may contain diamonds.

Magnetism For centuries, scientists have struggled to understand the mysteries of **magnetism,** a force exerted through space by certain substances such as iron. Magnets have both positive and nega-

Mount Katahdin Henry David Thoreau

In the morning, after whetting our appetite on some raw pork, a wafer of hard bread, and a dipper of condensed cloud or waterspout, we all together began to make our way up the falls, which I have described; this time choosing the right hand, or highest peak, which was not the one I had approached before. But soon my companions were lost to my sight behind the mountain ridge in my rear, which still seemed ever retreating before me, and I climbed alone over huge rocks, loosely poised, a mile or more, still edging toward the clouds; for though the day was clear elsewhere, the summit was concealed by mist. The mountain seemed a vast aggregation of loose rocks, as if some time it had rained rocks, and they lay as they fell on the mountain sides, nowhere fairly at rest, but leaning on each other, all rocking-stones, with cavities between, but scarcely any soil or smoother shelf. They were the raw materials of a planet dropped from an unseen quarry, which the vast chemistry of nature would anon work up, or work down, into the smiling and verdant plains and valleys of earth. This was an undone extremity of the globe; as in lignite we see coal in the process of formation.

At length I entered within the skirts of the cloud which seemed forever drifting over the summit, and yet would never be gone, but was generated out of that pure air as fast as it flowed away; and when, a quarter of a mile farther. I reached the summit of the ridge, which those who have seen in clearer weather say is about five miles long, and contains a thousand acres of tableland, I was deep within the hostile ranks of clouds, and all objects were obscured by them. Now the wind would blow me out a yard of clear sunlight, wherein I stood; then a gray, drawing light was all it could accomplish, the cloud-line ever rising and falling with the wind's intensity. Sometimes it seemed as if the summit would be cleared in a few moments, and smile in sunshine: but what was gained on one side was lost on another. It was like sitting in a chimney and waiting for the smoke to blow away. It was, in fact, a cloud-factory,—these were the cloud-works, and the wind turned them off done from the cool, bare rocks. Occasionally, when the windy columns broke in to me, I caught sight of a dark, damp crag to the right or left; the mist driving ceaselessly between it and me. It reminded me of the creations of the old epic and dramatic poets, of Atlas, Vulcan, the Cyclops, and Prometheus. Such was Caucasus and the rock where Prometheus was bound. Aeschylus had no doubt visited such scenery as this. It was vast, Titanic, and such as man never inhabits. Some part of the beholder, even some vital part, seems to escape through the loose grating of his ribs as he ascends. He is more lone than you can imagine. There is less of substantial thought and fair understanding in him, than in the plains where men inhabit. His reason is dispersed and shadowy, more thin and subtile, like the air. Vast, Titanic, inhuman Nature has got him at disadvantage, caught him alone, and pilfers him some of his divine faculty. She does not smile on him as in the plains. She seems to say sternly, why came ye here before your time? This ground is not prepared for you. Is it not enough that I smile in the valleys? I have never made this soil for thy feet, this air for thy breathing, these rocks for thy neighbors. I cannot pity nor fondle thee here, but forever relentlessly drive thee hence to where I am kind. Why seek me where I have not called thee, and then complain because you find me but a stepmother? Shouldst thou freeze or starve, or shudder thy life away, here is no shrine, nor alter, nor any access to my ear.

The tops of mountains are among the unfinished parts of the globe, whither it is a slight insult to the gods to climb and pry into their secrets, and try their effect on our humanity. Only daring and insolent men, perchance, go there. Simple races, as savages, do not climb mountains,—their tops are sacred and mysterious

From THE MAINE WOODS, by Henry David Thoreau, 1864. Thomas Y. Crowell Company, 1961.

Field Trip

tracts never visited by them. Pomola is always angry with those who climb to the summit of Ktaadn.

Perhaps I most fully realized that this was primeval, untamed, and forever untameable *Nature*, or whatever else men call it, while coming down this part of the mountain. We were passing over "Burnt Lands," burnt by lightning, perchance, though they showed no recent marks of fire, hardly so much as a charred stump, but looked rather like a natural pasture for the moose and deer, exceedingly wild and desolate, with occasional strips of timber crossing them, and low poplars springing up, and patches of blueberries here and there. I found myself traversing them familiarly, like some pasture run to waste,

or partially reclaimed by man; but when I reflected what man, what brother or sister or kinsman of our race made it and claimed it, I expected the proprietor to rise up and dispute my passage. It is difficult to conceive of a region uninhabited by man. We habitually presume his presence and influence everywhere. And yet we have not seen pure Nature, unless we have seen her thus vast and drear and inhuman, though in the midst of cities. Nature was here something savage and awful, though beautiful. I looked with awe at the ground I trod on, to see what the Powers had made there, the form and fashion and material of their work. This was that Earth of which we have heard, made out of Chaos and Old Night. Here was no man's garden, but the unhandselled globe. It was not lawn, nor pasture, nor mead, nor woodland, nor lea, nor arable, nor waste-land. It was the fresh and natural surface of the planet Earth, as it was made for ever and ever,—to be the dwelling of man, we say—so Nature made it, and man may use it if he can. Man was not to be associated with it. It was Matter, vast, terrific,—not his Mother Earth that we have heard of, not for him to tread on, or be buried in,—no, it were being too familiar even to let his bones lie there,—the home, this, of Necessity and Fate. There was there felt the presence of a force not bound to be kind to man. It was a place for heathenism and superstitious rites,—to be inhabited by men nearer of kin to the rocks and to wild animals than we. We walked over it with a certain awe, stopping, from time to time, to pick the blueberries which grew there, and had a smart and spicy taste. Perchance where *our* wild pines stand, and leaves lie on their forest floor, in Concord, there were once reapers, and husbandmen planted grain; but here not even the surface had been scarred by man, but it was a specimen of what God saw fit to make this world. What is it to be admitted to a museum, to see a myriad of particular things, compared with being shown some star's surface, some hard matter in its home! I stand in awe of my body, this matter to which I am bound has become so strange to me. I fear not spirits, ghosts, of which I am one,—*that* my body might,—but I fear bodies, I tremble to meet them. What is this Titan that has possession of me? Talk of mysteries!—Think of our life in nature,—daily to be shown matter, to come in contact with it,—rocks, trees, wind on our cheeks! the *solid* earth! the *actual* world! the *common sense! Contact! Contact! Who* are we? *where* are we?   1864

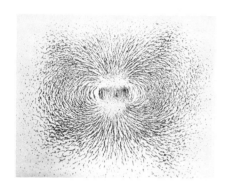

2.6 Iron filings on a sheet of paper distributed along magnetic lines of force around a small tabular magnet having its N pole on upper flat surface and its S pole on the bottom. (Nat Messik.)

tive poles, and lines of magnetic force extend from pole to pole as a banana-shaped *magnetic field* (Figure 2.6). The Earth itself behaves much like a giant bar magnet, generating lines of magnetic force that extend far into space. Like a bar magnet, the Earth also has two magnetic "poles": a positive pole toward the north and a negative pole toward the south. Although Einstein attempted to seek a common bond between gravity and magnetism, there are many differences between these two forces. For example, gravity is a part of an object forever, whereas magnetism may be lost. Gravity is also a single force; magnetism may be positive or negative.

The magnetic behavior of the Earth is complex. For one thing, the magnetic poles are not aligned with the geographic North and South Poles, which define the axis around which the Earth spins. The magnetic North Pole is located 11.4 degrees from the Earth's axis, or approximately the distance from Denver to Washington, D.C. (Figure 2.7). Also, the magnetic North Pole "wanders" slowly from place to place, and the strength of the Earth's magnetic field has dwindled about six per cent in the last 150 years. Furthermore, geophysicists have recently discovered that periodically the entire magnetic field reverses: the North and South Poles exchange places. During the last 71 million years, there have been 171 of those polar reversals. On the average, a reversal takes place about once each half million years. The last reversal took place 700,000 years ago, so we have reason to expect a polar reversal in the near future.

The reason these reversals are useful to the geologist is that they are "imprinted" in iron-bearing rocks. One kind of imprinting occurs when igneous rocks are cooling from molten liquid to solid form. As the cooling material passes through a temperature known as the **Curie**

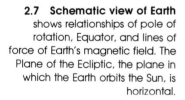

2.7 Schematic view of Earth shows relationships of pole of rotation, Equator, and lines of force of Earth's magnetic field. The Plane of the Ecliptic, the plane in which the Earth orbits the Sun, is horizontal.

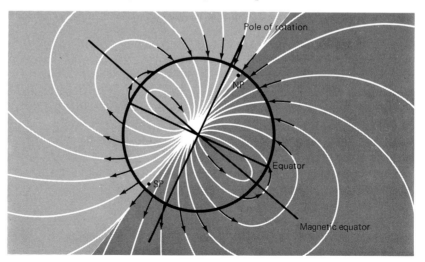

point, the iron in the rocks becomes magnetized by the Earth's magnetic field. Each tiny iron fragment takes on a weak magnetic field oriented the same as the Earth's magnetic field. The fragments maintain this field indefinitely—unless they are again heated above their Curie points, where they lose their magnetism. Therefore, if a geologist finds an igneous rock 100 million years old, it can be used to test whether the Earth's magnetic field was "reversed" or "normal" when the rock formed. Whether the rock has moved from its original north-south orientation can also be determined. In Chapter 14 we shall see how important this knowledge can be in relation to continental drift.

Minerals of the Earth's Crust

The Earth's crust is formed of a variety of raw materials known as minerals. A **mineral** is defined as a naturally occurring solid having a definite chemical composition and an ordered ionic arrangement resulting in a set of relatively uniform chemical and physical properties. Most minerals are formed by processes that do not involve living organisms, except for "hard parts" such as shells and skeletons. Glass (composed mostly of silicon and oxygen) is not considered a mineral because its ions are arranged randomly; quartz, also silicon and oxygen, is a mineral because its ions are locked in a characteristic structure (Figure 2.8). Naturally occurring elements, such as the metals gold, silver, and copper, whose ions are packed in orderly arrangements, are also considered minerals.

Composition and Structure of Minerals

Composition of minerals When one considers the size of the Earth and the complexity of its rocks, one might expect to find an infinite variety of minerals. In fact, there are only about 2000 different minerals (in this case, a small number), and most of them are very rare. The reason for this simplicity is that about 90 per cent of all the ions in minerals are either oxygen, silicon, or aluminum. Almost all rocks, therefore, are composed of silicate minerals (minerals composed of silicon, oxygen, and one or more metallic ions). Other kinds of minerals sometimes found as rock constituents are oxides, sulfides, and carbonates.

Structure of minerals All minerals consist of tiny building blocks called **unit cells.** A unit cell is a regular arrangement of ions held together by electrical forces. The unit cell of halite (common salt or sodium chloride), for example, contains four ions of chlorine and

68
Matter, Energy, and Minerals

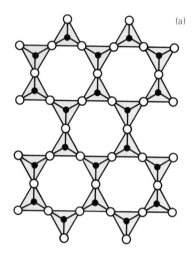

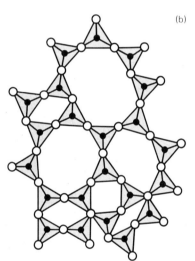

2.8 Particles in various solids, schematic sketches. In crystalline solid (a) particles are arranged in regular, symmetrical, repeating geometric pattern. In amorphous solid (b) particle arrangement is not regular.

four ions of sodium; in a single grain of table salt there are about 56×10^{17} (56 followed by 17 zeros) unit cells. These unit cells are stacked in a systematic way, like bricks in a wall, with virtually no spaces between them (Figure 2.9). The systematic ionic arrangement determines most of the properties of a mineral. These include: (1) the chemical characteristics, and (2) the beautiful geometric shapes of crystals. We define **crystals** as solid bodies bounded by natural plane surfaces; these surfaces are the external expressions of regular interior (ionic) structure. Because each crystal has its own distinctive unit cell, it follows that each crystal also has a distinctive external form.

This strict internal structure was not known in detail until 1912, when an experiment suggested by Max von Laue of the University of Munich revolutionized mineralogy. It was found that X-rays passing through mineral crystals are bent, or refracted, by the ions of the internal structure, or **lattice**. Because X-rays are close relatives of visible light, they can affect photographic film. By placing a specimen of a mineral between a source of X-rays (usually a metal target that is bombarded by high-speed electrons) and photographic film, the developed film shows spots where the ions are positioned (Figure 2.10). With their X-ray tools, crystallographers can calculate the sizes of the ions (Figure 2.11) of unit cells, as well as the distances and angles between the ions. These distances and angles are very important in determining such mineral properties as strength and hardness. These properties, in turn, govern the behavior of minerals in the geologic cycle.

The crystal lattice of any mineral has the same dimensions and angles each time that mineral forms. These forms can be defined by measuring the angles between the plane surfaces of natural crystals. These angles are always consistent, according to the Law of Constancy of Angles: *The angles between corresponding faces on different crystals of one substance are constant.*

Each unit cell is added precisely in the same way as the preceding unit cell, so that a crystal that has formed without interference is simply a giant version of the first unit cell. The growth of a crystal can be observed easily in the laboratory by slowly cooling a saturated solution of common salt; a large single crystal bounded by smooth surfaces will grow steadily as long as the cooling rate and supply of material are maintained.

The best-developed and most spectacular crystals grow where they are not impeded, as along open fractures or cavities in rock where the crystals can enlarge into an empty space (Figure 2.12). In the Ural Mountains in Russia, a whole mineral quarry was opened on a single crystal of feldspar that was a rectangular solid 10 meters square, of unknown length (depth).

Minerals of the Earth's Crust

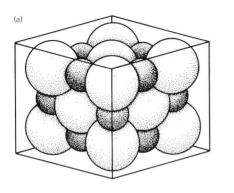
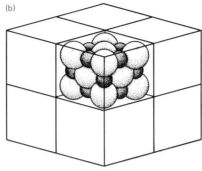

2.9 Unit cell and halite crystals
(a) Enlarged view of unit cell, consisting of small sodium ions (dark) and large chlorine ions (light).
(b) Crystal built of eight unit cells.

2.10 Study of crystals using narrow beam of X-rays. (Left) X-ray tube emits rays in all directions, but lead shield permits only a narrow beam of X-rays to strike crystal. Ions in crystal lattice deflect X-rays in a regular pattern, which is recorded on photographic film. (Right) Photograph made by pattern of X-rays reflected from a crystal of ice. (I. Fankuchen, Polytech. Inst. of Bklyn)

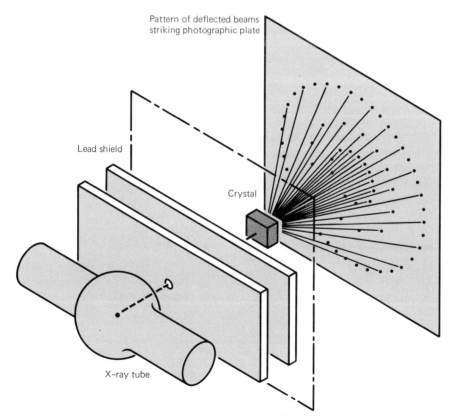

70
Matter, Energy, and Minerals

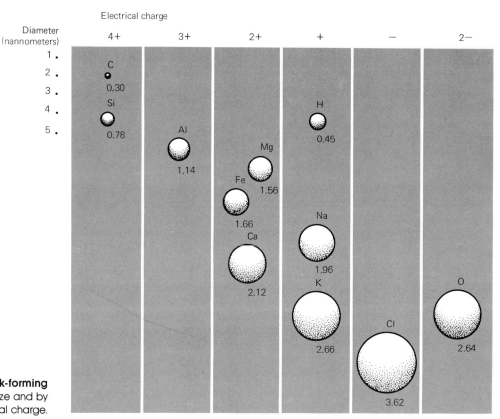

2.11 **Common ions of rock-forming minerals** arranged by size and by electrical charge.

2.12 **Tiny crystals of quartz.** Specimen viewed through scanning-electron microscope. (U. S. Geological Survey.)

Most natural crystals, however, are imperfect because of interference by neighboring minerals during formation. There is so much interference during the formation of most rocks that single crystals are not obvious without a microscope. Irregular **compromise boundaries** are formed when boundaries of crystals interfere with another crystal's growth (Figure 2.13).

Properties of Minerals

The arrangements of ions in crystal lattices determine the properties, or characteristics, of minerals. These properties vary considerably among different minerals, allowing geologists to utilize them in identification of minerals.

Color One of the most obvious and striking properties of a mineral—and one of the trickiest to use—is color. Some colors depend not upon which ions are present, but upon how these ions are arranged. A carbon mineral may feature the clear brilliance of diamond or the dull blackness of graphite. Or, a relatively small number of ions

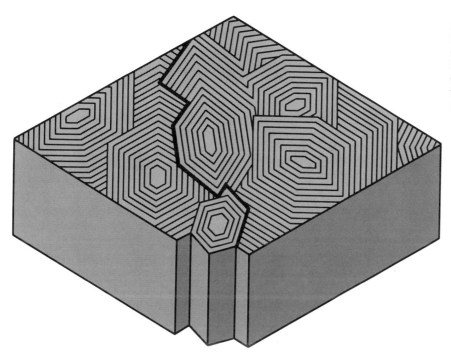

2.13 Compromise boundary (heavy line) results where enlarging crystals (six shown here) have interfered with one another's growth. Each enlarging crystal's true crystal faces shown by light lines.

may account for a large color change; a tiny amount of chromium is the only difference between a colorless crystal and an emerald. Clear quartz may become amethyst or rose quartz with a trace of silvery titanium or manganese. Other ions tend to produce their own color whenever they appear. Copper minerals are always green or blue, and uranium oxides, yellow.

Streak The color of a mineral that has been finely powdered is called **streak**. This property is more constant than color, and hence provides a somewhat more reliable diagnostic tool. Most transparent and translucent minerals have a white streak; nonmetallic minerals have a streak usually lighter than the color. Streak is especially useful in the study of metallic minerals.

Luster This property, largely independent of color, is related to the amount of light reflected by a mineral. Minerals with a *metallic* luster reflect most of the light falling on them, thus giving a shiny metal-like appearance. Nonmetallic minerals absorb more light than metals, reflect less, and have a glass-like appearance, or *vitreous* luster.

Crystal form Minerals are grouped into six systems of geo-

72 Matter, Energy, and Minerals

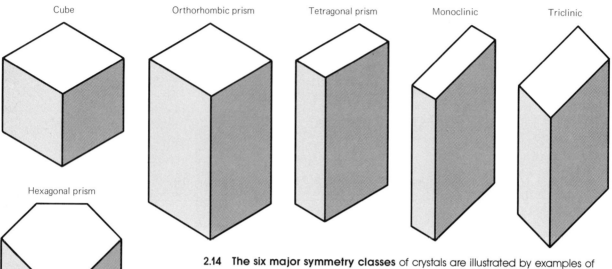

2.14 The six major symmetry classes of crystals are illustrated by examples of the solids that define them. In cubes, square faces meet at right angles. Orthorhombic prism is defined by two square faces and four rectangular faces, all meeting at right angles. In a tetragonal prism all sides are rectangles and all meet at right angles. A monoclinic shape consists of three pairs of sides, including two sets of rectangles and one of parallelograms. A triclinic shape consists of sides that are three pairs of parallelograms. Hexagonal prism is a six-sided figure with plane ends.

metric symmetry, according to the angles between the plane surfaces of crystals: isometric, orthorhombic, tetragonal, monoclinic, triclinic, and hexagonal (Figure 2.14).

2.15 Several flakes of muscovite split off along perfect cleavage in one direction. Vertical sides are crystal faces, which in this specimen are not as smooth as the cleavage planes. (Ward's Natural Science Establishment.)

Cleavage Many minerals tend to break along smooth planes that are parallel to zones of weakness in the lattice, a phenomenon known as **cleavage.** Some minerals, such as muscovite, have perfect cleavage, breaking easily along one plane, and breaking with great difficulty in any other direction (Figure 2.15). Other minerals, quartz, for example, lack distinct cleavage. Quartz breaks, or fractures, irregularly, in any direction. Cleavage may occur in more than one direction, and each may be perfect or imperfect. Halite, or common table salt, has three perfect cleavages at right angles. A broken grain of salt, examined by magnifying glass, will reveal a cubic shape or six plane surfaces joined at right angles (Figure 2.16). Calcite exhibits three perfect cleavages which are not at right angles. As a result, calcite breaks into rhombs (Figure 2.17).

Hardness A geologist uses many other clues to identify a mineral. One is **hardness,** or the resistance to scratching. The German mineralogist Friedrich Mohs in 1822 arranged 10 common minerals in a relative scale of hardness still used today. (See Appendix B, page 548, for a drawing of the Mohs scale of relative hardness.)

Specific gravity The term used to describe how heavy a mineral is in relation to water is **specific gravity.** Quartz, with a specific gravity of 2.65, is 2.65 times as heavy as the same volume of water. Specific gravity is usually figured by weighing a mineral sample in air, and again under water. The term **density** is also commonly used, referring to the mass per unit volume. Thus a rock of which a cubic centimeter weighs five grams has a density of five grams per cubic centimeter. The ions of dense minerals are more tightly packed than those of less dense minerals. Low-density matter can float on top of matter of higher density, as oil floats on water.

Other properties of minerals Some minerals can be identified by other properties, such as *magnetic behavior*. A magnet attracts some minerals, especially those bearing iron, and fails to attract others. Sensitive instruments called *magnetometers* can even locate iron-bearing mineral deposits from ships and airplanes. Another mineral property of great importance is *electrical conductivity*. Metals tend to be good conductors, because their electrons move freely through the crystal lattice, whereas nonmetals conduct poorly. A class of minerals called *semiconductors*, especially those containing silicon or germanium, have lately become extraordinarily useful as such modern electronic components as transistors. Other properties used by geologists are *feel* (talc is soft and slippery) and *taste* (halite is "salty").

2.16 Three directions of cleavage at right angles create tiny cubes when halite is broken. (E. J. Dwornik, U.S. Geological Survey)

2.17 Three directions of cleavage not at right angles create rhombs when calcite is broken. (Ward's Natural Science Establishment.)

Noncrystalline Solids

A few solids found in nature are not crystalline; such noncrystalline solids lack orderly geometric arrangements (Figure 2.8b). Such noncrystalline, or **amorphous**, solids have, as the name implies, "no shape." The best examples are glasses and gels. Natural **glasses** often form during volcanic activity, when molten material cools too quickly to form crystals. **Gels** are solids formed from colloids, which are halfway between true solutions (sea water) and suspensions (muddy water). The commonest natural gel solid is opal, which occurs in many colors. Opals contains 3 to 10 per cent water, and some opals are almost as precious as diamonds. Opals are usually found in cracks of volcanic rocks.

Chemical Groups of Minerals

Minerals may be arranged chemically into several major and minor groups. Major groups include silicates, sulfides, oxides, carbonates, and sulfates. Minor groups include nitrates, borates, chromates, and native elements.

Silicate Minerals

Minerals composed primarily of silicon and oxygen are so abundant and widespread that **silicates** have been nicknamed the "rock-forming minerals."

Kinds of silicates According to their chemical compositions silicate minerals may be divided into two groups: ferromagnesian and nonferromagnesian silicates. **Ferromagnesian** silicates are usually rich in the elements iron and magnesium, and dark in color. (The word ferromagnesian is derived from the Latin word for iron, *ferro*, and the word magnesian.) Common ferromagnesian silicates are olivine, pyroxene, amphibole, and biotite.

Nonferromagnesian silicates lack iron and magnesium. They are rich instead in the elements sodium, potassium, calcium, and aluminum, and are light in color. Common minerals of this group are the feldspars, quartz, and muscovite. The distinction between ferromagnesian and nonferromagnesian silicates is fundamental in rock classification, as we shall see in later chapters.

Silicate structure In all silicates, the ions of silicon and oxygen join one another in the shape of a tiny **tetrahedron** (plural, tetrahedra)

(Figure 2.18). One silicon ion is always bound to four oxygens by powerful ionic bonds. The tetrahedra, or groups of tetrahedra, are bound in a variety of structures by ions of magnesium, iron, or other common elements. Silicates may be classified by the way these tetrahedra are linked to one another as shown in Figure 2.19.

All minerals having chain structures cleave in the direction of the chains. In the single-chain mineral pyroxene cleavage occurs at 90-degree angles; in the double-chain mineral amphibole, at 60-degree and 120-degree angles (Figure 2.20).

Variations in silicate structure A solid grasp of silicate structures is essential to understanding how ferromagnesian and nonferromagnesian minerals may vary. Within many silicate structures, metallic ions of about the same size commonly replace one another during chemical reactions. Or, a single mineral may contain more than one element in a particular position in the structure. In the ferromagnesian mineral olivine, ions of magnesium and iron normally alternate as the metal ions that link individual, otherwise isolated silica tetrahedra.

The most common nonferromagnesian silicates are *feldspars* (minerals containing abundant silicon, oxygen, and aluminum). The two major classes of feldspars are distinguished by the alternation of the metallic ions sodium, calcium, and potassium. In *potassium feldspars*, potassium is more abundant than sodium or calcium. In *plagioclase feldspars*, sodium and calcium alternate with each other, and potassium is absent.

Many other building blocks find their way into silicate structures and help determine the nature of the resulting material—especially aluminum and other metallic ions, water, water-hydrogen groups, and extra oxygen. These other elements serve to balance the structure mechanically and electrically so that it is chemically stable.

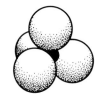
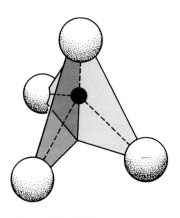

2.18 Tetrahedron formed by clustering of four oxygen ions around a silicon ion. (Top) Natural arrangement with oxygen ions touching one another. (Bottom) "Exploded" view to show all ions.

Oxide Minerals

This large class of minerals includes compounds in which ions, usually of metal, are combined with oxygen. **Oxides** have ionic bonds that are equal in every direction; that is, the oxygen ion is bound with equal strength to each of its neighbors. The oxides include *magnetite* (a compound of iron and oxygen), the only mineral that is attracted by a magnet (Figure 2.21) and may even be a magnet itself. Also called lodestone, magnetite provided a primitive tool for navigation, because it is magnetic, and can be made into a primitive compass.

Magnetite, hematite, and other iron oxides are mined worldwide

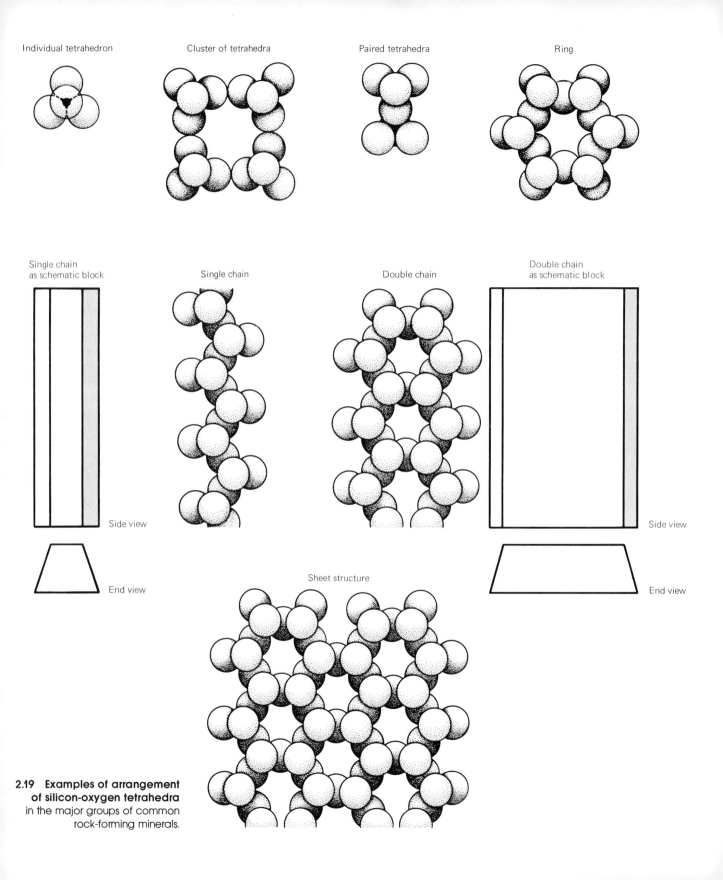

2.19 **Examples of arrangement of silicon-oxygen tetrahedra** in the major groups of common rock-forming minerals.

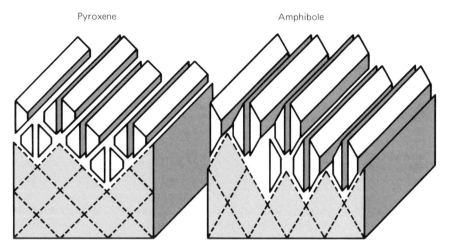

2.20 **Relationship of chains of tetrahedra** (shown by schematic blocks) to cleavages (dashed lines) in pyroxene and amphibole. Breakage between single-chain units creates two directions of cleavage at right angles in pyroxene. Breakage between double-chain units creates two directions of cleavage, not at right angles, in amphibole.

to make iron and steel. When exposed to air and moisture, they form ferric oxide (rust). Another common oxide, corundrum (aluminum oxide), is second only to diamond in hardness and, when formed with minute amounts of impurities, produces red rubies and blue sapphires. (A well-known exception to the definition of an oxide is quartz, which is chemically an oxide but has the structural characteristics of a silicate.)

Sulfide Minerals

One or more ions of sulfur combined with a metallic ion are sulfide minerals. Examples include galena (lead sulfide), and cinnabar (mercury sulfide). Sulfide minerals usually display metallic luster. The most common sulfide is pyrite (iron sulfide), sometimes called fool's gold for its gold-like appearance. A simple way to distinguish these minerals is hardness. A knife blade easily scratches gold but does not scratch pyrite. The mining of pyrite produces about 40 per cent of the world's sulfur; the rest comes from mining pure (native) sulfur. Sulfide minerals are formed in the absence of oxygen; when they come into contact with oxygen, they quickly break down to release pure sulfur or sulfuric acid.

Carbonate and Sulfate Minerals

In oxide minerals, oxygen behaves as an individual ion. In other oxygen-bearing minerals, the oxygen ions behave as clusters bound firmly to another ion. These link-ups form a complex ion called a **radical.** Examples are the carbonate radical, in which three oxygens are

2.21 **Magnetic characteristic** of massive specimen of magnetite shown by attraction of specimen to small hand magnet. (Nat Messik and J. E. Sanders.)

clustered around a carbon, and the sulfate radical, in which four oxygen ions are bound to a sulfur ion. Unlike molecules, which are electrically stable, radicals have positive or negative charges and thus are attracted to and combine with ions having appropriate opposite charges.

Some carbonate minerals form vast bodies of sedimentary and metamorphosed sedimentary rocks. Calcite (calcium carbonate) is the principal ingredient of limestone and calcite marble. Dolomite (calcium-magnesium carbonate) is the main component of dolomite rock and dolomitic marble. Because calcite is easily dissolved by rainwater bearing carbon dioxide, caves of great size and intricacy often form in limestones (Chapter 9).

The best-known sulfate mineral is gypsum, or hydrated calcium sulfate, found as extensive sedimentary deposits associated with limestone, red shales, sandstone, clay, and rock salt. Gypsum is mined worldwide for use in plasters, gypsum wallboard, roof tiles, cements, fillers for paper and paint, soil conditioner, and fertilizer.

Minor Groups of Minerals

Other radicals found in minerals are nitrate (nitrogen and oxygen), borate (boron and oxygen), and chromate (chromium and oxygen). Some native elements, such as gold and copper, are both easy to mine and economically valuable. Hence such deposits attracted primitive miners thousands of years ago.

Why Study Minerals?

The *rocks* of the Earth's crust consist of aggregates of minerals, many of which are of great importance and value to humans. From some minerals, for example, we extract valuable metals such as aluminum, nickel, and iron. Industrial minerals provide raw materials for ceramics, glass, fertilizers, and cement. Precious minerals that are prized for their color and hardness include diamond, emerald, and sapphire.

A summary of mineral structure may seem an excessively complex introduction to geology. But as we shall see, some understanding of minerals is necessary to discuss the major geologic processes. The formation of igneous rocks is a mineralogical subject, as is their conversion to sedimentary rocks or metamorphic rocks. Earthquake waves travel at different speeds through different minerals; geologists are investigating the minerals involved in the creation of new sea floor; and so on. Let us continue our study of geology with the fundamental processes of igneous activity and rock formation.

Chapter Review

1. Minerals and rocks are composed of minute particles called *atoms*. Atoms are composed of even smaller particles called *sub-atomic particles*.
2. The three most important sub-atomic particles are *protons, electrons,* and *neutrons*. Varying numbers of protons and neutrons make up the *atomic nucleus*.
3. Because almost all of the *mass* (amount of matter) of an atom is tightly packed into the nucleus, the atoms of even the heaviest rocks are mostly empty space.
4. The force that keeps the whirling arrangement of sub-atomic particles together is known as *strong interaction force*. Protons, which bear positive *electrical charge*, hold a strong attraction for electrons, which bear opposite, or negative electrical charge. Neutrons have no charge.
5. An individual atom is so small that if enlarged 100 million times, it would only be the size of a pea. We determine the *atomic number* of an atom by its number of protons; an atom with two protons has an atomic number of 2. All atoms having the same number of protons are considered the same kind of matter, or *element*.
6. Electrons do not fly just anywhere around the nucleus. They are organized in regions called *energy-level shells*. Each shell is capable of holding a fixed number of electrons—and no more. The farther from the nucleus a shell is located, the easier it is for an electron in that shell to be knocked out of orbit.
7. *Ions* are atoms or groups of atoms which have gained or lost electrons; when positive and negative ions attract each other an *ionic bond* forms.
8. Ions of one element, or of more than one element, may combine to form *molecules*. When ions of different elements combine in a liquid, the molecules form a *compound*. Atoms, which are electrically non-charged, cannot enter into chemical reactions.
9. Although 92 elements are found in nature, just eight of them make up nearly 99 per cent of the Earth's rocky crust. Oxygen, locked in solid compounds with other elements, is by far the most common element in the crust.
10. *Energy* can be defined as the capacity to do work. In geology, enormous amounts of energy are involved in moving glaciers, building mountain ranges, and shaking the ground during earthquakes.
11. Matter can be converted into energy. Albert Einstein's famous Theory of Relativity explained how energy is emitted when a neutron changes to a

Matter, Energy, and Minerals

proton. In this change mass decreases and the "lost mass" is converted into energy.

12 *Magnetism*, as far as we know, is unrelated to *gravity* and must not be confused with it. The Earth behaves like a giant magnet, and geologists can learn much about the Earth's physical history by clues found in iron-bearing rocks which have been magnetized.

13 The Earth's crust is formed of a variety of raw materials known as *minerals*. The rocks of the Earth's crust consist of aggregates of minerals, many of which are of great importance to man. Minerals are defined as naturally occurring solids having a definite chemical composition and an ordered atomic arrangement.

14 The number of minerals is not infinite. Only about 2000 (in this case, a small number) different minerals are known, and of these only a few are common. The reason for this simplicity is that about 90 per cent of all the ions in minerals are either oxygen, silicon, or aluminum. The systematic ionic arrangement of minerals accounts for the consistent and beautiful qualities of *crystals* and the geometric structure of crystalline forms.

15 Minerals are usually identified by color, streak, luster, crystal form, cleavage, hardness, specific gravity or density, and other properties such as magnetic behavior.

16 Minerals composed primarily of silicon and oxygen are so abundant and widespread that *silicates* have been nicknamed the "rock-forming minerals." Silicate minerals are either *ferromagnesians* or *nonferromagnesians*. In all silicates, the ions of silicon and oxygen join one another in the shape of a tiny *tetrahedron*; silicates may be classified by the way these tetrahedra are linked to one another.

Questions

1. Define *matter*.
2. What are *elements*? What are *atoms*? Does any relationship exist between elements and atoms? Explain.
3. Define *proton*, *neutron*, and *electron*. When all three of these kinds of particles are brought together, do they form any kind of systematic arrangements? Explain.
4. What is an *ion*? Can both atoms and ions join with other atoms or ions? Explain your answer.
5. Three kinds of bonds among ions are _____, _____, and _____. Explain the principle responsible for each kind of bond.
6. Define *energy*. The six chief kinds of energy are _____, _____, _____, _____, _____, and _____.
7. Explain the difference between *kinetic energy* and *heat energy*. How does *chemical energy* differ from *nuclear energy*?
8. What is the relationship between mass and energy?

9   Compare and contrast the forces of *gravity* and *magnetism*.
10 What is the *Curie point* of a magnetic substance? Explain the significance of the Curie point for understanding why the Earth is surrounded by a magnetic field.
11 Define *mineral*.
12 What is the single diagnostic feature of all minerals?
13 What is a *crystal*? Do minerals always appear as crystals? Explain.
14 Define *cleavage*. Compare and contrast *cleavage planes* and *crystal faces*. How can you distinguish a cleavage plane from a crystal face?

Suggested Readings

Asimov, Isaac, *Building Blocks of the Universe*. New York: Abelard-Schuman, 1961.

Barnett, Lincoln, *The Universe and Dr. Einstein* (Revised Edition). New York: Bantam Books, 1968.

Lapp, Ralph E., *Matter*. New York: Time-Life Books, 1966.

Pearl, Richard M., *1001 Questions Answered About the Mineral Kingdom*. New York: Dodd, Mead & Company, 1968.

Wilson, Mitchell, *Energy*. New York: Time-Life Books, 1966.

Zim, H. S., and Shaffer, Paul, *Rocks and Minerals*. New York: Golden Press, 1957.

"Iron girders were twisted like licorice, burning rum ran in streams through the streets."

Chapter Three
Igneous Activity and Volcanoes

Igneous Activity and Volcanoes

Igneous activity designates the making of hot liquid material from the mantle, its rising into the crust or onto the surface of the Earth, and its cooling, forming solid igneous rocks. Such rocks make up a vast bulk of the Earth's crust. Volcanoes are the vents through which this material passes to the Earth's surface. Accordingly, igneous activity is a process of fundamental importance in the geologic cycle.

Why Study Igneous Activity?

Igneous activity refers to what happens to natural molten rock material. When this molten rock material is on the surface of the Earth it is called **lava**, and when it remains inside the Earth's crust it is named **magma.** As either magma or lava cools and solidifies it forms **igneous rocks;** this relationship between surface molten material, lava, and cooled rocks is clearly evident in the case of a volcanic eruption, where we can actually watch the formation of igneous rocks as molten lava cools and hardens.

Igneous Activity and the Geologic Cycle

Why do we want to study igneous activity? First of all, igneous activity forms igneous rocks, which provide the raw material for sedimentary and metamorphic rocks. Without igneous rocks there would be no rocks at all. Igneous activity makes up an important part of the geologic cycle, and on other planets the *only* part of the geologic cycle that amounts to anything is the cooling of molten materials to make igneous rocks. Therefore, whatever understanding we achieve of igneous rocks on Earth can be extended to other planets such as the Moon and Mars.

Igneous activity is continuous The so-called primordial crust of the Earth no longer exists. Such a crust still remains on the Moon, and presumably on other planets, but on the Earth the geologic cycle of weathering and other processes have destroyed any primordial crust. All the bedrock on the Earth has been through an entire geologic cycle at least once, and perhaps many times, and so the Earth's original crust has disappeared.

The Causes of Igneous Activity

Originally, the Earth's crust was thought to be the cooled, scum-like surface floating on the hot liquid underneath, like the crusty scum that forms on cocoa as it cools in a cup. With the help of research related to earthquakes we now know that the material under the crust (it is called the *mantle*) is not liquid, but solid rock. Our problem is to explain how this solid material of the mantle melts and becomes the liquid we can see coming out of volcanoes.

Heat, pressure, and generation of magma Igneous activity is an expression of the Earth's heat. We have learned something about the Earth's heat from many sources such as deep South African gold mines, where the temperature increases as the mines get deeper. The mine workers are already so uncomfortably hot that they would be unable to work in mines that were any deeper. But even the deepest oil wells and gold mines penetrate only about 10 km into the Earth's crust, whose thickness beneath the continents averages 30 to 50 km. Obviously, much of what we know about the mantle is still conjectural.

But with the help of laboratory experiments we can establish the relationship between the melting points of materials and their temperature and pressure. If you have a substance near its melting point at a high temperature and high pressure and you reduce the pressure, the material will suddenly be above its melting point and will immediately change from a solid state into liquid.

Geologists think such a process takes place in the mantle. We can speculate that under great pressure and high temperature the Earth's mantle is solid, but if for some reason the pressure were to be released the material would be above its melting point at the new pressure, and hence it turns into liquid magma; this liquid magma may work its way upward all the way to the surface (Figure 3.1).

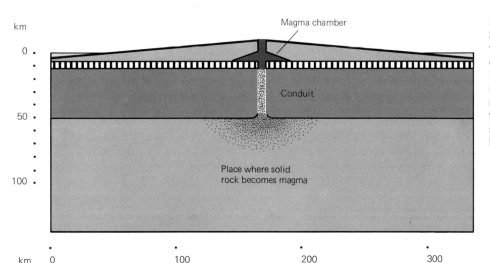

3.1 Melting and rise of magma. Schematic profile and section through Earth's lithosphere showing change from solid mantle rock to liquid magma at depth and upward shift of the liquid to a shallower magma chamber, from which it can feed a volcano. (Based on subsurface data from Hawaii; G. A. Macdonald, 1961.)

It is so tricky to collect the gases dissolved in lava that reliable measurements became possible only in the 1960s. Nonetheless, geologists can now say that lava erupts at around 1000°C, and that lava releases such gases as steam, carbon dioxide, sulfur dioxide, and nitrogen. Such clues, combined with chemical analyses of the lava, are helping geologists to understand the origin of magma.

Translated literally from the Latin, the word *igneous* means "fire-formed." No one who has witnessed the spectacular eruption of a volcano, or who has even seen films of an eruption, can doubt that igneous rocks seem to be "born of fire."

This chapter concerns itself with volcanoes and other expressions of the Earth's heat, such as geysers. These are the surface displays of igneous activity, and in the next chapter we will discuss the internal processes leading to the formation of igneous rocks.

Volcanoes

The most obvious effect of igneous activity is a volcano. A **volcano** is the place of emission of molten material or gas, and generally both, from the Earth's interior onto its surface. Commonly, the molten material accumulates as conical hills or mountains. Our purpose in studying volcanoes is to examine the products that come out, the features they build, the kinds of activities, and to become acquainted with some famous volcanoes.

Products of Volcanoes

During the Middle Ages, the "smoke and ashes" of volcanoes were thought to be caused by the fires of Hades. As late as the eighteenth century, some scientists believed that volcanoes were the vents of great fires burning within the Earth (Figure 3.2). Only gradually did early geologists discover that the "smoke" of volcanoes is not smoke at all, and the "ashes" are not products of combustion. The complex and variable products of volcanoes include liquids, solids, and gases.

Liquids Lava, the liquid ejected by a volcano, contains the same silicate materials as magma, but magma contains gases that are not present in lava. Lavas can be classified into two groups, (1) mafic, and (2) felsic. **Mafic** (MAY-fic) lavas, a term derived from the chemical symbols for magnesium and iron (Mg, Fe), tend to be liquid and free-flowing. **Felsic** lavas, rich in the raw materials of feldspars, are sticky and slow-flowing, or *viscous*. Volcanic activities vary with the

3.2 **Eighteenth-century artist's view** of fire and volcanoes. (New York Public Library.)

viscosity (resistance to flow) of the lava. In liquid mafic lavas the gases can escape continuously. In sticky felsic lavas the gases escape only after they have built up great pressures and cause violent explosions.

Volcanologists have divided lava flows into three broad categories (Figure 3.3). **Pahoehoe** (pah-hoey-hoey), a Hawaiian word that has come to be accepted universally for fluid mafic lavas, describes smooth-flowing lava with a crust that may be curved like folds of heavy cloth or coils of rope. **Aa** (ah-ah) is another Hawaiian word for more viscous mafic lavas with jagged, uneven surfaces. As the surface of aa lava cools and hardens, it breaks continuously by the slow but constant flow beneath it. For reasons that are still not clear, aa is also characterized by spines that grow mysteriously on the rough blocks. **Block lava,** rich in silica, resembles aa without the spines. Block lava is so viscous that it may move only a few meters a day, breaking itself into blocks by the motions of flowing over the ground.

Solid-gaseous mixtures A mixture of solid particles suspended in a gas behaves much like a fluid. By following this principle industrial engineers have learned to transport such solid materials as cement and coal dust through pipes. Much the same thing happens in some volcanic eruptions, when fine particles of frothy magma erupt in a stream of gases. Such an eruption is called an **ash flow,** even though we now know that the particles are not really ash at all but silica-rich (60 to 75 per cent) rock fragments.

3.3 Kinds of lava based on physical appearance of resulting extrusive igneous rock. (Left) Variety known as pahoehoe, Footprint Trail, Hawaiian Volcano National Park. (U. S. Geological Survey.) (Center) Rough-surfaced aa lava flow with pointed spines. Newberry Crater, Oregon (R. L. Nichols) (Right) Block lava. Yatsugatake volcano, Japan. (H. Takeshita)

Ash flows resemble fluid, fast-moving lava flows in many ways, spreading many kilometers from the vent. Ash flows moving downhill from a volcano commonly coast quietly over the surface of the Earth at 30 to 100 km/hr. If they are ejected laterally instead of upward the added push can boost the speed of a flow to around 150 km/hr, as in the case of the eruption that destroyed the city of St. Pierre in Martinique in 1902. That cloud contained white-hot fragments that glowed eerily at night; fragmental flows of this type have since been called **glowing avalanches** or **nuées ardentes** (noo-ay zar-DAHNT) (Figure 3.4).

The fragments ejected in ash flows are so hot (usually hotter than 1000°C) that they char trees and wooden buildings, and even after

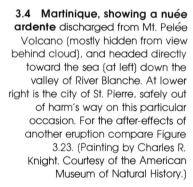

3.4 Martinique, showing a nuée ardente discharged from Mt. Pelée Volcano (mostly hidden from view behind cloud), and headed directly toward the sea (at left) down the valley of River Blanche. At lower right is the city of St. Pierre, safely out of harm's way on this particular occasion. For the after-effects of another eruption compare Figure 3.23. (Painting by Charles R. Knight. Courtesy of the American Museum of Natural History.)

long journeys through the air they may melt together as they settle. Rock made of melted, welded ejecta is **welded tuff.**

Solids The sudden release of gas in a volcano results in the explosive production of countless **ejecta** of rock, mineral particles, and glass. Ejecta are known collectively as *pyroclastic* ("fire-broken") materials or **tephra** (TEF-ra), a term coined by Aristotle. Sometimes the gas is released when magma comes in contact with groundwater or sea water to produce steam; this causes a *phreatic* eruption.

Tephra come in all shapes and sizes, but three general classes are recognized. Depending upon their shapes the largest fragments are called bombs or blocks. **Bombs** are ejecta larger than 32 mm that were thrown out while molten (Figure 3.5). During their flight through the air, they are usually shaped by air-caused friction into the form of a sweet potato or almond. The shapes of bombs are analagous to the ropy surface of pahoehoe lava, as blocks are analagous to aa. Bombs that are still liquid when they strike the ground are called *cow-dung bombs,* and an accumulation of these welded together is called *spatter.*

3.5 Volcanic bombs at cone north of Kalepeamoa, Mauna Kea, Hawaii. These bombs show the typical twisted shape. (C. K. Wentworth, U. S. Geological Survey, Volcano Observatory at Kilauea.)

Blocks are the same size as bombs, but emerge from a volcano in solid form (Figure 3.6). Accumulations of blocks, called **volcanic breccia** (BREH-cha), may lie loosely or they may be gradually cemented into a solid mass. Most bombs and blocks are fist- to football-sized, although blocks of more than 50 tons have been found.

Ejecta measuring from 0.06 to 4 mm are known as **cinder** or *lapilli* (laa-PILL-ee), from the Latin for little stones. Volcanic cinder bears no relation to the cinder produced by burning coal. Ejecta having diameters smaller than 4 mm constitute **ash,** which, like cinder, has nothing to do with combustion. Ash (or volcanic dust) is pulverized rock, and some of it is so fine that it may travel great distances in the atmosphere. Dust from the eruption of Krakatoa (Indonesia) in 1883 circled the world for several years, producing spectacular red sunsets on every continent.

3.6 Volcanic blocks are usually smaller than this one at the base of Makaopuhi, Hawaii. (U. S. Geological Survey, Hawaiian Volcano Observatory.)

Explosion products may take on strange forms depending upon the condition of the magma and the nature of the environment during eruption. Frothy, air-filled fragments of mafic lava are **scoria,** and of felsic lava are **pumice** (Figure 3.7). **Vesicles** are small cavities made in volcanic rock by expanding gas. If very liquid lava is caught in a strong wind, it may form thin glass threads known as *Pele's hair,* named after the Hawaiian fire goddess. Pele's hair occasionally falls thickly in the streets of Hilo, Hawaii, where local birds use it to build their nests.

As tephra settle after an explosion, they are sorted by size. Deposits from individual episodes of an eruption may be neatly graded,

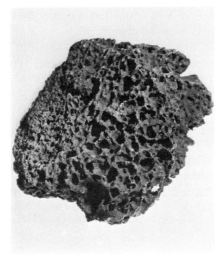 

3.7 Porous volcanic rock made because gases were expanding as lava solidified. (Nat Messik and J. E. Sanders.) (a) Jagged scoria with large, irregular vesicles in mafic extrusive rock. (b) Light-colored pumice with so many tiny vesicles (scarcely visible) that the rock will float on water.

with blocks and lapilli at the base and ash at the top. At greater distances from the vent, particle sizes become smaller. Some tephra deposits provide useful markers in time: a single layer of tephra covering tens of thousands of square kilometers, produced in a mere instant of geologic time, creates a layer which is everywhere the same age, providing a useful reference for dating widespread rocks.

Gases Virtually all the gases of the Earth's atmosphere and all the water of the Earth's oceans are thought to have been released by igneous activity. The release of these gases provides the driving force for volcanic action. Although gases make up only 1 or 2 per cent of the total weight of erupting magma, they expand into bubbles that provide the power to expel lava, tephra, and other volcanic products. When pressure on the magma is reduced, in a way not well understood, the effect is much like removing the cap from a bottle of soda pop: gas that was in solution suddenly forms bubbles that rush to the top of the bottle and escape into the air. The kinds of gases released in volcanoes vary widely and rapidly; during the 1963 eruption of Stromboli, Italy, the water content of escaping gas ranged from 0 to 45 per cent in only three minutes. The 1945 eruption of Parícutin (pah-REE-koo-teen) Volcano in Mexico liberated 15,000 tons of steam per day.

Volcanic Landforms

The popularized concept holds that a volcano is a conical mountain, with a circular base and a pointed top. Indeed, many volcanoes have built features matching this image. However, depending upon the nature of the products and how they accumulate, volcanoes build a variety of landforms. These products include domes, cones, craters and calderas, volcanic shields, lava plains, and composite cones.

Domes Cone-shaped structures, formed by the accumulation and hardening of lava, are called **domes.** Most domes grow by repeated additions of magma from beneath, as if they were being blown up like a balloon. A dome may also grow when lava oozes repeatedly onto the surface through an opening near the top (Figure 3.8). A few domes result from the upthrusting of a plugged conduit by magma pushing from below, as though a cork were being forced out of the neck of a bottle; these are called *plug domes.*

The shape of the dome depends largely upon the viscosity of the lava. Highly viscous felsic lavas tend to resist flowing away from the vent, and form high, steep-sided domes. Highly fluid mafic lavas travel farther from the vent and form lower, broader domes. Extremely fluid lavas often spread so thin that there is no noticeable dome, both on the sea floor and on land.

Dome formation varies widely. Some domes may be only half a meter high and a few meters across. By contrast, Lassen Peak, in California, is one of the largest domes in the world, 600 meters high and more than 1600 meters wide at the base. It is estimated that the dome of Lassen Peak took about five years to build. During its famous eruption of 1902 the dome of Mt. Pelée (pay-LAY), on the island of Martinique, was observed to be rising as fast as 25 meters a day. The shape of the dome may be changed even as it is forming, when expanding gases suddenly explode through the bottom or sides.

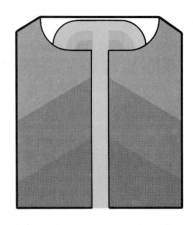

3.8 Volcanic cone and lava dome, schematic profile and section. The kind shown here has formed by successive oozing of pasty lava from the central vent of a cone. (Based on F. A. van Bemmelen, 1949.)

Cones The most familiar of all volcanic structures is the **cone,** named according to its eruption products. **Cinder cones** are made out of small, usually uniform-sized tephra that are ejected from a volcanic vent and then fall back nearby (Figure 3.9). The shape is highly characteristic, in fact, it is so consistent among different volcanoes that it can be calculated mathematically. The angle of the slope of nearly every cinder cone in the world is about 30 degrees. The cinders may eventually be cemented together by minerals carried by steam or other water. If the pool of lava rises through the vent after the cone has been built, it may coat the sides of the cone and leave a smooth lava veneer after subsiding again. Cinder cones may rise to their maximum height, usually about 450 meters, in as little as a few weeks, but usually their piling-up time occupies months.

Eruptions of small clots of liquid lava (spatter) often produce smaller structures called **spatter cones** (Figure 3.10). Rarely as high as

3.9 Tephra ("cinder") cone and lobate horizontal lava flows seen in oblique view from an airplane, Cerro Negro Volcano, west-central Nicaragua, erupting in November 1968. This tephra cone is not yet strong enough to support a column of lava that could fill the conduit to the top and thus to flow from the crater where the cloud of tephra is being released. Instead, the lava lifts part of the cone and oozes out from beneath its base. Old lava flows in foreground; new flows, still steaming, at left. Eventually the lava that issues from beneath a tephra cone may build up a broad, conical volcanic shield. A tephra cone atop a conical volcanic shield forms a composite cone. (Compare Figure 3.16.) (U. S. Geological Survey.)

30 meters, spatter cones look as though they were created by someone standing in a deep hole and hurling wet cement up onto the ground. Spatter tends to stick and become welded to other pieces, so that the structure is more rigid than cinder cones. Welded spatter may form walls that are nearly vertical.

Craters and calderas Once igneous activity begins, the structure of the volcano often changes dramatically. The eruption may be focused straight upward with great violence through the summit of a dome. If the explosions are violent, they may blow the top off the volcanic dome, creating an **explosion crater** as much as a kilometer in diameter. Following a large eruption, the downward retreat of magma

3.10 Spatter cones built at bases of two active lava fountains in August 1966 at Surtsey, Iceland. People at lower right indicate scale. (W. A. Keith, Iceland Tourist Bureau.)

within the chamber or reservoir may cause the top of the dome to subside along generally circular cracks. Such subsidence over an area smaller than 1½ km in diameter is called a **collapse crater,** or, if more than 1½ km, a **caldera** (cahl-DAY-ruh) (Figure 3.11).

Early observers noticed that a volcanic mountain is sometimes lower after an eruption than before. The eruption of Vesuvius, near Naples, Italy, in 79 A.D. is a famous case in point; after the eruption the top of the old volcano was completely missing. In its place was a huge depression 2 km across. This was taken to mean that the volcano had "blown its top," blasting the old peak into fragments. Likewise, in 1815 Mount Tamboro, in Indonesia, lost 1400 meters in altitude: the entire peak of the mountain was replaced by a caldera 11 km in diameter.

94
Igneous Activity and Volcanoes

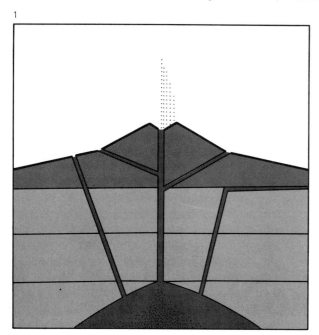
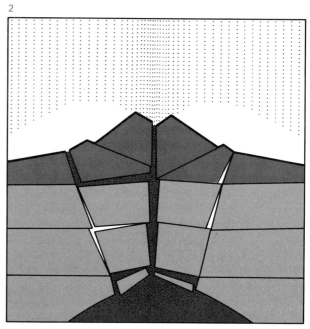

3.11 Stages in final explosive eruption and collapse of Mt. Mazama, Oregon, to form Crater Lake. (1) Magma chamber fills with gas-charged liquid. Gas separates vigorously from liquid and begins to erupt from central vent, spraying tephra into the air. (2) Eruption becomes more intense; gas and tephra escape violently from the magma chamber, partially draining this chamber. (3) After gas pressure has been reduced, the weight of the roof of the partially empty magma reservoir causes collapse, creating a caldera. (4) Water collects in caldera, forming Crater Lake (to be technically correct its name should be "Caldera Lake"). A second charge of magma, containing much less gas than the first, enters the reservoir; chiefly tephra is extruded to form Wizard Island, a small tephra cone. (Compare this sequence with the history of Vesuvius, Figure 3.20.) (Slightly modified from Howel Williams, 1941.)

By examining the products of such explosions, however, geologists learned what really happened. The fragments scattered over the landscape are not from the old rock of the mountain peak, but from freshly-created lava that must have come from the reservoir of molten

3.12 Profiles through Mauna Loa Volcano, Hawaiian Islands, and Mount Everest drawn to same scale.

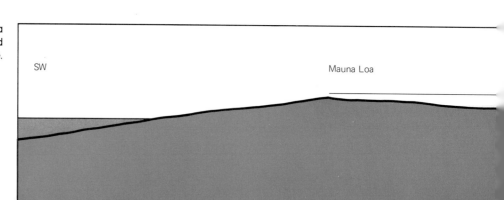

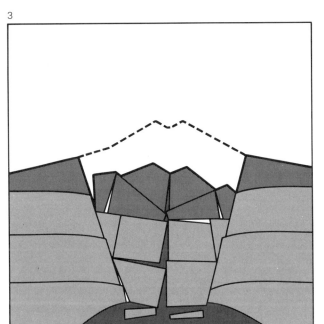
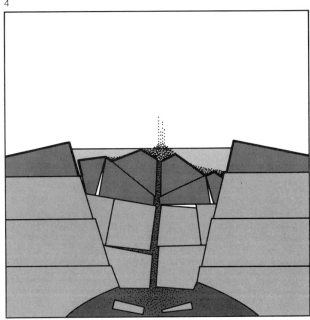

magma. After an eruption, the level of magma drops in the conduit and underlying reservoir. If the level drops far enough, it recedes from the roof of the magma reservoir itself, allowing a very large, generally circular block of the Earth's crust to sink downward. The caldera of Ngorongoro in Tanzania (tan-zuh-NEE-uh), Africa, is 18 km long, 16 km wide, and its floor lies 600 meters lower than its rim. The largest caldera in the world, at Aso Mountain in Japan, measures 16 by 22 km.

Volcanic shields Repeated quiet eruptions of liquid mafic lava create a broad, gently sloping hill of volcanic rock. There may be a small mound at the vent, usually a spatter or cinder cone. Because of the resemblance to the shields of early Germanic warriors, which had a pointed knob at the center, these structures are called **volcanic shields**. Although volcanic shields were originally named in Iceland, the best examples are the Hawaiian Islands and other mid-ocean is-

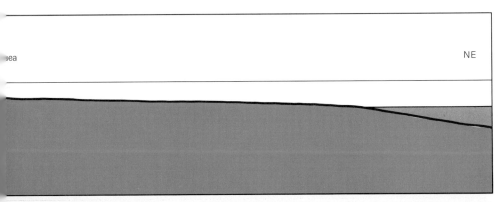
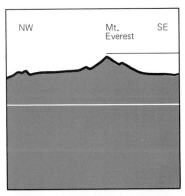

3.13 Fissure eruption, schematic block diagram. Lava spraying up along cracks forms fountains. After filling low block which subsided between parallel fissures the lava has begun to spread laterally beyond the fissures.

lands. Mauna Loa (mau-NUH LO-uh) built slowly upward from the sea floor over a million years or so, accumulating a towering mass of basalt 100 km wide at the base. The summit of Mauna Loa stands 4000 meters above the surface of the Pacific, and more than 9000 meters above the sea floor—higher from its base than Mount Everest is above sea level (Figure 3.12). So large is this mountain that even Kilauea (kee-lah-WAY-uh) and Hualalai (hwa-luh-LIE), full-sized structures by most measures, are mere flank craters on Mauna Loa. These giant shields are usually constructed not by one central vent but by zones of fissures called **rift zones** (Figure 3.13).

Lava plains As volcanic shields are larger and flatter than domes, cones, and calderas, lava plains are far larger even than shields. **Lava plains** (or *plateaus* if they exist at high altitude) are flat sheets of extrusive igneous rocks. Such sheets were built by repeated effusions of freely flowing mafic lava. The slope of a lava plain is so slight, often less than one degree, that it is usually not discernible by eye. Whereas a shield may be 10 km or so across, a lava plain may cover more than 250,000 square kilometers. After thousands of eruptions, the thickness of extrusive rock may reach about a kilometer. Thus, the volume of extrusive rock underlying a lava plain can reach 400,000 cubic kilo-

3.14 Columbia basalt plateau in Washington and Oregon. Textured areas shown in map are underlain by basalt that spread out in hundreds of successive fluid flows and eventually filled a former lowland. The maximum aggregate thickness of the basalt layers is about 1100 m. Photo shows sectional view of at least four basalt flows.

meters. The volume of the Columbia and Snake River basalts (the most common volcanic rock) of the northwestern United States (Figure 3.14) is so great that in places they have changed mountainous landscapes into level plateaus. We are only beginning to explore the largest lava plains of all—those found on the ocean floors. These thin layers of basalt are thought to be produced continuously along vast oceanic volcanic ridges.

Composite cones Obviously, the lavas of both shield volcanoes and lava plains are too fluid to build the steep slopes we associate with picturesque mountains; such lava would not build a pile but would flow out the same way a castle of too-wet sand on a beach subsides. Likewise, a cone composed only of loose pyroclastic material seems to be unstable if more than 300 to 450 meters high. Most of the spectacular volcanic mountains on Earth are combinations of both solid igneous rock and loose pyroclastic debris called **composite cones.** These are built by alternating layers of tephra and lava, the latter acting as cement for the former (Figure 3.15). Mountains such as Fuji in Japan, Vesuvius in Italy, Rainier in Washington, Mayon in the Philippines, and Shasta in California, all famed for their beauty, are composite volcanic cones (Figure 3.16). The secret of such symmetry is eruption through a single central vent. If the vent should move, or if lateral eruptions should break out, secondary cones would form and the symmetry is lost.

Kinds of Eruptions

Volcanoes may be defined not only by the landforms they build but also by the kind of eruption that shapes them. Most eruptions from

98
Igneous Activity and Volcanoes

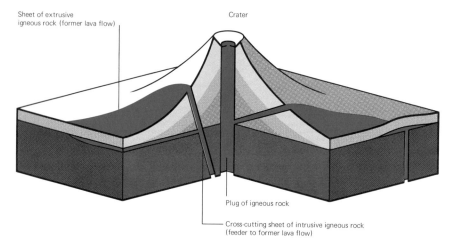

3.15 Volcanic cone shown in cutaway view of block diagram. Summit eruption and flank eruptions are illustrated in an idealized version. Further explanation in text.

cones are **summit eruptions,** which eject volcanic material from the central vent at the mountain peak. If the lava breaks through the sides of cones it is called a **flank eruption** (see Figure 3.15). Other eruptions emerge through cracks on land or at sea as **fissure eruptions.**

Most eruptions are combinations of these types, such as the activity in Puyehue (poo-YAY-way) Volcano in Chile in 1960. Puyehue, thought to be a summit type, began with a flank eruption. Later, lava poured from 28 craters along a new fissure 13 km long.

Another classifying feature is the violence of the eruption.

3.16 Composite cone of Mt. Mayon, Philippines, considered to be the most symmetrically formed composite cone in the world. (Peabody Association.)

Effusive eruptions Quiet eruptions of fluid mafic lava that may flow over the ground as fast as 20 to 30 km/hr are **effusive eruptions**. Because the viscosity of this lava may be very low, dissolved gases usually escape gradually, like the bubbles of boiling soup, before they create enough pressure to cause explosions. The lava flows out smoothly, spreading far from the vent or fissure. It builds up vast plateaus on the continents, such as the Columbia Plateau and the Deccan Plateau of India, and flat expanses of the floor of the ocean.

Explosive eruptions Highly viscous felsic lava produces explosive eruptions, usually accompanied by dense clouds of ash. Repeated explosions may send tons of lava and clouds of gas high into the air. The abundant silica, often up to 60 per cent, raises the viscosity, making it more difficult for gases to escape. Gas pressures rise until they reach a critical point, when they may shatter masses of lava, or even part of a mountain. Gas, especially water vapor, is thus the driving force behind volcanic explosions. Without the dissolved gases, silicic magma becomes inert; it can neither flow nor explode.

Fumarolic eruptions *Fumarolic* eruptions, or **fumaroles** (from the Latin *fumo* for smoke) are vents that issue only hot gases for long periods. A well-known example is Mount Fuji in Japan.

Distribution of Volcanoes

During the last 400 years, about 500 volcanoes are known to have erupted. There is probably no part of the world that has not been the scene of volcanic activity at some time. However, some regions of the Earth are far more active than others. As we shall see, there are good geologic reasons for this.

Geographic distribution Today, more than three-quarters of the world's subaerial surface volcanoes (volcanoes that erupt into the atmosphere) are distributed around the edges of the Pacific Ocean in a pattern sometimes called the "Ring of Fire" (Figure 3.17). An extremely active offshoot of this ring extends westward through Indonesia, where fully 14 per cent of the world's volcanoes are located. Most of the Pacific volcanoes form festoons of islands, called *island arcs*, such as the Aleutians, Kuriles, Bonins, Marianas, New Guinea, and the Solomons-New Hebrides.

Another cluster of volcanoes is in Iceland, where the eruption of Laki in 1783 produced the largest lava flow in history. The entire mid-Atlantic ridge, extending north and south along the floor of the

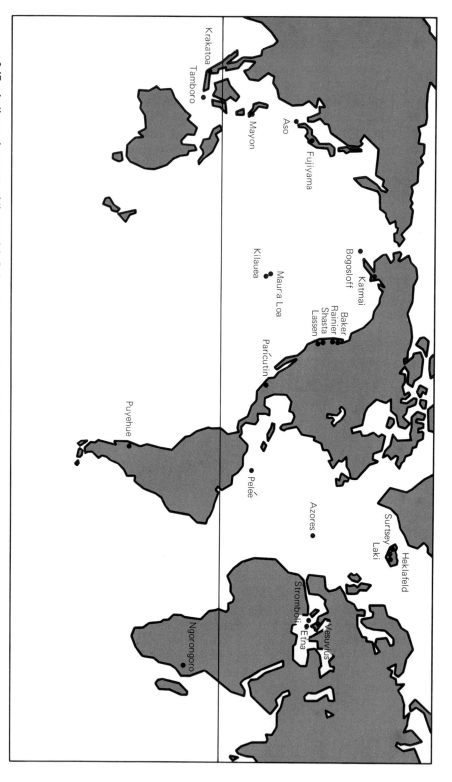

3.17 Active volcanoes of the world. Only volcanoes mentioned in text are labeled with their names.

Atlantic basin, quivers with volcanic activity (Figure 3.17). Other concentrations are located around the Mediterranean and in East Africa, along the Rift Valley.

Although more than 80 per cent of known volcanoes are on the continents, this figure does not really illustrate the true picture. The estimated total volume of volcanic rocks that have accumulated on the floor of the Pacific Ocean is 2½ times that erupted on all continents throughout geologic history. The only answer to this paradox is that most of the world's volcanism is occurring in secret—hidden and muffled by the great depths of the oceans.

Important differences exist between continental and submarine volcanoes. Perhaps the most obvious is the difference between types of lava, and therefore violence of eruption. The lava of continental volcanoes tends to be viscous and felsic, like the continental rock from which the lava melted. Highly viscous, felsic lava results in explosive eruptions. Oceanic lava is more fluid and mafic, allowing magmatic gases to escape freely before they build up enough pressure to explode.

Geologic distribution   For centuries, geologists have looked for a reason for this distribution of modern subaerial volcanoes. Why are there none in North America except on the West Coast? Why none in Greenland? In Australia? In most of Asia? Why are most of the subaerial volcanoes in the Pacific situated around the rim, whereas those in the Atlantic are in its middle?

During the last 15 years, answers to such questions have begun to appear. Volcanoes, and other major geologic activities such as earthquakes and mountain building, seem to be linked to a worldwide system known as **plate tectonics** (Figure 3.18), which we shall be discussing in Chapter 14 and elsewhere throughout the book.

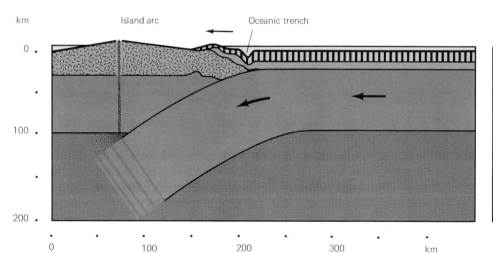
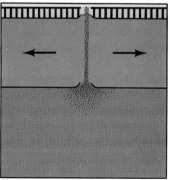

3.18 Relationship of volcanoes to plate tectonics. Schematic profile through outer part of Earth. At mid-oceanic ridge, where plates move apart, material rises to form volcanoes, most of which are underwater. Striped part of upper crust represents contrasting magnetic properties of rocks. Island arc volcanoes lie above zones where descending plate is thought to melt.

Some Classic Volcanic Eruptions

Over the years, some volcanoes have gained a special place in volcanology. Their eruptions were so distinctive that their names have been used to describe other eruptions. Examples include Vesuvius, Krakatoa, Mt. Pelée, Hawaii, and Stromboli.

Vesuvius Ever since Roman times, the activity of Vesuvius has been observed and recorded. After a long period of quiescence, Mt. Vesuvius erupted in 79 A.D. After the eruption, the Roman historian Pliny the Younger wrote what may be called the first scientific description of a volcanic eruption. Pliny had a strong personal reason for writing this description: his uncle, Pliny the Elder, died observing the activity.

Geologists have sorted out the details of the eruption by analyzing the thick layers of tephra. The eruption began by clearing out the vent. This was followed by the rapid expansion of gases in the magma. Then came a long and frightening series of violent explosions that threw 250 cm of pumice on nearby Pompeii. The pumice buried the city and suffocated its residents as they sat or lay (Figure 3.19). Then a blast of gas blew out the sandy sides of the conduit. During a lull in the violence, the crater walls collapsed. The material that collapsed into the vent was blown out by further explosions. In addition, the rising heat over the vent and the steam coming from it created tremendous updrafts that triggered local thunderstorms. The torrential rainfall saturated the tephra deposits on the western flank of the cone. Finally, this gooey mass slid loose and buried the nearby city of Herculaneum. When the magma reservoir was depleted, the entire top of the mountain collapsed, leaving a huge caldera that required many further eruptions to fill (Figure 3.20).

The word *Plinian* has since come to stand for eruptions consisting of the explosive emission of large amounts of lava, on the order of several cubic kilometers, which causes the collapse of the upper mountain to form a caldera. A Plinian eruption may or may not be accompanied by glowing avalanches.

Krakatoa The eruption in 1883 of Krakatoa (Figure 3.21), the most famous Plinian eruption of recent times, was so violent that "Krakatoan" may also be used to describe particularly energetic volcanism. On May 20, 1883, after 200 years of calm, Krakatoa, a large composite cone located in the Sunda Strait between Java and Sumatra, began a series of moderate and weak eruptions. Following three months of such activity the climax came suddenly on August 26 at 1 p.m. Explosions, heard throughout Java, more than 150 km away, con-

3.19 **Plaster casts of former residents of Pompeii,** Italy, who became victims of the explosive eruption of Mt. Somma in 79 A.D. The shapes of these bodies were preserved as molds in the tephra that engulfed them. Plaster casts were made after the soft tissue had oxidized and the bones were removed, and these specimens were the result. (SCALA.)

3.20 Collapse of Mt. Somma and later building of Vesuvius, a large composite cone. The collapse of Mt. Somma caldera followed the explosive eruption of 79 A.D. Later eruptions have built a composite cone large enough to bury all but the northeast part of the walls of the caldera. (Photo courtesy of Italian Government Travel Office.) (Bottom left drawing modified from Alfred Rittmann, 1962.)

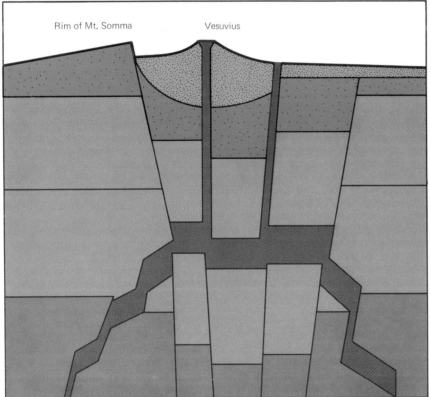

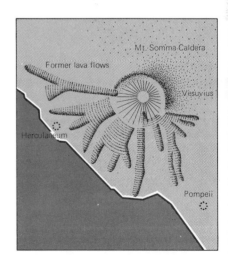

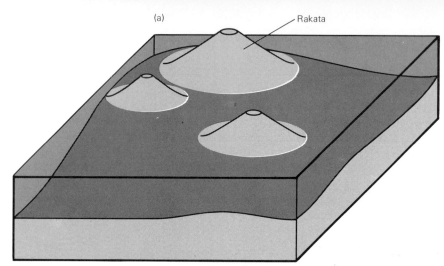

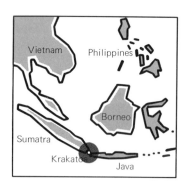

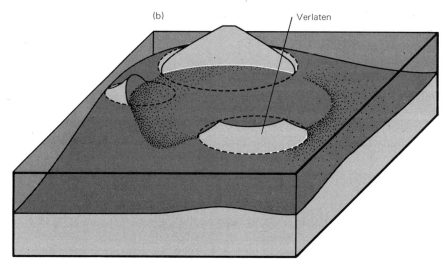

3.21 Stages in history of Krakatoa, Indonesia. (a) Three small conical islands, as restored prior to great explosion. (b) Collapse to form caldera after violent eruption of August 27, 1883 removed parts of all three island cones. (c) Later growth of small central cone creates a new island in the center of the group of three. (Bottom, left) Island of Mang, formed by the walls of a caldera closely resembling that of Krakatoa, in northern Marianas, Pacific Ocean, as photographed obliquely from an airplane. (American Museum of Natural History.)

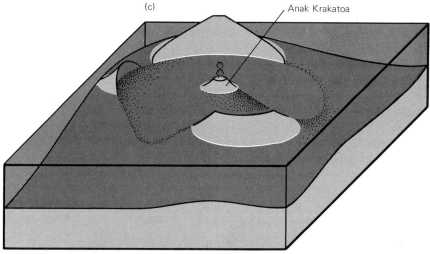

tinued all afternoon and all night, bringing a "rain" of ash and pumice that made it almost impossible to see, even with lamps. At 10:02 the next morning came a blast that has been called the loudest noise in history, shooting ash 80 km into the air and covering 800,000 square kilometers with tephra. The noise was heard 3500 km away in central Australia, and even residents of Rodriguez Island, nearly 5000 km distant in the Indian Ocean, reported sounds like heavy gunfire. Weak activity continued until February. When it was all over, nearly 18 cubic kilometers of tephra had been spread through 4 million square kilometers. Floating piles of pumice as much as 3 meters thick made navigation in the Sunda Strait hazardous.

Far more destructive to human life were the huge sea waves, or tsunami, caused by the eruption. A **tsunami** (tsoo-NAH-mee) is an unusual wave unrelated to tide or wind that is set off by the displacement of the sea floor by a volcano, earthquake, or undersea avalanche. The first tsunami hit Java and Sumatra a few hours after the explosions, damaging low-lying villages. The next morning, a giant wave carried a Sumatran gunboat into the center of the town of Telok Betong; another wave extended the boat's overland journey a kilometer and a half farther. Some of these waves reached heights of more than 30 meters and traveled as much as 150 km inland. Altogether, the waves of Krakatoa killed more than 36,000 persons! As for the island itself, more than two-thirds of it disappeared; it was replaced by three small islands, a tiny rock, and a caldera 275 meters deep.

Mt. Pelée The 1902 eruption of Mt. Pelée (pay-LAY), on the Caribbean island of Martinique (Figure 3.22), is considered by many volcanologists to have been the most dramatic in history.

The first signs of activity were observed by a high-school teacher on April 2. On April 23, ash fall and sulfur fumes began to spread and to grow more intense daily. By the end of April, falling ash was choking the streets in St. Pierre, a city 10 km from the summit. Businesses began to close. Horses and birds dropped dead in the city, and residents began to flee. However, the French governor wanted the people to remain so they could vote in elections scheduled for May 10. He appointed a commission which reported no danger from the volcano. When this move failed to halt the evacuation, he stationed troops around town on May 6 to keep people home. On May 7, the newspaper *Les Colonies* wrote: "Mount Pelée is no more to be feared by St. Pierre than Vesuvius is feared by Naples. Where could one be better off than at St. Pierre?" The article in *Les Colonies* was not purely a political statement, because all previous eruption clouds had traveled southwestward from the peak straight to the sea (Figure 3.22).

The city was bulging with several thousand panic-stricken refu-

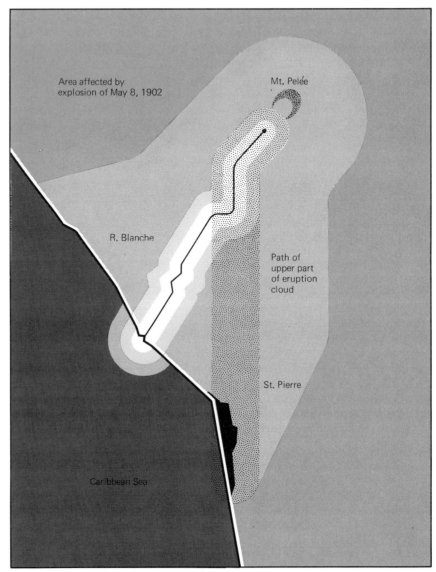

3.22 **Northwestern Martinique**, West Indies, showing area devasted by explosive eruption of Mt. Pelée on May 8, 1902. The densest part of the mass of hot tephra discharged by the explosion flowed directly to the sea down the valley of River Blanche (as shown in Figure 3.4). Such a large amount of hot tephra was discharged on May 8, however, that much of it did not make the sharp bend in the valley but spilled over the valley wall and headed straight south toward St. Pierre. Because all previous avalanche clouds from Mt. Pelée had followed the direct route to the sea via the valley of River Blanche, the residents of St. Pierre had every reason to believe that they would be safe. They did not imagine that any eruption cloud could overflow the valley sides. As it happened, the fatal eruption was so big that the stretch of the valley of River Blanche which trends N-S served as a rocket launcher pointed right at St. Pierre. (After G. A. MacDonald, 1972, Fig. 8-1, p. 144.)

gees from the countryside when, on May 8, at 7:50 a.m., four loud explosions rocked Mt. Pelée. One of them shot a glowing avalanche sideways through a V-shaped notch in the caldera. It started down its usual track toward the sea. However, it was so big, that most of it did not make the sharp turn in the valley, but headed straight south directly toward St. Pierre. The speed of this avalanche was later calculated to have been around 150 km/hr. Two minutes later the cloud hit the city, stopping the clock on the Hospital Militaire at 7:52 and killing an estimated 30,000 persons (including the governor).

Although hurricane winds in the Caribbean often move as fast as the glowing avalanche of Mt. Pélee, hurricanes are not so destructive. The reason is that the blast from the volcano consisted of a cloud of superheated steam filled with even hotter tephra particles. Thus the volcanic blast was denser than normal atmospheric gases, so that the fiery cloud hugged the ground as it traveled and hit buildings and people with greater impact. Observers who saw it from the side or from ships in the harbor described it as a black, rolling cloud that moved as though shot from a cannon.

Destruction to the city was almost total (Figure 3.23). Only two persons are known to have survived, one a prisoner in a dungeon. The rest appear to have died in seconds from inhaling the gas, which was hot enough (650 to 700°C) to have softened glass. Of the city itself, which had been the most important commercial center on Martinique, no building remained standing. Iron girders were twisted like licorice, burning rum ran in streams through the streets, one-meter-thick walls of cement and stone were torn apart, and a three-ton statue of the Virgin Mary was carried 15 meters from its base. The entire town was ignited as if by a giant match. During the next few months, at least five more explosions added more than 2000 persons to the number of fatalities.

The eruption of Mt. Pelée added a new term to the volcanologist's vocabulary: **nuée ardente** (noo-ay ar-DAHNT), French for glowing cloud. Although the term glowing avalanche is more accurate, a

3.23 Total devastation of St. Pierre shown in photograph made with camera facing due north toward Mt. Pélée Volcano. The avalanche of hot tephra which wiped out the city flowed south, directly toward the camera. Notice that the only walls left standing are aligned N-S, parallel to the flow of the glowing avalanche. Harbor is at left. Compare with Figure 3.4. (American Museum of Natural History.)

Peléean eruption is often defined as an eruption in which all of the magma mixes with gas and is expelled in the form of *nuées ardentes.* In such an eruption, almost no lava flows as a liquid. In the final stages of eruption, a mass of viscous lava may rise as a bulge in the crater. This happens when there is no longer enough gas in the magma to explode. In the case of Mt. Pelée, the lava erected a giant spine from the vent, towering above the crater like a huge plug.

Other volcano types Two other major types of volcanoes are named after specific eruptions.

Hawaiian eruptions have already been described as those involving fluid lava, usually of great quantity, in which the gases are liberated continuously and more or less quietly. Occasionally, fountains of lava, projected by jets of escaping gas, may reach heights of 300 meters or more. More commonly, lava oozes from vents or fissures quietly and persistently, building some of the largest mountains on Earth.

Strombolian eruptions are named after Stromboli, a volcano in the Aeolian Islands off the coast of Sicily (Figure 3.24). Stromboli has been known since ancient times as the "lighthouse of the Mediterranean" because of the red glow which can be seen almost all the time. Eruptions occur more or less regularly as moderate explosions that throw incandescent cinder, lapilli, and bombs to heights of tens or hundreds of meters, accompanied by a white cloud signifying low ash content. Stromboli has been exceptionally important to volcanologists both because it is easily accessible for study and because of its continuous discharge of hot clots of lava.

3.24 Stromboli at night, the "lighthouse of the Mediterranean," (Giorgio Gualco/B. Coleman)

3.25 Major peaks of Cascade Range were formed by growth of composite cones during Pleistocene and Holocene volcanic activity.

Volcanism in the continental United States In the United States, as elsewhere, it is difficult to calculate the number of volcanoes that have been active during geologic time, because most volcanic structures older than about 10 million years have been eroded beyond recognition by wind and water. Some older volcanoes can be identified by the rocks composing their conduits or vents, but most ancient volcanoes have been buried. Geologists estimate that during the past 60 million years there have probably been several thousand eruptions in the Cascade Range (Figure 3.25)—the only region with such recent volcanism. About 15 volcanoes have erupted during the last few tens of thousands of years, and seven have erupted since 1800, including Mt. Baker, Mt. Rainier, Mt. St. Helens, Mt. Hood, Mt. Shasta, Cinder Cone, and Lassen Peak.

Mt. Baker, in Washington, is the site of the most recent volcanic activity in the continental United States. In mid-1975 steam issued steadily from the newly reactivated vent on the southeast side. Heat

3.26 **Snow-capped peak of Mt. Baker,** Washington, showing steam cloud issuing from newly reactivated vent on southeast side. (Stephen Malone, University of Washington, courtesy Sigrid Schroeder.)

from the vent has been melting the snow and ice which blankets the summit. The possibility of catastrophic avalanches from the melted snow and ice caused park officials to close large areas of the Mt. Baker North National Forest in the summer of 1975.

In 1926, the U.S. Geological Survey set up an observatory at the town of Mineral, Oregon. Other stations throughout the Cascade Range have since been installed by the Survey and by the Center for Volcanology at the University of Oregon. Recent evidence provided by the U.S. Geological Survey has indicated that Mt. St. Helens has been "more active and more violent" during the past 2000 years than any other volcano in the continental United States. The scientists compare Mt. St. Helens' activity with that of Vesuvius in Italy, and predict another eruption before the end of this century.

Underwater Volcanoes

Modern methods of investigating the ocean depths have shown tens of thousands of volcanic cones beneath the surface of the sea. The study of these underwater volcanoes is a frontier area of science that is only now being fully explored. In this section we will describe some of the unique features of underwater volcanoes.

The Discovery of Yellowstone Park

Nathaniel Pitt Langford

Monday, August 29. At about one mile below our camp the creek runs through a bed of volcanic ashes, which extends for a hundred yards on either side. Toiling on our course down this creek to the river we came suddenly upon a basin of boiling sulphur springs, exhibiting signs of activity and points of difference so wonderful as to fully absorb our curiosity. The largest of these, about twenty feet in diameter, is boiling like a cauldron, throwing water and fearful volumes of sulphurous vapor higher than our heads. Its color is a disagreeable greenish yellow. The central spring of the group, of dark leaden hue, is in the most violent agitation, its convulsive spasms frequently projecting large masses of water to the height of seven or eight feet. The spring lying to the east of this, more diabolical in appearance, filled with a hot brownish substance of the consistency of mucilage, is in constant noisy ebullition, emitting fumes of villainous odor. Its surface is covered with bubbles, which are constantly rising and bursting, and emitting sulphurous gases from various parts of its surface. Its appearance has suggested the name, which Hedges has given, of "Hell-Broth springs"; for, as we gazed upon the infernal mixture and inhaled the pungent sickening vapors, we were impressed with the idea that this was a most perfect realization of Shakespeare's image in Macbeth. It needed but the presence of Hecate and her weird band to realize that horrible creation of poetic fancy, and I fancied the "black and midnight hags" concocting a charm around this horrible cauldron. We ventured near enough to this spring to dip the end of a pine pole into it, which, upon removal, was covered an eighth of an inch thick with lead-colored sulphury slime. . . .

Friday, September 2. To-day we have occupied ourselves in examining the springs and other wonders at this point. . . . Through a little coulee on the other side of the hill runs a small stream of greenish water, which issues from a small cavern, the mouth of which is about five feet high and the same dimension in width. From the mouth, the roof of the cavern descends at an angle of about fifteen degress, till at the distance of twenty feet from the entrance it joins the surface of the water. The bottom of the cavern under the water seems to descend at about the same angle, but as the water is in constant ebullition, we cannot determine this fact accurately. The water is thrown out in regular spasmodic jets, the pulsations occurring once in ten or twelve seconds. The sides of the mouth of this cavern are covered with dark green deposits, some which we have taken with us for analysis. About two hundred yards farther on is another geyser, the flow of which occurs about every six hours, and when the crater is full the diameter of the surface is about fourteen feet, the sides of the crater being of an irregular funnel-shape, and the descending at an angle of about forty-five degrees. At the lowest point at which we saw the water it was about seven feet in diameter on the surface. One or another of our party watched the gradual rise of the water for four or five hours. The boiling commenced when the water had risen half way to the surface, occasionally breaking forth with great violence. When the water had reached its full height in the basin, the steam was thrown up with great force to a height of from twenty to thirty feet, the column being from seven to ten feet in diameter, at the midway height of the column, from bottom to top. The water was of a dark lead color, and those portions of the sides of the crater that were overflowed and then exposed by the rise and fall of the water were covered with stalagmites formed by the deposit from the geyser. . . .

Monday, September 19. When we left Yellowstone lake two days ago, the desire for home had superceded all thought of further explorations. Five days of rapid travel would, we believed, bring us to the upper valley of the Madison, and within twenty-five miles of Virginia City . . . We had within a distance of fifty miles seen what we believed to be the greatest wonders on the continent. We were convinced that there was not on the globe another region where within the same limits Nature had crowded so much of grandeur and majesty with so much of novelty and wonder. Judge, then, of our astonishment on entering this basin, to see at no great

From THE DISCOVERY OF YELLOWSTONE PARK by Nathaniel Pitt Langford, 1905. University of Nebraska Press, 1972.

Field Trip

distance before us an immense body of sparkling water, projected suddenly and with terrific force into the air to the height of over one hundred feet. We had found a real geyser. In the valley before us were a thousand hot springs of various sizes and character, and five hundred craters jetting forth vapor. In one place the eye followed through crevices in the crust a stream of hot water of considerable size, running at nearly right angles with the river, and in a direction, not towards, but away from the stream. We traced the course of this stream by the crevices in the surface for twenty or thirty yards. It is probable that it eventually flows into the Firehole, but there is nothing on the surface to indicate to the beholder the course of its underground passage to the river.

On the summit of a cone twenty-five feet high was a boiling spring seven feet in diameter, surrounded with beautiful incrustations, on the slope of which we gathered twigs encased in a crust a quarter of an inch in thickness. On an incrusted hill opposite our camp are four craters from three to five feet in diameter, sending forth steam jets and water to the height of four or five feet. But the marvelous features of this wonderful basin are its spouting geysers, of which during our brief stay of twenty-two hours we have seen twelve in action. Six of these threw water to the height of from fifteen to twenty feet, but in the presence of others of immense dimensions they soon ceased to attract attention.

Of the latter six, the one we saw in action on entering the basin ejected from a crevice of irregular form, and about four feet long by three wide, a column of water of corresponding magnitude in the height of one hundred feet. Around this crevice or mouth the sediment is piled in many capricious shapes, chiefly indented globules from six inches to two feet in diameter.

We gave such names to those of the geysers which we saw in action as we think will best illustrate their peculiarities. The one I have just described General Washburn has named "Old Faithful," because of the regularity of its eruptions, the intervals between which being from sixty to sixty-five minutes, the column of water being thrown at each eruption to the height of from eight to one hundred feet. . . .

Near by is situated the "Giantess," the largest of all the geysers we saw in eruption. Ascending a gentle slope for a distance of sixty yards we came to a sink or well or an irregular oval shape, fifteen by twenty feet across, into which we could see to the depth of fifty feet or more, but we could discover no water, though we could distinctly hear it gurgling and boiling at a fearful rate afar down this vertical cavern. Suddenly it commenced spluttering and rising with incredible rapidity, causing a general stampede among our company, who all moved around to the windward side of the geyser. When the water had risen within about twenty-five feet of the surface, it became stationary, and we returned to look down upon the foaming water, which occasionally emitted hot jets nearly to the mouth of the orifice. As if tired of this sport the water began to ascend at the rate of five feet in a second, and when near the top it was expelled with terrific momentum in a column the full size of the immense aperture to a height of sixty feet. The column remained at this height for the space of about a minute, when from the apex of this vast aqueous mass five lesser jets or round columns of water varying in size from six to fifteen inches in diameter shot up into the atmosphere to the amazing height of two hundred and fifty feet. This was without exception the most magnificent phenomenon I ever beheld.

1905

Differences from Subaerial Volcanoes

The average depth of the world's oceans is about 4000 meters. For a volcano to take place on the deep sea floor the pressure must exceed the weight of all that water. Furthermore, before it becomes visible, the volcano must build a mountain 4000 meters high. Needless to say, a great deal of submarine volcanic activity never meets the eye—particularly the gentle but persistent eruptions from such zones as the mid-Atlantic ridge that are thought to be creating new sea floor (Figure 3.18, right).

Hydrostatic pressure Eruptions that occur under water more than about 600 meters deep are never explosive, no matter what kind of lava is emerging. The reason is that the pressure of the water at such depths is so great that gas is prevented from expanding. Because an explosion is essentially a rapid expansion of gases, deep submarine eruptions are quiet ones.

Vertical movement and rebuilding The vertical build-up involved in creating a submarine mountain is tremendous. The end result is the tallest mountains in the world, measured from sea floor to crater. The island formed by the top of a submarine volcano, however, is seldom permanent. Additional eruptions often change the shape, size, or location of such an island, or even replace it altogether with a new one.

New and Disappearing Islands

The formation of volcanic islands Very few submarine eruptions go on long enough to build a mountain that reaches to the surface of the ocean. If such a mountain does break the surface, the chances are good that the pounding waves will eventually slice off its top, leaving only a submerged shoal to mark the site. This happened to Falcon Island, in the Tonga group, which was first reported in 1867 as a shoal and in 1885 as a volcanic island about 100 meters high. By 1892 it had shrunk to less than 10 meters high, and by 1898 had completely succumbed to wave action.

Some volcanic islands come and go several times. The most famous volcano in the Aleutian Islands of Alaska is the "disappearing" island of Bogoslof. The first island, a small peak called Ship Rock, was discovered in 1768. Since then, the island has ducked up and down several times. Thirty-six of the 76 volcanoes in the Aleutians have been active since 1760 but Bogoslof has been the most active. In 1796 a second peak called "Castle Rock" was sighted. In 1883, "New Bogoslof" emerged, and in 1906 and 1907 two new cones appeared. These both vanished. New eruptions followed in 1910, 1926, 1931, and these alternately removed and replaced the elusive Bogoslof.

Surtsey The best-known "new" island of recent years is Surtsey, born November 14, 1963, within sight of the coast of Iceland (Figure 3.27). Eruption began early on the morning of the 14th, and by the 16th an island had appeared. By the 19th it was 60 meters high and 600 meters long. Hydroexplosions (explosions caused by the contact of molten lava with water) continued for several months. Occasionally bombs were thrown as high as 1000 meters. By April 1964, the cone had grown high enough so that water no longer reached the vents. The eruption became more similar to the Hawaiian type. Lava poured gently into the sea and armored the island against the waves. It is this armor that has made it possible for Surtsey to survive.

A little more than a year later, on May 7, 1965, the Surtsey eruption ended. Two weeks later another ash cone, Little Surtsey, was built about 600 meters to the northeast. By September, it had built up to 70 meters above sea level and by October 24 had washed away. Soon more eruptions began to the southwest of Surtsey. On December 26, these produced a visible island named Jølnir, or Christmas Island. Jølnir disappeared and reappeared five times. In the summer of 1966 it was washed away for good.

Underwater Eruptions and Lava

Pillow structure When lava erupts underwater, it cools more quickly than when it pours out on land. This rapid cooling often creates distinctive shapes and textures. Commonly, lavas erupted underwater are molded into pillow- or sausage-shaped lumps with a glassy but plastic skin. When they cool, such lavas create **pillowed rock.** The sack-like blobs, the pillows, usually accumulate in piles. The bottom of each pillow conforms to those beneath; the top remains rounded (Figure 3.28). Most pillowed rocks are thought to have formed from thick, pahoehoe lava flows. Pillowed lavas seem to form during shallow submarine eruptions, or subaerial eruptions that pour into the sea. Not until 1971 was there a well-authenticated observation of underwater formation of lava pillows. A team of divers saw them budding off from flowing tongues of lava pouring from Kilauea Volcano on the ocean floor off Hawaii.

3.27 Stages in the growth of Surtsey, a new volcanic island that appeared off the south coast of Iceland on November 14, 1963. All photos by Sigurdur Thorarinsson. (1) Clouds of tephra on November 15, 1963 conceal the island, which is 500 m long and 37 m high. (2) View from airplane one week later. Clouds of steam and tephra billow from the enlarging cone, clearly visible in foreground. (3) Tephra cone, almost 700 m in diameter. Lava appears for the first time and flows into sea through breached wall of cone; November 30, 1963. (4) Clouds, chiefly of steam, from reaction between lava and sea water. Small, dark tephra cloud in center; December 16, 1963. Island is 800 m in diameter, 87 m high. (5) View north into vent. Declining activity indicated by small steam cloud; April 1, 1964. Area of island, 1.15 km². (6) View west of lava reaching sea to form protective terrace against wave attack; April 11, 1964. (7) Distant aerial view of whole island; hot lava flowing into sea creates steam cloud. Protective terrace of igneous rock fully encircles island; June 18, 1964. Area of island, 1.41 km².

115
Underwater Volcanoes

3.28 Sectional view of three complete small pillows and partial view of two large pillows formed by extrusion of lava onto ancient sea floor. Of the three complete pillows shown the lower two, having circular sections in this view, evidently had hardened before the third (upper) one was formed. Before the third pillow had hardened, the two large ones were piled on top. This additional weight flattened the third pillow somewhat and pressed it down against the older two below, creating a rounded "bump" on its lower side. (Brad Hall.)

Broken-glass rocks The rapid cooling effect of water creates another volcanic product, masses of almost completely glassy particles which become cemented to form "broken-glass rocks." The glass is usually altered by oxidation and the absorption of water into a waxy-looking, yellow-brown substance. Broken-glass rock usually forms in oceans or lakes, but may also form when an eruption occurs under a glacier. Such an eruption may create a lake under the glacier, or a lake that melts its way to the top of the glacier. The lake may slowly drain, leaving deposits of broken-glass rocks.

Subglacial eruptions Subglacial eruptions, particularly in Iceland but also in Antarctica and in the Andes, are a variation of underwater eruptions that may cause enormous floods. If a volcano melts enough ice to form a lake, the lake may eventually break through the ice surrounding it, sending water and icebergs cascading over the countryside. Two eruptions of the Volcano Grimsvøtn in Iceland in 1934 and 1938 discharged water at about 50,000 cubic meters per second. This is five times the flow of the world's largest river, the Amazon. These *jokulhlaups*, as they are called, are responsible for shaping much of the southern coast of Iceland.

Fumaroles, Hot Springs, Geysers

In volcanic regions, evidence of the Earth's heat often continues to surface for thousands of years after volcanic activity has ceased. Magma cools very slowly, giving off hot gas as it does. This gas is

Fumaroles, Hot Springs, Geysers

mostly steam, but also contains carbon dioxide, hydrochloric acid, ammonia, and many other gases which rise from the mantle into the crust. In most places, the outer part of the crust contains pore spaces filled with water, much of which has seeped downward from the rainfall. Rising gases heat and react with this water, possibly changing some of it to steam. An easy method has not yet been developed for differentiating superheated water from the ground, or new water coming up from the mantle.

Underground Activity

As a result of the movement of steam described above, fumaroles, hot springs, or geysers may reach the surface.

Fumaroles issue gases which may be foul-smelling or even dangerous. When wind conditions are right, fumarolic carbon dioxide may accumulate in valleys, killing mammals and birds. **Hot springs** emit liquids that have been heated to temperatures above the average air temperature of the region. **Geysers** are periodically spouting hot springs.

Causes of underground activity Fumaroles, hot springs, and geysers are closely related to one another and often occur together. Occasionally, they spout only water in the ground that was derived from rain and that circulated to great depths where the Earth's natural heat is high. Most of them, however, are distributed above cooling magma. In both cases, deep water is heated above the normal boiling point (100° C). This heated water, which is lighter than cooler water, then rises toward the surface.

As the heated water rises, it comes to places having less pressure from the weight of the overlying rock and water. Because of the lower pressure, boiling begins without any change in the temperature. Expansion of the boiling water causes overflow at the surface.

Locations of underground activity Yellowstone National Park (Figure 3.29) is one of the world's outstanding examples of an area having many hot vents. The 3000 or so hot springs, boiling mudpools, and geysers of Yellowstone have been seething since the first eruptions began early in the Cenozoic Era (about 50 million years ago). A slowly cooling mass of felsic magma just below the surface guarantees that the activity will continue for thousands of years to come. Although other major regions of steam and hot-water activity are found in Italy, New Zealand (Figure 3.30), Iceland, and elsewhere, Yellowstone stands out for the number and variety of its hot-water vents.

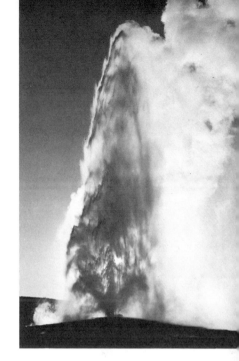

3.29 Old Faithful geyser, Yellowstone Park, Wyoming, at peak of one of its regular hourly eruptions. Height of water column is 20 m. (U.S. National Park Service)

3.30 Boiling mud pit marked by small circular holes where steam and hot water escape. Whakarewarewa Thermal Reservation, Rotorua, New Zealand. (Consulate General of New Zealand, New York City.)

Igneous Activity and Volcanoes

Old Faithful and Big Geysir Yellowstone National Park is probably best known for Old Faithful Geyser, which has erupted "faithfully" about once every hour for many years. The Park contains about 200 active geysers, compared with about 30 geysers in Iceland and a somewhat smaller number in New Zealand and Kamchatka, U.S.S.R., the other areas in the world where geysers are known.

For geologists, the most famous spouting hot spring is Big Geysir in Iceland, named in 1647 after the Icelandic word for "gush." In 1846 the German chemist Bunsen made a careful study of Geysir. His explanation for its intermittent eruptions is still generally accepted for Big Geysir, Old Faithful, and many other geysers. A tube leading to the geyser vent (more than 175 meters long in the case of Old Faithful) fills with water which is heated by magma or hot bedrock. Much of the water is heated above the temperature at which surface water boils, but, being trapped in the tube, the water is under high pressure and

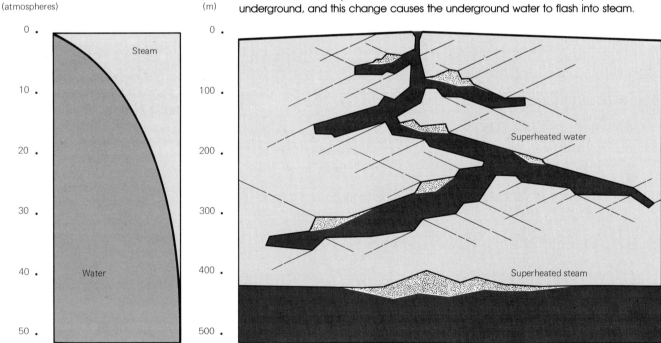

3.31 Schematic profile of Great Geysir, Iceland, alongside graph of change of boiling temperature of water with increasing hydrostatic pressure downward. Given the appropriate network of underground openings, a supply of water, and plenty of heat, a geyser can form. Eruption begins when all water reaches boiling temperature, 100°C at surface, increasing downward to 130°C at a depth of 30 meters. When steam pushes out water near the surface, it reduces the pressure underground, and this change causes the underground water to flash into steam.

cannot boil at that temperature. Eventually, continued heating from below finally starts some water boiling, forming steam. This steam forces the overlying column of water upward (Figure 3.31). As some of the water spills out, pressure on what is left below is reduced and more of the water is freed to flash into steam. This continues until the tube has been emptied and the eruption is over. The water from the eruption trickles back underground, and a new cycle begins.

Igneous Activity and Seismic Activity

One of the most reliable signals preceding an eruption is an abnormal vibrating of the Earth, caused by the passage of tremors through the ground. These vibrations are a form of seismic activity. Trembling may begin weeks or months before an eruption. The number and intensity of the tremors or small earthquakes usually increase sharply during the last few days.

Seismic Activity and Volcanoes

It is logical that seismic tremors should accompany volcanoes. The immediate cause of a volcano is probably some change in the upper mantle that reduces pressure on the mantle rock. Such a change would cause some "creaks and groans" that could be detected at the surface by sensitive *seismographs* (Figure 3.32). In the Hawaiian Islands, volcanic eruptions are preceded by two kinds of earth-

3.32 Portable seismograph carried in a jeep is prepared to record seismic waves created by movements of magma underground during flank eruption of Kilauea Volcano, Hawaii, in 1956. (U. S. Geological Survey.)

quakes. The shallow kind originate within about 8 km of the surface, and these are thought to be associated with the bulging and sagging of the volcanic cone before and during eruption. The deep kind come from 50 to 60 km below the ocean. These are thought to be generated by the creation and collection of magma. Although the early movements of this magma are poorly understood, it is possible that seismic activity may result from the gradual readjustment of solid rock as the magma moves out of it.

Examples of seismic activity Exceptionally large eruptions may be preceded by many years of seismic activity. Before the destructive eruption of Mt. Vesuvius in 79 A.D., seismic tremors shook Naples off and on for 16 years. Eruptions at Kilauea in Hawaii are signalled by as many as thousands of shallow tremors. One of the best-studied series of pre-eruption earthquakes is reported by Gordon A. Macdonald, former director of the Hawaiian Volcano Observatory. In 1942, a flurry of deep vibrations began, 40 to 50 km below Mauna Loa. On February 21 and 22 a swarm of shallow quakes occurred high on the flanks of the mountain along the rift zone where lava had previously erupted. This swarm was followed by others that gradually migrated up the rift zone. By March 7 the location of the vibrations had reached the summit. Based upon a slight tilting of Kilauea westward and the knowledge that the summit of Mauna Loa had erupted in 1940, R. H. Finch predicted that an eruption would take place along the flank of Mauna Loa within a few months, at an altitude of between 2500 and 3000 meters. Right on time, the eruption came on April 28 at 2800 meters. This is the most accurate prediction yet made of a volcanic eruption.

Prediction and Use of Igneous Activity

Each year, roughly several dozen volcanoes erupt on Earth, causing varying amounts of damage. Damage is done by direct igneous activity such as lava flows, suffocation by gases, showers of tephra, and glowing avalanches. Some indirect hazards are mudflows, created by torrential downpours related to the hot updrafts near the vent, and tsunami related to the collapse of calderas on volcanic islands. About 30 persons were killed by mudflows when Villarica Volcano in Chile erupted in December 1971. Two shepherds and hundreds of animals were buried in 1973 when Mt. Hudson in Chile erupted and caused mudflows. The ability to predict the eruption of volcanoes would allow evacuation of humans and portable property from threatened areas.

Forecasting Igneous Activity

So far, geologists have been very cautious about making specific predictions about times, places, and characteristics of volcanic eruptions. Only the most general predictions are really reliable.

Seismic, magnetic, heat indicators Volcanologists have found a number of clues that warn of an impending eruption to some extent. As we have already mentioned, one is seismic activity. Because seismic activity may persist for years or only weeks before an eruption, however, earthquakes alone are not enough to pinpoint the time of eruption. Another clue to rising magma seems to be increased heating at the volcano surface. This might be measured in three ways: (1) Temperature differences in the ground show up clearly on infrared film, so that camera-equipped aircraft can monitor heat changes in volcanoes (Figure 3.33); (2) Being hot, and above its Curie temperature, the magma has lost its magnetism. This change, too, can be detected from airplanes towing sensitive magnetometers above a volcano; (3) Finally, there is evidence that changes in the temperature of fumarole gases may warn of an eruption. Several interesting observations have been made of heat changes before an eruption, but volcanologists are still not sure whether these changes will be recognizable early enough to give advance warnings.

3.33 **Image formed by flying over Surtur I,** Iceland, during 1965 eruption with instrument sensitive to energy waves lying in the infrared range of the electromagnetic spectrum. White areas are hot; gray areas, cool. (U. S. Air Force Cambridge Research Laboratory, courtesy U. S. Geological Survey.)

Tiltmeters and satellites As the magma approaches the surface it inflates the volcanic cone. Thus, it causes the ground to tilt slightly before an eruption. This very slight tilting can be detected by extremely sensitive instruments called *tiltmeters,* modern solid-state instruments the size of a quarter, designed by the National Aeronautics and Space Administration (NASA) to monitor the attitude of satellites.

Thanks to tremendous improvements in radio communications and satellite technology, the Earth Resources Technology Satellite (ERTS – 1) is now relaying information from instruments placed on and near volcanoes. During the last few years, extensive arrays of radio-connected instruments have been deployed in the Cascade Range and also on several Alaskan and Central American volcanoes.

Center for Short-Lived Phenomena Keeping abreast of eruptions all over the world is a task performed by a group known as the Center for Short-Lived Phenomena in Cambridge, Massachusetts. Through a worldwide network of correspondents, the Center is informed by telegram of any eruptions (or other "short-lived phenomena," such as earthquakes or avalanches). In 1973, for example, the Center notified subscribing volcanologists that Santiaguito in Guatemala was producing the first known *nuées ardentes* since 1934.

Living with Igneous Activity

The bright side of volcanoes From a human point of view, volcanic eruptions are not always destructive. On the good side, volcanoes build land for us to live on, constructing all of the oceanic islands (probably all within the last 25 million years) and much of the continents. The 1960 eruption of Mt. Kilauea added 1½ square kilometers to the Island of Hawaii, and already this land is usable for housing and industry. Small sprinklings of volcanic "ash" or "tephra" provide farmers with "instant fertilizer." In Indonesia, most rural farmers live in areas where active, or very recently active, volcanoes have added ash to the soil. Even solid volcanic rock weathers rapidly, converting to rich soil. Rock formed from lava erupted in 1840 in eastern Hawaii is already densely covered with tropical forest.

Reducing volcanic damage Ever since the first farmers built stone walls trying to divert a lava flow, people have been thinking about and experimenting with ways to reduce the damage caused by an eruption. Perhaps the best-known early attempt at lava diversion was made in 1669 when a molten flow of aa lava from Mt. Etna was threatening Catania, Sicily. A group of men covered with wet cowhides dug a channel through the wall of hot lava, creating an escape route. This was successful—for a while. Unfortunately, the new flow headed toward the town of Paterno. Infuriated townsfolk from Paterno then drove away the men from Catania, allowing the lava to destroy part of Catania.

More modern techniques are not much different, except that they employ explosives instead of picks and shovels to break new channels. In Hawaii, volcanologists have tried dropping bombs from airplanes to breach a new channel in a lava flow, or break down the wall of the cone around the vent to allow lava to flow away from a populated area. In 1940, 12-meter-high diversion barriers were actually constructed by bulldozer in Hawaii to protect the city of Hilo and its harbor; this system has not yet been tested.

Fighting Helgafell The largest-scale attempt to modify a lava flow was made in Iceland in 1973, when the volcano Helgafell erupted, threatening the fishing port of Vestmannaeyjar. The volcano had been quiet for a long time, and the eruption caught the population by surprise. The people were quickly evacuated as tephra and lava

3.34 **Houses on Heimaey, Iceland,** nearly engulfed by black tephra emitted during 1973 eruption of Helgafell Volcano. (Consulate General of Iceland, New York City.)

burned or covered many buildings (Figure 3.34), and toxic gases accumulated to dangerous levels. Famed Icelandic volcanologist Thorbjorn Sigurgeirsson arrived to direct efforts to minimize damage to the important fishing harbor. Barriers were quickly constructed and the lava was sprayed with sea water from high-pressure hoses at carefully planned locations. Sigurgeirsson's strategy was to slow the flow, deplete the volume of the flows farthest from the vent, and promote the spread of lava on the upper slopes of the new cone that was rising. The efforts were at least partly successful, maintaining a narrow but usable harbor entrance around the new flows and saving enough of the town to rebuild and revitalize.

Geothermal Energy

Energy produced by the Earth's heat is known as **geothermal energy.** As coal, petroleum, and other conventional energy sources dwindle, geologists are seeking new ways to make use of this tremendous storehouse of heat.

Importance of Geothermal Energy

Hot springs have been used for many centuries for cooking, laundering, and bathing; Roman emperors maintained elaborate bathhouses at the hot springs of Italy. Only in the twentieth century, however, have there been serious efforts to use heat trapped below the Earth's surface for more ambitious purposes. Steam is the most useful carrier of this energy. If a reservoir of trapped steam is punctured by a drill, the steam will escape through the drill hole. Properly controlled, underground steam can be forced to turn a turbine and generate electricity. The first electricity generator powered by geothermal steam was installed in Italy in 1905 and since then many hundreds of steam wells have been drilled. Today, geothermal steam seems more appealing than ever before as an inexpensive and relatively clean power source. Most important, geothermal sources of energy can be relied upon to produce power for thousands of years—as long as it takes the underlying magma to cool.

World Locations of Geothermal Energy

So far, only four locations in the world have produced meaningful amounts of geothermal energy: Tuscany, Italy; Iceland; New Zealand; and California. Other countries now attempting to develop steam resources include Mexico, Russia, Japan, Indonesia, and Chile.

Italy pioneered the use of geothermal energy, and has served as a kind of working laboratory for others interested in the field. By 1923, Italian engineers had devised a system for removing most of the contaminating gases from the steam.

Iceland, built up entirely by volcanic processes, has abundant geothermal energy. Indeed, Iceland might better be called Fireland: it literally straddles the mid-Atlantic ridge, where hot mantle magma is thought to be rising continuously through the sea bed. The most abundant geysers and hot springs are in the vicinity of the capital of Reykjavik, where since 1925 natural hot water has been used to heat dwellings and greenhouses where tropical fruits are grown. Many Icelanders bake bread by burying the loaves in hot dirt. They also cook boiled food by lowering containers into hot springs. Some 90 per cent of the houses in the city are heated by water from hot springs 15 km away. The first plant to convert geothermal steam to electricity was inaugurated at Hveragerdi in 1964.

The only region producing geothermal power in the United States is The Geysers, California, 150 km north of San Francisco. Despite the name, the area has no geysers, but does seethe with 50 acres of fumaroles (the best-known are the Smokestack, Steamboat, and Safety Valve) and hot springs. In 1852, the first health resort was opened in The Geysers, and 1921 the first wells were drilled for steam. The project was abandoned until 1955, when a 180-meter experimental well was sunk, and in 1960 the Pacific Gas and Electric

Company opened the first commercial geothermal power station (12,500 kw) in this country.

Sources of Geothermal Energy

There are still few clues to guide geologists to new deposits of high-pressure steam. Each of the four major regions now producing electricity is quite different geologically. In most regions, steam is trapped by nonporous cap rock. There must be an underground chamber capable of storing steam, as well as a heat supply and adequate amounts of water. Prospectors are hunting hard for more steam reservoirs to provide much-needed electric power. More exciting still are attempts to tap the much larger power potential of hot rock formations and magma reservoirs that have no steam. It now appears possible to generate electricity from dry, hot rock by pumping down liquid water under pressure, which causes the rock to fracture. The water invades the cracks and turns to steam. This steam is then used to turn turbines as it returns to the surface. This process could open up vast new regions—including the interiors of live volcanoes—to geothermal energy production.

Chapter Review

1. *Igneous activity* designates the making of hot liquid material from the mantle, its rising into the crust or onto the surface of the Earth, and its cooling, forming solid igneous rocks.
2. When the molten rock material is on the surface of the Earth it is called *lava*, and when it remains inside the Earth's crust it is named *magma*. As either magma or lava cool and solidify they form *igneous rocks*.
3. Igneous rocks make up a vast bulk of the Earth's crust, and provide the raw material for sedimentary and metamorphic rocks.
4. Igneous activity is an expression of the Earth's heat. A combination of high temperature and a sudden reduction of the great pressure in the Earth's mantle may produce liquid magma from the ordinary solid material of the mantle. This liquid magma may work its way upward all the way to the surface. Much of what we know about the mantle is still conjectural.
5. The most obvious effect of igneous activity is a volcano. A *volcano* is the vent through which the hot liquid material or gas, and generally both, pass from the Earth's interior onto its surface.
6. Volcanoes produce several types of *lava*, solid-gaseous mixtures such as *ash flows* and *nuées ardentes*, various solids collectively named *tephra*, and varied *gases*.

7   Some volcanic landforms are *domes, cones, craters* and *calderas, volcanic shields, lava plains,* and *composite cones.*

8   Types of volcanic eruptions include *summit, fissure, flank, effusive, explosive,* and *fumarolic.* The first three produce distinctive identifying shapes, whereas the last three types describe categories of violence of the eruption.

9   There is probably no part of the world that has not been the scene of volcanic activity at some time. Today, however, more than three-quarters of the world's subaerial surface volcanoes are distributed around the edges of the Pacific Ocean in a pattern known as the "Ring of Fire." Most subaerial Atlantic volcanoes are located in the middle of the ocean rather than on the rim. These patterns are explained by the theory of plate tectonics.

10 Some classic volcanic eruptions have been so distinctive that their names have been used to describe other eruptions. Examples include Vesuvius, Krakatoa, Mt. Pelée, Hawaii, and Stromboli.

11 Most of the world's volcanic activity probably occurs underwater, and so we only see evidence of some of it. Besides creating submarine mountains which are the tallest mountains in the world, submarine eruptions may also produce new islands if these mountains actually reach the ocean surface. Such islands are rarely permanent, but the best-known "new" island of recent years is Surtsey, born in 1963 off the coast of Iceland.

12 Besides volcanoes, *fumaroles, hot springs,* and *geysers* are indications of the Earth's heat, and each emits specific gases or liquids. Yellowstone National Park is one of the world's outstanding examples of an area having many hot vents.

13 Volcanoes are usually preceded by tremblings of the Earth known as *seismic activity*, which are detected on delicate instruments called *seismographs*. Geologists continue to explore the relationship between earthquakes and volcanic activity.

14 Energy produced by the Earth's heat is known as *geothermal energy*. Geologists are seeking new ways to make use of this tremendous storehouse of heat, and by 1923 Italy had pioneered its use as a major source of energy. So far, only four locations in the world have produced meaningful amounts of geothermal energy: Italy, Iceland, New Zealand, and California. Several other countries are attempting to develop steam resources.

Questions

1   Define *igneous rock.*

2   Do the "smoke" and "ashes" mean that volcanoes are burning in the same sense as wood burns in a fireplace? Explain.

3   Define *lava.* Based on their physical appearances the three common kinds of lava are _____, _____, and _____.

4   List and describe the kinds of solids that may be propelled out of volcanoes.
5   What is a *volcanic dome*? Of what is a volcanic dome composed?
6   Compare and contrast *tephra* ("cinder") *cones, volcanic shields,* and *composite cones.*
7   Define *crater.* How does a crater differ from a *caldera*?
8   What activities take place during a volcanic eruption that is described as *Hawaiian*?
9   Explain the principles that regulate the operation of a *geyser.*
10 Discuss some of the effects of hydrostatic pressure on the eruption of a volcano on the sea floor.
11 What are *pillowed volcanic rocks*? How do pillows form?
12 Describe some of the methods used for making predictions about where and when a volcano will erupt.
13 What is *geothermal energy*? Is such energy available everywhere?

Suggested Readings

Bullard, Frederick, *Volcanoes: In History, In Theory, In Eruption.* Austin: University of Texas Press, 1962.

Macdonald, G. A., *Volcanoes.* Englewood Cliffs, N.J.: Prentice-Hall, Inc., 1972.

Tazieff, Haroun, *Volcanoes.* London: Prentice-Hall International, 1961.

Thorarinsson, Sigurdur, *Surtsey: The New Island in the North Atlantic.* New York: The Viking Press, 1967.

Vitaliano, Dorothy B., *Legends of the Earth: Their Geologic Origins.* Bloomington: Indiana University Press, 1973.

Wilcoxson, Kent H., *Chains of Fire: The Story of Volcanoes.* Philadelphia: Chilton Books, 1966.

Williams, Howel, "Volcanoes." Scientific American, November, 1951, pp. 45–53. (Offprint No. 822. San Francisco: W. H. Freeman and Company.)

"Igneous rocks compose roughly 90 per cent of the Earth's crust. Yet fundamental questions persist."

Chapter Four
Igneous Rocks

Having seen the more dramatic aspects of igneous activity we will now look at igneous rocks, the products of igneous activity, and try to reconstruct the processes that operated during their formation. In effect, in Chapter 3 we were watching living experiments in the making of igneous rocks. We were able to see the processes and the end products. Now in this chapter, when we study igneous rocks, we shall use the geologist's method of starting with the end products and trying to answer the question: What happened when we were *not* able to see the activity?

We will start by looking at the shapes of these rock bodies in relation to their surroundings. After we have seen large bodies of igneous rocks we will then focus closely on small specimens as in a laboratory. As we examine the general principles that govern crystallization and the formation of magma we will see that these principles can be applied not only to Earth, but to other planets as well.

Plutons

Volcanically formed rocks are known as **extrusive** igneous rocks. Magma that solidifies within the crust forms **intrusive** igneous rocks. Later, intrusive igneous rocks may reach the surface, but they do so only after they have been exposed by the erosion of formerly overlying crust.

Geologists have given the name **pluton** to any body of intrusive igneous rock. This name is in honor of Pluto, the Roman god of the underworld. Plutons are classified not on the basis of the kinds of rock of which they are composed, but according to their sizes, shapes, relationships, and placement with regard to the surrounding rock. The boundaries of **concordant** plutons are parallel to the layers of surrounding rock; the boundaries of **discordant** plutons are not parallel with the layers of surrounding rock (Figure 4.1). Plutons are also divided on the basis of their shapes into **tabular** (slab-like, or table-shaped); **lenticular** or dome-shaped; and **massive** varieties. Many plutons are quite complex, and include combinations of tabular, lens-shaped, and massive bodies.

Tabular Plutons

Dikes A **dike** is a discordant, tabular pluton which fills a fracture in older, previously formed rocks (named "country rock"). The margins of dikes typically are steeply inclined.

Dikes may cut through any type of older rocks—igneous, sedi-

4.1 Tabular plutons of mafic igneous rock that were intruded into nearly horizontal layers of limestone. Most of these plutons are discordant, hence are dikes, but a few sheets are concordant, and are sills. Montreal, Canada. (J. E. Sanders)

mentary, or metamorphic. Thus it is common to see, along a road cut or cliff, a distinctly contrasting strip of foreign-looking rock angling sharply through an otherwise uniform mass (Figure 4.2). Where the country rock erodes more easily than the dike rock, the margins of the dike may form vertical natural walls. Occasionally the reverse happens. If the dike rock is less resistant than the country rock, the dike itself is eroded into the form of a natural ditch.

Dikes range in size from mere slivers, 1 cm thick (Figure 4.3) and 1 to 2 meters long, to huge bodies 1 km thick and many kilometers long. Although one great dike in the north of England is more than 150 km long, most dikes are much smaller. In the United States, dikes are exposed in both eastern and western regions (for example, in Maine, Massachusetts, the Crazy Mountains of Montana, the Spanish Peaks of Colorado) but rarely in the center of the country. This distribution of visible dikes is consistent with zones of tectonic activity since the beginning of the Paleozoic Era (about 500 million years ago) in the United States. Evidence of this activity may be seen from the Rocky Mountains westward and from the Appalachians eastward, but rarely in between.

Sills When magma finds an easy opening upward, or a zone of reduced pressure, it may move in that direction to form a dike. In some cases, further upward movement is blocked and the direction of least pressure is parallel to layers of older rock, particularly sedimentary rocks such as shales and sandstones. When magma spreads later-

4.2 Dike of dark-colored mafic rock (basalt) cutting through white granitic rock. North side Route I-95, Branford, Connecticut. (J. E. Sanders)

4.3 Dikes of various sizes. (Left) Thin dike composed of mafic rock cutting through a hand specimen of light-colored granite. (B. M. Shaub, J. E. Sanders) (Right) Wall-like sheet of dark-colored rock in left foreground is a large dike several meters thick and hundreds of meters long. The steep-sided mass of rock forming the isolated, irregular hill is Shiprock (New Mexico), a small massive pluton which solidified in the conduit of a former volcanic cone, now completely eroded away. Such a body of rock is known as a volcanic plug. (Courtesy Barnum Brown Collection, American Museum of Natural History)

ally between layers of weak rock, it forms concordant tabular plutons named **sills** (see Figure 4.1).

Like dikes, sills vary greatly in size. Mount Royal, in Montreal, Canada, contains some sills only 5 to 10 cm thick and 2 meters across. In Glacier National Park, Montana, visitors can study sills 30 meters thick (Figure 4.4) that once extended at least 5500 square kilometers. Other large sills extend across parts of New Jersey and New York. Bare rock standing in regular columns along the west bank of the lower Hudson River constitutes the impressive Palisades. This striking example of the eroded edge of a tilted sill extends for 100 km at thicknesses ranging from 100 to 300 meters (Figure 4.5). The parent magma of the Palisades sill solidified far underground as a great horizontal slab. The rock came to the surface only after regional tilting of the layers and prolonged erosion. An even larger sill in Ontario, Canada is 1500 meters thick.

Superficially, some sills look much like ancient lava flows, but there are several sure ways to tell them apart (Figure 4.6). The most reliable is that sills cause heat-induced changes ("baking" or contact metamorphism; see Chapter 6) in the rock both above and below them, whereas a tongue of lava, extruded upon the Earth's surface, can affect only rock beneath it at the time of eruption. Commonly, such baking alters the chemical composition not only of adjacent rocks but also of igneous rock forming the pluton itself.

Plutons

4.4 Sill visible high in the cliffs underlain by ancient sedimentary rocks of Precambrian age. Mount Gould, Glacier National Park, Montana. (David W. Corson, from A. Devaney, N. Y.)

4.5 Steep slope underlain by tilted edge of thick sill composed of resistant dolerite forms the Palisades along the west shore of the Hudson River, from Hoboken, New Jersey, north to Haverstraw, New York. Columns of rock are bounded by network of cooling joints. (United Press International)

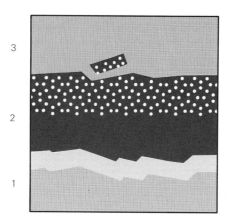 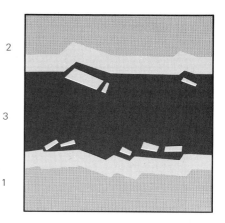

4.6 Sill and buried sheet of extrusive igneous rock formed by cooling of a former lava flow, schematic sections. (Left) Buried sheet of extrusive igneous rock displays vesicles in upper half; piece of vesicular igneous rock has been broken loose and incorporated in non-baked overlying sedimentary strata. (Right) A sill has broken loose pieces of the country rock at both top and bottom and has baked them, as well as the strata, next to the sill at both top and bottom contacts. Numbers show order of age of strata: 1, oldest; 3, youngest.

Lens-Shaped Plutons

Laccoliths A **laccolith** is a concordant mass of igneous rock with a flat floor and an arched roof that has intruded between older layers of rock, forcing the top layers upward (Figure 4.7). Laccoliths are commonly lens- or dome-shaped, initially creating a hill or mountain where they arched the Earth's surface upward. The name is derived from the Greek *laccos*, meaning reservoir, and *lithos*, meaning rock. If the magma is injected from below faster than it can spread parallel to the layers of the country rock, a developing sill may change into a laccolith. The composition of the magma, too, may determine the shape of the pluton. Thin, fluid magma spreads easily between layers of the country rock, forming a sill. By contrast, pasty, viscous magma tends to pile up in one place, thus forming a biscuit-like laccolith.

Viewed from the air, laccoliths appear to be roughly circular or oval. They may range in diameter from several hundred meters to several kilometers, and in maximum thickness from a couple of hundred meters to almost 2 km. Laccoliths were first identified in the Henry Mountains in the deserts of southeastern Utah. These laccoliths are oval domes from 1 to 6 km in diameter, about seven times as wide as they are high. Laccoliths are most common, however, in regions of the Rocky Mountains where the igneous rocks have intruded weak beds of shale, sandstone, and limestone. In many places the upper layers of country rock can be seen to have been stretched, thinned, and broken as the top of a laccolith forced its way upward beneath them.

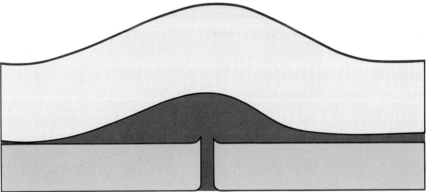

4.7 Laccoliths. (Top) Distant view of several peaks of Henry Mountains, Utah, where laccoliths were first described. (Union Pacific Railroad) (Bottom) Idealized sectional view through laccolith shortly after intrusion and before roof has been eroded.

Massive Plutons

Stocks Massive discordant plutons with an exposed area of less than 100 square kilometers are **stocks.** Most stocks are circular or oval when viewed from above and somewhat cylindrical in shape (Figure 4.8). When the rocks around them are worn away by erosion, stocks form steep-sided hills called **bosses.** One of the most familiar bosses in the United States is Stone Mountain, near Atlanta, Georgia. Others are the Great Snake and Little Snake Hills projecting above the

4.8 Stock and batholith, schematic block diagram. If upper third of block showing stock at left were eroded away, the appearance would be the same as in the batholith at right.

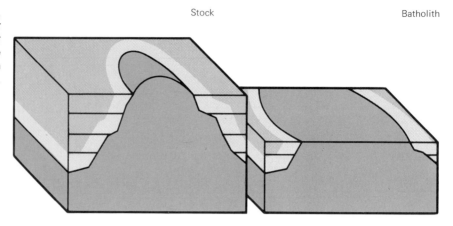

Stock   Batholith

New Jersey marshes near Newark. These bosses were fed by magma from the huge sill of the Palisades.

Batholiths Like stocks, **batholiths** differ from laccoliths in that they extend deep into the Earth's crust, and may be defined as massive, discordant plutons occupying exposed areas of more than 100 square kilometers. Whereas laccoliths are usually emplaced between layers of rock near the surface, batholiths, and also stocks, may extend downward indefinitely (Figure 4.9).

Because the rocks composing batholiths originated at great depth where cooling is extremely slow, they are massive. Many of these huge plutons compose the central cores of mountain ranges. The Idaho batholith, extending through much of that state and part of Montana, occupies about 40,000 square kilometers. The Coast Range batholith of British Columbia and southeastern Alaska occupies an exposed area of about 250,000 square kilometers. Batholiths also form the Sierra Nevada, in California, and the granitic mountains of New England, Scotland, and many other areas. Batholiths composed of Precambrian rock (older than about 600 million years) underlie about 5 million square kilometers of northeastern Canada and Greenland.

Plutons cool most rapidly along their margins (called *contacts*) where they are in contact with older, cooler rock. Crystals that grow near contacts usually are small. In the centers of plutons, where heat is lost more slowly, large crystals are found. One geologist studying the batholiths of California has calculated that hundreds of thousands or even millions of years are required for complete crystallization of batholiths.

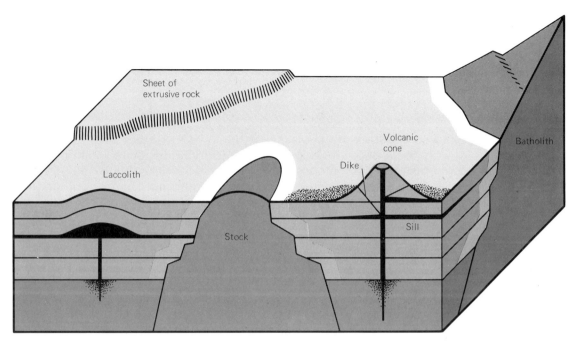

4.9 Major kinds of plutons and various bodies of extrusive igneous rocks shown on a schematic block diagram. White zones at margins of stock and batholith show prominent contact-metamorphic aureoles. Thinner zones of altered rock exist at margins of smaller intrusives, but are not shown here.

Geologic Dating of Plutons

One way to date the plutons is to make a radiometric measurement using the radioactive isotopes in the rock-making minerals. Radioactive isotopes of potassium are present in biotite and the feldspars. Radioactive isotopes of uranium occur in zircons and other minerals. These isotopes provide a direct answer to the time of crystallization in millions of years. If we are to use this information fully we need to relate it to the surrounding rocks.

We relate plutons to the surrounding country rock by examining the contact relationships using two important principles: (1) A pluton is younger than the rock it cuts or bakes (Figure 4.9), and (2) a pluton is older than the erosion surfaces that cut it. If the contacts are not exposed we still can use one other approach. A pluton is older than any gravel that contains pebbles or minerals derived from the igneous rock composing the pluton.

Textures of Igneous Rocks

When an igneous rock forms, two things happen: (1) The chemical composition of the magma or the lava determines what minerals are present in the rock, and (2) the history of the cooling determines the textures. (Fast cooling results in small particles; slow cooling creates large particles. If the cooling is uniform, the texture is uniform; if cooling is not uniform, the texture is also not uniform.) Therefore, the texture and mineral composition tell all that is needed about the formative processes of the rock. It also happens that texture and composition are used for the classification of igneous rocks, and so the classifying of these rocks ties them closely to their history.

In order to show how the classification works we will examine closely texture and composition to demonstrate how classification and history work together. We shall first consider the **texture,** the term that describes the size, uniformity of size, and mutual relationships of the particles of rocks.

Kinds of Textures

The sizes and uniformity of crystals in magma or lava depend largely upon the conditions of cooling and the presence of volatile matter, especially water. If magma cools slowly, rocks with large crystals form. If it cools very rapidly, as may happen when magma erupts onto the Earth's surface or the ocean floor, crystallization may be so rapid that the rocks will be wholly or partly glass.

Equigranular texture When the crystals of a rock are all about the same size, its texture is **equigranular.** The particles of equigranular rocks vary greatly in size. Plutonic bodies tend to cool slowly and, as a result, produce rocks whose crystalline particles are larger than 1 mm, which we shall call **granular.** Magma that cools rapidly produces particles smaller than 1 mm, which we shall call **aphanitic.** Extremely rapid cooling creates *glassy* rocks.

The presence of volatile matter increases the fluidity of a magma, allowing ions to flow more easily to growing crystals and therefore producing rocks with larger particles. The largest crystals grow in rocks called *pegmatites,* igneous rocks with particles larger than 10 mm.

4.10 Interlocking particles displaying crystalline texture in granite. White particles in center and right are quartz; gray particles at upper left are feldspars. The full width of the photograph represents about 6 mm. Paleozoic granite, southwest end of Lewis island, Branford, Connecticut. (J. E. Sanders)

Porphyritic texture When we studied minerals we examined crystals that were specimens of single minerals, such as olivine, biotite, hornblende, pyroxene, muscovite, various feldspars, and quartz. The crystal form was distinctive to that mineral. These same minerals form nearly all igneous rocks. In most igneous rocks, however, individual particles do not display crystal shapes. Instead, most rocks are collections of mineral particles having compromise boundaries (see Figure 2.13). The mineral particles are all about the same size and they all grew at once. The result was an interlocking aggregate resembling a jig-saw puzzle (Figure 4.10). In some igneous rocks only a few crystals began to grow while the rest of the material remained liquid. Because these crystals grew in the midst of a liquid, they could grow their true crystal shapes (Figure 4.11). The sizes of some of these early-growing crystals reached several millimeters or more. Later, their molten surroundings solidified. The later-formed particles grew all at once, and thus formed compromise boundaries. The resulting rock contains a scattering of large, beautifully shaped crystals (phenocrysts) set in the midst of smaller particles (the groundmass), all about the same size, having compromise boundaries. An igneous rock having a non-uniform texture in which some crystals are distinctly larger than others is a **porphyry,** and the texture is said to be **porphyritic.**

As with other igneous rocks, the textural characteristics of porphyry vary widely. The groundmass may be granular, aphanitic, or even glassy. The phenocrysts may be perfect or imperfect, from 2 cm

4.11 Porphyry in microscopic view. Oblong particle in diagonal position at lower center is a plagioclase phenocryst about 4 mm long. Groundmass consists of equal parts of plagioclase and pyroxene (irregular dark gray shapes). (J. E. Sanders)

140
Igneous Rocks

or so in diameter. They may be light-colored feldspar or quartz, or dark ferromagnesian minerals such as biotite, hornblende, or olivine. The important feature of porphyritic texture is a distinct contrast in particle sizes between the phenocrysts and the groundmass.

Knowing that large crystals form by slow cooling and small ones by fast cooling, it is easy to understand why many extrusive igneous rocks are porphyritic. Before magma pours out of the Earth's crust, thus becoming lava, it may have been crystallizing underground. Before erupting, these crystals may have reached considerable size. As the magma becomes lava, it cools suddenly and the remaining liquid portion hardens into finely crystalline rock or even glass. The first-formed crystals are the phenocrysts. They are trapped in the solidifying groundmass and a porphyry is formed.

Mineral Composition of Igneous Rocks

As we have said, two factors govern the classification of igneous rocks: (1) cooling history, and (2) the composition of the molten material. Rocks that are chemically unlike, and which had identical cooling histories will be texturally alike, but will contain different minerals.

We shall recognize three families of igneous rocks, **felsic, intermediate,** and **mafic** (MAY-fic), and will subdivide each of these according to texture.

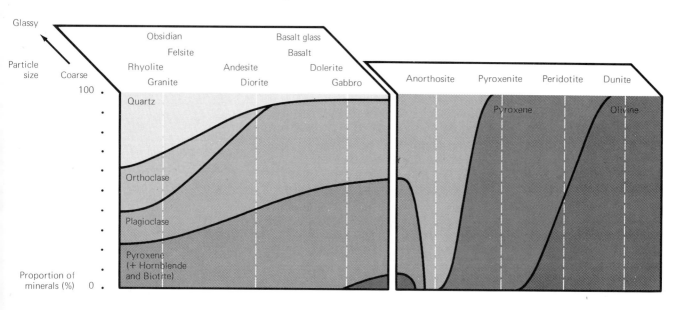

The two factors of texture and composition are shown in Figure 4.12. Felsic, intermediate, and mafic rocks range through all possible textural combinations from pegmatitic to glassy. All these rocks can be either intrusive or extrusive. However, a special group of granular igneous rocks are never fine grained or glassy. Rocks in this special group tend to consist of only one or two minerals, chiefly olivine, pyroxene, or very calcium-rich plagioclase. Collectively these rocks form a group known as **ultramafic.**

Felsic Rocks

Felsic rocks include three members: (1) granite, the coarse one, (2) rhyolite, the aphanitic or fine-grained, and (3) obsidian, the glassy.

    Granite The term *granite* has traditionally been applied by nongeologists to almost any coarse-grained granular igneous rock. (In this chapter we consider that granite is an igneous rock. As we shall see in Chapter 6, some granites are metamorphic rocks.) This usage persists in popular language, both because granite is such a common ingredient of the Earth's crust, and because its coarse texture is so distinctive.

    In more formal usage, **granites** are defined as equigranular coarse-grained igneous rocks consisting principally of the light-colored minerals, feldspars and quartz. Because of the dominance of these light-colored minerals, most granites are whitish, and flecked with some greenish or blackish minerals such as biotite or hornblende. Granites contain more orthoclase than plagioclase, and the plagioclase is sodium-rich (albite) rather than calcium-rich (anorthite).

    Sometimes the quartz and feldspars are gray, as in the granite of Quincy, Massachusetts, widely used as a building stone. In other localities, such as Pike's Peak, Colorado, the feldspars may be pink or reddish. This creates a colored variety of granite that is much in demand for building purposes. Because of its great strength and because it can be easily cut and polished, granite is used more extensively

4.12 Chart for classifying igneous rocks (opposite page). The vertical panels show the proportions of minerals, with the height of the panel equal to 100 per cent. The horizontal panels at top show particle sizes, ranging from coarse in front to glassy in the rear. The dashed lines on the vertical panels show the proportions of minerals in the granite, diorite, and gabbro families of igneous rocks. Panel at right is for ultramafic rocks, which form by settling of crystals. All ultramafic rocks are coarse-grained; therefore no divisions according to particle size appear on the horizontal panel, as in the other igneous rocks.

4.13 Varieties of granite, based on texture. (a) Orbicular variety, characterized by circular features of unknown origin. (Frank Dunand, J. E. Sanders) (b) Uniform, or ordinary variety. (Nat Messik, J. E. Sanders) (c) Porphyritic variety. (d) Graphic granite, a variety in which the quartz (dark gray) becomes aligned in rows, suggesting cuneiform writings. (Nat Messik, J. E. Sanders) All specimens one-half natural size. (e) Pegmatitic coarse grained. (J. E. Sanders)

than any other stone for buildings, bridges, sea walls, monuments, and ornaments.

Common varieties of granite are *orbicular granite,* containing unusual round or egg-shaped clusters of mineral grains; *porphyritic granite,* with phenocrysts of quartz or feldspars as much as 5 cm in diameter; and *graphic granite,* with irregular cleavage faces showing crystals of quartz that resemble the characters of Arabic writing and so give the rock its name (Figure 4.13).

Rhyolite A fine-grained equivalent of granite is called **rhyolite.** Rhyolites are usually so fine grained that their dominant mineral, feldspar, blends together as a single mass. Scattered in the feldspar are usually phenocrysts or quartz, and biotite. Extrusive rhyolites are particularly important because they prove that felsic magmas do exist. (As we shall see in Chapter 6, the existence of felsic magmas has been questioned by some geologists.) If quartz phenocrysts are not present it is not easy to separate rhyolite from similar-appearing fine-grained intrusive igneous rocks. The name **felsite** is a general term for any fine-grained felsic igneous rock.

Obsidian Solid, natural glass without crystals is called **obsidian**. Most obsidians are dark colored, usually black, and their finish normally bears brilliant luster. It is not surprising that obsidian looks much like artificial glass, because artificial glass is also made by cooling a molten solution of quartz and other minerals. This lustrous appearance has given obsidian high value among primitive peoples, who learned to fashion tools, ornaments, and weapons from it. Obsidian has conchoidal fracture, named for its resemblance to the inside of a clam (conch) shell (Figure 4.14). A conchoidal fracture is smooth and can occur in any direction. This makes obsidian easier to carve than other rocks, which may break irregularly in undesired directions. Ancient Mexicans were especially skilled obsidian carvers, able to shape long, thin knife blades. Perhaps the best-known locality to see obsidian is Obsidian Cliff in Yellowstone National Park, Wyoming.

4.14 Obsidian, showing typical conchoidal fracture. One-half natural size. (Nat Messik, J. E. Sanders)

Intermediate Rocks

Under this heading we include rocks that are generally light colored but which contain much less quartz than granites (some contain no quartz), and in which plagioclase is the dominant feldspar and ferromagnesians are more abundant than in the felsic group (see Figure 4.12). We shall discuss two intermediate rocks, diorite and andesite.

Diorite Coarse-grained granular igneous rocks, generally darker than granites, are **diorites.** In texture, diorites are so like granites that these two rocks may be difficult to distinguish. Some diorites can be distinguished from granites by their darker color. Between one-eighth and three-eighths of the minerals in diorite are dark colored. The commonest dark minerals of diorite are biotite, hornblende, and pyroxene. Unlike granites, diorites contain more plagioclase than orthoclase feldspars, and the plagioclase is more calcic (calcium-rich) than sodic (sodium-rich). The fabric is equigranular, although quartz phenocrysts are occasionally present. Although diorites are distributed throughout the world, they are bypassed by builders because of their dark color.

Andesites The fine-grained equivalents of diorites are **andesites,** which closely resemble rhyolites but lack quartz phenocrysts. As mentioned earlier, if it is not possible to divide fine-grained felsic rocks into rhyolites and andesites, the collective term **felsite** is used.

Mafic Rocks

Mafic rocks are generally dark gray, greenish, or black. For our purposes we recognize four textural varieties: (1) gabbro, (2) dolerite, (3) basalt, and (4) basalt glass (see Figure 4.12).

Gabbro Granular igneous rocks composed chiefly of ferromagnesian minerals, an equal or smaller amount of plagioclase (both sodium and calcium feldspar), and negligible quartz are called **gabbros.** Like granite and diorite, gabbro displays a variety of colors, ranging from dark gray to black to greenish. Gabbros may feature a banded or layered texture, produced by variations in chemical composition. Like granites, gabbros are widely distributed; like diorites, gabbros are also well suited for use in construction but largely ignored because of their sooty appearance. Unlike granites and diorites, however, some gabbros are mined extensively for iron, nickel, titanium, copper, and other elements that are found in many of the large gabbro intrusive bodies.

Dolerite A medium-grained mafic rock is **dolerite.** In color, dolerite resembles gabbro, ranging from gray to black to dark green. Bodies of dolerite form notable features along the east coast of the United States, such as the Palisades along the Hudson River and several ridges in the Connecticut Valley.

Basalt Aphanitic mafic rocks, are **basalts,** the dark-colored counterparts of felsites. In color, basalts range from greenish or purplish black to pure black. Compared to felsites, basalts contain more ferromagnesians and less feldspar. Thus the edges of thin chips of basalt are usually opaque instead of translucent, as they are in felsites. Most basalts are extrusive; they have formed by solidification of lava that was erupted through volcanoes or fissures to form such huge piles of extrusive rock as the Deccan Plateau in India and the Columbia River Plateau in Washington and Oregon (see Figure 3.14).

Basalt glass Glassy mafic rock resembling obsidian is **basalt glass.** Basalt glass and obsidian are differentiated by examining their thin edges. The thin edges of obsidian are translucent, and the thin edges of basalt glass are opaque.

4.15 Layered intrusive body of ultramafic rock, with settled crystals arranged in definite layers comparable to sedimentary strata. Duke Island complex, southeastern Alaska. (T. N. Irvine from "Petrology of the Duke Island Ultramafic Complex, Southeastern Alaska." Reprinted with permission of Geological Society of America, Memoir 138, p. 240, 1974)

Ultramafic Rocks

Ultramafic rocks are thought to originate by the settling out of a mafic magma of crystals of olivine, pyroxene, or plagioclase. Thus these rocks typically form layers, much resembling the strata of sedimentary rocks (Figure 4.15). In some ultramafic rocks olivine is the dominant or only mineral. Such varieties consist of 40 per cent silica and 60 per cent magnesium and iron oxides. Others are mixtures of olivine and pyroxene and still others contain only plagioclase.

Dunites consist wholly of olivine; **pyroxenites** are ultramafic rocks composed wholly or pyroxene. **Peridotites** are ultramafic rocks containing both olivine and pyroxene. Associated with peridotites are concentrations of valuable metals such as nickel, platinum, and chro-

Fuji-No-Yama Lafcadio Hearn

The most beautiful sight in Japan, and certainly one of the most beautiful in the world, is the distant apparition of Fuji on cloudless days—more especially days of spring and autumn, when the greater part of the peak is covered with late or with early snows. You can seldom distinguish the snowless base, which remains the same color as the sky: you perceive only the white cone seeming to hang in heaven; and the Japanese comparison of its shape to an inverted half-open fan is made wonderfully exact by the fine streaks that spread downward from the notched top, like shadows of fanribs. Even lighter than a fan the vision appears—rather the ghost or dream of a fan;—yet the material reality a hundred miles away is grandiose among the mountains of the globe. Rising to a height of nearly 12,500 feet, Fuji is visible from thirteen provinces of the Empire. Nevertheless it is one of the easiest of lofty mountains to climb; and for a thousand years it has been scaled every summer by multitudes of pilgrims. For it is not only a sacred mountain, but the most sacred mountain of Japan—the holiest eminence of the land that is called Divine—the Supreme Altar of the Sun;—and to ascend it at least once in a lifetime is the duty of all who reverence the ancient gods.

I arrived too late to attempt the ascent on the same day; but I made my preparations at once for the day following, and engaged a couple of gōriki ("strong-pull men"), or experienced guides. I felt quite secure on seeing their broad honest faces and sturdy bearing. They supplied me with a pilgrim-staff, heavy blue tabi (that is to say, cleft-stockings, to be used with sandals), a straw hat shaped like Fuji, and the rest of a pilgrim's outfit;—telling me to be ready to start with them at four o'clock in the morning.

August 25 Sky clears as we proceed;—white sunlight floods everything. Road reascends; and we emerge again on the moorland. And, right in front, Fuji appears—naked to the summit—stupendous—startling as if newly risen from the earth. Nothing could be more beautiful. A vast blue cone—warm-blue, almost violet through the vapors not yet lifted by the sun—with two white streaklets near the top which are great gullies full of snow, though they look from here scarcely an inch long.

Now there are no more trees worthy of the name—only scattered stunted growths resembling shrubs. The black road curves across a vast grassy down; and here and there I see large black patches in the green surface—bare spaces of ashes and scoriæ; showing that this thin green skin covers some enormous volcanic deposit of recent date. As a matter of history, all this district was buried two yards deep in 1707 by an eruption from the side of Fuji. Even in the far-off Tōkyō the rains of ashes covered roofs to a depth of sixteen centimetres. There are no farms in this region, because there is little true soil; and there is no water. But volcanic destruction is not eternal destruction; eruptions at last prove fertilizing; and the divine "Princess-who-causes-the-flowers-to-blossom-brightly" will make this waste to smile again in future hundreds of years.

The black openings in the green surface become more numerous and larger. A few dwarf-shrubs still mingle with the coarse grass. The vapors are lifting; and Fuji is changing color. It is no longer a glowing blue, but a dead sombre blue. Irregularities previously hidden by rising ground appear in the lower part of the grand curves. One of these to the left—shaped like a camel's hump—represents the focus of the last great eruption.

The land is not now green with black patches, but black with green patches; and the green patches dwindle visibly in the direction of the peak. The shrubby growths have disappeared. The wheels of the kuruma, and the feet of the runners sink deeper into the volcanic sand.

Fuji has ceased to be blue of any shade. It is black—charcoal-black—a frightful extinct heap of visible ashes and cinders and slaggy lava.

6.40 a.m. We reach Tarōbō, first of the ten stations on the ascent: height six thousand feet.

There is nothing very difficult about this climbing, except the weariness of walking through sand and cinders: it is like walking over dunes. We mount by zigzags. The

From EXOTICS AND RETROSPECTIVES, by Lafcadio Hearn. Boston: Little Brown and Company, 1898.

Field Trip

sand moves with the wind; and I have a slightly nervous sense—the feeling only, not the perception; for I keep my eyes on the sand—of height growing above depth. The wind has suddenly ceased—cut off, perhaps by a ridge; and there is a silence that I remember from West Indian days: the Peace of High Places. It is broken only by the crunching of the ashes beneath our feet. I can distinctly hear my heart beat.

We are out of the fog again. All at once I perceive above us, at a little distance, something like a square hole in the face of the mountain—a door! It is the door of the third station—a wooden hut half-buried in black drift. How delightful to squat again—even in a blue cloud of wood-smoke and under smoke-blackened rafters! Time, 8.30 a.m. Height, 7085 feet.

The ascent is at first through ashes

and sand as before; but presently large stones begin to mingle with the sand; and the way is always growing steeper. I constantly slip. There is nothing firm, nothing resisting to stand upon: loose stones and cinders roll down at every step. If a big lava-block were to detach itself from above! Enter the fourth station, and fling myself down upon the mats. Time, 10.30 a.m. Height, only 7937 feet;—yet it seemed such a distance!

Off again. Slope has become very rough. It is no longer soft ashes and sand mixed with stones, but stones only—fragments of lava, lumps of pumice, scoriae of every sort, all angled as if freshly broken with a hammer. All would likewise seem to have been expressly shaped so as to turn upside-down when trodden upon. Ah! sixth station!—may all the myriads of the gods bless my gōriki! Time, 2.07 p.m. Height, 9317 feet.

With the stones now mingle angular rocks; and we sometimes have to flank queer black bulks that look like basalt. On the right rises, out of sight, a jagged black hideous ridge—an ancient lava-stream. The line of the left slope still shoots up, straight as a bowstring. Rocks dislodged by my feet roll down soundlessly;—I am afraid to look after them. Their noiseless vanishing gives me a sensation like the sensation of falling in dreams. Time, 4.40 p.m. Height, 10,693 feet.

August 26, 6.40 a.m. Start for the top. Hardest and roughest stage of the journey, through a wilderness of lava-blocks. The path zigzags between ugly masses that project from the slope like black teeth. The trail of castaway sandals is wider than ever. Reach another long patch of the snow that looks like glass beads, and eat some.

Twelve thousand feet, and something—the top! Time, 8.20 a.m. Stone huts; Shintō shrine with tōrii; icy well, called the Spring of Gold; stone tablet bearing a Chinese poem and the design of a tiger; rough walls of lava-blocks round these things—possibly for protection against the wind. Then the huge dead crater—probably between a quarter of a mile and half-a-mile wide, but shallowed up to within three or four hundred feet of the verge by volcanic detritus—a cavity horrible even in the tones of its yellow crumbling walls, streaked and stained with every hue of scorching. I perceive that the trail of straw sandals ends in the crater. Some hideous overhanging cusps of black lava—like the broken edges of a monstrous cicatrix—project on two sides several hundred feet above the opening; but I certainly shall not take the trouble to climb them. Yet these—seen through the haze of a hundred miles—through the soft illusion of blue spring weather—appear as the opening snowy petals of the bud of the Sacred Lotus! No spot in this world can be more horrible, more atrociously dismal, than the cindered tip of the Lotus as you stand upon it. 1898

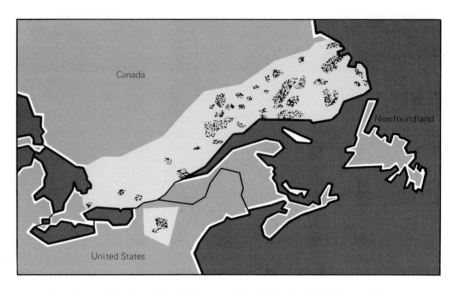

4.16 Anorthosite bodies (stippled) exposed in belt of ancient Precambrian rocks (white) that stretches from the Adirondacks, New York State, northeastward to Labrador, in Canada. (Redrawn from J. Martingnole and K. Schrijver, 1970, Tectonophysics, v. 10, pp. 403–409)

mium. Peridotites are commonly altered into other rocks, usually *serpentinite*, a soft, greenish rock thought to make up much of the Earth's lower crust.

Anorthosites are coarse-grained igneous rocks consisting predominantly or entirely of plagioclase. Some anorthosites contain small amounts of pyroxene. As the amount of pyroxene increases, the rock grades into a gabbro. The number of occurrences of anorthosite rocks is not great, but their masses are enormous, particularly in Norway and North America (Figure 4.16).

Magma and the Formation of Igneous Rocks

Considering the overwhelming abundance of igneous rocks (they compose roughly 90 per cent of the Earth's crust to a depth of about 15 km), magma formation and behavior are clearly among the Earth's most important processes. Yet fundamental questions about magma persist. Most of what we know about magma is derived from the products of volcanoes and the composition of igneous rocks, and from laboratory experiments. However, our accounts of magma behavior in its sub-crustal "habitat," where fiery temperatures and crushing pressures prevail, still depend upon inferences.

The Relation of Igneous Rocks to Magma

So far in this chapter we have been discussing igneous rocks. How are igneous rocks related to magma, and vice-versa? We have seen four major families of igneous rocks: (1) felsic, (2) intermediate, (3) mafic, and (4) ultramafic. Does this mean there are necessarily four different kinds of magmas? How would one go about answering these questions? Because no one has ever seen magma we can do three things: (1) study lava, (2) analyze the varieties of igneous rocks, and (3) make small-scale laboratory experiments at high pressures and temperatures.

Kinds of lava Obviously, the lavas that come out of volcanoes are closely connected to magmas underground. We actually have quite a spectrum of lavas. The exclusively mafic composition of oceanic volcanoes contrasts with the greater variety of lavas from continental volcanoes. This suggests that at least two kinds of lava exist, and therefore presumably there are two or more kinds of magmas.

Varieties of igneous rocks We can also look at the kinds of igneous rocks, because they obviously must be related to magma. The texture of igneous rocks tell us a little about the order of crystallization and the formation of igneous rocks from magma. From the textural evidence of the minerals in igneous rocks we can infer the order of crystallization. It has been found that crystals of magnetite and ilmenite in granites almost never have other crystals inside them. The importance of this observation is that magnetite and ilmenite must crystallize before other minerals. From the textural evidence of relative growth an order of crystallization has been worked out. The order is: oxides of iron, ferromagnesian minerals, calcium feldspars, alkali feldspars (sodium and potassium feldspars), and finally quartz.

It is important to remember that the order of crystallization is not as straightforward as it seems. The melting points of all these minerals change according to pressure, the amount of water present, the kinds of other magma ingredients, and other conditions (Figure 4.17).

Laboratory experiments Experimental evidence about igneous rocks has been available since late in the eighteenth century, when Sir James Hall, a friend of James Hutton, melted granite in a blacksmith's forge. More sophisticated experiments are now possible that re-create

4.17 Interlocking particles in mafic igneous rock illustrate how shapes of particles in crystalline texture can be used to infer order of crystallization. Elongate plagioclase crystal, 2mm long (at right of center) displays crystal shape and indicates first growth. Next came pyroxene (as in large particle at upper left and just above center). Last to crystallize was magnetite (black, in center), which wraps around all particles that had solidified previously. View through polarizing microscope of paper-thin slice of dike at Pine Rock, New Haven, Connecticut. (J. E. Sanders)

temperatures and pressures from well inside the Earth's mantle. Tiny thimblefuls of ground-up rock material are placed in piston-type squeezers, heated to high temperatures, and then cooled. Small specimens of igneous rocks crystallize when the apparatus is cool, and these specimens can be studied.

A striking series of laboratory results were obtained by N.L. Bowen early in the 1900s. He discovered a fundamental principle of the behavior of silicate minerals.

Bowen's Reaction Principle Bowen experimented with two kinds of materials, plagioclase feldspars and ferromagnesian minerals. He found that these materials behaved quite differently. In the plagioclase experiments the first crystals to form contained calcium and no sodium. As the temperature dropped, the plagioclase crystals reacted with the liquid (the laboratory "magma"). Calcium left the lattices and sodium entered. Each time this exchange took place an aluminum ion exchanged with a silicon ion. This continuous exchange of ions between feldspar lattices and the "magma" was named a **continuous reaction series** (the right side of Figure 4.18).

The ferromagnesian minerals displayed different behavior than the plagioclase. Two kinds of things took place: (1) Growing olivine crystals reacted with the "magma" to exchange iron and magnesium, just as sodium and calcium exchanged in the plagioclase. (2) As the temperature dropped, the olivine crystals surprisingly disappeared and pyroxene crystals began to grow. These pyroxene crystals exchanged ions with the liquid; presently the pyroxene vanished and in its place amphibole appeared. Therefore, as cooling took place, minerals appeared and disappeared. Such a relationship was called a **discontinuous reaction series** — it is represented by the discrete steps on the left side of Figure 4.18.

Origins of Ultramafic Rocks

Bowen's experiments showed that the first minerals to crystallize from the mafic magma are calcium-rich plagioclase and olivine. Because the density of olivine is greater than that of the magma, olivine crystals should sink to the bottom of the magma chamber and accumulate there as dunite. Similarly, plagioclase could accumulate

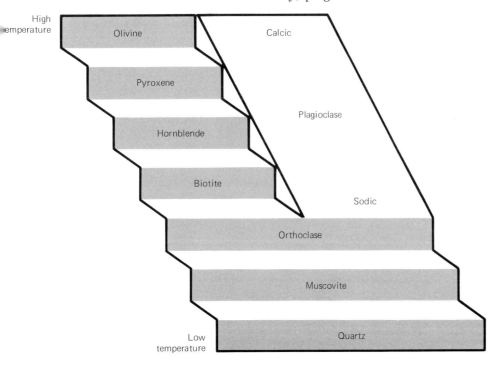

4.18 Bowen's reaction series shown schematically. Continuous reaction in plagioclase feldspars shown by sloping plane at right. Discontinuous reaction in ferromagnesian minerals shown by stair-step arrangement of treads and risers at left. In the ferromagnesian minerals the vertical risers indicate continuous reactions between individual minerals and the magma. The treads mark discontinuous steps where the reaction creates a new mineral lattice.

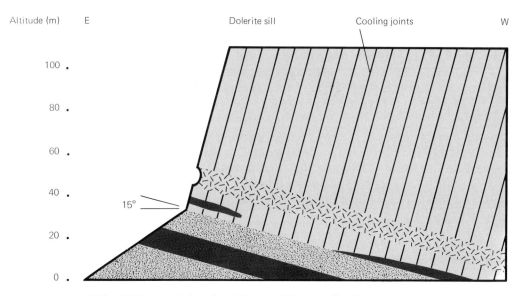

4.19 Olivine crystals in the Palisades sill have settled through the magma and stopped about 15 m above the basal contact. Evidently the lower 15 m had already become stiff enough to keep the olivine crystals from sinking any farther. Sinking took place before sill was tilted.

as anorthosite and pyroxene crystals as pyroxenite. Such accumulations of settled crystals explain the known kinds of ultramafic rocks. Examples are known from the Palisades sill, in New Jersey, where 15 meters above the bottom the heavy olivine minerals are found in a layer 3 meters thick (Figure 4.19). Most of the dolerite lies above the olivine layer. This happened because the minerals forming dolerite crystallize at lower temperatures than does olivine, and their specific gravities are less than those of olivine. Minerals that are metallic oxides, or elements whose specific gravities may be as high as 5.0, are usually found near the bottoms of plutons. Light minerals such as quartz (specific gravity 2.65), tend to rise as they crystallize. This fractionization creates *layered intrusives* that may look very much like sedimentary strata (see Figure 4.15).

The Igneous Rocks of the Moon

During the late 1960s and early 1970s, "Moon rocks" became a familiar phrase to people all over the world. To most of them, the phrase implied the excitement of viewing and handling exotic material from an alien "planet." To geologists, the rocks were hard mineralogical clues to the Moon's history.

Basic Composition of the Moon

Before the Apollo 11 spacecraft landed at the Sea of Tranquility in July of 1969, geologists had little specific information on the composition of the lunar surface. Speculation abounded. As recently as a few years previously, a leading American scientist had predicted that the lunar landing craft and its crew might disappear in as much as a kilometer of fine powder.

This, of course, did not happen (Figure 4.20), but the differences between the Earth and the Moon proved great enough to guarantee enormous differences in the rocks of the surface. Perhaps most important is the total lack of weathering on the Moon. There is no rainfall or wind to cause erosion; no transport of rocks of soil from place to place by streams, rivers, or glaciers; and no water to alter the chemical compositions of the minerals. The most important process affecting surface rocks has been meteorite bombardment, which has ground them into a rough regolith and fused some of the regolith into glass (Figure 4.21).

4.20 Footprint of the first human on the Moon, U. S. Astronaut Neil A. Armstrong. This footprint shows the powdery texture on the surface of the Moon is only a few cm thick. (NASA)

Results of Apollo landings During five lunar landings between 1969 and 1974, American astronauts retrieved more than 100 kg of Moon rocks and soil for earthbound geologists. The astronauts landed on a rough, crumbly layer of regolith above bedrock. Compared to Earth rocks, the lunar material proved to be simple mineralogically. This simplicity has resulted in part from the absence of water that on Earth is responsible for countless hydrations and other changes of rocks. The lunar samples consisted entirely of mafic igneous rocks. As millions of people around the world watched the lunar excursions on television, most of them heard for the first time the words "basalt" and "gabbro"—the most common lunar rocks.

A major surprise was the discovery on the Moon's surface of anorthosite, a relatively rare rock on Earth. As we mentioned previously, on Earth anorthosite occurs in a few areas of Precambrian rocks in the Adirondack Mountains, in Labrador, and in Quebec (see Figure 4.16).

The lunar basalts consist of approximately half pyroxene, a little more than one quarter plagioclase, and significant amounts of ilmenite (far more than on Earth), orthoclase, olivine, and quartz. Glass is far more abundant than on Earth. This glass is thought to have formed when meteorite impacts melted the lunar regolith into drops of liquid that quickly hardened into glass spheres. Once formed, the glass spheres, in turn, became pitted with micro-craters. These smooth craters are thought to have been created by the penetration of hypervelocity micrometeorites into the glass. Such micrometeorites vaporize on

4.21 Tiny spheres of glass in the size range of 150 to 250 microns forming the lunar regolith Taurus-Littrow site, Apollo 17 mission, collected by Astronaut Schmitt. (NASA)

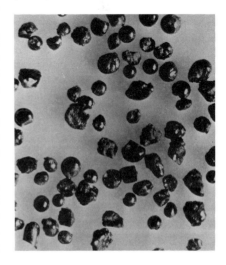

4.22 Lunar breccia, collected during Apollo 11 mission, (NASA)

4.23 Vesicular lunar basalt, collected during Apollo 17 mission. Specimen measures 10 by 18 by 14 cm. (NASA)

impact and melt a small, smooth pit. In addition to crystalline and glassy rocks, the third common component of the lunar regolith is breccia, a mixture of all other rock types (Figure 4.22), including glass spherules, meteorite fragments, and mineral fragments. One breccia was dated at four billion years—two hundred million years older than any rock yet found on Earth.

Moon Mysteries

Many people hoped that the Moon landings would answer the questions about the Moon that have nagged at scientists for centuries. This hope was only partly fulfilled. This is not surprising. If astronauts were to land at five sites on the Earth, they would surely come away with more geologic questions than answers.

Volcanism on the Moon During the 1960s a fierce debate raged among lunar experts between those who believed there are volcanoes on the Moon and those who thought the Moon is a cold, inert, dusty sphere. The abundance of igneous rocks collected by the astronauts conclusively proved that although no evidence of recent eruptions exists, lunar rocks were once molten and cooled rapidly, as do volcanic rocks. Their fine-grained structure is evidence of this, as are the many tiny vesicles and vugs (irregular cavities) in the rocks (Figure 4.23).

Origin of the Moon Before Apollo, four theories on the formation of the Moon were defended with approximately equal vigor: (1) The Moon was torn from the belly of the Earth, probably from the Pacific Ocean; (2) the Moon was a "planet without an orbit" that was "captured" by the Earth's gravitational field; (3) together as twin planets the Earth and Moon formed out of the same primordial dust cloud; and (4) the Moon coalesced from a ring of "planetesimals" (solid bodies smaller than planets) that circled the Earth as the Earth formed.

Enthusiasm for the first three theories has faded considerably. The chemical composition of the lunar surface differs sufficiently from that of the Earth (more ilmenite, more of the metal titanium, basalts of distinct composition) so that a common Earth-Moon origin seems out of the question. Similarly, the likelihood that the Earth could capture a fast-moving Moon seems remote. Only Theory 4 remains plausible, particularly when we look at the over-all chemical situation on the two planets. The Moon has been depleted (has unexpectedly low amounts) of the so-called volatile elements (including mercury, zinc, cadmium, lead, sulfur, chlorine, and bromine). These are the elements that would have escaped under high-temperature conditions before the so-called "heavy" elements (such as scandium, titanium, strontium, and zirconium). This is the situation that would be expected if the Moon had gradually agglomerated from a hot cloud of planetesimals. More volatile elements would tend to fly off and the heavier ones would remain.

The debate has not been resolved, but this is not surprising considering our relatively limited knowledge of the Moon's surface. Most geologists agree that they would be unable to decipher the history of the Earth if they were allowed only brief visits to a half dozen points on its surface. Indeed, fundamental questions about the Earth remain unanswered even though man has inhabited and studied it for thousands of years.

Chapter Review

1. Igneous rocks are classified as *extrusive* (volcanically formed) and *intrusive* (formed from magma that solidifies within the Earth's crust); together they make up approximately 90 per cent of the bulk volume of the Earth's crust.

2. *Plutons* represent any body of intrusive igneous rock. According to their sizes, shapes, relationships, and placement with regard to the surrounding rock, plutons are classified as either *concordant* or *discordant*. Plutons are also divided on the basis of their shapes into *tabular, lenticular,* and *massive* varieties.

3. Tabular plutons consist of *dikes* and *sills*. *Laccoliths* are lens-shaped plutons. *Stocks* and *batholiths* are classified as massive plutons.

4. The classification of igneous rocks involves both *texture* and *mineral composition*. Texture is determined by the size, uniformity of size, and the mutual relationships of the particles of rocks. The size and uniformity of crystals depend largely upon the conditions of cooling. Because rocks of different texture may be formed from the same magma it is essential to consider the chemical, or mineral, composition of rocks as well as their texture.

5. In terms of mineral composition we may classify igneous rocks as *felsic, intermediate, mafic,* and *ultramafic*.

6. Magma formation and behavior are among the Earth's most important processes. Magma seems to be generated by partial melting of mostly crystalline (solid) rock in the lower crust or upper mantle.

7. *Bowen's Reaction Principle*, developed in 1915, helps us understand the crystallization of magma by demonstrating how a crystallizing mineral adjusts its composition to remain in balance with the still-liquid portion of magma.

8. Apollo Moon landings have detected a total lack of weathering on the Moon. Moon rock samples returned by astronauts were virtually all

mafic igneous rocks, and basalt and gabbro appear to be the most common lunar rocks.

9 The abundance of igneous rocks on the Moon proves conclusively that although no evidence of recent eruptions exists, lunar rocks were once molten and cooled rapidly, as do volcanic rocks.

Questions

1 *Extrusive igneous rocks* form (at, beneath) the Earth's surface by the solidification of _____. *Intrusive igneous rocks* form (at, beneath) the Earth's surface by the solidification of _____.

2 What is a *pluton*? Are plutons classified according to the kind of rock composing them? Explain.

3 Two kinds of *tabular plutons* are _____ and _____. They can be distinguished because _____ are _____ whereas _____ are _____.

4 Explain how to distinguish a sill from a buried sheet of extrusive igneous rock formed by solidification of an ancient lava flow. List two points of dissimilarity.

5 Define *laccolith*. Sketch the cross section of a laccolith.

6 Compare and contrast *stocks* and *batholiths*.

7 List the two factors that control the formation of igneous rocks.

8 Explain how the cooling history affects the appearance of igneous rocks.

9 What is a *porphyry*? The particles composing a porphyry are named _____ and _____.

10 The four chief families of igneous rocks are _____, _____, _____, and _____.

11 Which family of igneous rocks typically forms layers that are analogous to sedimentary strata?

Suggested Readings

Bayly, Brian, *Introduction to Petrology.* Englewood Cliffs, N.J.: Prentice-Hall, Inc., 1968.

Ernst, W. G., *Earth Materials,* Englewood Cliffs, N.J.: Prentice-Hall, Inc., 1969.

Pirsson, L. V., and Knopf, Adolph, *Rocks and Rock Minerals,* Third Edition. New York: John Wiley and Sons, Inc., 1947.

Spock, L. E., *Guide to the Study of Rocks,* Second Edition. New York: Harper & Row, Publishers, Inc., 1962.

Tuttle, O. Frank, "The Origin of Granite." *Scientific American*, April 1955. pp. 77–82. (Offprint No. 819. San Francisco: W. H. Freeman and Company.)

Zim, H. S., and Shaffer, Paul, *Rocks and Minerals*. New York: Golden Press, 1957.

"Cleopatra's Needle weathered slightly after 3000 years in Egypt. In 100 years in New York City's air its hieroglyphics are obscured."

Chapter Five
Weathering and Soils

Weathering and Soils

One of the important lessons emphasized by James Hutton concerned the things that happen to bedrock when it becomes exposed to the atmosphere. Everywhere he looked Hutton saw the activities and effects of "a universal system of decay and degradation." In modern terms we would say that Hutton was talking about weathering and erosion. We use **weathering** as a general term that describes all the changes in rock materials that take place as a result of their exposure to the atmosphere. "Exposure to the atmosphere" means that the rock is subject to a series of conditions which collectively characterize the subaerial environment. Weathering is a part of a larger geologic activity known as **erosion.** In general, the term weathering means the effects of processes that happen to rock materials without moving them visibly. By contrast, erosion indicates that significant *motion* has taken place. We discuss erosion in later chapters and concentrate on weathering in this chapter.

Weathering and the Geologic Cycle; Everyday Importance of Weathering

Weathering is important both in geologic theory and in many ways that affect our everyday lives. In the first place, weathering offers visible proof that the geologic cycle is operating today. No matter where you are, you can see the effects of weathering. All you have to do is look carefully around you and you will be convinced that one of the key parts of the geologic cycle is taking place. Secondly, weathering illustrates how geologic materials and certain kinds of energy react to bring the materials into adjustment with conditions in the subaerial environment, or, as we shall also refer to it, the *environment of weathering*. Thirdly, weathering creates important materials, the products of weathering. These include the solid framework particles of the regolith and ions dissolved in water. By reacting with bacteria and other microorganisms the upper part of the regolith becomes capable of supporting rooted plants. When this has happened the upper part of the regolith is said to be a *soil*.

The solid particles of the regolith include three kinds of materials. These are: (1) rock fragments and mineral particles that have been broken loose from the solid bedrock but which otherwise have not been changed much, (2) secondary alteration products that formed during weathering, and (3) resistant substances that were released from the bedrock but survived weathering without being changed.

Without soil there would be very few land plants and almost no land-dwelling animals. Thus, it is no exaggeration to state that without soil we would not be here. Hence, because weathering affects soil, weathering affects us all. Moreover, some individual deposits formed by weathering are mined to recover iron oxides, bauxite (a source of aluminum), or kaolinite (a pure clay). Some valuable deposits of resistant minerals such as platinum, gold, monazite (a rare silicate mineral containing thorium), and ilmenite (an iron oxide containing titanium) are accumulations of particles that were released from bedrock by weathering. *Kerogen,* a complex solid hydrocarbon found in so-called oil shales, survives weathering. Kerogen released during weathering may be the raw material from which certain oil and gas deposits were derived.

Weathering is important for its effects on the engineering properties of bedrock. Zones of "rotten rock" created by weathering have to be avoided in building dams and other heavy structures. Weathering enlarges cracks and dissolves cavities out of carbonate rocks. The spaces thus created by weathering may later serve as one of the several kinds of reservoirs for oil and gas (Chapter 16). Finally, weathering can increase the concentration of sulfide mineral deposits. Nearly all of the world's copper is mined from places where the proportion of copper has been increased as a result of weathering. From all these examples it is clear that many aspects of weathering are involved in modern industrial and engineering activities.

The effects of weathering are related to climate. This means that the discovery in the geologic record of an ancient weathered residue not only proves that the area was formerly part of the land, but in addition, the residue may contain clues about the ancient climate.

Now let us examine various aspects of weathering one at a time. We begin with the environment of weathering. Then come the processes and products of weathering, rates of weathering, and finally, soils.

The Environment of Weathering

In discussing an environment we can be specific about what we mean by including the pressure, the temperature, the kinds of solutions, the kinds of free gases, and the kinds of organisms present. For example, in the subsurface environment where intrusive igneous rocks form, the pressure is enormous. The temperature is high and is more or less constant. Rainwater, with its dissolved chemical load, is not present. No free oxygen is available; all oxygen is bound in the silicate mineral lattices. And organisms are absent.

By contrast, at the surface of the Earth, in the environment of weathering, the pressure is only that of the atmosphere. The temperature ranges from below the freezing point of water (0°C) to as high as 50°C or so. Rainwater may be plentiful. Free gases include abundant

Weathering and Soils

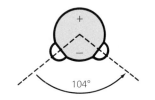

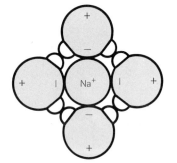

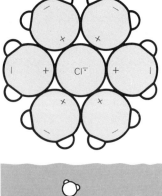

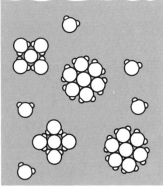

oxygen, nitrogen, and carbon dioxide. Many kinds of organisms are present.

By far the most important thing to know about weathering is how water behaves in the environment of weathering. We shall explore how water affects weathering by discussing: (1) the molecular structure of water, (2) the chemical composition of rainfall, and (3) the effects of climate.

Other aspects of the environment of weathering to be discussed individually include the slope situation at the Earth's surface, the pore spaces within bedrock and regolith, and the kinds of organisms and organic materials.

Structure of Water Molecules

For many reasons, water is a more important factor in weathering than is any other single substance or process. The first reason is the sheer abundance of water. Secondly, water is the only compound that occurs naturally at the Earth's surface as a solid, as a liquid, and as a gas. Finally, water is highly active chemically.

The molecular structure of water (H_2O) is responsible for many properties which make water both stable and strongly reactive. In water molecules, two hydrogen ions, each having an "extra" electron, and one oxygen ion, and each needing two electrons to fill its outer shell, form what is termed a hydrogen bond. The two hydrogen ions line up on the same side of the oxygen ion. The result is the profile as shown in Figure 5.1. Because of the concentration of the hydrogen ions on one "side," the molecules of water are not charged electrically by the same amount in all directions. One "side" contains an excess of positive charge, whereas the other side contains a negative excess.

This uneven distribution of electrical charges causes each water molecule to act much like a dipole magnet. One side attracts positively charged ions and the other side, negative ions. Water is able to wedge its way between two ions that are held by weaker ionic bonds,

5.1 Water molecule, schematic sketch (a). Water dissolving salt (sodium chloride, the mineral halite): (b) Negative sides of four water molecules surround positively charged sodium ion; (c) positive sides of six water molecules surround negatively charged chlorine ion; (d) schematic view of water containing both sodium and chlorine ions in solution.

as in halite. The positive side of the water molecule "steals" a negative chlorine ion, and the negative side latches onto a positive sodium ion. The separated ions from the halite crystal are then surrounded by water molecules, or dissolved (Figure 5.1d). Because of its molecular structure water can both detach ions from mineral lattices and carry these ions away. Water is so effective at dissolving other substances that pure water is rarely found anywhere in nature.

The shape of water molecules likewise is responsible for water's great power to expand as it freezes. When it crystallizes (when it freezes), solid water forms characteristic lattices. These include familiar six-pointed snowflakes (Figure 5.2). The structure of ice forms an open network. Hence, ice, unlike most solids, takes up more space than does the same mass of its parent liquid, water. As the temperature of water drops from 5° to 0°C, randomly arranged water molecules organize themselves into the open crystal pattern of ice. As the water freezes its volume increases by about 9 per cent. Freezing water expands with such force that it can break apart solid masses of bedrock.

Chemical Composition of Rainwater

If you have been taught to expect that rainwater is as pure as distilled water, then you may be surprised to see a heading such as "chemical composition of rainwater." But careful microchemical analyses have proved that rainwater contains many dissolved materials. First and foremost, it contains ordinary salt (sodium chloride, or in the mineral term, halite). This salt comes from sea water; it is picked up by the wind where the surface of the sea features whitecaps and sheets of spray. Secondly, rainwater includes dissolved gases, including nitrogen, oxygen, sulfur dioxide, carbon dioxide, and carbon monoxide. Finally, rainwater contains various organic compounds. These derive from films of organic matter on the surface of the sea that were picked up with the spray, and from organic compounds released into the atmosphere by land plants. Because of these and other substances, rainwater is a slightly acid solution, is very active chemically, and contains natural fertilizers and plant food. Rainwater contains a microflora consisting of bacteria and tiny spores and seeds of plants.

Effects of Climate

We can define **climate** as the time-averaged weather in a given region. The climate is a result of the kind of air masses that flow into the

5.2 Snow crystals, seeded by nuclei of silver iodide, as seen through an electron microscope. Despite their great diversity, all snowflakes display a six-sided (hexagonal) symmetry. (State University of New York at Albany, Institute for Atmospheric Research, courtesy Roger J. Cheng)

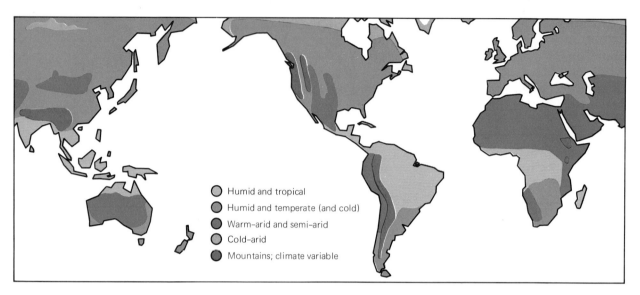

5.3 Climates of the world.

area. The environment of weathering depends intimately upon climate, particularly upon temperature, precipitation, and evaporation and on the way these factors influence the distribution and kinds of organisms. We can illustrate how these factors interact by summarizing four climate zones (Figure 5.3). These are: (1) humid tropics, (2) humid-temperate zones, (3) warm-arid regions, and (4) cold-arid regions.

Humid tropics In general, the most intensive weathering on Earth takes place in the humid tropics. The steady rainfall (commonly more than 250 cm a year) and high temperatures accelerate chemical reactions. For each rise of 10°C in the temperature chemical activity doubles or triples. Biological activity increases as well. The cover of vegetation is dense and complete. A typical example is the tropical rainforest. The bedrock and regolith are continuously moist. The air temperature rarely if ever drops below about 20°C. This means that snow and ice are never seen.

Humid-temperate zones In humid-temperate regions, such as much of North America, rainfall usually is less than in the humid tropics. Temperatures are more variable; freezing during the winter creates ice and snow. The natural vegetation forms a complete cover, usually in the form of woodlands.

Warm-arid regions In warm-arid regions temperatures are high and span a considerable range. Rainfall is scanty. Evaporation exceeds rainfall. As a result, water is drawn upward out of the regolith instead of trickling steadily downward through it, as in the humid regions. Without water, chemical processes of all kinds are inhibited and only a few specialized plants can survive. These are scattered and do not form a continuous cover. In desert regions, therefore, rocks may endure for many thousands of years without changing very much. South of Aswan, in Egypt, granite blocks used for building 4000 years ago are still quite fresh.

Cold-arid regions According to popular imagination the polar lands of the Earth are swept by frequent blizzards and are buried beneath kilometers of ice and snow. This may be true of Antarctica, but in the Northern Hemisphere, many of the coldest places are also among the driest in the world. Their rain-equivalent precipitation is only a few centimeters per year (a centimeter of rain is equivalent to 10 cm of snow). The little precipitation that falls remains frozen most of the time. In addition the water in the top few hundred meters or so of deep regolith is frozen the year around, forming permafrost. Little liquid water is available for chemical reactions or for plant growth. Summer thawing of the top part of the permafrost layer and later refreezing of the thawed parts pushes stones to the surface along great polygons (Figure 5.4).

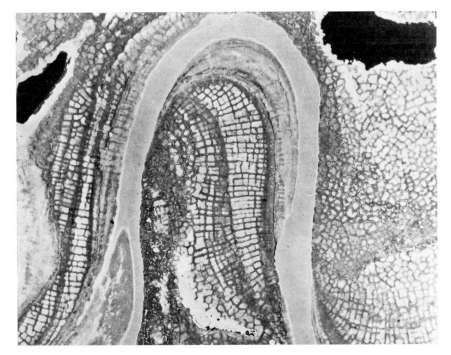

5.4 Polygonal and rectangular patterns in regolith created by freezing and thawing of top of permafrost layer, in the vicinity of Meade River, about 56 km southeast of Barrow, Alaska, as seen from an airplane. (United States Geological Survey.)

Slope Situation

The slope situation refers to the height of a slope, its shape, and its direction of inclination with respect to sunlight and to rain-bearing winds. The slopes form parts of the Earth's surface. The shape of the Earth's surface, including slopes, is known technically as the Earth's **morphology. Relief** expresses the numerical value of the difference in altitude (height above or distance below sea level) between any two points. **Topography** is the study and mapping of morphologic features.

Elevated rocks are subject to more intense effects of rain, of wind, and of changes of temperature than are low-lying rocks. This is true for two reasons: (1) at high altitudes the climatic conditions include greater extremes, and (2) bits of weathered rock are more likely to fall away from elevated rocks, thus exposing additional surfaces of fresh bedrock to the effects of weathering. In regions of low relief, weathered residues tend to remain in place. If weathering products are not moved away, they may form thick deposits. Effects of weathering have been observed to depths of 100 meters in Nigeria, Uganda, and Czechoslovakia. Construction excavations and borings for a hydroelectric project in Australia revealed weathered granitic rock at depths of 170 meters, weathered schist at 180 meters, and weathered gneiss at 350 meters below the surface. Russian scientists have reported examples where weathering has reached a depth of more than 1200 meters.

Altitude, or lack of it, is not the only factor that affects weathering on slopes. Rocks on the north slope of a mountain range weather at a different rate than those on the south slope. In the Northern Hemisphere, the Sun's rays strike south-facing slopes far more directly than they do north-facing slopes. Solar energy evaporates more

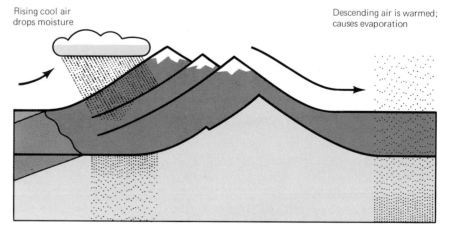

5.5 Rain-shadow effect of coastal mountains, schematic profile. Uneven stippling on front face of block indicates movement of water in regolith, downward at left (where rainfall is abundant), and upward at right, where evaporation is great and desert conditions prevail.

water from bedrock and regolith underlying south-facing slopes. On north-facing slopes, more ice is likely to form and to last longer during a thaw. Slopes receiving more rainfall tend to support more plant cover than other slopes. For example, the northeastern sides of sand ridges on Fire Island, New York, are thickly vegetated, whereas southwestern slopes are barren.

In some mountainous regions, vastly different weathering environments can be found on the east and west slopes of the same range. In western United States, the western slopes of the Sierra Nevada, northern Coast Ranges, and some ranges in the Rocky Mountains are forested, whereas the eastern slopes lack trees. This is explained by the effects of the mountains on the flow of air. As moisture-bearing air moves landward from the Pacific Ocean, the mountains deflect it upward. As it rises, the air cools and drops most of its moisture in the form of rain or snow along the western slopes. When the air reaches the eastern side of the mountains, it descends, is warmed, and thus increases its capacity to hold moisture (Figure 5.5). This descending air dries out the eastern slopes and may even form local deserts.

Pore Spaces in Bedrock and Regolith

The environment of weathering is located in the pore spaces in bedrock and regolith. In bedrock, this means the spaces between crystals or in cracks in the bodies of rock. These cracks in bedrock are named according to whether or not the blocks on the opposite sides have been displaced. Cracks along which the opposite sides have not been relatively shifted are **joints**. Cracks along which the opposite sides have been relatively displaced are **faults**. Some joints result from large-scale fracturing of the bedrock (Figure 5.6a), which is related to the shifting of plates of the lithosphere. Other joints result from the shrinkage that accompanies cooling of bodies of igneous rock (Figure 5.6b). In regolith, pore space exists among the framework particles. Water tends to remain longer in pores than on surfaces exposed to sunlight.

Organisms and Organic Materials

Organisms affect weathering in many ways. They cycle elements into and out of the regolith. The kind of vegetation growing in an area determines the kind of organic litter which falls onto the ground. Various wild animals and insects rearrange the regolith by

5.6 Joints in bedrock form spaces along which water is able to break the rock further. (Top) These joints, in massive limestone, resulted from the effects of deformation caused by earth movements. The Burren, County Clare, Ireland. (Courtesy Irish Tourist Board) (Left) Systematic pattern of columnar cracks that formed during cooling and decrease in volume of this body of igneous rock. Devil's Postpile, California. (U.S. National Park Service photograph by C. W. Stoughton, July 1971) (Right) Effects of weathering along joints in massive sandstone, Canyonlands, Utah. (American Airlines)

burrowing into it. (This discussion does not include the effects of one organism, *Homo sapiens*, whose impact on regolith can top them all.)

Visible Evidence of Weathering

How can you tell if something has been weathered? The most obvious way is to find some object whose original condition you know and then to see how it looks after being out of doors for a time. Stone sculptures (Figure 5.7) and tombstones (Figure 5.8) provide clear proof of weathering. Dated tombstones have been the basis for study-

5.7 Accelerated weathering during twentieth century of porous sandstone sculpture carved in 1702, Herten Castle, near Recklinghausen, Westphalia, Germany. In 206 years, until 1908 (photo on left) only minor weathering took place. In the next 61 years rapid weathering has nearly destroyed the statue (right, 1969). (Dr. K. Schmidt-Thomsen, Landesverwaltungsdirektor, Der Landeskonservator von Westfalen-Lippe, Muenster, Germany)

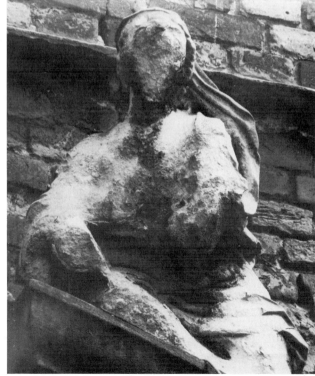

5.8 Comparative weathering of different rocks shown by preservation of tombstones. (Left) Inscriptions completely destroyed by weathering in humid climate; red sandstone gravestones erected at unknown date after 1766 in New York City. (Rhoda Galyn, May 1973) (Right) Negligible weathering in dry climate of marble monument erected near Little Bighorn, Montana. (U. S. National Park Service)

ing the effects of weathering within a single area on rocks of different kinds. One study in Great Britain showed that one kind of tombstone developed and lost a weathered surface layer 2.5 cm thick. Another kind required 500 years for this same effect.

Processes of Weathering

Weathering processes are so complexly interlocked that it is not easy to organize them in a simple way as a basis for a coherent discussion. As a first approximation weathering processes can be divided into those which involve only physical activities and those which include chemical activities. Basically the physical processes do the breaking and the chemical processes, the alterations. However, as we shall see, in nature things are not so clearly separated as would be suggested by these two categories. All weathering processes work together. Breaking rocks creates more surface area (Figure 5.9), which becomes available for chemical attack. Chemical activities cause the physical release of minerals from some rocks. Crystallization of salt helps loosen rocks physically. The effects of organisms are both chemical and physical. In order to include all these variables, we shall discuss the processes of weathering under three main headings. These are: (1) physical weathering, (2) chemical weathering, and (3) complex weathering. We shall also look at the weathering of some common rocks.

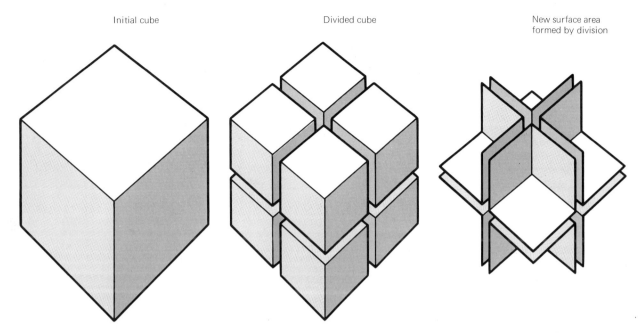

5.9 Increase of surface area results from dividing cube in half along three mutually perpendicular planes. In reducing the length of the cube sides by one half, the surface area has been doubled.

Physical Weathering: Disintegration

Physical weathering is the name given to processes which break apart rocks without altering their chemical compositions. Other names for physical weathering are *mechanical weathering* and *disintegration*.

The chief effect of physical weathering is to break rocks into smaller fragments and eventually into individual mineral particles. We discuss the particle-by-particle separation first and then describe near-surface fractures.

Granular disintegration The physical separation of the individual mineral particles of a rock from one another is a process named **granular disintegration.** A rock undergoing granular disintegration crumbles easily (Figure 5.10).

Near-surface fractures Bedrock that is brought up to the Earth's surface is subjected to uneven conditions which cause the rock to fracture. The chief processes involved are pressure release from unloading and thermal effects.

5.10 Granular disintegration is causing these massive granite blocks to crumble. Algeria. (United Nations, FAO photo by P. Keen)

5.11 Sheeting joints parallel to land surface in massive granite, Liberty Cap, Yosemite National Park, California. (U. S. National Park Service)

Any rock formed deep inside the Earth is subject to uniform pressure which results from the weight of the overlying rocks. This pressure amounts to about 0.26 kilograms per square centimeter for each meter of depth. At a depth of 15 meters the pressure equals the air pressure inside a high-pressure bicycle tire. Bedrock exposed at the surface has been relieved of the weight of its former overburden. As a result the bedrock expands upward slightly. This upward expansion can be great enough to form **sheeting joints** parallel to the ground surface (Figure 5.11). The massive granite "domes" of Yosemite National Park, California, are superb examples of landforms shaped chiefly by sheeting joints (Figure 5.12). The spalling off of bodies of bedrock along sheeting joints is **exfoliation.**

Thermal effects, frost wedging, heating and cooling of rocks in deserts The effects of temperature on rocks are most pronounced where temperatures are extreme and fluctuate. We can illustrate this by comparing what happens at the cold end of the scale with what happens at the warm end.

Where air temperatures reach the freezing point of water (0°C) liquid water is transformed into solid ice. As mentioned previously, when water freezes its volume expands by 9 per cent. Such freezing does two things: (1) it lifts overlying particles in regolith, a process known as **frost heaving** (to be discussed further in Chapter 7), and (2) it can split apart bodies of solid bedrock. The prying apart of solid bedrock by growth of ice crystals is **frost wedging.** The most vigorous frost wedging takes place at high altitudes, where the temperatures go above freezing during the day but drop into the freezing range at night. The result of these repeated freeze-thaw-refreeze cycles is to create extensive accumulations of angular blocks called **boulder fields.**

Other mechanical breakage results from crystallization of salts out of water solutions. We discuss this process under the heading "complex weathering."

If you have ever built a campfire on solid bedrock you probably have seen sheets of rock break off when the rock was hot and when you doused the fire by drenching it with water. The rock spalls off in sheets because the fire causes it to expand upward. More breaking takes place when cool water causes an abrupt temperature change in the heated rock.

In deserts boulders have been found fractured as though by frost wedging. In addition, desert travelers have reported hearing rocks break with sharp cracking sounds similar to gunshots. (It has been reported that upon hearing these sounds French Foreign Legionnaires in the Sahara have gone to battle alert.)

The dry heating and cooling of a small cube of granite was repeated for 89,400 cycles in the laboratory, a routine thought to be equivalent to 244 years of daily heating by the Sun and cooling at night. The heating was for five minutes to 140°C; the cooling was to 30°C. At the end of the three-year test, no cracks were found in the specimen, even under a microscope.

How are the field observations and laboratory experiments to be reconciled? No one really knows, but some scientists suspect that the laboratory conditions did not fairly duplicate what happens to rocks in their natural desert settings.

5.12 Granite dome formed by weathering along sheeting joints, Half Dome, viewed from Washburn Point, Yosemite National Park, California. (American Airlines)

Chemical Weathering: Decomposition

Chemical weathering, which causes the composition of rock materials to change, is known as **decomposition.** In order to understand how decomposition works we need to consider two factors: (1) chemical processes and (2) geologic materials. Because a given chemical process affects some materials and not others we shall have to discuss these two topics together. At the end we shall summarize the responses to various chemical weathering of some of the common rocks. The

Weathering and Soils

chemical processes we shall consider are: (1) oxidation, (2) dissolution, (3) hydrolysis, and (4) carbonation.

Oxidation The addition of oxygen to another element is a process known as **oxidation.** The elements in rocks that are most easily oxidized are iron (found in ferromagnesian and sulfide minerals) and sulfur (found in sulfides, such as pyrite). The rusting of a nail is a familiar example of oxidation. The iron of the nail combines with oxygen from the atmosphere to form a reddish iron oxide (the "rust").

As with other kinds of processes in chemical weathering, oxidation proceeds more rapidly if water is present. Oxidation in the desert is slow compared with oxidation where the rocks are exposed to water.

The iron-oxide minerals that form during chemical weathering are colored red, yellow, orange, or brownish. Many rock surfaces underground are covered with a thin film of pyrite, an iron sulfide, which is easily oxidized. As a result, the iron forms limonite, and the released sulfur may collect as an element or may combine with water to create sulfuric acid, a powerful chemical agent.

Also present in some rock materials is organic matter. Most forms of organic matter have such a great affinity for oxygen that they will unite with any oxygen that is available and may even use up the total supply. The end products of the oxidation of organic matter are carbon dioxide gas and water. When carbon dioxide is dissolved in water it forms a weak acid (carbonic acid), which can attack minerals.

5.13 **Clay minerals (kaolinite) crystals** seen through electron microscope. (K. M. Towe, Smithsonian Institution)

Dissolution As we have described previously, water is capable of dissolving some solid rock materials. Layers of rock salt (halite rock) are so easily dissolved that they are exposed only in the driest regions. Proof of water's ability to dissolve materials has been provided by measurements of new-fallen snow in the Sierra Nevada, California. As soon as the snow melted and seeped into the mountain regolith its chemical load increased seven and a half times. Almost immediately the content of silica increased nearly 100 times. This indicates that dissolution of rock materials begins almost as soon as water comes into contact with the minerals of the regolith and bedrock. If the water remains in the ground for several months, its chemical load doubles again. The removal of materials in water solution is a process named **leaching.**

Hydrolysis An important process in the weathering of silicate minerals is their combination with water, or **hydrolysis.** In this process water attacks the crystal lattices of feldspars. Some of the OH ions from the water become parts of the crystal lattices of clay minerals (Figure 5.13). Clay minerals are stable solids under all but the most humid tropical conditions. Therefore, these minerals are nearly universal ingredients of regolith; they are the chief secondary-alteration product of weathering. Feldspars weather so readily into clay minerals that only rarely do they appear as survival products in the regolith. When they do, we can be certain that the mechanical weathering of breaking the feldspars loose from the bedrock took place faster than chemical weathering could convert them into clay minerals. During the weathering of feldspars to clay minerals, sodium, calcium, and potassium are leached from the feldspar lattices and go into solution. The potassium is taken up immediately by any plant rootlets; the sodium and calcium are transported in solution to the sea.

Carbonation Carbon dioxide gas, which is present in the atmosphere and is a byproduct of the oxidation of organic matter, readily dissolves in water to form a weak acid, carbonic acid. The reaction between carbonic acid and minerals is **carbonation.** Carbonate minerals, especially calcite, dissolve easily in carbonic acid. Because calcite is the chief ingredient of limestones, such rocks are greatly affected by carbonation.

Complex Weathering

Under this generalized heading we discuss the crystallization of salts, the combined effects of chemical and physical weathering, and the effects of organisms.

Crystallization of salts As we have seen, the water circulating in the regolith contains many elements in solution. When this water evaporates, its dissolved materials crystallize as solids. These crystal lattices grow and exert large forces on their surroundings. When large amounts of solids are precipitated, they may cement sediments into sedimentary rocks. When only small amounts are precipitated the crystals pry rocks apart just as ice does.

The two common crystal salts that wedge apart rocks are gypsum and halite. Gypsum, a hydrated calcium sulfate, is particularly troublesome where limestone or marble is exposed to sulfur oxides, which are common in coal smoke. In such places the rainwater becomes a dilute solution of sulfuric acid that reacts with the calcite in the limestone and marble to form gypsum. As gypsum crystals enlarge in the cracks, they cause pieces of solid rock to flake off from buildings and monuments.

Halite, ordinary rock salt, crystallizes from the salty under-

5.14 Salt weathering possibly formed the recesses surrounding these Indian dwellings. Mesa Verde National Park, Arizona. (American Airlines)

ground water in deserts. Where evaporation is intense, salt-bearing water is pulled upward by capillary forces to the surface. As the water evaporates from porous rock, such as sandstone, halite is precipitated. The growing halite crystals help make the rock crumble. Salt weathering created many of the caves in the American southwest used by cliff-dwelling Indians as homes and shelters (Figure 5.14). In

5.15 Cubes of rock become spheres by attack at corners and edges. Cube at left shows effects only at corners. In middle sketch weathering has proceeded along sides as well. At right the rock splits apart along concentrically curved shells.

Egypt, halite crystals have forced plaster away from the walls of buildings in Cairo and have severely weathered the temples at Luxor and Karnak. In London, the Houses of Parliament have been damaged by crystallization of magnesium sulfate.

Combined effects of chemical and physical weathering As feldspars are converted to clay minerals by hydrolysis, the mineral fabric of the rock swells. Such swelling may be strong enough to break the rock.

If a cube of rock is undergoing hydrolysis and breaking, the effects appear first at the corners, where the three surfaces join. The next place to show the effects is along the edges, where two planes meet. The last places affected are the sides, where only one plane is exposed. As the corners and edges are attacked and break more rapidly than the plane sides, the original cube becomes more and more spherical (Figure 5.15). Later, weathering proceeds inward evenly from all directions. The end result is that a series of more or less regular sheets or shells peel off like the layers of an onion. These concentric partings have been named **spheroidal joints** (Figure 5.16).

Effects of organisms The weathering created by organisms is both physical and chemical. Earthworms turn over vast amounts of soil and many animals dig burrows into the regolith (Figure 5.17). The effect of all this turning over and burrowing is to bring rock particles to the surface, where they can be weathered, and to admit gases and water into the regolith where they might not otherwise penetrate. Plant roots follow fractures in rocks; they grow where they can find water (Figure 5.18). Growing roots widen the cracks, just as they cause considerable damage to sidewalks, walls, and foundations.

Lichens and bacteria break down organic material and this liberates acids which attack silicate minerals. Other organic material affects the acidity of the waters and thus greatly influences the chemical environment of weathering.

Weathering of Some Common Rocks

We can bring together our discussion of weathering by reviewing how a few common rocks respond to weathering. We include granite, mafic igneous rocks, limestone, and sandstone.

Weathering of granite In considering the weathering of granite we need to check what happens to the chief minerals in it: feldspars, quartz, and the minor (or accessory) minerals, including biotite, zir-

5.16 Spheroidal joints in weathered mafic volcanic rock, Kanai, Hawaii. (American Museum of Natural History)

5.17 Vast underground networks of shafts and tunnels are built by these prairie dogs outside the entrances of their burrows. Around the entrances they heap up levees about 30 cm high. The vertical shaft, 8 to 10 cm in diameter, descends 3 to 4 m and then connects to vast networks of horizontal tunnels. Mackensie State Park, Lubbock, Texas. (U. S. Department of Agriculture, Soil Conservation Service, 1963)

con, ilmenite, and magnetite. In a humid climate the feldspars undergo hydrolysis and become clay minerals. Much or all of their potassium, sodium, and calcium is removed in solution. The quartz may not be much changed, but simply loosened. Thus quartz tends to accumulate as a regolith of sand-size particles. Along with the quartz in this sandy regolith are the resistant minor minerals zircon, ilmenite, and magnetite. In the humid tropics the magnetite may be oxidized to hematite. Any biotite present may be altered to chlorite in a cool climate, and to illite in warmer climates. In some tropical locations the conditions of chemical weathering cause quartz to be dissolved.

Weathering of mafic igneous rocks The chief minerals in mafic igneous rocks are plagioclase and pyroxene; the chief minor mineral is magnetite. Chemical weathering can destroy all of these minerals. Both plagioclase and pyroxene undergo hydrolysis and are leached. The result: clay minerals and ions in solution (chiefly calcium from the plagioclase and magnesium and calcium from the pyroxene). Oxidation changes the iron in the pyroxene and magnetite to hematite and limonite, which are not very soluble.

Weathering of limestone The chief mineral of limestone is calcite. Other minerals include quartz, various other forms of silica (SiO_2), and pyrite. In moist regions the calcite undergoes carbonation, and the calcium is removed in solution. The pyrite is oxidized. The iron from the pyrite forms limonite and the sulfur may become sulfuric acid that can react with other calcite to create gypsum. The quartz and other silica materials accumulate as a sandy residue. In moist climates the areas underlain by limestone are lowered rapidly and become valleys and lowlands. By contrast, in arid regions limestone is one of the most resistant rocks. In dry climates limestones cap the highest mountain peaks.

Weathering of sandstone Because sandstone consists mostly of minerals that have already survived one or more cycles of weathering, not many more changes are likely to happen when sandstone is weathered. The chief mineral in sandstone is quartz. Others may include feldspar (which escaped destruction during a previous cycle of weathering) and various materials in the cements, such as calcite, iron oxides, pyrite, and even quartz. When sandstone is subjected to chemical weathering, usually the chief effect is on the cement. If the cement is calcite, the effects parallel those just described for limestone. In moist climates the calcite is destroyed and the quartz particles form a sandy residue. In dry climates the calcite persists and the

5.18 Tree roots growing along joint helped wedge apart the large piece of rock that has recently fallen away, exposing the roots to view. North side of Connecticut Turnpike, W. of overpass for Todds Hill Road, Branford, Connecticut. Bedrock is a volcanic breccia of Late Triassic-Early Jurassic age. (Joanne Bourgeois, 1974)

rock is resistant. Iron oxides and quartz cements are resistant nearly everywhere. Pyrite is oxidized, with results as explained previously.

Now that we have examined the chief processes of weathering and have seen how a few rocks react, let us proceed to rates of weathering. After that we can consider soils.

Rates of Weathering

The rates of weathering depend upon the interplay among three factors. These are: (1) susceptibility to weathering of the various minerals, (2) amount of surface area initially available, and (3) intensity of the weathering processes (which depends on all the factors that influence the environment of weathering). The rate of chemical weathering can be increased greatly in areas where the air has been polluted by chemicals.

Selected Examples of Weathering Rates

As we have seen, the various rock-making minerals vary in their responses to chemical weathering. Comparative study has resulted in the establishment of a "weathering potential index" (WPI). In this index quartz is assigned a value of 1. Minerals having numbers higher than 1 are more weatherable than quartz (Table 5.1).

Tephra deposited in warm, moist regions exemplify the combination of conditions that create the fastest weathering: susceptible minerals, large initial surface areas, and environment of maximum chemical activity. For example, in only 50 years, the tephra that were ejected during the great explosion of Krakatoa in 1883 had lost about 5 per cent of their initial silica, had formed clays, and had made a soil more than 2.5 cm thick.

Table 5.1
RELATIVE RESISTANCE TO WEATHERING OF ROCK-MAKING SILICATE MINERALS

Mineral	Resistance
Olivine	Least resistant
Pyroxene	
Hornblende	
Biotite	
Plagioclase feldspars	
Muscovite	
Orthoclase	
Quartz	Most resistant

Weathering and Soils

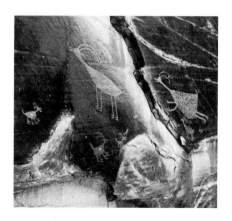

5.19 Hieroglyphics on sandstone, Monument Valley, Arizona. Although these Indian drawings may be thousands of years old they are clearly legible. (American Airlines)

The slow end of the weathering scale is found in dry regions. In polar areas having slight precipitation many exposed rock surfaces remain nonweathered for thousands of years. Similarly, in desert regions such as the Nile Valley and the southwestern United States, inscriptions carved on rock faces thousands of years ago are still legible (Figure 5.19).

Removal of material during weathering (and by related processes) lowers the land surface. If rocks are uniform to begin with and are being weathered uniformly, the land surface is lowered more or less equally everywhere. The effect can be compared with the melting of an ice cube. In some places, however, a particularly resistant material is present. Then a tough surface rock may protect the underlying layers and locally retard the general lowering. Such a "cap" rock may create a pedestal (Figure 5.20). The uneven wearing away of rock material is known as **differential weathering.**

Effects of Air Pollution

Air-borne chemicals that are the waste materials of many industrial activities can greatly increase the rate of weathering. The chief corrosive agent is sulfuric acid that forms from the sulfur dioxide created

5.20 Pedestals created by protective effect of hard-to-erode caps. (Left) Caprock formed by precipitation of sodium salts around shores of Mono Lake, California. (R. and J. Spurr/Bruce Coleman, Inc.) (Right) Large boulders atop tall pillars of glacial sediments, near Evolena, Val d'Herens, Valais, Switzerland. (Swiss National Tourist Office)

5.21 Differential weathering of Cleopatra's Needle, Central Park, New York City. Sides that face west and south (left) have weathered much more than the side that faces east and north (right). (Ann McCaughey, 1975)

by the burning of certain coals and fuel oils. In London, Venice, Rome, and many other cities, stone structures have weathered much more rapidly since the beginning of the Industrial Revolution than they had previously.

A typical example of pollution-induced weathering (that is combined with a change from a dry to a moist climate) is provided by one of the Egyptian obelisks known as Cleopatra's Needle. The original monument, 20 meters high, was erected in the dry, clean air of Egypt in 1475 B.C. In 1879, it was moved to Central Park, in New York City. During more than 33 centuries in Egypt, it weathered only very slightly. In less than one century in New York City's moist air, however, its hieroglypics have been almost obscured by the effects of weathering (Figure 5.21). The stone is wearing away unevenly. On the northeast side, which is subjected to the most rainfall, relatively little damage has been done. On the southwestern side the greatest decomposition is evident. In New York, the southwest winds prevail during fair weather; however, such winds bring with them sulfur dioxide and other chemical pollutants that were discharged into the air from factories and refineries across the Hudson River in New Jersey and in areas farther southwest.

Soils

As we have emphasized previously, weathering is a process that is essential to life as we know it on Earth. Without weathering the surface of the Earth would be much like that of the Moon, Mars, Mercury, or Venus. It would be rocky, barren, and lifeless. Instead, we live on a planet whose continents bear extensive forests, grasslands, and farms, and these support a wonderful complexity of animal life. Neither plant nor animal communities could exist unless the Earth's bedrock could be converted into regolith and the regolith, in turn, could be converted into soil.

We shall examine soils by starting with a definition. Then we shall discuss the relationship of soil to regolith, describe some examples of modern soils, briefly consider the management of soils, and conclude with some remarks about ancient soils in the geologic record.

Definition of Soil

As we noted previously, **soil** is defined as the upper part of the regolith that is capable of supporting the growth of rooted plants. Soils consist of solids, liquids, and gases, all of which must be in proper balance to support plant growth. Soil is a dynamic product; it is alive with complex chemical, physical, and biological activities.

From our previous discussion you may recall that rainfall contains many things necessary for plant life: inorganic and organic substances and tiny spores, seeds, and bacteria. In order for these ingredients to interact to grow plants, they need to find shelter in the framework of the regolith, where water persists between rain showers. Thus, the important contribution of the regolith is to provide the required shelter and moisture. This is done by the mineral particles that range in size from gravel to tiny colloids.

The *liquid portion* of soils consists of complex chemical solutions with their rich variety of dissolved ions. The gases in the pore spaces include those of the atmosphere (chiefly nitrogen and oxygen) along with those liberated by biological and chemical activity, chiefly carbon dioxide.

After plants have started to grow, their residues are added to the soil to make **humus,** which consists of finely divided and partially decomposed organic matter which imparts a dark color. Humus forms when dead plants, fallen leaves, and other organic litter are oxidized very slowly and are attacked by soil bacteria.

Relationship of Soil to Regolith: Soil Zones and Soil Profiles

The formation of soils can best be understood by considering two factors: (1) the kind of regolith, or parent material of the soil; and (2) how the water moves within this regolith and what chemical changes the water creates.

Kinds of regolith We have discussed regolith repeatedly, but before we take up its relationship to soil in detail it is appropriate to emphasize that regolith can be divided into two major categories: (1) **residual regolith,** which is formed by the deep weathering of the underlying bedrock, and (2) **transported regolith,** which has come from someplace else and may be altogether unlike the underlying bedrock. Whether a transported regolith is or is not like the underlying bedrock, the important point is that it is a regolith and, as such, is ready to become a soil.

Important regolith-forming processes other than weathering include glacial grinding of bedrock, downslope gravity transport of particles that broke loose from the bedrock, and volcanic explosions, which create tephra, a kind of "instant regolith." In addition, sediments deposited during floods on the bottoms of valleys and the dust blown by the wind may blanket the bedrock with a regolith having a composition unlike that of the bedrock. Therefore, although in the previous parts of this chapter and elsewhere we have mentioned the importance of the weathering of bedrock into residual regolith as an important part of the geologic cycle, when we analyze soils, we need to include all kinds of regolith. Exact figures are not available, but it is likely that many soils important to agriculture have formed on transported regolith.

Soil zones and soil profiles The conversion of the upper part of the regolith into soil creates layers. These are characterized by distinctive textures, colors, and compositions. The major layers are named **soil zones.** (In the literature of soil science these are referred to as "horizons," but in geology "horizon" implies a surface having zero thickness. Therefore, we shall refer to these layers as "zones.")

The development of a zonal soil profile requires time. A mature soil is one that displays two distinctive zones, designated by the capital letters A and B (with numbered subdivisions, if any are pres-

Realms of the Soil
Rachel Carson

The thin layer of soil that forms a patchy covering over the continents controls our own existence and that of every other animal of the land. Without soil, land plants as we know them could not grow, and without plants no animals could survive.

Yet if our agriculture-based life depends on the soil, it is equally true that soil depends on life, its very origins and the maintenance of its true nature being intimately related to living plants and animals. For soil is in part a creation of life, born of a marvelous interaction of life and nonlife long eons ago. The parent materials were gathered together as volcanoes poured them out in fiery streams, as waters running over the bare rocks of the continents wore away even the hardest granite, and as the chisels of frost and ice split and shattered the rocks. Then living things began to work their creative magic and little by little these inert materials became soil.

Life not only formed the soil, but other living things of incredible abundance and diversity now exist within it; if this were not so the soil would be a dead and sterile thing. By their presence and by their activities the myriad organisms of the soil make it capable of supporting the earth's green mantle.

The soil exists in a state of constant change, taking part in cycles that have no beginning and no end. New materials are constantly being contributed as rocks disintegrate, as organic matter decays, and as nitrogen and other gases are brought down in rain from the skies. At the same time other materials are being taken away, borrowed for temporary use by living creatures. Subtle and vastly important chemical changes are constantly in progress, converting elements derived from air and water into forms suitable for use by plants. In all these changes living organisms are active agents.

Perhaps the most essential organisms in the soil are the smallest—the invisible hosts of bacteria and of thread-like fungi. Statistics of their abundance take us at once into astronomical figures. A teaspoonful of topsoil may contain billions of bacteria. In spite of their minute size, the total weight of this host of bacteria in the top foot of a single acre of soil may be as much as a thousand pounds. Ray fungi, growing in long threadlike filaments, are somewhat less numerous than the bacteria, yet because they are larger their total weight in a given amount of soil may be about the same. With small green cells called algae, these make up the microscopic plant life of the soil.

Bacteria, fungi, and algae are the principal agents of decay, reducing plant and animal residues to their component minerals. The vast cyclic movements of chemical elements such as carbon and nitrogen through soil and air and living tissue could not proceed without these microplants. Without the nitrogen-fixing bacteria, for example, plants would starve for want of nitrogen, though surrounded by a sea of nitrogen-containing air. Other organisms form carbon dioxide, which, as carbonic acid, aids in dissolving rock. Still other soil microbes perform various oxidations and reductions by which minerals such as iron, manganese, and sulfur are transformed and made available to plants.

Besides all this horde of minute but ceaselessly toiling creatures there are of course many larger forms, for soil life runs the gamut from bacteria to mammals. Some are permanent residents of the dark subsurface layers; some hibernate or spend definite parts of their life cycles in underground chambers; some freely come and go between their burrows and the upper world. In general the effect of all this habitation of the soil is to aerate it and improve both its drainage and the penetration of water throughout the layers of plant growth.

Of all the larger inhabitants of the soil, probably none is more important than the earthworm. Over three quarters of a century ago, Charles Darwin published a book titled *The Formation of Vegetable Mould, through the Actions of Worms, with Observations on Their Habits*. In it he gave the world its first understanding of the fundamental role of earthworms as geologic agents for the transport of soil—a picture of surface rocks being gradually covered by fine soil brought up from below by the worms, in annual amounts running to many tons to the acre in most favorable areas. At the same time, quantities of organic

From SILENT SPRING by Rachel Carson, 1962. Houghton-Mifflin Company. Abridged with permission.

Field Trip

matter contained in leaves and grass (as much as 20 pounds to the square yard in six months) are drawn down into the burrows and incorporated in soil. Darwin's calculations showed that the toil of earthworms might add a layer of soil an inch to an inch and a half thick in a ten-year period. And this is by no means all they do: their burrows aerate the soil, keep it well drained, and aid the penetration of plant roots. The presence of earthworms increases the nitrifying powers of the soil bacteria and decreases putrification of the soil. Organic matter is broken down as it passes through the digestive tracts of the worms and the soil is enriched by their excretory products.

This soil community, then, consists of a web of interwoven lives, each in some way related to the others—the living creatures depending on the soil, but the soil in turn a vital element of the earth only so long as this community within it flourishes.

The problem that concerns us here is one that has received little consideration: What happens to these incredibly numerous and vitally necessary inhabitants of the soil when poisonous chemicals are carried down into their world, either introduced directly as soil "sterilants" or borne on the rain that has picked up a lethal contamination as it filters through the leaf canopy of forest and orchard and cropland? Is it reasonable to suppose that we can apply a broad-spectrum insecticide to kill the burrowing larval stages of a crop-destroying insect, for example, without also killing the "good" insects whose function may be the essential one of breaking down organic matter? Or can we use a nonspecific fungicide without also killing the fungi that inhabit the roots of many trees in a beneficial association that aids the tree in extracting nutrients from the soil?

The plain truth is that this critically important subject of the ecology of the soil has been largely neglected even by scientists and almost completely ignored by control men. Chemical control of insects seems to have proceeded on the assumption that the soil could and would sustain any amount of insult via the introduction of poisons without striking back. The very nature of the world of the soil has been largely ignored.

From the few studies that have been made, a picture of the impact of pesticides on the soil is slowly emerging. It is not surprising that the studies are not always in agreement, for soil types vary so enormously that what causes damage in one may be innocuous in another. Light sandy soils suffer far more heavily than humus types. Combinations of chemicals seem to do more harm then separate applications. Despite the varying results, enough solid evidence of harm is accumulating to cause apprehension on the part of many scientists.

We are therefore confronted with a second problem. We must not only be concerned with what is happening to the soil; we must wonder to what extent insecticides are absorbed from contaminated soils and introduced into plant tissues. Much depends on the type of soil, the crop, and the nature and concentration of the insecticide. Soil high in organic matter releases smaller quantities of poisons than others. In the future it may become necessary to analyze soils for insecticides before planting certain food crops.

As applications of pesticides continue and the virtually indestructible residues continue to build up in the soil, it is almost certain that we are heading for trouble. This was the consensus of a group of specialists who met at Syracuse University in 1960 to discuss the ecology of the soil. These men summed up the hazards of using "such potent and little-understood tools" as chemicals and radiation: "A few false moves on the part of man may result in destruction of soil productivity and the arthropods may well take over."   1962

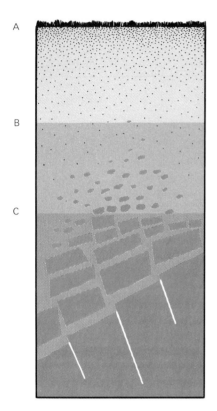

5.22 Zonal soil formed by weathering of residual regolith overlying decomposing limestone bedrock.

ent). These two zones are collectively designated as the solum; they overlie the subsoil, designated zone C (Figure 5.22). The subsoil is the parent material which has been altered to form the soil. Thus, zone C consists of regolith, which, as noted previously, may be residual or transported. If zone C consists of residual regolith, then the lower boundary of this zone is extended downward to include the bedrock. Other special designations are: zone O for the upper organic zone that is rich in decaying organic matter, and zone R for the bedrock, whether or not it is related to the overlying regolith.

Humus, bacteria, and organic acids The humus in some soils forms a continuous, dark-colored surface layer. Such an organic layer at the top (the O zone) can exist where the rate of accumulation of plant litter exceeds the rate of its destruction by bacterial activity, oxidation, and other processes. In the tropics, where the mean annual temperature exceeds 25°C, plant material is abundant, but bacteria are so extraordinarily abundant that they consume all the plant litter and no humus forms. The happy home of humus is on the floor of a forest in a humid-temperate region. Rainwater which moves downward through the humus becomes charged with organic acids and thus becomes an especially effective leaching agent. Beneath the layer of humus the color of the soil in the A zone typically is gray. All the iron minerals, which impart brownish colors, have been leached by the organic acids and carried downward to the B zone.

Time Because the reactions involved in the development of a soil do not happen all at once but take place in a progressive sequence, time is a factor when needs to be considered in discussing the relationship of soil to regolith. The time required for a soil to develop varies greatly and depends to a large degree on the kind of regolith. Only a few years are required to convert a layer of newly deposited mafic tephra from a tropical volcano into a soil. If the regolith itself is residual and is being developed slowly by decomposition of the underlying bedrock, then the time required to form a mature soil may be thousands of years. The silts deposited by the yearly floods of some rivers, such as the Nile, are "instant soils" that can be planted as soon as the flood waters dry off.

Examples of Modern Soils

Many kinds of modern soils have been recognized. Because of the three factors of (1) time; (2) the variability of soil-forming processes within different environments of weathering, which is particularly

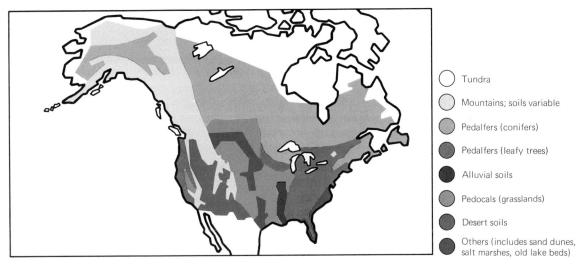

5.23 Soil groups in North America. (Redrawn from C. B. Hunt, 1972)

responsive to climate; and (3) the different kinds of parent regolith materials on which soils form, it is not surprising to discover that soil scientists have been able to map hundreds of varieties of soils. For our purposes we shall consider four groups, starting with the extreme kind found in the humid tropics and progressing to less-humid regions (Figure 5.23). Our four groups are: (1) latosols, (2) pedalfers, (3) pedocals, and (4) others.

Latosols The reddish soils of the humid tropics are **latosols**. Because of intense bacterial activity, latosols typically lack humus; they contain abundant iron oxides and aluminum oxides and have been leached of nearly all other metallic ions and even of quartz. Under conditions of extreme leaching iron oxides and aluminum oxides may be so concentrated that they become valuable sources of iron or aluminum. Regolith rich in aluminum oxide is **bauxite**; regolith rich in iron oxide is **laterite**. Lateritic regolith may become so hard that it can be used as a building material. The ancient temples of Angkor Wat, in Cambodia, have been constructed from carved laterite (Figure 5.24).

Pedalfers In moving away from the humid tropics into the humid-temperate regions one leaves behind the latosols and encounters a different kind of soil, named **pedalfer** (pe-DAL-fur) soils. The name comes from *pedo*, Greek for soil, and the chemical symbols for

5.24 Bricks carved from laterite were used to build ancient Cambodian temples. Bayon temple, Angkor, 1960 (United Nations)

aluminum (Al) and iron (Fe), the two most characteristic components. (Soil scientists do not seem to be troubled by the fact that iron oxides and aluminum oxides are also the main ingredients of latosols; one would never guess this from the name.) In a pedalfer soil the dominant movement of water is downward. The water leaches away calcium carbonate and other easily soluble materials from the A zone. Pedalfer soils tend to be sandy in the A zone and clay-rich in the B zone, because clay formed by decomposition of feldspars trickles downward from the A zone and becomes concentrated in the B zone.

Pedalfer soils predominate in the eastern half of the United States, where rainfall exceeds about 60 cm per year. Based on vegetation, pedalfer soils are divided into three varieties: (1) conifer soils, (2) leafy-tree soils, and (3) prairie (or grassland) soils.

Pedocals In areas having progressively less rainfall than in the humid-temperate regions, the balance shifts between downward percolation and upward movement of water as a result of evaporation. So far we have been discussing soils in which the dominant movement of interstitial water was downward. Therefore, soluble ions were removed from the A zone and deposited in the B zone. When, in response to intense evaporation, the water-movement balance shifts in favor of upward movement, any soluble constituents that start downward can be drawn back upward again and, when the interstitial water evaporates, can be precipitated at or near the surface of the soil. As a result, calcium, though soluble, tends not to be removed from the A zone. Because of the presence of calcium the name for these soils is **pedocal** (PE-doh-cal; from the Greek root for soil and cal, short for calcium). A pedocal soil is one in which calcium has been precipitated as calcite in the A zone. In the United States pedocal soils are located in the western part, where annual rainfall is less than 60 cm. Again, as with pedalfers, we can use vegetation present to divide pedocals into three varieties: (1) grassland soils, (2) forest soils, and (3) soils of sparse vegetation (desert soils).

In soils where evaporation is responsible for appreciable upward movement of interstitial water, evaporite minerals are deposited within the soil and at its surface. Crusts of calcite known as **caliche** are found in semi-arid regions. In still drier climates the entire surface may become cemented with various evaporite minerals, including silica, dolomite, gypsum, and halite.

Other soils A few other specialized soils should be added to our list of examples. These include *bog soils*, which consist of partly decomposed plant matter that becomes peat. Such soils are common in the roofing over by plants of small ponds and lakes in low-lying sites of North America and Europe. *Tundra soils*, widespread in the Arctic, consist of sandy clay and raw humus. Many tundra soils are underlain by permanently frozen ground, or *permafrost*. (See Figure 5.4.)

Management of Soils

Granting the constant pressure of an ever-increasing population, much effort has been given, and will continue to be given, to the possibility of expanding the area of land available for agriculture. In this connection, desert areas and the humid tropics are repeatedly considered for agricultural development. The main difficulty in dry regions is evaporation. Even where water is available for irrigation, its incessant evaporation through the soil will eventually cause precipitation of evaporite minerals. Optimistic schemes for extensive agricultural development of such sites as the South American rainforest generally do not take account of the fact that in the tropics leaching is so intense that latosols cannot support agriculture longer than a few years. Because the soil is such a precious asset for any country, great efforts have been expended to study the behavior of the various kinds of soils and to adopt agricultural practices that do not cause depletion or erosion of the soil.

Ancient Soils

As we have seen, both weathering and soil formation depend upon climate. Therefore, an ancient soil buried in the rock record contains clues to the nature of ancient climates. Such buried soils have permitted geologists to infer, for example, that cold periods of glacial advance have alternated with mild interglacial periods. Soils that formed during cold periods differ from those that developed during interglacials. Other ancient climates did not involve glaciers. From the presence of ancient laterites in Ireland, we can conclude that millions of years ago Ireland enjoyed balmy tropical temperatures. Other evidence from ancient soils indicates that England was largely a desert land during the Triassic Period (about 200 million years B.P.) and South Africa had glaciers in the late Paleozoic Era (about 300 million years B.P.). (B.P. stands for *before present*.)

Chapter Review

1 *Weathering* is a general term that describes all the changes in rock materials that take place as a result of their exposure to the atmosphere. Weathering is a part of a larger geologic activity known as *erosion*.

2 Weathering is important both in geologic theory and in many ways that affect our everyday lives. Weathering offers visible proof that the geologic cycle is operating today; weathering illustrates how geologic materials and certain kinds of energy react to bring the materials into adjustment with the *environment of weathering;* weathering creates important products including regolith and soil.

3. Water is a more important factor in weathering than is any other single substance or process. Water is the only compound that occurs naturally at the Earth's surface as a solid, liquid, and gas; water is highly active chemically.

4. *Climate* can be defined as the time-averaged weather in a given region. The environment of weathering depends intimately upon climate, particularly upon temperature, precipitation, and evaporation.

5. The *slope situation* refers to the height of a slope, its shape, and its direction of inclination with respect to sunlight and to rain-bearing winds. Elevated rocks are subject to more intense effects of climate than are low-lying rocks.

6. The environment of weathering is located in the pore spaces in bedrock and regolith. In bedrock, this means the spaces between crystals or in cracks in the bodies of rock. In regolith pore space exists among the framework particles. Bacteria and lichens create organic acids that can break down silicate minerals.

7. Organisms affect weathering in many ways. Plants drop organic litter onto the ground, and various wild animals and insects rearrange the regolith by burrowing into it.

8. The processes of weathering include *physical weathering*, or disintegration, *chemical weathering*, or decomposition, and *complex weathering*, which consists of the crystallization of salts, the combined effects of chemical and physical weathering, and the effects of organisms.

9. The rates of weathering depend upon the interplay among the susceptibilities to weathering of the various minerals, the amount of surface area initially available, and the intensity of the weathering processes. The rate of chemical weathering can be increased greatly in areas where the air has been polluted.

10. Neither plant nor animal communities could exist unless the Earth's bedrock could be converted into regolith and the regolith, in turn, could be converted to soil. The upper part of the regolith that is capable of supporting the growth of rooted plants is *soil*.

11. Soils consist of solids, liquids, and gases, all of which must be in proper balance to support plant growth. Soil is alive with complex chemical, physical, and biological activities.

12. Regolith can be divided into two major categories: *residual regolith*, formed by the deep weathering of the underlying rock, and *transported regolith*, which has come from someplace else and may be altogether unlike the underlying bedrock.

13. The conversion of the upper part of the regolith into soil creates layers, which are characterized by distinctive textures, colors, and compositions. The major layers are named *soil zones*, and are lettered from the top down A, B, and C according to the effects of migration of soluble materials during chemical weathering.

Weathering and Soils

Questions

1. Why is weathering important? What is the significance of weathering in the geologic cycle?
2. Explain two ways in which the structure of water molecules influences weathering processes.
3. What is meant by *environment of weathering*? How do conditions in the environment of weathering compare with those in the environment where igneous rocks form?
4. When "it rains" what kinds of materials are likely to accompany the water?
5. How could you convince your roommate who has not taken a geology course that weathering is a reality?
6. The three chief processes of weathering are _____, _____, and _____.
7. What is the most important process in the weathering of silicate minerals? When a granite is weathered what products are formed? What happens to these products in the geologic cycle?
8. Compare and contrast the weathering of a limestone in a humid climate with that in a dry climate. What landscape changes might take place after a long time in a region underlain by layers of limestone and other kinds of rocks if the climate changed from moist to arid?
9. How does air pollution affect weathering? Can you find any examples where you live to illustrate your answer?
10. Define *soil*. How are soils related to *regolith*? How is regolith related to *bedrock*?
11. What is a *soil profile*?
12. How do plants, climates, and soils interact?
13. What is the significance of an ancient soil buried in the geologic record?

Suggested Readings

Basile, R. M., *A Geography of Soils*. Dubuque, Iowa: W. C. Brown, 1971.

Bloom, A. L., *The Surface of the Earth*. Englewood Cliffs, N.J.: Prentice-Hall, Inc., 1969.

Bridges, E. M., *World Soils*. New York: Cambridge University Press, 1970.

Carroll, Dorothy, *Rock Weathering*. New York: Plenum Press, 1970.

Hunt, C. B., *The Geology of Soils: Their Evolution, Classification and Uses*. San Francisco: W. H. Freeman and Company, 1972.

Kellogg, Charles E., "Soil." *Scientific American*, July 1950. (Offprint No. 821. San Francisco: W. H. Freeman and Company.)

Shimer, John, *This Sculptured Earth: The Landscape of America*. New York: Columbia University Press, 1959.

Winkler, E. M., *Stone: Properties, Durability in Man's Environment*. New York-Wien: Springer-Verlag, 1973.

"Under conditions of metamorphism, rocks that are otherwise rigid can be forced to behave like silly putty."

Chapter Six
Sedimentary Rocks and Metamorphic Rocks

Sedimentary Rocks and Metamorphic Rocks

At this point, turn back to page 10 and look closely at the rock-cycle diagram. You will see that we have already discussed igneous rocks and some aspects of weathering, which convert rocks to sediments. But we have not yet explored the entire left-hand section of the diagram, which shows sediments becoming sedimentary rocks and sedimentary rocks becoming metamorphic rocks.

To a geologist, the rock cycle is the very foundation of the entire subject of geology. But, by contrast, to someone just starting the study of geology, it may be difficult to relate the rock cycle to things that are important to you. However, consider the following simple facts: (1) most of the energy sources we have used in the past and will continue to use in the immediate future come from the products of sediments and sedimentary rocks, and (2) metamorphic rocks being formed from sedimentary rocks are constantly creating new foundation rocks for continents. In this way, metamorphic rocks keep the continents from being worn away by erosion.

In this chapter, we will isolate four broad sections of the rock cycle in order to complete our tour around the cycle. The main sections will be (1) sediments and their conversion to sedimentary rocks, (2) a description of how sedimentary rocks are classified, (3) a description of how metamorphic rocks are classified, and (4) a discussion of surface and subsurface environments to show how sedimentary materials respond to higher and higher temperatures up to the point of melting. Once a rock melts it becomes magma, and the complete cycle may start again.

Sediments and Their Conversion to Sedimentary Rocks

Sediment is broadly defined as regolith that has been transported at the surface of the Earth and deposited in low places as layers (strata). In this section we will examine the particles that make up sediment, discuss their conversion to sedimentary rocks, and look at some features of strata which are characteristic of both sediments and sedimentary rocks. We will start by discussing sedimentary particles.

Sedimentary Particles

In the simplest case, sedimentary particles consist of solids—they are transported regolith which was derived from solid bedrock (which may have consisted of any one of the three great groups of rocks: igneous, metamorphic, and sedimentary). As long as a particle remains a solid it carries with it some information about where it came from, and perhaps something about what happened to it during its travels. As we recall from our discussion of igneous rocks, once a rock melts it becomes magma, and all previous history is erased. When it cools, it crystallizes, and once solid it starts a new history all over again. All particles of igneous rocks are alike in this respect.

Sedimentary rocks would be simple to understand if all their particles were derived from a single group, as with igneous rocks. However, sedimentary rocks are complicated because they contain four major groups of particles: (1) solid particles eroded from bedrock and transported as solids to a basin of deposition, (2) solid particles whose "raw materials" traveled as ions in solution, but which became solid particles in the waters of the basin of deposition, either by chemical or biochemical activities, (3) solid particles of volcanic origin, and (4) particles and materials formed from the tissues of plants and animals. The particles in each of these groups are distinctive in appearance and in significance.

6.1 **Large rock fragment** still displaying clearly the characteristics of the source rock (a coarse-grained intrusive igneous rock cutting across a fine-grained metamorphic rock). (J. E. Sanders)

Relationship of sedimentary particles to bedrock Bedrock becomes sediment as particles that survive weathering, known as **survival products,** or by special means that bypass weathering. Weathering creates three kinds of solids, plus ions in solution. The solids are: (1) rock fragments, (2) resistant minerals, such as quartz, which survive weathering, and (3) secondary alteration products, chiefly clay minerals, created during the decomposition of feldspars.

Description of Sedimentary Particles

Detritus Many of the particles derived from the breaking up of bedrock were transported and deposited physically as solids of varying sizes. All these particles are collectively designated **detritus** (deh-TRY-tus). Detritus is easy to recognize. It consists of one or both of two kinds of materials: (1) *rock fragments*, that is, particles which are large enough to retain recognizable minerals and textural characteristics of the original bedrock (Figure 6.1), and (2) particles that contain only one mineral; most of these are quartz, and include other rock-forming silicate minerals. Quartz is so abundant that it serves as an "index mineral" for recognizing detritus (Figure 6.2).

All the kinds of sedimentary detritus have managed to escape being destroyed by chemical weathering.

A different kind of detritus, also solid and derived from bedrock, includes the secondary alteration products that formed during decomposition. Chief among these are the clay minerals that formed by the

6.2. **Quartz particles** in sand-size sediment, view through microscope. (B. M. Shaub.)

198
Sedimentary Rocks and Metamorphic Rocks

hydrolysis of feldspars. Because feldspars are so abundant, this type of detritus is more prominent than all the other kinds (Figure 6.3). Other alteration products include chiefly iron oxides.

Particles derived from ions carried in solution The chief ions in solution are calcium, sodium, potassium, and magnesium. Because these ions have been completely separated from their bedrock sources they no longer retain any clues to where they came from. Thus, the ions which traveled in solution to the basin where they were deposited are not regarded as detritus. Instead, these ions form particles that contain information about how they came out of solution and got back into solid form once again. In the waters of the basin of deposition, the two chief processes of the return of dissolved ions to the solid state are: (1) secretion of **skeletal particles** in the hard parts of organisms (fossils), and (2) crystallization which took place when the water containing these ions was evaporated. Particles that crystallize as the result of evaporation are known as **evaporites.**

Skeletal particles are mostly *calcareous* (composed of calcium carbonate in the form of calcite or of aragonite), *siliceous* (composed of silica, SiO_2), or *phosphatic* (composed of calcium-phosphate compounds). Calcareous skeletal particles include whole or broken shells,

6.3 Clay minerals, enlarged 10,000 times as seen through an electron microscope. (Technische en Fysische Dienst voor de Landbouw, Wageningen, The Netherlands)

6.4 Various sedimentary particles. (a) Calcareous skeletal material secreted by organisms that live attached to rocks near shore, washed up on the beach at Gardiners Island, New York. (J. E. Sanders) (b) Pleistocene diatoms, New Zealand. Width of square specimen is $1/10$ mm. (U.S. National Museum, Smithsonian Institution)

(a) (b)

 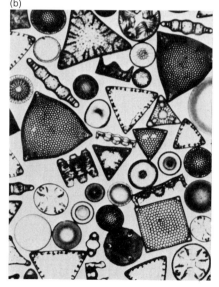

bones, plates, spines, and other parts of organisms (Figure 6.4a). All were secreted within the soft tissues of organisms, either as external housing or as some kind of internal support.

Siliceous skeleton makers include single-celled marine and nonmarine plants, the *diatoms* (Figure 6.4b), and various single-celled marine organisms, the *radiolaria*, and more complex marine organisms, the sponges. (Some sponges secrete calcareous materials; others secrete complex materials.)

Fish scales, the shells secreted by a few kinds of marine invertebrate organisms, and peculiar jaw-like remains known as *conodonts* consist of phosphatic material. The teeth of many organisms are composed of tough, complex material that includes much calcium.

Evaporite particles consist of well-formed crystals (Figure 6.5, top) or of small spherical particles known as *oöids* (OH-oh-ids) (Figure 6.5, bottom).

Organic material Some sedimentary particles and other sedimentary materials are the work of plants or animals; these are strictly organic in the sense that they consist chiefly of the elements carbon, hydrogen, and oxygen. The particles include the woody tissue of plants, such as tree trunks, branches, twigs, and leaves (Figure 6.6, left). Other particles derived from plants include spores and pollen. These tiny particles are spread through the atmosphere and are so resistant to chemical alteration that they persist in sediments (Figure 6.6, right).

The fatty and waxy tissues of woody land plants, and the soft tissues of aquatic plants, such as algae, and of animals of all kinds, do not form particles. Nevertheless, such soft "organic matter" is closely associated with modern sediments and is considered to be the ultimate raw material from which petroleum forms underground. We

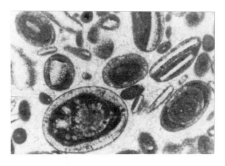

6.5 Two kinds of precipitates that are deposited from salt water that has been concentrated as a result of evaporation. (Top) Tiny cubic crystals of halite that grew along a twig. (B. C. Schreiber). (Bottom) Oöids composed of calcium carbonate. (G. M. Friedman)

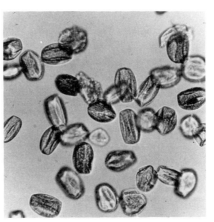

6.6 Organic particles formed by plants. (Left) A leaf. (U.S. Geological Survey). (Right) Tiny pollen grains, 0.01 mm long, seen through a microscope. (Atmospheric Research Center, State University of New York at Albany, courtesy of R. J. Cheng)

200
Sedimentary Rocks and Metamorphic Rocks

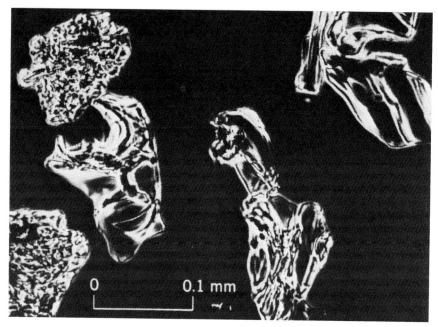

6.7 Volcanic particles, represented by tiny fragments of glass. Debris exploded from Mount Mazama, Oregon, and collected in sediments underlying Creston Bog, Washington. (R. B. Taylor, U. S. Geological Survey)

shall be discussing organic matter further in a later section of this chapter and again in Chapter 16.

Volcanic particles The materials ejected from volcanoes consist of tiny bits of glass, of crystals, or of volcanic rock fragments (Figure 6.7). Volcanic particles join the sedimentary materials in a basin of deposition either straightaway from the volcano (in the form of pyroclastic debris) or by being eroded and transported at a later time.

After this summary of sedimentary particles let us turn to the conversion of sediments to sedimentary rocks, or lithification.

Lithification

After sediment has been transported and deposited, one step remains in the production of rock. The over-all term for all processes of converting loose sediment into a coherent aggregate, or rock, is **lithification** (from the Greek *lithos*, stone). The chief processes of lithification are cementation, recrystallization, and replacement. Compaction of clay minerals is a minor factor of lithification.

Cementation The key process in the production of a sedimentary rock is **cementation,** the crystallization of minerals from solution in the pore spaces of the framework particles. The **framework particles**

are those that are of sand size or coarser. The minerals that form as cements may or may not be the same as those of the framework particles. The most common cements are calcium carbonate and silica. Others are iron carbonate (siderite) and iron oxide (hematite). These cements may eventually make up a large part of the rock itself. Hematite, in particular, may fill sandstone to such an extent that it is mined as iron ore. Fine particles which become cemented among coarser particles are called a **matrix**.

Recrystallization and replacement As water circulates slowly through the remaining pore spaces of compacted sediment, a variety of chemical reactions may occur. Minerals are easily dissolved, transported, and precipitated elsewhere to form new crystals. The particles of limestone may grow larger as calcium carbonate from muds or shells dissolves and recrystallizes.

Having now seen how sediments become sedimentary rocks, let us examine a feature common to both sediments and sedimentary rocks, strata.

Sedimentary Strata

In the Introduction we mentioned that an outstanding feature of both sediments and sedimentary rocks is that they form layers, known as **strata** (singular, *stratum*). The individual strata can be distinguished easily because they differ in some visual feature, such as color, thickness, degree of cementation, size of particles, or mineral composition. Some sedimentary strata are 30 meters or more thick (Figure 6.8). Others are paper thin. The thick layers are sometimes called

6.8 Thick stratum of sandstone (left) extends from bushes along bank of Green River to top of cliff. Dinosaur National Monument, Utah. (Dick Frear, U.S. National Park Service). (Right) **Millimeter-thin laminae** of calcite and gypsum, Castile Formation, west Texas. (J. E. Sanders)

beds and the thin layers (thickness less than 1 cm) are designated as **laminae** (singular, *lamina*). The attribute of sediments and sedimentary rocks in which they are arranged in layers is called **stratification.**

Origin of strata Strata exist because something changed. If there are no changes in color, in size of particles, or in some other features, a single stratum may become so thick that it might not appear to be stratified (Figure 6.8). Within a stratum the sizes of the particles may be uniform throughout. Another possibility is that a systematic gradation in size of particles exists, with the larger particles at the bottom and progressively smaller particles higher up. A bed having such a gradient in the sizes of its particles is a **graded layer** (Figure 6.9). A graded layer forms when a moving current carrying particles of many sizes begins to drop the largest ones first, and as the speed of the current decreases, deposits successively finer particles.

Strata accumulate and build sequences that are parallel to the surface on which they were spread out. In most cases this surface is almost perfectly horizontal. Some sediment is transported along the bottom, and becomes deposited at the front of a growing embankment that is advancing into deeper water. As the successive sheets of sandy sediment fall down the inclined front of the embankment they build it forward. The successive layers are added at the angle of slope of the front of the embankment (Figure 6.10). Similarly sandy bottoms are shaped by winds and water currents into various regularly spaced mounds (dunes are an example). As these migrate they deposit sediment in inclined layers.

6.9 Graded layer, containing pebbles at base and fine sand at top. Pliocene of Ventura Basin, exposed along Santa Paula Creek, California. (J. C. Crowell)

6.10 Cross-strata formed by water current transporting sand, and forming an embankment which grows forward into deeper water. Schematic sectional view.

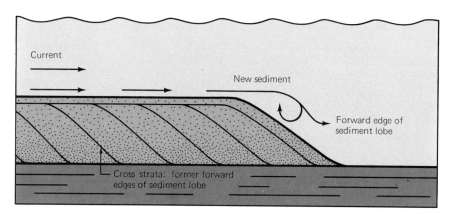

Sediments and Their Conversion to Sedimentary Rocks

6.11 **Cross strata in modern sediments and ancient sedimentary rocks.** (Top) Cross strata forming where sand at front of large dune is deposited at angle of repose. Upper Medano Creek, Great Sand Dunes National Monument, Colorado. (Bob Hangen, U. S. National Park Service, 1961). (Bottom) Cross strata of Navajo Sandstone (Jurassic) seen in section perpendicular to stratification. Zion National Park, Utah. (Edith Vincent, September 1968; courtesy of photographer and with permission of Elsevier Publishing Company)

Sloping layers of the embankments or mounds become sandwiched between originally horizontal layers formed where a flat bottom built upward. The inclined layers are named **cross strata;** they are common features of both recent sediments and ancient sedimentary rocks (Figure 6.11).

An important and fundamental principle applies to all cross-strata: they are inclined in the direction toward which the current was flowing. Recording and mapping of the directions of inclination of cross strata are important in the analysis of sedimentary formations.

Surface features of strata The surfaces of modern sediments contain many features that are identical to those found on the bedding surfaces of ancient sedimentary rocks. On the tops of many strata are

6.12 Features of bedding surfaces of sediments and sedimentary rocks. (a) Ripples created by waves in shallow water (G. M. Friedman) (b) Ripple marks in ancient sandstone; Devonian (G. M. Friedman) (c) Cracks formed by drying out of mud (FAO photo by H. Null) (d) Footprint of dinosaur in ancient sandstone of Triassic-Jurassic age. (J. E. Sanders) (e) Bird tracks and snail trails in modern sediments. (G. M. Friedman)

ripple-marks, polygonal cracks made by the shrinkage of fine-grained sediment that was dried out by the Sun, footprints of various walking animals, and the tracks and trails of crawling animals (Figure 6.12).

How Sedimentary Rocks Are Classified

In this section we discuss the basis for classifying and naming sedimentary rocks. Photographs of the most common kinds of sedimentary rocks within each of the major groups are found in Appendix C.

Basis for Classification

As with igneous rocks, the fundamental properties used for classifying and naming sedimentary rocks are mineral composition and texture. These properties reflect the origin of the materials and the processes to which they have been subjected. Recall that with igneous rocks the mineral composition told us about the magma and the lava, and the texture enabled us to determine the cooling history of the rock.

In sedimentary rocks, the composition of the particles tells us where the particles came from, and their texture contains clues about the sedimentary processes that deposited them.

Mineral composition The mineral composition of sedimentary rocks is determined by the output of weathering processes. Weathering yields three types of materials that may become sedimentary rocks: (1) *survival products*, minerals and rock fragments that were not altered chemically during weathering; (2) *alteration products*, minerals formed by the chemical alteration of other minerals, usually feldspars; (3) *dissolved products* of weathering that reach their sites of deposition in solution and there are converted back to solid material and deposited as sediment.

The first two groups of materials are transported and deposited in solid form by some "mechanical" process, such as running water or wind. Materials of the third group are not transported mechanically. Instead, they are changed from dissolved to solid form in the environment of deposition. This change is usually performed by plants and animals, which absorb ions from solution to build shells and skeletons of minerals, such as calcite. Less commonly, ions are converted to solids by evaporation or by precipitation.

Textures of sedimentary rocks Sedimentary rocks display two major kinds of textures: (1) clastic, and (2) nonclastic, which includes crystalline and organic texture. Each results from a distinctive process.

Sedimentary rocks having **clastic textures** (clastic means broken) can be identified by their broken and worn appearance (Figure 6.13). When rocks having clastic textures break, the fracture surface

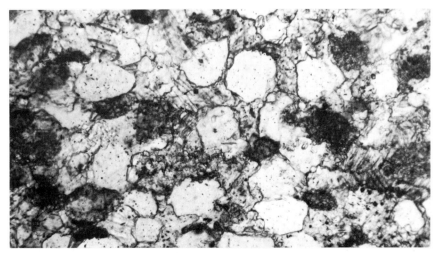

6.13 Clastic texture of sedimentary rock seen in view through polarizing microscope of paper-thin slice of sandstone. White particles with jagged edges (from corrosive effects of solutions that deposited the cement) are quartz about 0.5 mm in diameter. Cement is mostly calcite (light gray area with darker parallel lines). Mississippian, northeast Tennessee. (J. E. Sanders)

206
Sedimentary Rocks and Metamorphic Rocks

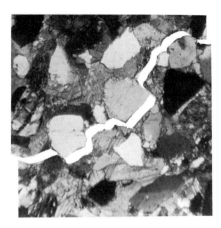

6.14 Breaking of sandstone around particles illustrated by cutting out photograph made of view through polarizing microscope of paper-thin slice of sandstone. (J. E. Sanders)

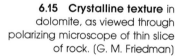

6.15 Crystalline texture in dolomite, as viewed through polarizing microscope of thin slice of rock. (G. M. Friedman)

and the mineral cement (Figure 6.14). On a freshly broken surface, therefore, a look through a hand lens will show particles standing in relief. This relief causes the broken surface to feel rough; if the particles are of sand size, the surface feels like sandpaper.

Sedimentary rocks displaying clastic textures are products of the cementation of pre-existing sediments. Three important aspects of the particles in both sediments and clastic sedimentary rocks derived from them are: (1) the sizes of the individual sedimentary particles, (2) the uniformity of sizes of aggregates of sedimentary particles, and (3) the shapes of the particles.

Sedimentary rocks with **nonclastic textures** may look more like igneous rocks than sedimentary rocks because their minerals do not appear broken and worn, and the particles are all interlocked and closely grown.

Rocks with *crystalline textures* display tightly joined crystals (Figure 6.15); the textures of such rocks resemble those of igneous rocks but can be distinguished from igneous rocks by their distinctive sedimentary minerals, such as calcite, halite, and gypsum. When rocks having crystalline textures break, the fracture passes randomly across the particles. The broken surface typically consists of many tiny segments that are cleavage planes of the individual minerals in the rock. These planes reflect light and are clearly seen with a hand lens. To the touch a broken surface of a sedimentary rock having crystalline texture feels like an igneous rock. The surface is rough, but not as abrasive as the broken surfaces of sedimentary rocks having clastic textures.

Sedimentary rocks having *organic textures* are easily identified by the characteristic appearances of the remains of shells, skeletons, or other structures secreted by animals or plants. These remains are most obvious on broken surfaces. Such surfaces may feel smooth or highly irregular, depending on the sizes of the organic structures. The pattern revealed along some broken surfaces may resemble the inside of an apartment building with the outside walls removed; see especially the right side of Figure 6.16.

Sedimentary Rocks Formed by Cementation of Sediments

Three chief groups of sedimentary rocks are formed by the cementation of sediments. These are: (1) detrital group, (2) volcanic group, and (3) carbonate group. Within each group the main kinds of rocks are recognized by sizes and shapes of framework particles.

6.16 Organic texture shown by broken piece of coral in limestone of Pleistocene age. (G. M. Friedman)

Rock Formations: The Missouri
Prince Maxmilian

At break of day the weather was extremely cool and disagreeable; the thermometer at half-past seven was only at 58°, and a bleak wind prevailed, which enabled us to use our sails. The part of the country called The Stone Walls, which now opened before us, has nothing like it on the whole course of the Missouri, and we did not leave the deck for a single moment the whole forenoon. Lewis and Clarke have given a short description of this remarkable tract, without, however, knowing the name of Stone Walls, which has since been given it. In this tract of twelve or fifteen miles, the valley of the Missouri has naked, moderately high mountains, rounded above, or extending like ridges, with tufts of low plants here and there, on which the thick strata of whitish coarse-grained friable sandstone, which extends over all this country, are everywhere visible. As soon as we have passed Judith River this white sand-stone begins to stand out in some places, till we have passed Bighorn River, and entered the narrower valley of the Stone Walls, where the strata extend, without interruption, far through the country, and lie partly halfway up the mountain, and partly form the summits. They are the continuation of the white sand-stone which occurs in such singular forms at the Blackhills. At all the places which are bare of grass, they are visible, and there we see horizontal or perpendicular angles and ledges resembling walls, some of which contain caverns. This sandstone formation is the most striking when it forms the tops of more isolated mountains, separated by gentle valleys and ravines. Here, on both sides of the river, the most strange forms are seen, and you may fancy that you see colonnades, small round pillars with large globes or a flat slab at the top, little towers, pulpits, organs with their pipes, old ruins, fortresses, castles, churches, with pointed towers, etc., etc., almost every mountain bearing on its summit some similar structure.

Towards nine o'clock the valley began to be particularly interesting, for its fantastic forms were more and more numerous; every moment, as we proceeded along, new white fairy-like castles appeared, and a painter who had leisure might fill whole volumes with these original landscapes. As proofs of this we may refer to some of these figures, which Mr. Bodmer sketched very accurately. In many places the clay formed the summits of the hills; in these parts there were patches of *Juniperus repens*, and on the bank of the river, small and narrow strips covered with artemisia and the thorn with flesh-coloured leaves (*Sarcobatus nees.*). Long tracts of the sand-stone strata perfectly resembled a large blown-up fortress, because the stratification everywhere gave these walls a certain regularity, while, at the same time, they bore marks of having been destroyed by violence. In several places where the sandstone summit appeared plainly to represent an ancient knight's castle, another remarkable rock was seen to traverse the mountain in narrow perpendicular strata, like regularly built walls. These walls consist of a blackish-brown rock, in the mass of which large olive-green crystals are disseminated. They run in a perfectly straight line from the summits of the mountain to the foot, appearing to form the outworks of the old castles. The surface is divided by rents or furrows into pretty regular cubic figures like bricks, which renders their similarity to a work of art still more complete. The breadth of these perpendicular strata seldom exceeds one or two feet. One of these walls was particularly striking, which ran, without interruption, over the tops of three mountains, and through the clefts between them, and connected the three masses of white sand-stone on the summits in so regular a manner, that one could hardly fancy they were natural, but that they were a work of art. All these eminences are inhabited by numerous troops of the wild mountain sheep, of which we often saw thirty or fifty at a time climbing and springing over the sand-stone formation.

Early in the afternoon we came to a remarkable place where the Missouri seems to issue from a narrow opening, making a turn round a dark brown rugged painted towerlike rock on the south, to which the traders have given the name of the Citadel Rock. This singular isolated rock seems to consist of clayslate, grauwacke, and a conglomerate of fragments or rock in

From TRAVELS IN THE INTERIOR OF NORTH AMERICA, 1839

Field Trip

yellowish clay, and is joined to the south bank by a ridge. On the bank opposite to it the white sand-stone runs over the ridge of the hills, which Mr. Bodmer has very accurately represented. After we had doubled the Citadel Rock we lay to on the south bank, and our people took their dinner. A herd of wild sheep looked down upon us from these heights. We had, however, not yet taken leave of the extraordinary sand-stone valley, on the contrary, we now came to a most remarkable place. The stratum of sand-stone, regularly bedded in low hills, runs along both banks of the river, which is rather narrow, like a high, smooth, white wall, pretty equally horizontal above, with low pinnacles on the top. At some distance before us, the eye fell on an apparently narrow gate, the white walls in the two banks approaching so near to each other, that

the river seemed to be very contracted in breadth as it passed between them, and this illusion was heightened by the turn which the Missouri makes in this place to the south-west. Looking backwards, the high, black, conical rock rose above the surrounding country; and on our right hand, there were, on the bank, dark perpendicular walls, seemingly divided into cubes, in the form of an ancient Gothic chapel with a chimney. Some pines grew singly about these walls, where there appear to be regular gateways formed by art. A little further on there was, on the north bank, a mass which much resembled a long barrack or some other considerable building, the corners of which were as regular as if they had been hewn and built up by a skilful workman. Beyond the rocky gate a herd of buffaloes were grazing on a small lateral valley; our hunters contrived to get near them and to kill four. As evening was come, and the people had to cut up the buffaloes, we lay to for the night on the north bank. I took this opportunity to ascend the remarkable eminences. I found the sand-stone so soft that it crumbled in my hand; whereas the yellowish-red sand-stone, which, in some places, formed the tops of roofs of the strange white masses, were of a rather harder grain. Extremely stunted and often strangely contorted cedars (*Juniperus*) grew among these rocks; but the pines (*Pinus flexilis*) were well grown and flourishing, though not above forty feet high. When standing among the remarkable masses of the sand-stone, we fancied ourselves in a garden laid out in the old French style, where urns, obelisks, statues, as well as hedges and trees clipped into various shapes, surround the astonished spectator. The balls and slabs, often of a colossal size, which rested on the above-mentioned pedestals, were likewise soft and friable, but not so much as the white sand-stone, and there were in them many round holes. Stratification could be perceived in all these stones, for even round spherical blocks were easily divided into regular plates, nearly an inch thick.

We looked with impatience for the following day, the 7th, in order to reach what is called the Gate of the Stone Walls. We soon came to a dark brown rock, like a tower, rising in the middle of the white wall, the front of which had fallen down, and had a great number of boulders about it. From this tower it is between 600 and 800 paces to the place which appeared to us yesterday to form a narrow gate; before reaching it, there is, on the north bank, a stream called, by Lewis and Clark, Stonewall Creek, which is about fifty paces broad at the mouth, and its banks are bordered with high poplars. A cold wind blew from the gate, beyond which there was another tower-like dark brown rock, not so large as the other, while the white sand-stone walls decreased and became less regular. The hills became gradually lower, the sand-stone partly disappeared, and was only seen occasionally.   1839

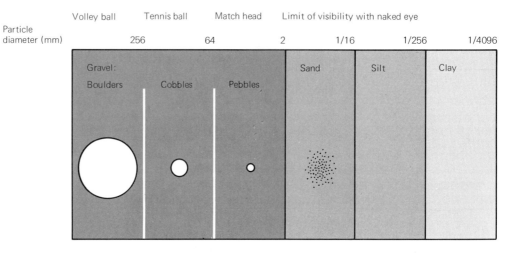

6.17 Sizes of sedimentary particles, based on standard scale, with comparison of familiar objects (at top).

The *sizes of clastic sedimentary particles* are grouped and named on the basis of a standard scale, using as diagnostic measurement the "diameter" of the particles. Obviously, unless the particles are spheres, they do not really have a "diameter." Instead, nonspherical particles possess three internal dimensions of length, width, and breadth. Unless otherwise specified, the general term "diameter" as applied to sedimentary particles designates the largest of these three internal dimensions. The main groups of sedimentary particles based on size are gravel, sand, silt, and clay (Figure 6.17).

The degree of uniformity of sizes of aggregates of sedimentary particles is known as **sorting**. The single most important aspect of sorting is based on what happened to the fine particles, the silt and clay. These particles are readily transported by currents of water or the wind. Hence, moving water or air nearly always separates these fine particles from the sand and larger sizes. Accordingly, we will consider that sediments are **well sorted** if the silt and clay have been separated from coarser material, and **poorly sorted** if they have not.

Size is not the only basis of the sorting of sediment particles; *specific gravity* is another factor. Dense minerals sink to the bottom faster than less dense minerals. Therefore, small but dense minerals, such as magnetite, for example (density of 5.2 g/cm^3), tend to be deposited along with larger particles of quartz (density of 2.65 g/cm^3).

The *shapes of sedimentary particles* result from the initial shapes of the particles in their present rocks, and from the way these initial shapes are changed during transport. If a particle starts out with sharp edges, transport by water and wind can change its shape. This change

How Sedimentary Rocks Are Classified

6.18 Angular and rounded rock particles, Yellowstone National Park, Wyoming. (Top) Angular fragments in a fresh rock fall along gravel road. (Bottom) Rounded boulders in Tower Creek, viewed from bridge at Tower Falls. (Photos by U. S. National Park Service)

results from **abrasion,** the rounding and smoothing of particles as a result of collisions with other particles (Figure 6.18). The rate of abrasion depends both on the composition and on the size of the particles. Soft rocks, such as limestone (composed of calcite having a mineral hardness of 3), are smoothed quickly. Hard rocks (such as sandstone cemented by quartz) became rounded slowly. Large particles become rounded after only a few kilometers of transport in a stream. By contrast, small particles may never become rounded. A

6.19 Conglomerate containing rounded boulders. Roadcut on Route 128, south of Boston, Massachusetts. (J. E. Sanders)

single particle of silt may move all the way from the Rocky Mountains to the Gulf of Mexico without changing shape.

Detrital sedimentary rocks The four main detrital sedimentary rocks correspond to the four major classes of sediments. We merely list rock types here; further descriptions are located in Appendix C. Cemented gravels form **conglomerates** (Figure 6.19) and **sedimentary breccia** (Figure 6.20), cemented sand forms **sandstones, arkoses,** and **graywackes.** Lithified silt forms siltstone; and hardened clay, claystone. In addition, the fine-grained rock, **shale,** designates materials of silt-size and/or clay-size that have been changed as a result of deep burial (Figure 6.21).

6.20 Sedimentary breccia, showing angular particles. North Side of U.S. 1, East Haven, Connecticut. (J. E. Sanders)

6.21 In fissile shale the rock splits into thin chips. (J. E. Sanders)

Sedimentary rocks composed of volcanic material Rocks in this group are classified according to the sizes of their particles. The coarse-grained variety are called **agglomerates** and the fine-grained variety are called **tuffs.** The boundary between these two varieties is not drawn at 2 mm, as with detritus. Instead, the boundary is placed at 32 mm.

Carbonate rocks The rocks in this group include the most abundant kinds of limestones, namely those derived from the cementation of skeletal debris (Figure 6.22). These kinds of limestones closely correspond with the detrital group of sedimentary rocks. In fact, such limestones differ from detrital sedimentary rocks chiefly because the particles in limestone came from the waters of the depositional basin, whereas detritus came from outside the basin.

Sedimentary Rocks Not Formed by Cementation of Sediments

In this category are four kinds of sedimentary rocks: (1) "instant" limestones, (2) evaporites, (3) cherts, and (4) dolomites.

6.22 Cementation of calcareous skeletal materials forms limestone known as coquina. Pleistocene, eastern Florida. (U.S. National Museum, Smithsonian Institution)

"Instant" limestones A few limestones were never calcareous sediments. Instead, they were secreted by the corals in a reef, or by the escape of carbon dioxide from the waters of a cave or hot spring. As long as the carbon dioxide is held in solution, the calcium, which ultimately may make calcite, remains dissolved. When carbon dioxide escapes, calcite is crystallized as rocky deposits. These include the speleothems of caves and travertine around springs (Chapter 9).

Evaporites The two chief minerals in evaporite rocks are halite and gypsum. These evaporites crystallize when the waters of the sea or groundwater in arid regions evaporate. Because the evaporite minerals form tiny particles that sink to the bottom, these may be moved by currents before they are finally deposited.

Cherts A tough rock composed of silica is a **chert.** It may occur as isolated nodules or as continuous layers. The silica of many cherts came from the shells of siliceous organisms that were dissolved and reprecipitated. In some cases, this change has been so complete that no traces of the original shell materials remain. (See page 555, number 11.)

Dolomites The rock, **dolomite,** is defined as a carbonate rock containing more than 50 per cent of the mineral dolomite (a double carbonate of calcium and magnesium). Most dolomite rocks form by the replacement of calcium-carbonate sediments and limestones.

Metamorphic Rocks

In appearance and origin, metamorphic rocks are distinctive from igneous and sedimentary rocks. Commonly, metamorphic rocks display strange, twisted forms (Figure 6.23) whose origin is not fully known, and exotic-sounding minerals found nowhere else. Although metamorphic rocks are derived from ordinary rock-making minerals that were originally part of older rocks, complex processes have changed these minerals, often physically as well as chemically, into new and distinctive materials.

The Concept of Metamorphism

The concept of metamorphism was introduced in 1832 by the English geologist Sir Charles Lyell. He described the fundamental concept that igneous or sedimentary rocks can be converted, under the influence of temperature, pressure, and circulating fluids, to different kinds of rock. To explain this conversion, he used the word **metamorphism,** from the Greek "change of form."

6.23 Contorted layers in metamorphic rock, exposed in cuts along I-684, Bedford, New York. (Charles Haberman)

Today we define metamorphism as the mineralogical and structural adjustment of solid rocks in response to chemical and physical conditions intermediate between those that produce sedimentary rocks and those that produce igneous rocks. Under conditions of metamorphism, rocks that are otherwise rigid can be forced to behave like silly putty. Metamorphic changes create new rock textures. If the original rock was a sedimentary with clastic texture, its minerals can be crystallized so that the resulting metamorphic rock is crystalline. Similarly, a crystalline rock can be crushed into particles that are relithified into a rock with a clastic-like texture. **Metamorphic rocks** are rocks that have been formed by metamorphism.

Parent rocks Any pre-existing rock that has been changed into metamorphic rock is referred to as a **parent rock.** Here, as elsewhere in geology, gradations exist between parent rock and metamorphic rock. For example, limestones grade imperceptibly into marbles and mudstones grade into schists. Metamorphic rocks are divided into many suites, or groups, based upon the kind of parent material from which they were derived. Chemically there are six broad groups of parent rocks: (1) shales, (2) sandstones and cherts, (3) carbonate rocks, (4) coal, (5) mafic igneous rocks, including volcanics, and (6) felsic igneous rocks. Depending upon the conditions of metamorphism, the same parent material may produce different kinds of metamorphic rocks. If new minerals are introduced during metamorphism, the final product may vary even further.

Metamorphic change True metamorphism includes changes that affect rocks in the solid state. Some changes that affect igneous and sedimentary rocks are not considered to be processes of metamorphism. Surface changes, brought about by such agents as rainwater and the growth of salt crystals, are treated as weathering (Chapter 5). True metamorphism occurs below depths of about one kilometer. By contrast, some processes which take place beneath the surface, such as melting of rocks in zones of high temperatures, are igneous processes (Chapter 3). True metamorphism is a process intermediate between weathering and igneous activity (or melting).

The Basis of Classification of Metamorphic Rocks

All the rocks we have discussed have been classified according to their mineral composition and texture, and the same is true for metamorphic rocks, except that we add a different feature (foliation), which is characteristic of metamorphic rocks. In this section we will

216
Sedimentary Rocks and Metamorphic Rocks

talk about the principles of classifying metamorphic rocks, and we will be discussing the chief kinds of metamorphic rocks. Photographs of these rocks and discussion on how to recognize them are found in Appendix C.

Mineral composition The minerals of metamorphic rocks are determined by the composition of the parent rock and the intensity of the metamorphic processes. Some metamorphic rocks consist entirely of quartz. These clearly are metamorphosed sandstones that contained nothing but quartz to begin with. Other metamorphic minerals are aluminum-rich silicates, unlike anything we have seen so far. These rocks were formed by the metamorphism of shale, containing abundant clay minerals rich in aluminum.

Texture Textures of metamorphic rocks reflect two important metamorphic processes: (1) recrystallization, during which the particles enlarge and interlock because they are all enlarging simultaneously (Figure 6.24), and (2) smashing and grinding, during which solid particles are broken into smaller pieces. We use the two textures that result from these processes to recognize the two great subdivisions of metamorphic rocks, the crystalline metamorphic rocks and the cataclastic metamorphic rocks.

Presence or absence of foliation in small specimens Because of their derivation from sedimentary rocks or formation under unequal pressures within the Earth, many metamorphic rocks display a distinctive layered aspect known as **foliation** (Figure 6.25). Foliation is a parallel or nearly parallel fabric of the mineral in a metamorphic

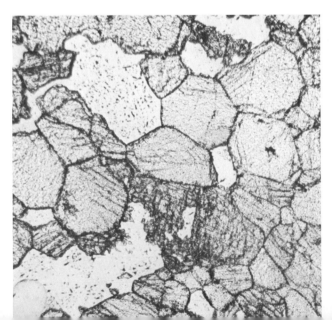
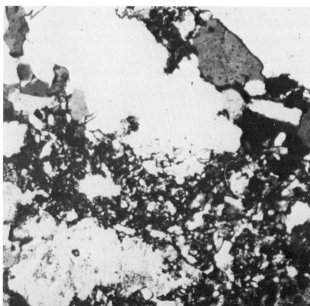

6.24 Textures of metamorphic rocks as seen in view through polarizing microscope of paper-thin slices. Pre-Triassic rocks, near Branford, Connecticut. (J. E. Sanders) (Left) Crystalline texture in form of a mosaic pattern created by growth of particles during metamorphism. (Right) Result of breakage of particles by intense grinding during metamorphism.

6.25 Foliation of metamorphic rocks seen on large scale in roadcut along Westchester Avenue, White Plains, New York. (J. E. Sanders)

rock. It is expressed as alternating layers of different minerals or parallel arrangements of the same mineral. Although almost all metamorphic rocks on a large scale are foliated, on the scale of hand specimens many are not foliated. Thus, we can use foliation as a basis for classifying metamorphic rocks as foliates or nonfoliates.

Accordingly, our classification of metamorphic rocks begins with two major groups mentioned previously, the crystalline and the cataclastic; within each group we make further subdivisions based on the presence or absence of foliation.

Metamorphic Rocks Having Crystalline Textures

The common metamorphic rocks having crystalline textures in the foliate group include slate, phyllite (FILL-ite), schist, and gneiss (NICE). The nonfoliates include marble, quartzite, and hornfels. The two other crystalline metamorphic rocks, amphibolites and greenstones, may be either foliated or nonfoliated.

Foliates Both the particle size of the rocks and the spacing of their parting surfaces (the surface along which a rock splits) are arranged in a gradational series that extends from slate to schist. The

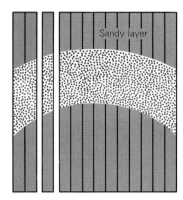
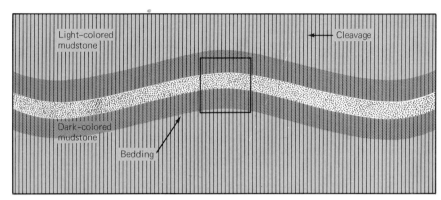
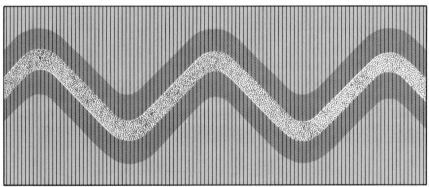

6.26 Closely spaced foliation in slates. Sketches showing relationship between strata (folded and apparent because of changes of color and particle size) and slaty cleavage (seen as parallel lines in edge view), which tends to be parallel to planes of symmetry of folds. The drawing in the left-hand margin is an enlargement of the rectangular section shown in the top drawing.

foliation of slates results from the compression and alignment of flake-like micas (Figure 6.26). When the micas have become aligned, the rock will split into paper-thin sheets along parallel surfaces. The breaking of slates along smooth surfaces is known as **slaty cleavage** (Figure 6.27). The breaks are so clean that slate is used widely to make roofing shingles, billiard tables, and blackboards. The natural parting surfaces parallel to the slaty cleavage are so smooth that they seem to have been polished.

The foliation of schist is determined by the principal schist minerals, which are micas, amphiboles, chlorite, and talc. In fact, schists display such a great variety that they are commonly identified by the dominant mineral, such as talc schist, or hornblende schist. A schist will break more readily and smoothly if the rock is rich in flaky min-

6.27 Slaty cleavage that is nearly horizontal and strata that are steeply inclined. Moe River, Quebec. (K. C. Bell, Geological Survey of Canada)

erals (talc, mica) which are tightly organized in parallel fashion. If platy minerals are few and not well aligned, foliation may be scarcely recognizable. Commonly, foliation surfaces are slightly to highly irregular; this causes the rock to appear wavy or wrinkled.

The foliation in gneiss is produced by alternating layers of different minerals. Because the foliation is a result of coarser crystals and fewer micas, gneisses tend to break less regularly than schists. Commonly, gneisses break completely across the layers. In the layers, light-colored minerals (usually quartz and feldspars) alternate with dark-colored minerals, such as hornblende and mica, producing the characteristic "zebra" pattern of some gneisses. Like some schists, gneisses display wierdly contorted patterns that reflect the tremendous forces by which they have been shaped (see Figure 6.23).

Nonfoliates The chief nonfoliates are marble, quartzite, and hornfels. Marbles are metamorphosed limestones, quartzites are metamorphosed sandstones, and hornfels designates a variety of fine-grained rocks that have been subjected to high temperatures. The principal difference between quartzite and sandstone is that sand-

Sedimentary Rocks and Metamorphic Rocks

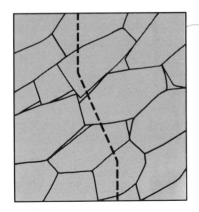

6.28 Quartzite breaks across the particles, with surface of breakage shown by dashed line. Compare with Figure 6.14.

stone fractures along the boundary between cement and quartz particles (see Figure 6.14). By contrast, quartzites are bonded so tightly that fracture takes place indiscriminately through quartz particles (Figure 6.28).

Amphibolites and greenstones are metamorphic rocks containing abundant mafic minerals, such as chlorite, hornblende, and epidote. Sometimes amphibolites and greenstones are foliated, and sometimes they are not.

Cataclastic Metamorphic Rocks

Cataclastic metamorphic rocks may be foliated or nonfoliated, and all are given the name **mylonite**. Mylonites are products of abrupt and powerful forces on dry, brittle near-surface rocks. The typical metamorphic change is straightforward crushing and pulverizing that is not complicated by chemical changes or recrystallization.

Surface and Subsurface Environments

The whole concept of metamorphism is that something has changed. The basis of this change is a response to an environment where materials are being put back together, instead of being broken apart as in the environment of weathering. In Chapter 5 we contrasted the environment of weathering and that of igneous activity. Now we must look at the environment between these two, the intermediate range between the surface domain of weathering and the deep subsurface domain of igneous activity. We will return to the near-surface environment of sedimentation in later chapters.

Conditions of Subsurface Environments

The temperature range between the surface and the subsurface environment can be as much as 1000°C. For most materials, temperature is the dominant factor in subsurface environmental changes. Other factors include pressure and salinity.

We have two sources of information about temperature gradients underground. One is deep borings and mine shafts, and the second is observation of contact zones of plutons. Bore holes and mines take us down to about 10,000 meters, and to temperatures up to 200°C. At these temperatures the only changes we can see involve the change of organic materials to the various grades of coal and petroleum. The contact zones surrounding plutons allow us to study the range of the country rock. In this range we can see evidence of changes involving not only organic materials but the less-reactive rock-forming silicate minerals.

We will examine the factors of subsurface environments, and go on to trace the changes these organic materials and silicate minerals undergo.

Pressure The subsurface environments exhibit increases of pressure downward. The downward increase in pressure results from the weight of overlying materials. This pressure depends on the densities of the materials involved. The range of densities lies between about 1.0 grams per cubic centimeter for water, and about 2.7 grams per cubic centimeter for dolomite. The pressure exerted by the weight of overlying water is **hydrostatic pressure.** The pressure exerted by the weight of overlying rock is **lithostatic pressure.**

Temperature Subsurface environment is also affected by the **geothermal gradient,** the regular increase of temperature toward the center of the Earth. The temperature that would be recorded in a bore hole starts at the top with the mean annual air temperature and increases steadily with the depth of the hole.

Salinity Deep borings made in search of petroleum show that underground fluids become progressively saltier with depth. A typical example of this change is shown in Figure 6.29. Notice that the salinity ranges in surface and subsurface environments are comparable. However, in the subsurface environment, the changes from any given depth to another are negligible.

The conditions of pressure, temperature, and salinity at any one depth level tend to remain constant. Conditions change only as the depth changes. A layer of sediment spread out at the surface of a subsiding area sinks, and another layer is spread on top of it. With continuing sinking, covering, and spreading of new layers the sediment progresses downward. As these materials descend they pass through specific depth levels which are important environmental boundaries. The reaction between these levels and materials is the critical factor that changes sediments to sedimentary rocks, and many sedimentary rocks to metamorphic rocks.

To see how this process of change works we must first become acquainted with various sedimentary materials, described in the next section.

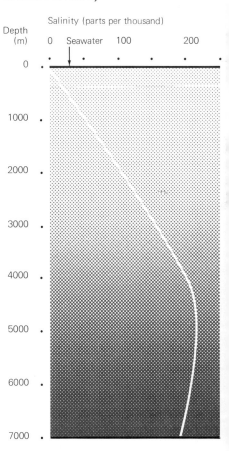

6.29 **Changes of salinity** of water at depth beneath the Earth's surface in the Gulf-coast region of Louisiana. (J. S. Bradley, 1975, American Association of Petroleum Geologists Bulletin, v. 59, no. 6)

Sedimentary Rocks and Metamorphic Rocks

Subsurface Changes in Organic Materials

At this point, we shall consider how organic materials are affected by the increased temperatures they encounter when being buried. We shall select two sample progressions. One starts with peat and ends with graphite. The second starts with organic matter and ends with graphite.

Peat to coal to graphite Figure 6.30 is a graph showing the effects of temperature, depth of burial, and time on plant material. At the surface in the upper left corner is peat. With exposure to progressively higher temperatures, the plant matter gives up its volatile constituents. At first these are carbon dioxide and water. Later, methane gas is given off. The driving off of these volatile constituents leaves behind more and more carbon, which is the fuel that is burned in the various coals. The next product after peat is lignite, and the end product is anthracite (Figure 6.31). Heating of anthracite yields graphite.

Lignites are highly immature, brownish coals, containing 30 to 40 per cent water, and burning with a smokey, yellow flame and strong odor. **Bituminous** coals are harder, blacker, and richer in carbon, and burn better. **Anthracite** coals, found in the United States only in eastern Pennsylvania, are the hardest; they bear 80 to 90 per cent fixed carbon and burn with a pale-blue flame without smoke or odor. Anthracite is formed by the heating of bituminous coals to a temperature of 170°C for 1000 million years or to 250°C for 10 million years.

Organic matter to petroleum to graphite As with the peat-coal-graphite sequence, liquid organic matter undergoes progressive changes that lead to petroleum, and ultimately to graphite. These changes are illustrated on the graph shown in Figure 6.32. The first change is from raw organic matter to **kerogen,** a complex solid hydrocarbon composed largely of carbon and hydrogen, but also including nitrogen and sulfur. Kerogen is the most widespread form of natural organic carbon. It is found in rocks of all ages and on all continents. The fine-grained rocks known as "oil shales" actually do not contain oil, but kerogen. When the kerogen rocks are heated to a temperature of about 350°C, the kerogen becomes petroleum.

In nature, the progressive changes beyond kerogen are: first to asphalt-base ("heavy") crude oil, and then to paraffin-base ("light") oil. The next product is a light liquid called petroleum distillate, and the next, natural gas. The thermal destruction of methane leaves a pure carbon residue in the form of graphite.

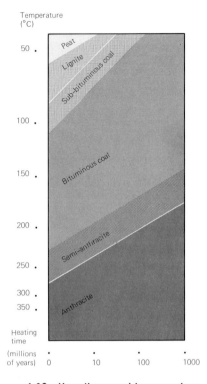

6.30 How time and temperature affect plant material and change it from peat (at upper left) to anthracite (lower right). With longer heating times the temperatures at which the various changes take place are progressively decreased. Thus, the change from high-volatile bituminous coal to low-volatile bituminous coal, which takes place at a temperature of about 230° C in laboratory experiments, requires a temperature of only about 150° C if the heating time is 1000 million years. (Replotted by J. E. Sanders from A. Hood, C. C. M. Gutjahr, and R. L. Heacock, 1975, American Association of Petroleum Geologists Bulletin, v. 59, no. 6)

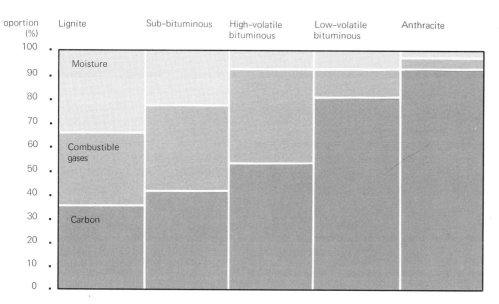

6.31 Proportions of fixed carbon, combustible gases, and moisture in various grades of coal. (U. S. Bureau of Mines)

6.32 How time and temperature change organic matter into kerogen, oil, and gas. As in Figure 6.30, the effect of long heating times is to decrease the temperature required to create a given end product. (Replotted by J. E. Sanders from Hood, Gutjahr, and Heacock)

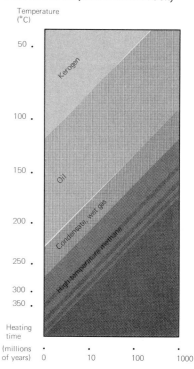

Subsurface Changes in Rock-forming Silicate Minerals

The progressive burial of clay minerals results in the formation of shale, and ultimately of mica schist. We will trace the first part of this progression in deep borings, and then turn to plutons for the final stages.

Changes in clay minerals The modern clay-bearing sediments contain abundant montmorillonite, a clay mineral containing iron and magnesium, and many layers of water molecules. Montmorillonite is formed by the alteration of volcanic rock, and has the remarkable power of swelling to many times its size when water is added to it. Montmorillonite is unknown in ancient shales. In its place is the mica mineral chlorite. The change from montmorillonite to chlorite takes place at a depth of burial of about 3000 meters, where the temperature is 150°C.

Once this change has taken place, the claystone has become shale. Further changes in the sheet silicate minerals take place at temperatures that exceed 500°C. From here on to the melting point of minerals, we are in the range of changes that affect all silicate minerals, and the changes of silicate minerals can no longer be illustrated by what we find out in deep boring. Instead, we are required to look at the contacts of plutons, and to see what happens in the interiors of mountain ranges.

Concept of Metamorphic Facies

The sequence to be discussed now is one you can observe as you move away from the contact where a pluton has invaded sedimentary rocks. The concept of **metamorphic facies** refers to assemblages of metamorphic minerals that formed under the same environmental conditions. The idea of metamorphic facies began by studying alterations of country rock by magma.

Contact-metamorphic aureoles The effects of the hot magma are greatest near the contacts, but some changes occur throughout a shell, or halo, which extends in all directions away from the contacts. These three-dimensional zones of alteration are **contact-metamorphic aureoles** (Figure 6.33). Around small dikes or sills, aureoles may extend for only a few centimeters into the country rock. Around the largest batholiths, the widths of aureoles may reach several thousand meters.

The most noticeable field expression of a contact-metamorphic aureole is a baking or hardening of rocks. Such hardened rock may be able to resist weathering and erosion better than both the intrusive igneous rock which solidifies from the magma that did the baking, and the surrounding country rock that was not baked. In the Crazy Mountains of Montana, rocks forming the aureoles of contact zones project as high ridges and peaks surrounding an eroded pit of igneous rock and stand above an eroded plain of unaltered shales and sandstones.

Contact-metamorphic minerals The types of country rock most susceptible to contact-metamorphic changes are sedimentary—sandstones, dolomites, limestones, and shales. Pure quartz sandstones are fairly resistant to change, except along the contact itself where they become quartzite. Limestones are easily recrystallized into marble or, if they contain quartz impurities, into wollastonite or pyroxene. Shales and slates, which vary greatly in composition, change easily into a wide range of metamorphic products. The most common metamorphic shale is hornfels. Farther from the contact, pure hornfels grades into a second zone of hornfels mixed with tiny crystals of andalusite in a groundmass of mica and quartz. Farther still from the contact this grades into slate with concentrations of carbonaceous matter. Finally there is a zone of practically unchanged shale. In all contact-metamorphic aureoles, the effects of metamorphism fade in this way as distance from the contact increases.

The same kinds of changes can be seen on a regional scale in mountain chains.

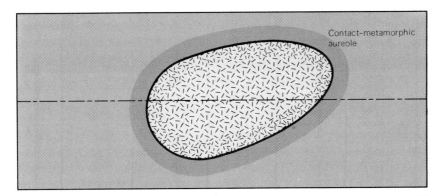

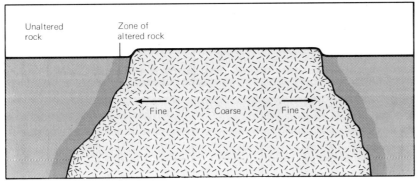

6.33 Contact-metamorphic aureole seen from above (map view, at top) and in vertical slice at right angles to Earth's surface, as might be found in the wall of a canyon.

The term **regional metamorphism** is somewhat confusing, because it does not describe a process. This term was adopted to deal with widespread environmental changes that cannot be accounted for by localized stresses or contact metamorphism. The conditions which give rise to regional metamorphism may pertain to areas of thousands of square kilometers, such as the Canadian shield, a relatively flat expanse of very old rock which covers much of eastern Canada; and the Fenno-Scandian Peninsula, which extends through most of Finland, Norway, and Sweden. Most of the rocks exposed in these and other large tracts are metamorphic, and this is taken to mean that most of the still-covered rocks likewise are metamorphic. This signifies such a widespread distribution that no single agent of metamorphism could account for the complex variety of rock types.

Most regional metamorphism has occurred in great linear belts hundreds of kilometers wide and thousands of kilometers long in which deformation has been concentrated. Within these elongated belts the conditions of metamorphism vary both vertically and laterally.

Sedimentary Rocks and Metamorphic Rocks

Within each belt, chemically similar parent rocks as unlike to the eye as shale, basalt, and dolomite can be converted to virtually identical metamorphic rocks. Three broad facies of regional-metamorphic rocks are (1) greenschist, (2) epidote-amphibolite, and (3) amphibolite. The amphibolite facies gives way to gneisses that resemble granite. Such a change defines the metamorphic granite facies.

The Metamorphic Granite Facies

Most metamorphism takes place in the upper part of the Earth's lower crust under moderate (in geologic terms, at least) temperatures. Little is known of the reactions that occur below these regions, where both temperatures and pressures are extremely high. At some point, these conditions become indistinguishable from those through which silicate minerals pass when a magma cools. At this zone of convergence, conditions match those which accompany the solidifying phases of metamorphic igneous activity. **Plutonic metamorphism** is a term used to describe the conversion of very deep sedimentary rocks into metamorphic rocks. Most geologists agree that even some granites, rocks traditionally considered to be the exclusive products of the solidifying of a magma, can be created by plutonic metamorphism (Figure 6.34). In fact, some geologists think that *most* granites are really products of plutonic metamorphism. The controversy over the origin of granite has raged since the nineteenth century and to this day is not altogether resolved.

The granite controversy is one in which both sides are at least partially correct. Many granitic rocks are clearly of magmatic origin. Some granitic batholiths are demonstrably surrounded by contact-metamorphic aureoles—proof that the material which solidified to form the intruded rock was emplaced when it was hotter than its surroundings. Blocks of country rock (called *xenoliths*, for "foreign rock") may have been wrenched loose and completely surrounded by igneous rock (Figure 6.35).

On the other side, those who argue that all granite is metamorphic, point to the so-called "space problem." If granite batholiths are igneous intrusives in which the magma came from below, they ask, what happened to all the country rock that the magma must have displaced? Recent measurements have indicated that batholiths are somewhat smaller than originally thought, but the space argument remains strong. Another point is that in some areas it is possible to follow a rock formation as it grades continuously from sedimentary rock to granite. Bedding features have also been found in granite, indicating

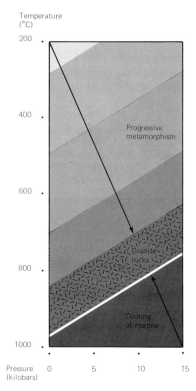

6.34 Origin of granite shown on graph of temperature versus pressure. Temperature on vertical scale increases downward. The field marked granitic rocks can be entered from either side. During progressive metamorphism entry is from the top. During the cooling of a magma entry is from the bottom.

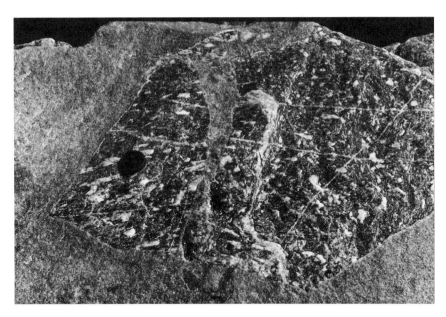

6.35 Xenolith of metamorphic rock surrounded by coarse-grained, light-colored, massive granite. (J. E. Sanders)

a sedimentary origin. One difficulty for the idea that granites originate by metamorphism is that enormous masses of sedimentary rock would have to be infiltrated and significantly altered by certain minerals to achieve *granitization*, or the transformation, by metamorphic processes, of sedimentary rocks into granite (Figure 6.36).

Because both sides of the granite controversy have such strong points in their favor, it seems clear that some granites may result from solidification of magma and others may form by granitization. One factor that illustrates how closely granites and metamorphic rocks are related is the existence of migmatite. **Migmatites** (or "mixed rock") may have both the banded or layered appearance of gneiss and, not far away, the non-oriented, random textural pattern of granite. In migmatite, thin stringers of schist or gneiss may fade imperceptibly into granite as if they had been dissolved. Wide zones of migmatites around granitic plutons seem to be truly intermediate between metamorphic and igneous rock.

Indeed, such an intermediate stage would be consistent with what we have learned about rock behavior so far. (See the discussion of the rock cycle in the Introduction.) We have now seen how, in many instances, one kind of rock is easily altered to another when environmental conditions change. Igneous rocks are weathered into

6.36 Migmatite formed during extreme metamorphism under conditions that created granitic rocks. Pre-Triassic rocks exposed in cuts along Route 95, Waterford, Connecticut. (J. E. Sanders)

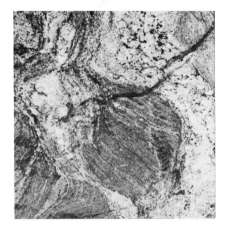

particles of sediment; these sediments may then be lithified to form sedimentary rocks; sedimentary rocks are transformed by heat and pressure into metamorphic rocks. If metamorphic rocks melt and subsequently solidify into igneous rocks, the great cycle starts over again. The previous history recorded by the rock is literally melted into oblivion.

Chapter Review

1. Most of the energy sources we use come from the products of sediments and sedimentary rocks. Metamorphic rocks being formed from sedimentary rocks are constantly creating new foundation rocks for continents.

2. *Sediment* is broadly defined as regolith that has been transported at the surface of the Earth and deposited as strata in low places.

3. Sedimentary rocks contain four major groups of particles: solid particles eroded from bedrock, solid particles whose "raw materials" traveled as ions in solution, solid particles of volcanic origin, and particles and materials formed from the tissues of plants and animals.

4. Many of the particles derived from the breaking up of bedrock were transported and deposited physically as solids of varying sizes. All these particles are collectively designated *detritus*.

5. The over-all term for all processes of converting loose sediment into a rock is *lithification*. The chief processes of lithification are *cementation*, *recrystallization*, and *replacement*. Compaction of clay minerals is a minor factor.

6. The attribute of sediments and sedimentary rocks in which they are arranged in layers is called *stratification*.

7. The fundamental properties used for classifying and naming sedimentary rocks are *mineral composition* and *texture*. The composition of the particles indicates where the rocks came from, and their texture results from the sedimentary processes that deposited them.

8. Sedimentary rocks can be classed in two major groups, depending on whether or not they were formed by cementation of sediments.

9. The concept of *metamorphism* states that igneous or sedimentary rocks can be converted to metamorphic rocks under the influence of temperature, pressure, and circulating fluids. Any pre-existing rock that has been changed into metamorphic rock is a *parent rock*. True metamorphism includes changes that affect minerals in the solid state; it excludes weathering or melting.

10. Metamorphic rocks are classified according to their *mineral composition*, *texture*, and *foliation*. Foliation is a distinctive layered aspect of many metamorphic rocks.

11. The basis of metamorphism is a response to an *intermediate subsurface environment* where materials are being put back together, instead of be-

ing broken apart as in weathering. For most materials, *temperature* is the dominant factor in subsurface environmental changes. Other factors include *pressure* and *salinity* of subsurface waters.

12 The concept of *metamorphic facies* refers to assemblages of metamorphic minerals that formed under the same environmental conditions. *Plutonic metamorphism* describes the conversion of very deep sedimentary rocks into coarse-grained metamorphic rocks resembling igneous rocks.

13 Geologists are uncertain about the origin of granite. Those scientists who do not believe that granite was formed by *granitization* argue that at least some granites may result from the solidification of magma.

Questions

1 Why are sedimentary rocks more difficult to understand than igneous rocks?
2 Describe the significance of *strata* as they are related to sedimentary rocks. How do strata originate?
3 What are the major considerations in classifying sedimentary rocks? How do these factors compare with classification of igneous rocks?
4 How is the size of particles used to classify sedimentary rocks?
5 Describe the concept of *sorting*.
6 Why are weathering and melting not considered to be agents of metamorphism?
7 Explain *foliation*.
8 What are the factors of the subsurface environment that affect metamorphism?
9 How is *plutonic metamorphism* related to the concept of *metamorphic facies*?

Suggested Readings

Bayly, Brian, *Introduction to Petrology*. Englewood Cliffs, N. J.: Prentice-Hall, Inc., 1968.

Ernst, W. G., *Earth Materials*. Englewood Cliffs, N. J.: Prentice-Hall, Inc., 1969.

Huxley, Thomas Henry, *On a Piece of Chalk*. New York: Charles Scribner's Sons, 1967.

Laporte, Léo F., *Ancient Environments*. Englewood Cliffs, N. J.: Prentice-Hall, Inc., 1968.

Spock, L. E., *Guide to the Study of Rocks*, Second Edition. New York: Harper & Row, Publishers, Inc., 1962.

Zim, H. S., and Shaffer, Paul, *Rocks and Minerals*. New York: Golden Press, 1957.

"Hundreds of kilometers away from the earthquake, avalanche debris flew over the tops of trees."

Chapter Seven
Stability of Slopes

Throughout geologic history, the surface of the Earth has been changing. In Chapters 7 through 12 we will be concerned with these changes. Almost everyone has observed some of the processes that change the Earth—shepherds have watched "landslides" (a popular term for any slope failure), fishermen have seen waves gnaw at coastlines, and alert farmers have struggled to prevent the erosion of topsoil by rainwater.

The continuous operation of the visible part of the geologic cycle is made obvious through our study of slopes, running water and erosion, groundwater, deserts and the wind, and oceans and the shoreline. We will see what these environments are, how they operate, how human beings have affected the surface of the Earth, and how the surface changes affect human activity. The landscape represents a momentary glimpse, a single frame of a motion picture of the Earth operating through time. Through many of these glimpses we will be able to predict the future of the landscape and understand its past.

The surface of the Earth is made up of low-lying, nearly horizontal plains underlain by sediment, and inclined areas, or **slopes,** which may be underlain by regolith or bedrock. Slopes include the sides of mountains, volcanoes, valleys, and isolated hills. These slopes are actively moving, and their stability is of vital importance in the process of the changing landscape. The force that makes slopes change is gravity.

The study of these changes and the landforms that result is **geomorphology**, from the Greek terms *geo*, meaning Earth; *morphe*, meaning form; and *logos*, discourse.

Introduction to Landscapes and Downslope Movement

All the diverse kinds of downslope movement, taken as a group, are known as **mass-wasting.** This is the displacement of a mass of bedrock or regolith adjoining a slope when the center of gravity of the mass advances outward and downward.

Why is Mass-Wasting Important?

Mass-wasting is a process that affects all slopes in all places all the time. Therefore, any house, for example, that is built on sloping ground is subject to the effects of mass-wasting. Because some slopes are stable (mass-wasting proceeds very slowly on them), years may pass without any problems. However, other slopes are notoriously unstable, with mass-wasting taking place fast enough to cause problems within a few years.

Introduction to Landscapes and Downslope Movement

In coastal California the hills are young, geologically speaking. They are so new that even though they consist of soft shale bedrock their slopes are steep and their relief is great. Hillsides that face the Pacific Ocean are vantage points with spectacular scenic views. Hence, these sites are considered prime spots for residential real-estate development. Many homes have been built on sloping lots where stability is not assured. The result has been a series of local disasters. After every rainy season newspapers are full of reports of houses that have slid down unstable slopes (Figure 7.1).

In other areas the results of mass-wasting are less spectacular than in coastal California, but they are no less real. Retaining walls tip over, boulders crash down on houses, and pieces of lawn or garden slip away. Therefore, knowledge of mass-wasting can be of great economic importance to you. In fact, by being knowledgable about potential hazards resulting from mass-wasting you could well avoid having to pay exorbitant damages. As taxpayers you should understand the vast costs which may arise from the effects of mass-wasting on highways and other public property.

7.1 House in danger of falling into Pacific Ocean as a result of mass-wasting of underlying materials. Pacific Palisades, California. (United States Geological Survey)

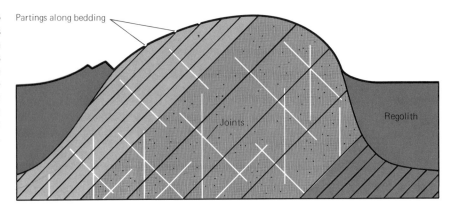

7.2 Joints and bedding-plane partings create large blocks in mass of bedrock. Under conditions shown in this schematic profile, blocks broken loose along joints that dip steeply to the right will fall down the right side of the hill. Blocks broken loose along bedding planes, which dip steeply to the left, will fall down the left side of the hill.

Some questions about slopes The most important factor affecting a slope is the kind of material underlying it. Is the slope underlain by bedrock? If so, what kind of rock? What kinds of parting surfaces are present? What is their spacing? Are these parting surfaces vertical, horizontal, or inclined? If they are inclined do they incline in the same direction as the slope (Figure 7.2) or in the opposite direction? Is regolith present? If so, how thick is it? What particle sizes are present? Can water seep through the regolith? These and other questions affect the outcome of a constant contest that takes place on and beneath a slope between the component of gravity that tends to move material downslope and the forces of friction (including another component of gravity) that tend to hold material in place.

Components of Gravity on a Slope

The Earth's gravity is the inward-acting force that tends to pull all particles toward the center of the Earth. This force acts along lines that are at right angles to the Earth's horizon (Figure 7.3). On a slope the force of gravity affects material in two ways: (1) One tends to push the material against the slope and thus to keep it in place. In effect, this push toward the slope keeps things from moving. We will call it a **resisting force.** (2) The other force of gravity acts along the slope and down the slope. We will call this the **pulling force.** These parts of the Earth's gravity are known as **components.** We shall refer to the first as the **slope-normal component** and to the second as the **slope-parallel component** (Figure 7.4).

As long as the resisting force is larger, the only energy involved is *potential* energy and the slope is stable. If the pulling force is larger,

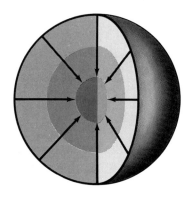

7.3 The Earth's gravity tends to pull all objects toward the center of the Earth, as shown in this schematic cutaway view. Arrows show directions along which the Earth's gravity acts.

Introduction to Landscapes and Downslope Movement

then potential energy is converted into *kinetic* energy and the slope is not stable—the material moves downward, either rapidly or slowly. The angle of the slope is the only factor that affects the strength of the pulling force. On gentle slopes, the resisting force is larger. On steep slopes, the pulling force is larger (Figure 7.5). The pulling force tends to make the materials on the slope slip downward along a series of parallel planes like cards in a deck (Figure 7.6). Such a tendency for solids to slip along parallel planes is called **shearing**.

We will now look into the relationship between the factors that affect these forces, starting with those which affect the resisting force.

Reduction of Resisting Force

Various things can happen to change the balance between the pulling force and the resisting force, thus reducing or even eliminating the resisting force altogether. Water and fluid pressure, and other factors, such as ice and earthquake waves, can enable even the slightest amount of pulling force to move material down very gentle slopes.

The presence of slippery material increases the chances for downslope movement. Small, flaky particles, such as the clay minerals present in silts and clays, are one kind of "slippery" material. Clayey slopes are notoriously unstable.

Water and fluid pressure Another factor commonly associated with mass-wasting is water. The effect of water depends on how abundant it is. As we said earlier, mass-wasting events are most likely to take place during, or shortly after, periods of heavy rain. Why

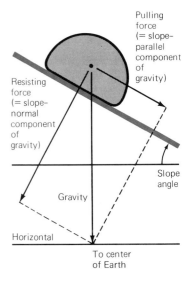

7.4 Contrasting components of the Earth's gravity on a slope.

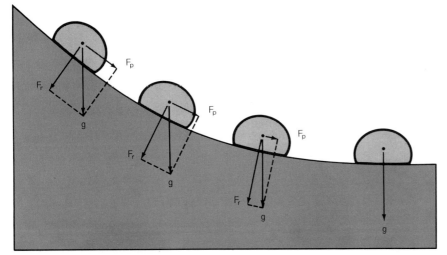

7.5 As slope angle changes the relative values of the two components of the Earth's gravity shift. On gentle slopes the component acting against the slope is larger. The component acting along the slope becomes larger as the slope angle increases.

236
Stability of Slopes

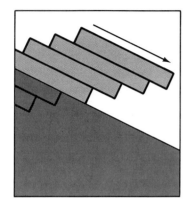

7.6 Effect of gravity component acting along the slope tends to create slippage on shear planes, which can be compared to the sliding that takes place between the individual cards in a deck of playing cards.

should a slope that is stable when it is dry become unstable when it is wet? Is water serving as a lubricant, or is something else happening? To answer these questions we need to examine the effects of water on regolith.

If the only water present in fine-grained regolith is a thin film, then the surface tension binds the particles together. If the spaces between particles become completely filled with water, the surface-tension effect is lost and the particles can move more easily. Thus, a film of water promotes stability, but a full load of water reduces stability.

The saturation of a slope by heavy rains or seepage adds considerable weight to the mass of debris. This extra weight increases the pulling force. It also increases the resisting force, but the effect on the pulling force is greater because water has no shearing strength. When the water becomes so abundant that its weight increases the pressure within the pore spaces, this pressurized water tends to "float" adjacent particles. This **pore pressure** pushes the particles apart, and thus overcomes the resisting force (Figure 7.7). When sand particles are pushed apart by such flow of pore-water they become "quicksand" and behave like a liquid. A "quick" condition develops when the pore-water pressure is great enough so that all grain-to-grain contacts are lost, and the grains are supported by the fluid.

The amount of space between particles, or **porosity**, increases abruptly in a process known as **spontaneous liquefaction.** When sandy regolith becomes liquefied its resisting force disappears, and the results can be catastrophic. The regolith flows down even the slightest slope until its surface becomes horizontal. Eventually the particles move closer together again and resume the appearance of a "solid" body of regolith. Spontaneous liquefaction can be caused by rapid jolts, such as the shocks of a pile driver, or by changes that follow the rapid use or fall of the level of underground water. The best known flows of this kind have occurred in The Netherlands on gentle

7.7 Upward flow of fluid increases pore pressure and lifts particles. When the particles no longer are in contact they can flow as a liquid.

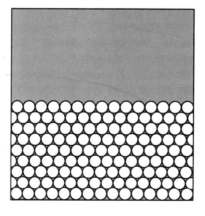
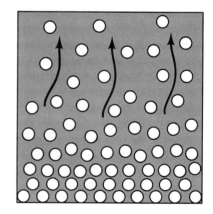

sandy slopes along the coast; between 1881 and 1946 229 such flows were reported.

The stability of a slope may change greatly if the water level adjacent to it falls by more than a few meters per day. This may happen to a river after a flood, or around a storage reservoir whose level is being lowered. In both cases, a volume of regolith that had been under water suddenly finds itself above water level. Unless this water can escape from the regolith almost as fast as the level of the river or reservoir drops, the trapped water exerts tremendous force on the slope near the dropping water level. The force exerted on the soil particles by the downward-percolating water is **seepage pressure.** This pressure acts in the direction of the flow, and increases in proportion to the speed of seepage. Both seepage pressure and speed are greatest at the foot of the slope. If these two factors become great enough, the slope may fail. Once the lower part of a slope has failed, the upper part soon follows. Rapid drawdown is a common cause of slope failures, especially in sediments of a size somewhere between sand and clay.

Other factors Ice can overcome the resisting force by the mere act of freezing, which causes water to expand and lift the particles. This lifting away from the slope overcomes the resisting force and the particles finally let go when the ice melts. Ice becomes slippery only when it is covered by a thin film of water, and ice crystals in regolith therefore act as a cement rather than as a lubricant.

Earthquake waves passing through regolith physically lift the particles, thus overcoming the resisting force. Once the particles start to move their motion gives them another lift, creating a self-perpetuating effect which is transferred to other particles along the way (Figure 7.8).

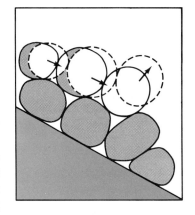

7.8 Shearing along slope causes particles to collide and to move away from the slope. Such movement away from the slope destroys the rigid framework formed by grains in contact and enables the entire body of dispersed particles to flow down the slightest slope.

Increase in Pulling Force

The pulling force upon a mass of material on a slope can be increased (1) by adding weight to the upper portion of the slope, (2) by increasing the steepness of the slope (Figure 7.5), and (3) by removing material from the lower part of the slope.

The placing of fill for roads and foundations may overload a slope. A massive slope failure in Hudson, New York, in 1915, was probably triggered by fill overloading in combination with increased water content after a month of heavy rains. Because the slope was old, it must be supposed that it had survived other periods of prolonged rainfall. Geologists investigating the event found an additional condition unprecedented in the slope's history: A stockpile of crushed

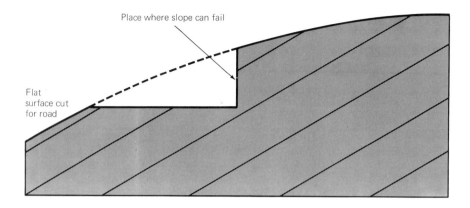

7.9 Removal of material from the toe of a slope, as in a roadcut, creates instability and invites slope failure.

rock, estimated to weigh about 25,000 tons, had been placed along the upper edge of the slope. The investigators concluded that the two factors—high pore pressure and overloading—acted together to trigger the slope's failure.

A common cause of slope failure that results from construction projects is removal of support at the toe of a slope (Figure 7.9). The same kind of result takes place where a stream undermines its banks or a sheet of water takes away sand from the top of a beach. The removed material has contributed to slope stability. When this material is taken away, the slope fails.

Now that we have summarized the mechanics of the forces involved in mass-wasting, we can examine the kinds of downslope movements.

Kinds of Downslope Movements

Mass-wasting may be sudden and catastrophic, or it may be extremely slow. The effects of slow mass-wasting are noticeable only after the material has moved during long periods of time. The most sudden movements occur in materials which stick stubbornly together. Irregularly jointed rocks, cemented sands, and certain fine-grained regolith are examples of materials that often cling to even steep slopes. When such cohesive debris does "let go," however, it is apt to do so all at once, with great violence and high speed. Other materials of low cohesiveness, such as wet clays, are highly unstable. These materials tend to spread or move slowly downward even on gentle slopes.

Rapid Movements

The most rapid mass-wasting events are those that involve single pieces or great bodies of bedrock (which, when they break loose, become "instant regolith") or regolith. The following processes involve one or more pieces of bedrock that have suddenly become regolith: rock fall (blocks moving as individuals), rock slide (several blocks sliding along a well-defined surface) and rock avalanche (great bodies of bedrock suddenly loosened and moving catastrophically). Other mass-wasting activities involve the shifting of bodies of stratified bedrock or the shifting of sand or clay along well-defined surfaces. The speed of such shifting may not equal that of rock slides or of rock avalanches, but is generally considered to be "fast" rather than slow. Once started, such a moving mass that began as a single block may lose its internal structure and continue as a chaotic, tongue-like flow.

Rock fall A common form of rapid mass-wasting on cliffs and steep slopes underlain by bedrock involves the relatively free falling of detached individual blocks of bedrock. These may be of any size. They can fall from a cliff (Figure 7.10), a steep slope, a cave, or an arch. The rock may fall directly down or make a series of bounds and rebounds over other rocks or regolith on a steep slope. There is little or no interaction among the falling fragments. Such movement of individual blocks is named **rock fall**. Rock falls are common along

7.10 Rock fall. (Left) Blocks of sandstone that fell away from a joint that is inclined steeply away from a lake, thus forming an overhanging cliff. (Dick Frear, U. S. National Park Service). (Right) Single large block of dolerite bounded by cooling joints has fallen from steep face of Palisades cliff, New Jersey. By granular disintegration dark material (at bottom of photo), which contains abundant olivine and lacks cooling joints, falls away one particle at a time. When a recessed area in the cliff face becomes deep enough, large blocks of the overlying dolerite are undermined, and fall down. (Ann McCaughey)

240
Stability of Slopes

7.11 Sliderock accumulating in fanlike talus cones at foot of steep slope underlain by massive igneous rock in northern Afghanistan. (United Nations)

the head walls of canyons and along rocky cliffs at the shore. Such falls may be caused by the wedging of jointed rocks by tree roots or ice. An accumulation of rock fragments that have fallen from above is a **talus** (TAY-lus) (Figure 7.11). The angular blocks in a talus are known as *sliderock*.

An active talus forms a steep slope, usually 30–40 degrees. Such an angle assumed by a body of loose particles is known as the **angle of repose.** The angle of repose is the natural slope which results from the interplay of the gravity components acting against the slope and along the slope (Figure 7.4). Loose particles cannot accumulate at any angle that is steeper than the angle of repose. Instead, they will spontaneously readjust by downslope movement.

It is important to realize that although loose particles do not accumulate at angles steeper than their angle of repose, once they have formed such an angle they may later move downslope. The angle might better be named the "angle of temporary repose."

Rock slide Some movements involve not single pieces but many pieces broken loose from solid masses of bedrock. The term **rock slide** describes the downward and usually rapid movement of many newly detached segments of former bedrock. The sliding takes place along some parting surface, as along a joint, along a fault, or between beds (Figure 7.12). Large, natural rock slides occur frequently in such rugged areas as the Rocky Mountains. Smaller but more frequent rock slides are produced by such human activities as road and railroad excavation which undercut rocky slopes.

One of the most intensively studied rock slides in the United States disrupted the Gros Ventre (Grow VAHN-truh) River Valley in Wyoming in 1925 (Figure 7.13). Some 38 million cubic meters of rocky debris broke loose from the side of the valley and traveled about 600 meters down a slope inclined from 18 to 21 degrees. The sliding mass lunged across the valley, climbed 100 meters up the opposite slope, and then settled back, forming a dam 70 to 75 meters high and more than half a kilometer long. The river backed up in a lake nearly 8 km long, and in 1927 heavy spring runoff washed away 15 meters of the dam, causing extensive damage downstream and killing several persons in the town of Kelly, Wyoming.

7.12 Rock slide along steeply inclined beds of sandstone, near Hole in the Rock, Utah. (A. J. Eardley)

Introduction to Landscapes and Downslope Movement

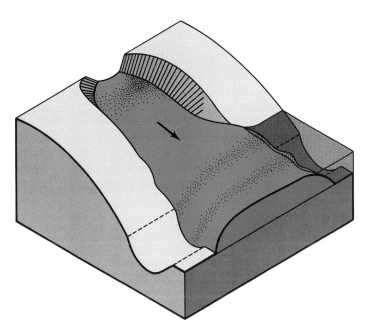

7.13 Slope failure along side of Gros Ventre River, Wyoming. (Drawing, above) Tongue of debris from valley side accumulated in axis of valley, blocking the river, and forming lake. (Photo, below) View from airplane of scar on valley side marking site where rock broke loose. Bedding surfaces in sedimentary rocks are inclined toward the axis of the valley at an angle of about 20 degrees. (Forest Service, USDA-353999)

Stability of Slopes

Rock avalanche The most awesome variety of downhill movement is a **rock avalanche** (Figure 7.14). The composition of "avalanches" may vary from pure ice and snow to rocky debris; only those containing notable amounts of rock debris are named rock avalanches. Avalanches, which move at more than 100 km/hr, are terrifying events. This great speed may result in part from a cushion of air trapped beneath the falling material. It has been suggested that this cushion of air works on the same principle which allows a Hovercraft to move so swiftly over rough ground or water. To reduce avalanche damage, artillery shells are fired into steep masses of snow and rock. The impact triggers small avalanches before enough material can accumulate for a more destructive one.

One of the most spectacular avalanches of modern times crashed down from the high Alaskan coast ranges and spread across the Sherman Glacier (Figure 7.15). The solid bedrock was shaken loose on March 27, 1964, during the Prince William Sound earthquake, whose source was hundreds of kilometers away. The avalanche debris literally flew over the tops of trees growing in a low hollow along its course. The avalanche finally came to rest as a blanket of debris 1 to 3 meters thick. The surface of this material contains long grooves and the edges are very irregular. These shapes indicate that the moving mass behaved as a series of fluid-like tongues. Viewed from a distance, this sheet of debris resembles a lava flow.

7.14 Avalanche debris triggered from north peak of Nevados Huascaran, Peru, by earthquake of May 31, 1970, includes everything from mud (beneath cracked surface at lower right) to giant boulders weighing an estimated 700 tons. (U.S. Geological Survey)

Kinds of Downslope Movements

7.15 Rock avalanche from Sherman Peak, Alaska, triggered by Prince William Sound earthquake of March 27, 1964 (Photo by Austin Post, U. S. Geological Survey). Drawing shows profile along main path of avalanche from Shattered Peak, the source, to Sherman Glacier. The avalanche literally flew over trees growing along the west side of Spur Ridge. (Plotted by J. E. Sanders from data in R. L. Shreve, 1966, Science, v. 154)

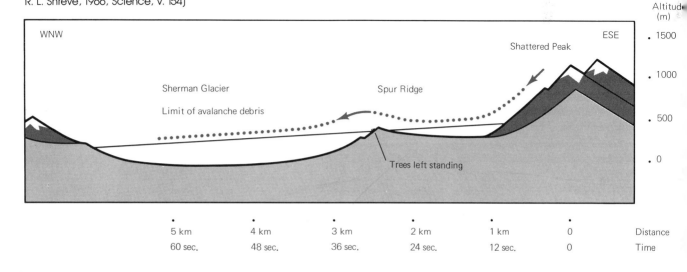

7.16 Slope failure by slumping. Material at lower end where the road is blocked has been converted to debris flow. Oakland, California, December, 1950. (William S. Young, San Francisco Chronicle)

7.17 Tilting toward slump surface of tops of slump blocks recorded by inclination of tree; small ponds commonly form in closed depressions that were created by slumping.

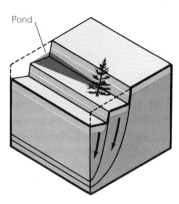

Slump When a mass of bedrock or of regolith breaks loose along a distinct surface of failure it rotates downward and outward from a slope. Such block-like failure along a distinct surface is **slump.** A slumped mass usually does not travel very far nor spectacularly fast. Nevertheless, slumping is one of the chief problems facing a highway engineer. Slumping takes place along an internal surface that is typically spoon-shaped (Figure 7.16). If a slump affects the steep edge of a flat-topped plateau, what was initially a horizontal surface commonly tilts backward into the hillside (Figure 7.17). At the bottom, the slumped mass bulges upward. Because the entire mass often slumps as a single unit, plants and even houses on the former horizontal surface may move downslope and tilt intact. In mountainous areas, lakes commonly form in these sunken regions.

Slumping usually occurs in regions where massive sedimentary strata (such as sandstones or limestones) or sheets of volcanic rock lie above weak formations of shale or clay. Such weak formations tend to erode away uniformly, and the resistant cap rock fails as large blocks break loose along fractures or along slump surfaces. The landscape of the Snake River Plains and the Columbia Plateau is broken by many small, step-like slumps where weak rock has eroded beneath the harder basalt flows.

Slumping in sand As we have just seen, the lower ends of slumped masses involving clay-rich regolith or soft shale commonly lose their internal structure and become chaotic liquid-like flows. These are a variety of *debris flow*, to be explained in the next section. By contrast, bodies of sand commonly slump along steep surfaces because particles have been moved away from the base—as along the bank of a stream (Figure 7.18) or a beach being eroded by thin sheets of water flowing along the shore. As soon as the body of sand breaks loose its internal structure disappears and it becomes a **debris slide.** Many debris slides—those that occur along undercut banks of streams, for example—are so small that they are not recorded in detail. During high water along a river such as the Mississippi, the collapse of sections of an undercut bank may be heard many times a day. Debris collapses along the Amazon River have been known to continue for several hours, affecting sections of bank more than a kilometer long and creating large waves.

7.18 **Vertical sandy bank of stream collapsing as debris slide** because high water eroded sediment from toe of slope. Colorado River, Grand Canyon, Arizona. (E. A. Heiniger, Rapho Guillamette/Photo Researchers)

Debris flow Wherever dry regolith on a slope becomes fully saturated, it is likely to begin to flow as a viscous, pasty mass. The general name for such movement is **debris flow.** If fine particles are especially abundant the term **mudflow** is appropriate. Debris flows are common in three places: (1) at the toes of slumped masses (Figure 7.16), (2) in stream valleys in dry regions subject to occasional torrential downpours, and (3) on volcanoes, where tephra and rainfall can mix together.

Mudflows are capable of transporting complex mixtures of rock debris, including boulders several meters in diameter. Mudflows are common where little or no vegetation anchors the soil, where slopes are moderately steep, where periods of intense rainfall alternate with dry periods, and where a deep layer of regolith contains enough clay or silt to aid in lubrication. Low-viscosity mudflows have been

Kinds of Downslope Movement

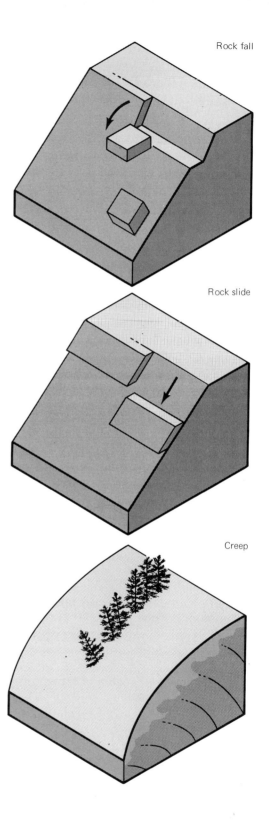
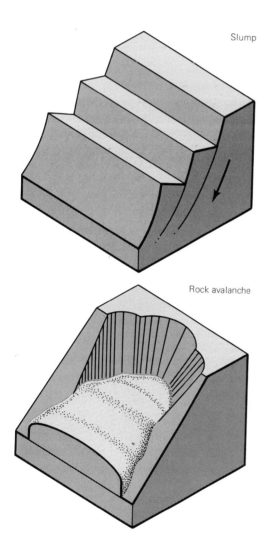

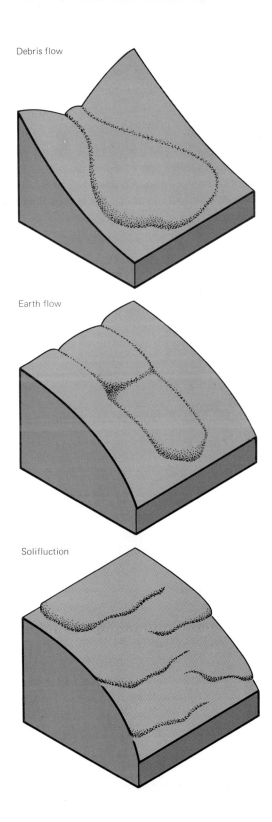
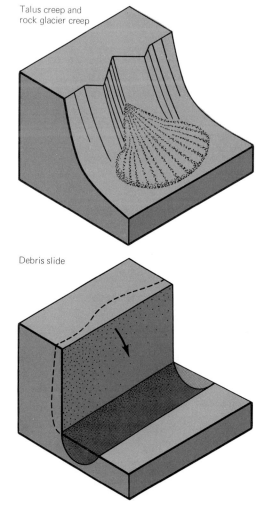

known to move so fast that they overtook railroad trains.

Mudflows vary in length from less than a meter to several kilometers. One of the largest recorded flows descended from the flanks of Mt. Rainier in Washington about 4800 years ago. Like many mudflows, this one is thought to have been associated with volcanic activity, partly induced by small eruptions on the east flank of Mt. Rainier. The flow originated above the Emmons Glacier; it was mixed with water as it came down the White River Valley. The movement of the flow accelerated with increasing water content, so that the mass, varying in thickness from 0 to 25 meters, moved all the way to the Puget Lowland, some 88 km from the source. Here it spread over about 170 square kilometers of lowland (Figure 7.19).

Another famous mudflow of volcanic origin accompanied the 79 A.D. eruption of Mt. Somma, the ancestor of modern Vesuvius in Italy (Figure 3.20). Heavy rains soaked the newly fallen fine tephra on the slopes of the volcano, converting the formerly dry tephra into a heavy unstable mass that began to move downslope rapidly. The mass, still hot when it eventually reached the Roman city of Herculaneum, at the base of Mt. Vesuvius, buried the city completely, just as the fallout of tephra from the atmosphere had snuffed out Pompeii.

7.19 **Exposure of Osceola mudflow** resting on Vashon stratified drift; about 50 km downvalley from Mount Rainier, Washington. Note the general decrease upward in number and size of stones. (D. R. Crandell, U.S. Geological Survey)

7.20 Effects of creep shown by inclinations of trees and by shifting of rock debris downslope from sources in bedrock.

Slow Movements

By slow movement of regolith we refer to movement that is so slow that most people do not know it is happening. It occurs on almost every moderate or steep slope, and greatly affects agriculture and soil fertility by removing topsoil faster than it can be replaced by natural soil-forming processes. Slow movements include creep, solifluction, earth flow, and talus creep and rock-glacier creep.

Creep The best-known and most widespread type of mass-wasting is **creep,** the imperceptible downslope movement of surface soil or rock debris. Creep affects all sizes of rock debris, from fine sands and silts to boulders tens of meters across. The debris may be dry or fully saturated with water. Although creep appears to be continuous, as the name implies, it is in reality a never-ceasing series of minute individual movements.

Normally, creep operates just in the top few meters of the regolith, but the mechanism is so effective that it carries vegetation and human-made objects with it. In fact, the best evidences for creep can be seen on almost any steep slope in the form of tilted fence posts, monuments, and telephone poles; displaced walls or foundations; misaligned roads; and inclined trees (Figure 7.20). Near-surface vertical strata of sedimentary rock may be tipped over so that they are inclined downhill. Perhaps the most striking demonstration of creep are tree trunks, which may bend uphill as they strive to counteract the downhill tilting.

Broken Country
George W. Kendall

As soon as we had given our horses and mules a short rest, and made a light meal of our half-cured meat, we re-saddled and resumed our journey. We were going forward at a rapid pace, the prairies before us presenting no other appearance than a slightly undulating but smooth surface, when suddenly, and without previous sign of warning, we found ourselves upon the very brink of a vast and yawning chasm, or *canon,* as the Mexicans would call it, some two or three hundred yards across, and probably eight hundred feet in depth! As the front ranks suddenly checked their onward course, and diverged at right angles, the rear sections were utterly at a loss to account for a movement so irregular; they could not see even the edge of the fearful abyss at a distance of fifteen yards from its very brink. The banks at this place were almost perpendicular, and from the sides projected jagged and broken rocks, with here and there a stunted, scrubby cedar. There was some appearance of a zigzag and precipitous trail down the sides of the canon at the point where we first reached it, and Mr. Hunt and Dr. Brenham took it with the intention of reaching the bottom if possible: they continued their winding path until they seemed mere pigmies, and only stopped when their progress was arrested by high and perpendicular bluffs. On their return, after an absence of some half an hour, they said they had not advanced half way to the bottom, and that to attempt crossing at this, or any other point within sight, would be useless. We travelled a mile or two along the banks, but finding it impossible to discover a crossing-place, we finally encamped in a little hollow of the prairie near the edge of the ravine. Here, finding that a large portion of our badly-cured meat was spoiling, we cooked what could still be eaten, and threw much of it away for the wolves and buzzards.

When morning came, which was bright and cloudless, we crawled out from under our wet blankets, and I doubt whether a more miserable, wobegone set of unfortunates, in appearance, have been since the passage of the Red Sea. Not a man among us who was not as wet as though he had been towed astern of a steamer from the Falls of St. Anthony to the Balize, and without the privilege of going ashore at any of the "intermediate landings."

The immense chasm we were upon ran nearly north and south, and by watching the current of the stream far below us — a furious torrent raised by the heavy rain — it was seen that it ran towards the former point. This induced Mr. Hunt to seek a crossing to the southward, and after saddling our horses we set off in that direction. We had gone but a few miles when large buffalo or Indian trails were seen, running in a southwest course, and as we travelled on, others were noticed bearing more to the west. We were obliged to keep out some distance from the ravine, to avoid the small gullies emptying into it, and to cut off the numerous turns, and in this way we travelled until about noon, when we struck a large trail running directly west. This we followed, and on reaching the main chasm found that it led to the only place where there was any chance of crossing. Here, too, we found that innumerable trails centered, coming from every direction; proof conclusive that we must cross here or travel many weary miles out of our way.

Dismounting from our animals, we looked at the yawning abyss before us, and the impression upon all was that the passage was impossible. That buffalo, mustangs, and very probably Indians with their horses had crossed here was evident enough, for a zigzag path had been worn down the rocky and precipitous sides; but many of our horses were unused to sliding down precipices as well as climbing them, and drew back repulsively on being led to the brink of the chasm. After many unsuccessful attempts, a mule was started down the path, then another was induced to follow, while some of the horses were fairly forced, by dint of much shouting and pushing, to attempt the descent. In some places they went along the very verge of rocky and crumbling ledges, where a false step would have precipitated them hundreds of feet to instant death; in others they were compelled to slide down pitches nearly perpendicular. Many of them were much bruised, but after an hour's hard work we all gained the bottom without sustaining any seri-

From NARRATIVE OF AN EXPEDITION ACROSS THE GREAT SOUTHWESTERN PRAIRIES, FROM TEXAS TO SANTA FE, 1845

Field Trip

ous injury.

I shuddered, on looking back, to see the frightful chasm we had so successfully passed, and at the time thought it almost a miracle that we had got safely across; but a few days afterward I was convinced that in comparison the undertaking we had just accomplished was as nothing.

The morning of September 3rd broke bright and cloudless, the sun rising from out the prairie in all his majesty. Singular as it may appear, nearly every shower, from the time we left Austin until we reached the settlements of New Mexico, fell during the night, generally commencing shortly after sundown.

We had scarcely proceeded six miles, after drying our blankets, when we suddenly came upon another immense rent or chasm in the earth, exceeding in depth the one we had so much difficulty in crossing the day

before. No one was aware of its existence until we were immediately upon its brink, when a spectacle, exceeding in grandeur anything we had previously beheld, came suddenly in view. Not a tree or bush, no outline whatever, marked its position or course, and we were all lost in amazement as one by one we left the double-file ranks and rode up to the verge of the yawning abyss.

In depth it could not be less than eight hundred or a thousand feet, was from three to five hundred yards in width, and at the point where we first struck it the sides were nearly perpendicular. A sickly sensation of dizziness was felt by all as we looked down, as it were, into the very depths of the earth. In the dark and narrow valley below, an occasional spot of green relieved the eye, and a small stream of water, now rising to the view, then sinking beneath some huge rock, was bubbling and foaming along. Immense walls, columns, and in some places what appeared to be arches, were seen standing, modelled by the wear of the water, undoubtedly, yet so perfect in form that we could with difficulty be brought to believe that the hand of man had not fashioned them. The rains of centuries, falling upon an immense prairie, had here found a reservoir, and their workings upon the different veins of earth and stone had formed these strange and fanciful shapes.

Before reaching the chasm we had crossed numerous large trails, leading a little more to the west than we were travelling; and the experience of the previous day led us to suppose that they all terminated at a common crossing near by. In this conjecture we were not disappointed, for a trot of half an hour brought us into a large road, the thoroughfare along which millions of Indians, buffalo, and mustangs had evidently travelled for years. Perilous as the descent appeared, we well knew there was no other near. The leading mule was again urged forward, the steadier and older horses were next driven over the sides, and the more skittish and untractable brought up the rear. Once in the narrow path, which led circuitously down the descent, there was no turning back, and our half-maddened animals finally reached the bottom in safety. Several large stones were loosened from their fastenings by our men, during the frightful descent; these would leap, dash, and thunder down the precipitous sides, and strike against the bottom far below us with a terrific and reverberating crash.

On leaving the main camp on the Quintufue it was thought by all that we could not be more than a hundred miles from San Miguel—we had now more than made that distance, and were still upon the immense prairie. To relieve ourselves from the horrible suspense we were in, to get *somewhere*, in short, was our eager aim, and hurriedly we pressed onward, in hope of finding relief. Our horses, in the meantime, had comparatively suffered less than ourselves, for the grazing of the prairie had been good. 1845

7.21 Freezing and gravity combine to shift pebble downslope. In left panel pebble lies on slope. In center panel the growth of ice crystals has pushed the surface outward parallel to itself, moving the pebble upward and outward at right angles to slope. In right panel, after thaw, gravity has moved the pebble vertically downward.

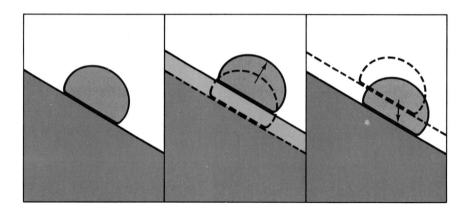

One of the most important causes of creep is the growth of ice crystals. As crystals enlarge, they grow perpendicularly away from a slope, pushing particles of rock debris ahead of them. When the crystals melt, the particles settle vertically downward for a net downhill movement (Figure 7.21), or they even topple downslope for a meter or more.

Frost may encourage creep not only by its upward heaving of rocks and soils, but also by its wedging action in fissures. The heaving power of frost has been measured at more than 1.5 tons per square meter, and uplifts of 15 cm or more are common. Many other causes of creep are known. These include (1) alternate saturation and drying of water in cracks in bedrock or in pores of regolith; (2) wedging by plant roots which open cracks—after the roots have been decayed, the cracks fill with creeping soil; (3) filling of animal burrows from the uphill sides; (4) trees swaying in the wind which aid movement by the prying action of the roots; and (5) the footsteps of humans and heavy animals, and the passage of automobiles on hillsides. Because movement may be as little as a few centimeters per year, grass may be able to maintain a continuous cover over an area that is creeping.

Solifluction A special type of creep, which occurs in high-latitude regions or at high altitudes where the ground freezes deeply, is **solifluction** (from the Latin for "flowing soil"). Solifluction is typical of subpolar areas where permanently frozen regolith ("permafrost") is covered by a shallow layer which thaws in the summer. This "active layer," a couple of meters thick, consists of tundra peat, rock debris, and other weathering products. Because water permeates this layer in summer, but cannot penetrate the zone of permafrost below, the surface mass may flow along slopes as gentle as two or three degrees. Solifluction is not altogether restricted to subpolar regions. It may occur in any place where loose regolith overlies packed clay or impermeable bedrock. In such settings, the water cannot escape downward, and the effect is the same as if permafrost were below.

The movement of solifluction is slower and more continuous than mudflow, and it affects whole slopes rather than just local areas. One of the earliest descriptions of solifluction, by Edward Belcher in 1855, lists conditions on Buckingham Island that are characteristic:

> As noon passed, the soil in all the hollows or small watercourses became semi-fluid, and very uncomfortable to walk on, or sink into. At the edge of the southern bank, the mud could be seen actually *flowing;* reminding one more of an asphalt bank in a tropical region than our position in 77°10' N. . . . About two P.M. . . . the entire slope, in consequence of the thaw, had become a fluid moving *chute of debris* for at least one foot in depth. (*The Last of the Arctic Voyages*)

Earth flow When the regolith on a steep slope becomes saturated with water, it may sag downward in a series of irregular terraces. This process is known as an **earth flow,** and its effects are seen most commonly in humid climate zones after a heavy rain, where the regolith has flowed down to form a bulging "toe" with large, smooth wrinkles. An earth flow may be confined to an area as small as a few square meters or it may cover many acres. Some earth flows are shallow, and affect only the regolith; others may extend much deeper, especially if the bedrock consists of unstable clay or shale, or of igneous rock that has been deeply weathered. Millions of tons of bedrock have been moved in huge, mud-like plastic deep earth flows. Earth flows are common along rivers such as the St. Lawrence, in Canada, where a layer of clay is buried by sand. The clay is normally stable unless disturbed by a shock, such as explosions or earthquakes. A shock may destroy the tenuous bonds between water and clay particles, changing the mass suddenly into a form that is more like a liquid than a solid. Blasting log jams in the St. Lawrence with dynamite has set off earth flows as much as 15 meters thick and thousands of meters wide. The powerful earthquake near Anchorage, Alaska in 1964 also set off destructive earth flows that caused more damage than the shaking. A water-saturated layer of clay 20 meters below the surface was liquefied by the tremors, so that upper layers of stiffer clay, sand, and gravel were literally floated seaward. One earth flow moved more than a kilometer across an intertidal mud flat into the ocean (Figure 7.22).

Talus creep and rock-glacier creep Creep may affect large rocks and boulders just as it affects soil. Typical accumulations of

7.22 Flow of regolith into the sea, as a result of triggering by Prince William Sound earthquake, March 27, 1964. (U. S. Geological Survey)

7.23 Rock glacier on Atlin Mountain, northern British Columbia (Information Canada Photo)

large, angular blocks are taluses (Figure 7.11). As long as a talus is fed by a continuing supply of rocks from above, it will maintain an angle of about 40 degrees. If the dribble of new rock should cease, however, the talus will gradually flatten out as the rocks shift farther downhill by **talus creep.** Talus creep is most rapid in cold regions where the rocks are moved by the alternating expansion and thawing of ice between the fragments.

In some alpine and arctic mountains, tongue-shaped masses of large rocks form features called **rock glaciers** (Figure 7.23). Some of the world's most impressive rock glaciers are in Alaska, where single masses may measure more than a kilometer long and half a kilometer wide.

The origin of rock glaciers is still not well understood. Some geologists think that interstitial ice is an important factor. This ice could be the remnants of the compacted snow that forms true glaciers. Others think the ice of rock glaciers derives not only from compacted snow but also from meltwater, rain, and rising groundwater. Still others think that such ice is not needed to cause rock glaciers to flow.

Mass-Wasting and Land Use

Wherever roads, buildings, or other structures are built in the vicinity of slopes, the builders must be aware of the potential danger of mass-wasting or else their projects may incur large damage costs. We shall review some of the kinds of damages and their costs, and also consider the prediction and prevention of slope failure.

Engineering Aspects of Mass-Wasting

Damages and costs   Highways, railroads, and pipelines are most susceptible to mass-wasting. In some areas farmlands, resorts, and homes are regularly damaged. The most damage-prone areas are located along rivers and lakes where undercutting and slippage occur. During the great Alaskan earthquake of 1964, slope failures triggered by the quake caused the most damage to structures and helped to cripple temporarily the economy of the whole state. At both Valdez and Seward, spontaneous liquefaction and slumping below sea level carried away the waterfronts of both towns. In the giant Turnagain Heights (Anchorage, Alaska) slump and earth flow, trees that had stood 20 meters above sea level were swept down and out to sea by the movement. Large areas of farmlands have been destroyed by slides in Switzerland, Czechoslovakia, and Quebec, Canada. Failure of a single slope in Oregon took 24 houses from one city. Another failure along Lake Michigan cost a railroad a quarter of a million dollars.

Engineering solutions   To anchor unstable slopes and to prevent slope failure, geologists and engineers use a number of techniques. One of the commonest and most effective approaches is to reduce the water content of the regolith underlying a slope. Large-scale drainage or drying lowers pore pressure and thus increases the strength of the regolith. On the surface of a slope, ditches of tile or concrete may be used to divert flowing water. More important is the reduction of subsurface water by deep drainage galleries or by drainage borings—excavations and pipes that draw many gallons of water per minute from the regolith. Engineers sometimes increase the stability of a slope by removing the load at the head of the slope or by adding to the toe, or both. Either technique reduces the pulling force along a slope without expensive construction.

If slopes are adjacent to railroads, highways, or buildings, retaining walls may be needed to provide insurance against slope failures. To be effective, such walls must be thick (sometimes several meters). Less-expensive structures that provide some slope stabilization are rock bolts, which are huge bolts sunk through the slope into underlying rock, and pilings, which are most useful on small slopes. Finally, the growth of vegetation increases the stability of a slope. Plants such as alders, poplars, willows, birches, and other trees that draw water rapidly from the regolith, serve to anchor it with deep roots. In regions where mass-wasting is a threat, extensive tree cutting on slopes is a poor practice.

Construction of the Grand Coulee Dam (Washington), the world's largest concrete dam, involved two unusual efforts at controlling and preventing mass-wasting. Silt beds had been deposited against the rock wall of the Columbia River Gorge at the downstream toe of the dam. In 1934, a slump of about 1.5 million cubic meters in the silt threatened lines of communication as well as vehicle access to the dam. This configuration was so unstable that geologists feared that tunneling through the silts to install drainage pipes would cause the very slide they were trying to prevent. This water was safely removed by excavating a tunnel in the bedrock. From the tunnel, holes were drilled to the pockets of water. This approach succeeded in reducing the water content of the mass without triggering further slumps. Later, a troublesome slope failed at another construction site of the Grand Coulee. This slope proved so difficult to control that the engineers elected an elaborate "last-resort" solution: artificially freezing the base of the slope. To the surprise of some observers on the scene, this icy dam successfully held back the silts until excavation could be finished.

Prediction and Prevention of Slope Failure

Early warnings To many people it may seem that a slope fails without warning. The material may rumble downhill with a suddenness that usually prevents the escape of anyone unfortunate enough to be trapped in its path. In reality, however, the only slopes that truly fail without warning are those triggered by earthquakes or by spontaneous liquefaction. All others are preceded by progressive deformation of the regolith or bedrock and by a downhill movement of the entire surface. Rather than saying a slope fails without warning, it is more accurate to say that casual observers fail to detect the warning signs. The slide at Goldau, Switzerland, in 1806, took the villagers by surprise, but several hours before the event, horses and cattle became restless and bees deserted their hives.

Prevention of slope failure Once a slope has started to fail, nothing in the way of prevention can be done, unless the movement is slow. The principal job of the engineering geologist is to prevent a recurrence of old failures and to anticipate future failures on obviously steep or otherwise unstable slopes. Preventive techniques are essentially the same as the engineering solutions described above. Many other methods of preventing slopes from failing have been tried, particularly in emergency situations such as leakage of water into a slope or overloading by unusually heavy rains. One is the artificial hardening of the regolith by the injection of chemicals or cement. Another means of increasing regolith's resistance to shearing is electro-osmosis. In this process water is induced to move out of soil pores by passing an electric current between electrodes driven into the ground. Also, highway cuts in nonsorted glacial sediments have been successfully stabilized by armoring them with a "pavement" of large rocks.

Prediction of slope failure Theoretically, if an engineering geologist knows the essential characteristics of a given slope—the kinds of regolith and rock present, the inclination, the internal structure, the location of all fracture surfaces, and the amount of pore water—he can calculate the chances of slope failure. Using such instruments as a pore-pressure gauge, it is possible to estimate the stability of a slope and the amount of corrective engineering work that should be done to prevent failure. In practice, the prediction of the behavior of slopes, which comes under an engineering subspecialty known as "soil mechanics," is still an inexact discipline. (Notice here the engineer's use of "soil" for what geologists call regolith.) The determination of slope stability by mathematics is only as precise as the information that goes into the equations, and that information is seldom perfect. Despite our best efforts, there is a good chance that slope failures will remain unpredictable for many years to come.

Chapter Review

1. The surface of the Earth is made up of low-lying, nearly horizontal plains underlain by sediment, and inclined areas, or *slopes*, which may be underlain by regolith or bedrock.
2. All the diverse kinds of downslope movement, taken as a group, are known as *mass-wasting*, the displacement of a mass of bedrock or regolith adjoining a slope, in which the center of the mass advances outward and downward.
3. One part of gravity tends to push the material against the slope and thus to keep it in place. This push toward the slope is a force called the *slope-normal component*, or the *resisting force*. The other part of gravity acts along the slope and down the slope. This force is the *slope-parallel component*, or the *pulling force*.
4. The angle of the slope is the only factor that affects the strength of the pulling force. On gentle slopes, the resisting force is larger. On steep slopes, the pulling force is larger. The pulling force tends to make materials slip along parallel planes, a tendency called *shearing*.
5. The resisting force can be reduced by the introduction of such factors as water and fluid pressure, ice, earthquake waves, and slippery materials such as clay minerals. *Pore pressure* can push particles apart and thus overcome the resisting force.
6. The amount of space between particles, or *porosity*, increases abruptly

in a process known as *spontaneous liquefaction*. The force exerted on the soil by downward-percolating water is *seepage pressure*.

7 The pulling force upon a mass of material on a slope can be increased by adding weight to the upper portion of the slope, by increasing the steepness of the slope, and by removing material from the lower part of the slope.

8 Downslope movements can be either rapid or slow. The rapid movements are *rock fall, rock slide, rock avalanche, slump, debris slide,* and *debris flow.* The slow movements are *creep, solifluction, earth flow, talus creep,* and *rock-glacier creep.*

9 Mass-wasting that directly affects humans can be a major source of damages and costs. Highways, railroads, and pipelines are most susceptible to mass-wasting damage, but in some areas farmlands, resorts, and homes are regularly damaged. Geologists and engineers use a number of techniques to increase the stability of slopes, and constant surveillance is underway in attempts to predict and prevent slope failures.

Questions

1 Explain how a stable slope can be made unstable by changing one of the *components of gravity*.
2 Describe how the saturation of a slope by heavy rains can push particles apart. Explain a "quick" condition.
3 Define *spontaneous liquefaction*.
4 How is a slope usually affected by the removal of the toe of a slope?
5 Describe three rapid mass-wasting movements.
6 Describe three slow mass-wasting movements.
7 How are slope failures related to earthquakes?
8 What kinds of areas are most vulnerable to slope failure and mass-wasting damage?
9 How can slope failure be prevented? How can it be predicted?

Suggested Readings

Bloom, A. L., *The Surface of the Earth.* Englewood Cliffs, N. J.: Prentice-Hall, Inc., 1969.

Crandell, D. R., "The Geologic Story of Mount Rainier." U. S. Geological Survey, Bulletin 1292, 1969.

U. S. Geological Survey. "Alaska's Good Friday Earthquake." Washington, D.C.: U.S. Government Printing Office, Circular 491, 1964.

Wilson, Mitchell, *Energy*. New York: Time-Life Books, 1966.

"The Emscher River in Germany contains so much chemical waste that its waters will develop photographic film."

Chapter Eight
Running Water As a Geologic Agent

Running water is nature's most pervasive agent of erosion. We can see the result of persistent running water most dramatically in the Grand Canyon. Whether by the flowing, trickling, or dropping of water, this massive ditch was carved out of rock—it took millions of years, but the work was accomplished by water. Later we shall see how even a tiny raindrop contributes to the reshaping of the landscape.

In the next three chapters we discuss various aspects of water. In Chapter 8 we concentrate on the flowing water of streams and rivers. In Chapter 9 the subject is underground water. In Chapter 10 we examine water in its solid form—ice and glaciers. Each of these chapters deals with specialized parts of the general topic of water. Before becoming involved with these special topics we will look closely at the hydrologic cycle and at the over-all problem of water supply. Then we shall see how running water affects people, how water transports and deposits its loads, and summarize running water's effects as an agent of landscape change.

The Hydrologic Cycle

As we have seen, solar energy keeps much of the Earth's water constantly moving. As a result, the rains keep falling and water from the rain directly and indirectly keeps streams supplied with water. The Earth's water and its circulation in the hydrologic cycle are essential to all life on Earth. Moreover, wherever it rains we can see the geologic effects of running water.

A fundamental concept of **hydrology,** the study of water, is that all water on the surface of the Earth is participating in the *hydrologic cycle* (Figure 8.1). The hydrologic cycle includes water in all three of its states: *solid* (polar ice caps and glaciers), *liquid* (both fresh and salt water), and *gas* (water vapor carried by the atmosphere).

The water of the hydrologic cycle is distributed in three reservoir systems. In order of importance, these are: (1) the oceans, (2) the continents, and (3) the atmosphere. About 97.3 per cent of the total water is in the oceans. The continents, including the glaciers of the Arctic and the Antarctic, hold only 2.7 per cent. Only one hundred-thousandth of the Earth's water is borne by the atmosphere at any one time. The total quantity of the Earth's water has recently been estimated at 1.5 billion cubic kilometers.

Water may enter the atmosphere either directly by evaporation or through the leaves of plants by **transpiration.** Water vapor may return to the earth as precipitation almost immediately, or it may

8.1 Hydrologic cycle, shown in schematic profile. Total evaporation from world's oceans exceeds precipitation back to the ocean by 9 per cent. This "extra" water falls on the land. Further movements of water explained in text. Not shown are evaporation from land and transpiration from plants.

travel thousands of kilometers with high-altitude air currents. Although only a tiny amount of water circulates in the atmosphere, this amount is of enormous importance in the formation of clouds, the determination of climate, and the maintenance of the entire atmospheric energy budget.

Running water depends upon precipitation—rain, snow, sleet, hail, and any other moisture which falls from the atmosphere to the ground. (Dew, frost, and clouds are not classed as precipitation because they do not fall.) Worldwide rainfall averages about 0.3 meter per year but this amount is not distributed evenly, either geographically or seasonally.

Precipitation may be disposed of in several ways. In its natural, undisturbed state typical regolith is capable of absorbing most or all of a light or moderate rainfall. The seepage of rainwater into the regolith is known as **infiltration.** Rainwater sinks into the Earth through natural passageways between particles of regolith or through cracks in bedrock.

If rainwater accumulates faster than it can infiltrate, the excess flows as a surface sheet or film in a downhill flow called **runoff.** Most of this runoff becomes organized into flows in definite channels, or **stream flows.** We shall use **streams** as a general term to include all water flowing in channels, however small or large, ranging from tiny rivulets to mighty rivers.

Water and the Biosphere

All one needs to do to appreciate the importance of water is to try going without it. In most parts of the world public water supplies are adequate and so reliable that most of us take water for granted. We naturally suppose that all we have to do is turn on the faucet and fresh, pure water will emerge. How many of us ever stop to think just how much water we use every day and where it all comes from?

Consumption of Water

In planning for the water supply of a small town lacking large industry, water engineers estimate that 100 gallons of water per person per day must be available. This means that each year 36,500 gallons of water are needed for each individual. Do you have any idea how much water that is? Suppose you were able to catch and store, without any losses, all the rain that fell during a year. How much area would you need? The answer, of course, depends on how much it rains. But if we take the world's average figure of 0.3 meter per year, then the area required is 12.5 square meters, about the size of an average room.

There is a slight "catch" in our experiment of collecting rainfall. Actually, we cannot trap all the rain that falls. In a natural setting about one-third of the water is "captured" by plants, another third seeps into the ground, and the last third flows over the surface of the land. The overland flow plus seepage out of the ground supplies the water to rivers.

Household uses of water The most obvious use of fresh water is for drinking. In order to maintain the human body's proportion of water at 72 per cent of body weight we must drink an average of 2.6 quarts per day, which is 241 gallons per year. Other household uses include laundry, bathing, cleaning, flushing toilets, garden and lawn use, cooking, and dish washing.

Industrial uses of water In the United States all industrial uses require 250 billion gallons of water per day. The largest amount of this water is used for cooling (90 to 95 per cent), and other uses include cleaning, conveying, and the manufacture of beverages such as soft drinks and beer. One to two tons of water are needed to manufacture a ton of bricks, 20 tons for a ton of gasoline, 200 tons for a ton of steel, and 600 tons for a ton of nitrate fertilizer (Table 8.1).

Agricultural uses of water Even greater than industrial needs are demands for water by farmers in dry regions of the world, where rainfall is inadequate for crop survival. To grow a ton of sugar or corn by irrigation, about 1000 tons of water are "consumed," or changed by evaporation from soil and transpiration through plants into water vapor. A ton of wheat may use 1500 tons of water, a ton of rice 400 tons of water, and a ton of cotton fiber 10,000 tons of water. Unfortunately, water losses during irrigation are highest where the water is needed most—in hot, arid climates.

Water Problems: Uneven Distribution

In the United States we are currently using one-quarter of the available water, even though much of this water has already been used

Table 8.1 QUANTITIES OF WATER REQUIRED TO PRODUCE GOODS AND SERVICES

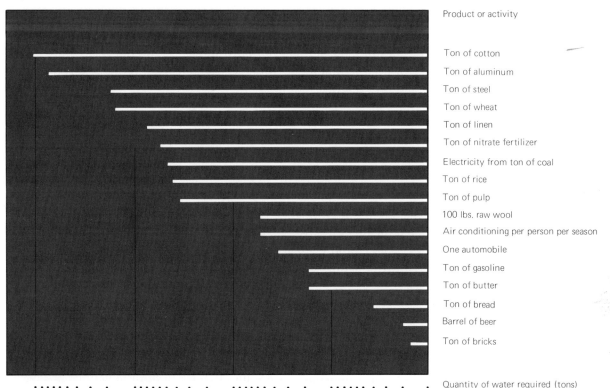

more than once through recycling. The rest of the water runs off to the sea. This illustrates that our total usage is nowhere near the total water available, and yet local problems of water shortage exist because of uneven distribution.

For example, the lack of sufficient rainfall in the southwestern United States could prevent further agricultural expansion. The alternative is industrial activities which consume less water.

Water, then, is very often in the wrong place in the wrong form. Acute shortages exist where the population density has exceeded the water supply. Such shortages will continue until either the people or the water can somehow be distributed more evenly. One of the major problems facing hydraulic engineers is the transportation of water from areas having abundant rainfall (such as the American northwest) to population centers lacking adequate water supplies (such as Los Angeles).

Rivers and Civilization

People have always settled close to rivers to use them for transportation and because they represent an easy supply of water. These advantages have to be offset against the tendency of rivers to flood. In recent times, because of our great impact on drainage systems, we are creating environmental problems which have become severe, as we shall see in later sections.

Uses of river water In addition to industrial, agricultural, and public water supplies, we use river water to generate electricity and to dispose of waste.

Until 1975 the consumption of electricity in the United States was increasing at the rate of about 7 per cent a year. In 1940, 36 per cent of our electricity came from hydroelectric generators, that is, generators turned by the kinetic energy of falling water. Most of the great dams in this country, such as Hoover Dam, Bonneville Dam (Figure 8.2), and Grand Coulee Dam, were built to produce electricity. However, it is estimated that in 1980 only 15 per cent of our electricity will come from hydroelectric sources. Most suitable river sites in this country are already being used to generate power, and thus we cannot easily expand hydroelectric power. We must use other means of generating electricity.

Nearly everyone is familiar with the extensive pollution of rivers by sewage in the United States. In 1936, the states of New Jersey, New York, and Connecticut were producing 275 million gallons of sewage per day; by 1971 this figure had climbed to 2.2 billion gallons per

day. Processing plants for this flood of waste cannot be built quickly or cheaply. With the passage of the National Environmental Policy Act of 1969, hopes rose that our rivers would soon be clean again, but this act has been slow to take effect.

8.2 Bonneville Dam, on the Columbia River, in Washington State, is the site of a great hydroelectric generating plant. (Department of the Army, North Pacific Division Corps of Engineers)

Floods and flood control River flooding, a natural process that is the largest factor in eroding watersheds and shaping river channels, is disastrous to human settlements (Figure 8.3). The low, flat areas next to the stream channel that are occasionally inundated with water are **flood plains.** Although people have consistently built on flood plains, they have done so without regard for the geologists' dictum that "the flood plain belongs to the river," and that floods are a normal part of the history of flood plains. In fact, flood plains are a natural storing place for flood water. By not building on flood plains, the river water is allowed to spread outward without causing distress to people.

8.3 Houses in the flood plain of the Meramec River, a tributary of the Mississippi River, were flooded by backwater from the Mississippi on April 5, 1973. On April 28 the Mississippi completely overpowered the Meramec and caused it to turn backward. (Charles B. Belt)

Many of the world's great rivers are subject to repeated floodings, causing heavy loss of life and property. One of the greatest floods in history struck China in 1887, when the Hwang Ho flooded 130,000 square kilometers, causing the direct death of a million people and the starvation of a million more because of crop destruction. Thus, geologists have long studied floods and their frequency of recurrence in order to minimize damage to population centers and agriculture. One of the most-employed means has been the construction of artificial levees (Figure 8.4), raising the natural banks of a river so that water is confined to its channel. So much extensive levee construction has taken place along the flood-prone Mississippi that only small parts of the lower valley now display natural conditions.

Environmental impact upon rivers Humans have affected both the quantity and quality of water flowing in streams. Each time a dam is built the downstream flow is reduced, and large areas are subjected to evaporation. Also, the current practice in Europe and North America of straightening and reinforcing river channels can seriously alter stream dynamics by speeding local river flow, and producing more erosion downstream. In southern California, this downstream effect of upstream change has forced engineers to build complete concrete channels for many streams from city to ocean; not to do so would increase the flood hazard and potential damage to each succeeding city downstream.

8.4 Artificial levees built to contain the Mississippi River. The levees are indicated on the east side by a white strip along the river. This is an aerial view looking north toward St. Louis, Missouri, 1973. (Charles B. Belt)

There are limits to how much water can be removed from a river. The increasing use of water from the Rio Grande River by farmers in the southwest has become the focus of an international dispute. As the flow has shrunk downstream, the relative amount of salt dissolved in the water has risen. Mexican farmers object that by the time it reaches them the river water is too salty for use on their farms.

Some streams have become industrial sewers. In the Ruhr district of Germany, the Emscher River contains so much sodium waste from chemical plants that its waters will develop photographic film! In the early 1970s the polluted waters of the Cuyahoga River in Cleveland were ignited and burned for a couple of days. Rivers like the Emscher and the Cuyahoga have obviously exceeded their self-purification capabilities. One of the things we need to know about rivers is the limit of their abilities to absorb waste products. (One proposal to allow rivers to perform their natural cleaning activities has been to demand that industrial users discharge their waste water upstream instead of downstream.)

Finally, at the same time that human demand for river water rises, the pressure to preserve wild rivers has grown. With the growth of the environmental movement in the late 1960s and 1970s, people became concerned over the loss of rivers that used to flow freely and cleanly without damming or diversion. Environmentalists fought what they regarded as unnecessary construction of dams and artificial channels, objecting to the obliteration of scenic beauty and to the destruction of rich plant and animal communities. Such rivers as the Tombigbee in Louisiana, and even the mighty Colorado became the subject of complex lawsuits between those who would modify them and those who would leave them wild.

The Mechanics of Streams

There are many interrelated factors about stream mechanics. They all interact simultaneously, but in order to understand them we must discuss them one at a time. The chief factors are energy relationships, discharge, speed, slope, load, and shape of the channel's cross section.

Energy Relationships

As we mentioned earlier, the basic energy relationships in streams begin with potential energy and its conversion to kinetic energy and heat. The potential energy is represented by the quantity of water and the distance through which it will fall. As soon as this water begins to move, the energy becomes kinetic, and the kinetic energy is used up in the flow of the water, its geologic work, and heat losses because of friction. Heat losses are scarcely measureable factors, but they have been shown to exist.

Discharge

The volume of water flowing through a cross section of the stream channel is its **discharge** (Figure 8.5). Stream discharge is measured in volume per unit of time (cubic meters per second, abbreviated m^3/sec). A small creek may discharge only 5 m^3/sec; a small river perhaps 50 m^3/sec. The largest rivers discharge hundreds of thousands of cubic meters per second. The mean discharge of the Mississippi River at New Orleans, Louisiana is 81,000 m^3/sec and the range is from 40,500 to 337,500 m^3/sec. The largest discharge of any river in the world was measured during 1963 and 1964 in the Amazon River by members of a joint U.S.-Brazilian expedition. The investigators found that at Obidos, Brazil, some 650 km upstream from the river's mouth, the Amazon discharges about 1.5 million m^3/sec. This discharge is four times that of the Congo River and 10 times that of the Mississippi. The investigation showed that the discharge of the Amazon is about twice as large as previously supposed; it represents about 15 per cent of the total world river discharge.

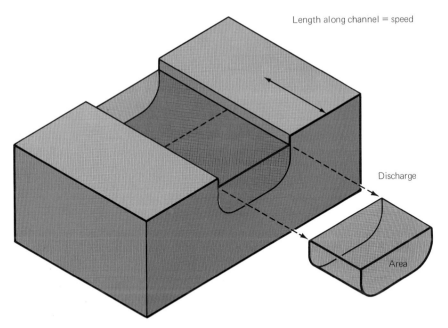

8.5 Discharge of a stream shown schematically by removing the amount of water that has flowed along it in one second. Each parcel of water cannot move until the parcel downstream from it has flowed onward. As each parcel passes down the channel its place is taken by another parcel from upstream.

In order to maintain a river flow of 1 m³/sec, an annual rainfall of 75 cm is required on an area of 130 square kilometers.

The discharge consists of water from two sources: (1) the *base flow*, which comes from underground water seeping into the stream valleys, and (2) *flood bulges*, long waves of water that are dumped rapidly into a stream system and that travel independently of the base flow. If the supply of base-flow water gives out, the river simply ceases to exist as a "permanent stream," but becomes an *intermittent stream*. When the flood bulge has flowed out of a given part of the stream, continued base flow maintains the discharge.

The base-flow water is backed up continuously all the way from the ocean (or a lake). Water that is locked into the base flow simply cannot flow any faster than the water downstream, which must "get out of the way" in order to let new water move downstream. Thus, the base flow can be compared to a horizontal conveyor belt at large airports. A flood bulge is analogous to a person walking on a conveyor belt in the same direction as it is traveling. The flood-bulge water moves at its own speed, but adds this to the speed of the base flow. Thus, if the speed of the base flow is 3 km per hour and that of a flood bulge 8 km per hour, the speed of flow of the water in the flood bulge is 11 km per hour (3 km/hr for base flow +8 km/hr for flood bulge). If the flood water did not make such a moving wave it could not travel faster than the base flow.

Running Water As a Geologic Agent

Speed

In order to calculate the discharge of a stream or river the speed of the water must be known. To make such measurements, the U.S. Geological Survey maintains more than 6000 stations on important streams.

The location of maximum speed of flow depends upon the channel shape, roughness, and sinuosity, but usually lies near the center of the channel, below the surface less than a quarter of the way to the bottom (Figure 8.6). Speed of flow decreases toward the sides of the channel and the bottom. Speed also increases downstream—a fact which surprises many people. Everyone knows that rivers grow wider and deeper downstream, but it is not so apparent that speed increases as well. A wild leaping mountain stream spends much of its energy creating turbulent eddies with almost as much backward as forward motion. For example, in a downstream direction speed increases from the tiny Owl Creek in Wyoming (0.6 m/sec) to the Bighorn River in Montana (0.75 m/sec) to the Mississippi River itself at Vicksburg, Mississippi (1.5 m/sec). One of the greatest speeds measured by the U.S. Geological Survey was 6.6 meters per second in a rocky gorge of the Potomac River during the flood of March 1936. The highest natural speed for running water is about 9 meters per second (32 km/hr).

8.6 Speeds of flow of water in various parts of the channel; transverse profile through a stream. Data from River Klarälven, southern Sweden, collected from current meters spaced at 10-meter intervals across the stream and at 0.5-meter intervals of depth. Lines show equal speeds in centimeters per second. (Åke Sundborg, 1956)

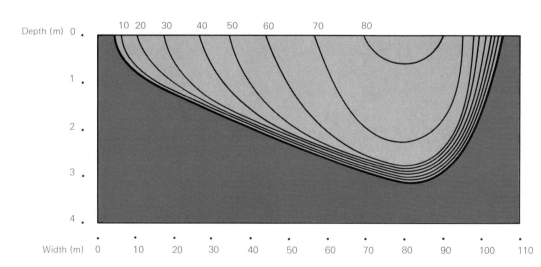

Slope, Profile, and Base Level

The slope on which a stream flows is its **gradient.** Near their sources the gradients of almost all streams and rivers are steeper than farther downstream. The aggregate of all the local gradients constitutes the **longitudinal profile** of a stream. The lower end of each stream profile is the sea or a lake. The level of water in the sea or a lake forms a **base level** for entering streams. Base level regulates all activities of the stream.

The longitudinal profile of a stream is concave up, or rounded upward like the inside of a bowl (Figure 8.7). The main reason for this concavity is that, in general, discharge increases downstream. As volume increases, the channel becomes more efficient.

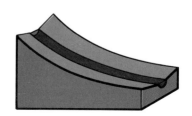

8.7 Longitudinal profile of a stream shown as a schematic block diagram.

Channels

The channel of a stream may be envisioned as a long, narrow trough. The shape of the trough is always the most effective design for moving the amount of water and sediment supplied from a specific drainage basin. To maintain this effectiveness, the river constantly redesigns its channel.

The shape or cross section of a channel is often expressed as the form ratio: depth divided by width. Thus a river 1 meter deep and 100 meters wide has a form ratio of 1/100. The Columbia River near the Canadian border has a form ratio of 1/19 in the summer; at the same time of year, the Platte River near Duncan, Nebraska, has a form ratio of 1/160. The Columbia River, relatively deeper, can carry its load with a gradient of only 0.66 meters per km, whereas the Platte River, with a wide, shallow cross section, needs a gradient of 1.7 meters per km to transport its load.

Load

The geologic "work" of a river includes picking up sediment and transporting it downstream. A stream's capacity to pick up and carry sediment depends upon several factors, such as speed, gradient, and discharge. Capacity to move sediment along the bottom of the channel (named the stream's bed) goes up as the third to fourth power of the speed; if a stream's speed doubles during a flood, it can move 8 to 16 times as much sediment along its beds. Members of Major John Wesley Powell's pioneer expedition (see the Field Trip in this chapter), which descended the Colorado River through the Grand Canyon in

1871–1872, wrote of the sounds of huge boulders being rolled over at night by the swift current.

Concept of a Graded Stream

The interaction of all the factors just discussed establishes a condition of dynamic balance which characterizes a **graded stream.** Grade is maintained by constant changes in slope, discharge, and cross-sectional area.

As we have mentioned, a stream is capable of reacting to changes in its environment, almost in the same way a living creature can react. An increase or decrease in load, the total material a river carries, causes it to make a corresponding change in gradient. If the river cannot move its load beyond a certain point on the profile, it will increase its gradient. The gradient can be steepened by depositing some of its load at that point, building up the channel bed and creating a steeper slope below the point. This build-up at the same time decreases the slope above the point, so that the river continues the process in the upstream direction until the entire slope has been adjusted.

The stream has many built-in self-adjusting mechanisms. They all operate simultaneously to maintain a graded condition, which may be thought of as a condition of dynamic balance.

Laminar and Turbulent Flow

In order to see how streams operate, we need to examine how water flows. Water, like other fluids, bends and flows easily in response to a force, such as gravity. Masses of water can move by either laminar flow or turbulent flow. **Laminar flow** (from the Latin for blade) can take place only in smooth, straight channels at extremely low speeds (less than a millimeter per second), as parallel layers of water shear past one another as smooth planes (Figure 8.8). Flow also approaches laminar type at the lip of a waterfall, where the top layers of water suddenly shoot past the bottom layers. In general, laminar flow is not found in natural streams. The nearest approximation occurs along the bed and banks, where speed approaches zero.

Far more common is **turbulent flow,** characterized by a variety of vortices and eddies that are continually forming and disappearing. If we could keep track of an individual water molecule in a turbulent flow, we would see it following a looping, forward-and-backward corkscrew kind of motion. The upward movement of flow paths within turbulent eddies provides the force for lifting (or suspending)

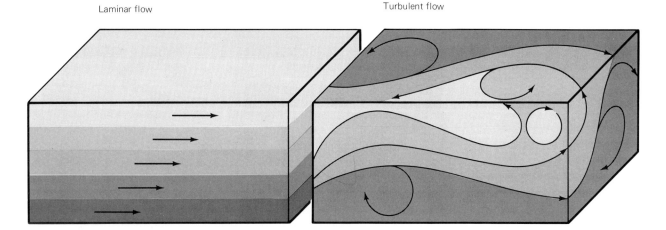

8.8 **Laminar and turbulent flow.** Compare smooth planes slipping one over another in laminar flow (at left) with complex curved flow paths (shown by arrows) in turbulent flow (at right).

fine sediments; without this "lift," such sediment would concentrate near the bottoms of streams.

The Geologic Work of Running Water

The flowing water does three things: (1) It erodes bedrock, (2) it carries chemical load in solution, and (3) it transports solid particles of sediment.

Erosion of bedrock Bedrock is eroded by a combination of drilling, sawing, and corroding. The drilling and sawing take place at waterfalls. The falling water drills the familiar structures known as **potholes** (Figure 8.9). Potholes form easily in weak or soluble rock where the hole is persistently abraded and smoothed. Most potholes are only a meter or so in diameter and less than half a meter deep, but giants 20 meters across and 6 meters deep or larger are occasionally found.

Any reactions between chemicals dissolved in stream water and the rocks of the stream bed are known as **corrosion.** In humid climates, stream channels are often determined by outcrops of soluble rock formations. Limestone, which is regarded as a hard and durable rock in arid regions, is particularly susceptible to corrosion in humid areas. The bicarbonate ions that are found in virtually all stream water easily dissolve calcium carbonate to produce deep valleys.

Dissolved load Although the load carried by a stream in solution is usually invisible, the dissolved load may weigh as much or more than the visible sediment load. An estimated 4000 million tons of

8.9 Potholes in limestone bed of Catskill Creek, northwest of Catskill, New York. (J. E. Sanders)

soluble matter is transported each year from the continents to the oceans. The dissolved load of the Delaware River at Trenton, New Jersey, is 830,000 tons per year; of the Colorado River near Cisco, Utah, 4.4 million tons per year; and of the Mississippi River at Red River Landing, Louisiana, 101.8 million tons per year. In a sample of about 70 rivers in different regions of the United States, approximately 20 per cent of the total load is carried in solution, varying from 1 per cent in the Little Colorado River in Arizona to 64 percent in the Juniata River in Pennsylvania.

Sediment load According to its way of being moved, a stream's sediment load is divided into suspended load and bed load. Particles dropped in still water sink at various *settling rates*. The Earth's gravity pulls the particles downward in proportion to their diameters, shapes, and specific gravities. A particle of coarse clay may settle through still water at a rate of 0.00023 cm/sec, whereas coarse sand may settle at 20 cm/sec (Figure 8.10). In order for these particles to be transported by the stream the water must exert a force on them that overcomes their tendencies to settle. Particles that are supported by water that follows the upward flow paths within turbulent eddies in a stream constitute the **suspended load** (Figure 8.11).

Small particles, such as clay and silt, are most easily kept in suspension. When these particles are abundant, as during floods, they cause the water to appear murky and turbid.

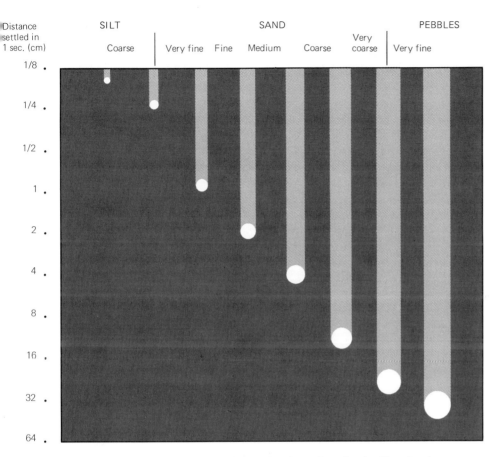

8.10 Settling speeds of quartz spheres of various diameters in still water shown by lengths of lines measured down from top of graph (to represent vertical distances particles would settle in 1 second). (Data from R. J. Gibbs, et al., 1971)

Most of the suspended sediment comes not from the channel walls and bed of a stream but from mass-wasting and gullying of slopes in the river's tributaries. In fact, the amount of suspended sediment is directly proportional to the amount of gullying on hillsides.

Particles that slide, roll, and bounce along the stream bed make up the **bed load.** The separation of suspended load and bed load is not always distinct: as conditions of flow change, particles of the bed load may suddenly become suspended, or vice-versa. Particles of sand, gravel, and small cobbles tend to move as bed load, either individually or in groups. Without being in the suspended load a particle may rise a few centimeters to as much as 30 cm above the bottom, travel downstream some distance, and then be pulled back by gravity (Figure 8.11). Such movement is known as *saltation* (Latin for leap).

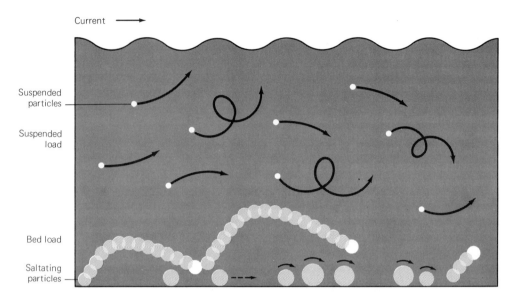

8.11 Sediment load of a stream shown by schematic longitudinal profile through water. Successive positions of saltating particles in bed load shown by overlapping circles.

Measurement of bed load is not easy, and even estimates are often far from actuality. Investigators generally agree that bed load makes up about 10 per cent of the total solid load, but in fast-moving rivers may exceed 50 per cent.

Once in motion, large particles will move faster than smaller ones, and rounded particles more easily than angular or flat shapes. The largest particle a stream can move is a measure of the stream's *competence.* Laboratory experiments indicate that the competence of a stream varies as the sixth power of the speed. Thus a slight increase in speed brings about a great increase in competence—another illustration of why most stream "work" is performed during floods. As large rocks or cobbles bounce along the stream bed, they may shatter, smooth, or dislodge other particles in the bed.

Particles in the bed load collide repeatedly. These collisions may produce smaller fragments, or round off the corners of angular particles. The rounding of particles is known as **abrasion.** The effectiveness of abrasion is apparent to anyone who has noticed the smooth, rounded pebbles along a stream (see Figure 6.18) or beach.

Having summarized the mechanics of streams and their loads, let us now examine the geologic effects of streams. There are two contrasting realms of streams. One is illustrated by streams that erode bedrock. In such streams the bedrock inhibits the sideways movement of the channels. Therefore, the main work of the stream is to cut downward. Downcutting is usually confined to upstream areas that are being uplifted. The contrasting situation is where streams are flowing entirely on their own deposits. These are usually subsiding areas downstream where the channel is free to shift sideways.

In the next sections we will isolate the eroding activities from the depositional activities. Finally, we will see how these activities respond to major geologic changes.

Erosion by Running Water

In this section we will look at the way in which running water and associated activities erode bedrock. We will first consider rivers and valleys, then patterns of rivers, and finally some ideas of what will happen if erosion continues through long periods of time.

Before the nineteenth century a debate was taking place over the subject of the relationship between rivers and valleys. One group argued that valleys were great cracks that had formed by splitting open the Earth's crust. The premise of this argument was that the valley was formed by the splitting, and the river merely found it and flowed through it. However, Playfair's law, discussed in the following section, provided a different viewpoint.

Playfair's Law

In 1802, James Hutton's friend, the English mathematician John Playfair, stated Hutton's ideas that streams do not simply occupy valleys, but create them. This principle, which has come to be known as **Playfair's law,** was expressed as follows:

> Every river appears to consist of a main trunk, fed from a variety of branches, each running in a valley proportioned to its size, and all of them together forming a system of vallies, communicating with one another, and having such a nice adjustment of their declivities, that none of them join the principal valley, either on too high or too low a level; a circumstance which would be infinitely improbable, if each of these vallies were not the work of the stream that flows in it.

Despite some exceptions, Playfair's law is now generally accepted and forms the basis for most modern studies of rivers.

Rivers create valleys by two contrasting processes: (1) headward erosion, and (2) changes in the valley-side slopes.

Running Water As a Geologic Agent

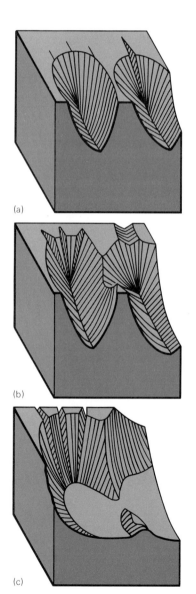

8.12 **Erosion of inactive volcanic cone,** Hawaii, by streams and mass-wasting. Rapid erosion of steep-sided gorge takes place by retreat of waterfalls. Later, side slopes retreat and valleys widen. (Based on G. A. Macdonald and A. T. Abbott, 1970)

Headward Erosion

The formation of a valley begins soon after "new" land is raised above sea level. The sequence has been studied in Hawaii on inactive volcanic cones (Figure 8.12). Rain collects in cracks and flows downhill. Tiny rivulets form and lead into larger trickles of water which come together in small streamlets. Eventually, at the lowest level, the streamlets come together in a master stream. The master stream cuts downward rapidly, high in energy because of its steep slope. Where the stream plunges steeply as waterfalls, it cuts backward, moving the waterfall as it goes. Other erosion is accomplished by drilling potholes, by loosening blocks bounded by fractures, by scouring, by dis-

8.13 **Steep gorge** formed by retreat of waterfall, leaving behind a row of potholes. Watkins Glen State Park, New York. (Nat Messik)

solving, and by hydraulically downcutting the stream bed. This leaves behind a steep-sided gorge (Figure 8.13).

After this gorge has formed, the main work of channel erosion is completed, and further activities take place on the side slopes.

Factors Shaping Valley-Side Slopes

Although rivers are responsible for carving valleys, they do not erode the entire volume of sediment removed from the cross profile of a valley. Most of the sediment travels from the valley slopes to the stream by mass-wasting (Figure 8.14).

The exact behavior of valley-side slopes has been the subject of some controversy. It is still not agreed what happens after the original slot has been cut. Everyone agrees that the next stage is the formation

8.14 Slot-like gorge cut by retreating waterfall is direct result of stream erosion; typical V-shaped valley profile results from transport of material down valley sides by mass-wasting.

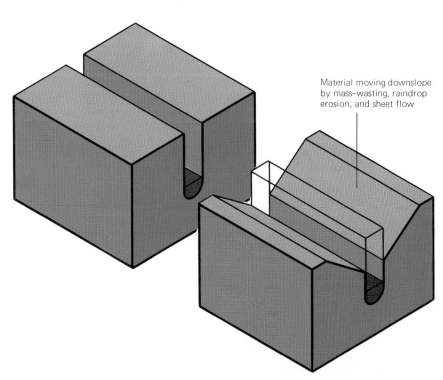

of a V-shaped cross section, which is typical of stream valleys. One opinion is that the V-shaped cross section gradually widens like a book being opened and laid flat (Figure 8.15a). The other viewpoint, which has gained support from recent field observations, is that the sides of the V retreat more or less parallel to each other (Figure 8.15b).

The important processes operating on valley-side slopes are the mass-wasting processes discussed in Chapter 7. In addition, there are several other processes we will now discuss: those caused by raindrops and thin sheets of water.

Raindrops Until recently, no one took the splashing of raindrops very seriously as agents of erosion. Before the 1940s, soil erosion was thought of as the washing of bare soil by flowing water. It was then discovered, however, that wash-off losses actually account for less than 10 per cent of soil removal by a rainstorm. The rest is attributable to raindrop splashes, which throw tiny silt-size particles into the air and drop them in new positions (Figure 8.16). If a board is

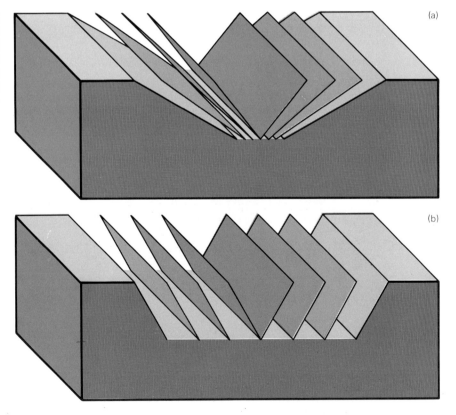

8.15 Contrasting ideas about behavior through time of valley-side slopes. (a) Concept of gradual lowering of slopes through time. (b) Concept of valley widening by retreat of slopes having essentially constant inclinations.

left lying on bare soil during a rainstorm, its upper surface will be coated with silt thrown up by raindrops. On sloping ground, soil splashed away from the slopes tends to fall back vertically, and thus to be displaced downhill.

Sheet flows After soil pores are clogged (a process known as surface sealing), puddles form and surface water begins to flow in rather uniform thin layers known as **sheet flows.** The thin sheets of water exert a dragging force over the soil. Depending upon their speeds of flow, the sheets pick up particles ranging in size from fine clay to sand. Sheet flows may also carry off valuable organic matter that has been dissolved from soil aggregates by rain splash. Sheet flows and rain splash are most severe on steep slopes where the water moves at greater speed.

Rills, gullies, and badlands On steep slopes, during heavy rainfall, sheet flows become more localized and eventually begin to carve countless tiny channels in the soil called **rills.** If rills continue unchecked by erosion-control practices such as tillage and mulching, rills may grow into even deeper channels called **gullies.** Gullies, in turn, may form a landscape of closely spaced V-shaped valleys known as **badlands** (Figure 8.17).

8.16 Impact crater of raindrop falling on wet soil. The tinier water droplets, formed when the large drop disintegrates, travel upward and outward, carrying along silt particles. (U.S. Soil Conservation Service, Food and Agricultural Organization)

8.17 Badlands created by stream erosion in an arid region. Death Valley, California. (U.S. National Park Service)

The Grand Canyon
John Wesley Powell

The Grand Canyon is a gorge 217 miles in length, through which flows a great river with many storm-born tributaries. It has a winding way, as rivers are wont to have. Its banks are vast structures of adamant, piled up in forms rarely seen in the mountains.

Down by the river the walls are composed of black gneiss, slates, and schists, all greatly implicated and traversed by dikes of granite. Let this formation be called the black gneiss. It is usually about 800 feet in thickness.

Then over the black gneiss are found 800 feet of quartzites, usually in very thin beds of many colors, but exceedingly hard, and ringing under the hammer like phonolite. These beds are dipping and unconformable with the rocks above; while they make but 800 feet of the wall or less, they have a geological thickness of 12,000 feet. Set up a row of books aslant; it is 10 inches from the shelf to the top of the line of books, but there may be 3 feet of the books measured directly through the leaves. So these quartzites are aslant, and though of great geologic thickness, they make but 800 feet of the wall. Your books may have many-colored bindings and differ greatly in their contents; so these quartzites vary greatly from place to place along the wall, and in many places they entirely disappear. let us call this formation the variegated quartzite.

Above the quartzites there are 500 feet of sandstones. They are of a greenish hue, but are mottled with spots of brown and black by iron stains. They usually stand in a bold cliff, weathered in alcoves. Let this formation be called the cliff sandstone.

Above the cliff sandstone there are 700 feet of bedded sandstones and limestones, which are massive sometimes and sometimes broken into thin strata. These rocks are often weathered in deep alcoves. Let this formation be called the alcove sandstone.

Over the alcove sandstone there are 1,600 feet of limestome, in many places a beautiful marble, as in Marble Canyon. As it appears along the Grand Canyon it is always stained a brilliant red, for immediately over it there are thin seams of iron, and the storms have painted these limestones with pigments from above. Altogether this is the red-wall group. It is chiefly limestone. Let it be called the red wall limestone.

Above the red wall there are 800 feet of gray and bright red sandstone, alternating in beds that look like vast ribbons of landscape. Let it be called the banded sandstone.

And over all, at the top of the wall, is the Aubrey limestone, 1,000 feet in thickness. This Aubrey has much gypsum in it, great beds of alabaster that are pure white in comparison with the great body of limestone below. In the same limestone there are enormous beds of chert, agates, and carnelians. This limestone is especially remarkable for its pinnacles and towers. Let it be called the tower limestone.

Stand at some point on the brink of the Grand Canyon where you can overlook the river, and the details of the structure, the vast labyrinth of gorges of which it is composed, are scarcely noticed; the elements are lost in the ground effect, and a broad, deep, flaring gorge of many colors is seen. But stand down among these gorges and the landscape seems to be composed of huge vertical elements of wonderful form. Above it is an open, sunny gorge; below, it is deep and gloomy. Above, it is a chasm; below, it is a stairway from gloom to heaven.

The Grand Canyon of the Colorado is a canyon composed of many canyons. It is a composite of thousands, of tens of thousands of gorges. In like manner, each wall of the canyon is a composite structure, a wall composed of many walls, but never a repetition. Every one of these almost innumerable gorges is a wall of beauty in itself. In the Grand Canyon there are thousands of gorges like that below Niagara Falls, and there are a thousand Yosemites. Yet all these canyons unite to form one grand canyon, the most sublime spectacle on the earth. Pluck up Mt. Washington by the roots to the level of the sea and drop it headfirst into the Grand Canyon, and the dam will not force its waters over the walls. Pluck up the Blue Ridge and hurl it into the Grand Canyon, and it will not fill it.

The carving of the Grand Canyon is the work of rains and rivers. The vast

From THE EXPLORATION OF THE COLORADO RIVER AND ITS CANYONS, 1895.

Field Trip

labyrinth of the canyon by which the plateau region drained by the Colorado is dissected is also the work of waters. Every river has excavated its own gorge and every creek has excavated its gorge. When a shower comes in this land, the rills carve canyons—but a little at each storm; and though storms are far apart and the heavens above are cloudless for most of the days of the year, still, years are plenty in the ages, and an intermittent rill called to life by a shower can do much work in centuries of centuries.

The erosion represented in the canyons, although vast, is but a small part of the great erosion of the region, for between the cliffs blocks have been carried away far superior in magnitude to those necessary to fill the canyons. Probably there is no portion of the whole region from which there have not been more than a thousand feet degraded, and there are districts from which more than 30,000 feet of rock have been carried away. Although, there is a district of country more than 200,000 square miles in extent from which on the average more than 6,000 feet have been eroded. Consider a rock 200,000 square miles in extent and a mile in thickness, against which the clouds have hurled their storms and beat it into sands and the rills have carried the sands into the creeks and the creeks have carried them into the rivers and the Colorado has carried them into the sea. We think of the mountains as forming clouds above their brows, but the clouds have formed the mountains. Great continental blocks are upheaved from beneath the sea by internal geologic forces that fashion the earth. Then the wandering clouds, the tempest-bearing clouds, the rainbow-decked clouds, with mighty power and with wonderful skill, carve out valleys and canyons and fashion hills and cliffs and mountains. The clouds are the artists sublime.

Besides the elements of form, there are elements of color, for here the colors of the heavens are rivaled by the colors of the rocks. The rainbow is not more replete with hues. But form and color do not exhaust all the divine qualities of the Grand Canyon. It is the land of music. The river thunders in perpetual roar, swelling in floods of music when the storm gods play upon the rocks and fading away in soft and low murmurs when the infinite blue of heaven is unveiled. With the melody of the great tide rising and falling, swelling and vanishing forever, other melodies are heard in the gorges of the lateral canyons, while the waters plunge in the rapids among the rocks or leap in great cataracts.

The glories and the beauties of form, color and sound unite in the Grand Canyon—forms unrivaled even by the mountains, colors that vie with sunsets, and sounds that span the diapason from tempest to tinkling raindrop, from cataract to bubbling foundation. But more: it is a vast district of country. Were it a valley plain it would make a state. It can be seen only in parts from hour to hour and from day to day and from week to week and from month to month. A year scarcely suffices to see it all. It has infinite variety, and no part is ever duplicated. Its colors, though many and complex at any instant, change with the ascending and declining sun; lights and shadows appear and vanish with the passing clouds, and the changing seasons mark their passage in changing colors. You cannot see the Grand Canyon in one view, as if it were a changeless spectacle from which a curtain might be lifted, but to see it you have to toil from month to month through its labyrinths. It is a region more difficult to traverse than the Alps or the Himalayas, but if strength and courage are sufficient for the task, by a year's toil a concept of sublimity can be obtained never again to be equaled on the hither side of Paradise.

1895

We have seen how valleys grow headward and spread sideways. Now let us look down at these valleys from above and see what patterns they form.

Drainage Patterns

A river and its tributaries make complex drainage patterns. These patterns depend upon the amount of water and the kinds of bedrock in the drainage basin. The chief drainage patterns, where water is plentiful, are dendritic, trellis-rectangular, radial, and annular (Figure 8.18).

Dendritic patterns The most common patterns are **dendritic.** They form where underlying rocks have no pronounced grain, so that stream flow can be equal in all directions.

Trellis-rectangular patterns Where the bedrock has long strips of rock that are of unequal resistance to erosion, the streams tend to flow along the weakest rocks. These strips may be caused by the spacing of joints or by the tilted edges of different kinds of strata. If the master streams are long and straight, following a belt of weak rock, and the tributaries enter at right angles, the pattern is called **trellis** drainage. If the master stream itself makes many right-angle bends the pattern is **rectangular.**

Radial patterns Drainage patterns in which the main streams flow away from a central point are **radial patterns.** Radial networks of this type are typical of volcanic cones and of circular domes.

Annular patterns In many places, sedimentary strata have been warped into broad, dome-like structures ranging from 160 to 320 km in diameter. In a few cases, master streams may flow in the curving belts of easily eroded strata. The resulting drainage pattern is chiefly concentric, or **annular.**

Rates of Erosion

In considering the rates of erosion we will want to know how erosion is measured. Furthermore, we will want to see what factors affect the rates of erosion.

8.18 Common kinds of drainage patterns formed by streams in regions having abundant rainfall (opposite page).

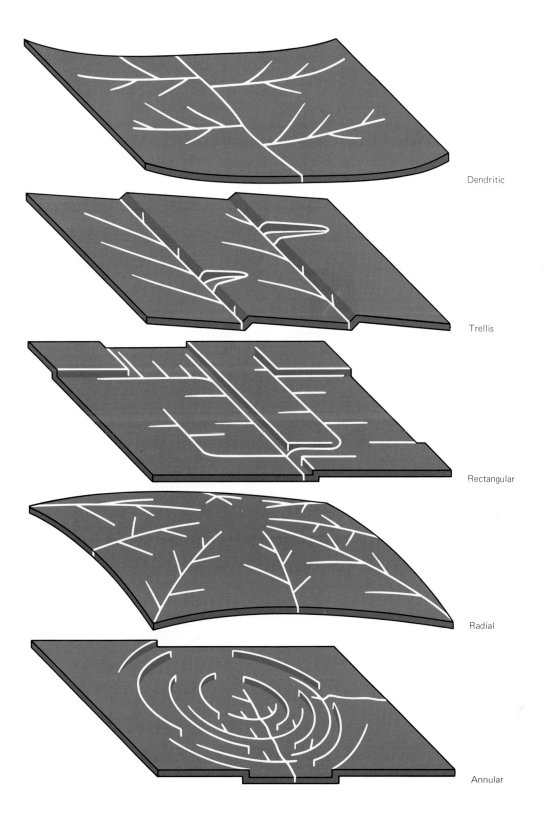

How erosion is measured Erosion can be measured in two ways: (1) by finding the rate at which the land surface is lowered, and (2) by measuring the amount of sediment and dissolved loads in streams. The easiest way to do this is to measure the stream loads and calculate rates of stream lowering based on amounts of material in the streams. The amount of sediment supplied per unit area is known as the *sediment yield*. At present, running water removes soil from the surface of the United States at an average rate of 1 cm per 300 years.

Factors affecting erosion The rates of erosion are influenced by such factors as the amount of rainfall, kinds and completeness of the vegetation cover, and kinds of geologic materials present. All of these factors are affected by the climate. In wet climates, where rainfall is high, regolith is protected from erosion by a complete cover of plants, including abundant trees. The leaves of the trees absorb the impact of the raindrops, and the roots hold the soil in place. Therefore, despite the great stream flow, sediment yields are small and thus rates of erosion are slow. If, however, the plants are taken away — for construction of houses, for example — the sediment yields can be very large.

While the land was laid bare during the recent construction of 89 houses on a 20-acre site in Montgomery County, Maryland, more than 3800 tons of sediment were eroded. This is the equivalent of 3 cm of soil over the entire area (1000 years' worth of normal erosion).

As we shall see in Chapter 11 (Deserts and the Wind), the maximum sediment yields come from semi-arid regions where plant cover is discontinuous, and the rain tends to fall in tremendous cloudbursts.

The Concept of the Erosion Cycle

An idealized erosion cycle is thought to begin with the uplift of a large land mass, ranging from an island to a mountain range to an entire continent. The cycle ends when this uplifted area has been eroded away and all that is left is an almost flat plain lying near base level (Figure 8.19). Such a flat surface is known as a **peneplain** (PEE-neh-plain). An erosion cycle takes so long that no one has ever seen a complete cycle. Geologists try to piece together the parts of the cycle by examining different regions in the modern world and by looking at the geologic record.

The idealized erosion cycle assumes that there are no interruptions. Acutally, there can be many interruptions, which we will look at in the last section of this chapter, Stream Response to Major Geologic Changes.

The Concept of the Erosion Cycle

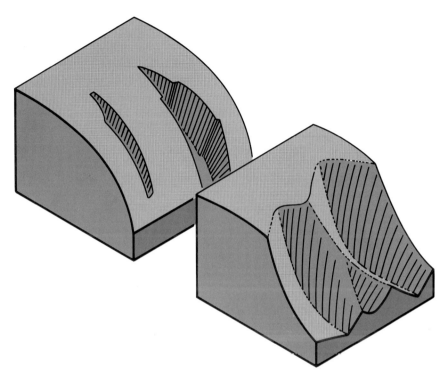

8.19 **First two stages of cycle of erosion** of a landmass by streams and by mass-wasting shown in schematic block diagrams. Youthful stage features narrow, steep-sided main streams and areas between streams that have not been dissected. Mature stage; divides are narrow and stream network extends to all parts of area.

The Three Stages of the Erosion Cycle

The cycle, which was originally formulated by William Morris Davis in the nineteenth century, has been broken down into three stages known as youth, maturity, and old age. These are defined on the basis of the results of stream erosion.

Youthful stage In the youthful stage of the cycle the streams are actively eroding headward. Valley walls are steep, even vertical (Figure 8.20). Waterfalls and rapids are numerous, and large areas in between streams have not yet been dissected.

Mature stage After the young valleys have been cut into the uplifted block, and new networks of tributaries have formed, and all valleys show typical V-shaped profiles, the stage of maturity is

8.20 **Steep-walled gorge** carved by stream in massive bedrock. Tamina "canyon," northeastern Switzerland. (Swiss National Tourist Office)

8.21 Image of mature landscape dissected by V-shaped stream valleys, seen from above, and based on side-looking radar scans. Near Sandy Hook, Kentucky. Width of photograph equals 10 km. (Autometric Division of Raytheon Company, courtesy U.S. Army Engineers, Topographic Laboratories, Fort Belvoir, Virginia)

achieved. In the mature stage the network of streams has extended itself completely across the area (Figure 8.21).

Old age It is not easy to show a photograph of a landscape that everyone would agree is of old age. From the geologic record it is known that there are ancient peneplains, but there are no peneplains in the modern world. Therefore, the details of the shift from the stage of maturity to old age are conjectural. Part of this problem is related to the history of valley-side slopes, shown in Figure 8.15. Some descriptions of old-age landscapes have been based on regions where stream sediments are being deposited. This is incorrect, because such regions are not subjected to systematic changes as in eroded valleys, but always look the same. We will next take up these depositional realms.

Streams and Stream Sediments

Regions that are sinking, on the one hand, and being built up as streams deposit their sediments, on the other, may be underlain by thick stream sediments. All stream-deposited sediments are known as **alluvium**. Streams and stream sediments are separated into two groups depending on the continuity of the supply of running water. Where water is plentiful, channels form great sweeping curves, and flow on wide lowlands **(flood-plain rivers)**. Where the water supply is discontinuous streams are intermittent and their channels are braided into short interconnected segments.

Features of Flood-Plain Rivers

The three major features of flood-plain rivers are the curving channel, the low sediment ridges which border the channel, and the wide, flat, low-lying flood plains. We shall describe how these features form and how they affect stream sediments.

Channels, natural levees, flood-plain basins A profile through a flood-plain river shows a well-defined **channel.** Along the banks of these channels are **natural levees,** broad, low ridges of fine sand or silt that has been deposited by repeated overflow (Figure 8.22). Levees range in width from about 20 meters on small streams to several kilometers on the lower Mississippi. Heights range from about 10 cm along small streams to several tens of meters along the Mississippi. Because levees are the highest points in a flood plain, they are the safest places to be during a flood.

As flood waters rise over the banks of the channel, the speed of

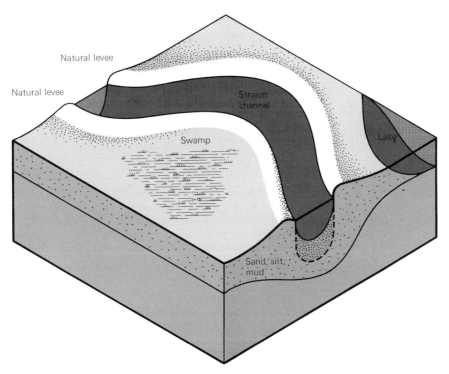

8.22 Features of a flood-plain river, shown in schematic block diagram. Top of block shows channel meander flanked by natural levees (composed of sandy sediment), which stand higher than flood plains, on which are found swamps and local lakes. Dotted line on right-front face shows depth of channel during flood.

the water drops abruptly. This slowed-down water can no longer carry all of its load, and deposition on natural levees takes place. The largest particles settle first; smaller particles, such as fine sand and silt, may travel long distances from the channel in suspension. Hence, the banks themselves receive the thickest deposits of sand and other coarse sediment, and rise above the adjacent flood plain.

Flood-plain basins are sites of swamps and temporary lakes. In them clay and organic matter accumulate. If flood-plain basins dry out between floods the fine sediments become mud-cracked. During the 1952 flooding of the Missouri River, areas of the flood plain near Kansas City received as much as 15 cm of sediment. Because flood-plain deposits are so rich in organic matter, many ancient civilizations, in Egypt, India, China, and other countries, concentrated their farms on flood plains.

8.23 Meanders and cutoff, forming ox-bow lakes, as seen from above. Hay River, northwestern Alberta, Canada. (Geological Survey of Canada)

Streams and Stream Sediments

Meanders and point bars There are probably no truly straight rivers in the world except those following faults, and it is almost certain that no river is straight for a distance greater than 10 times its width. Most stream channels are parts of great sweeping curves known as **meanders** (Figure 8.23), the most efficient route for a river to follow. Water meanders because its surface does not remain level (Figure 8.24).

Meanders form most commonly in loose alluvium which erodes easily. The alluvium is eroded from the outside, or *concave*, portion of a bend. The reason erosion occurs here is that as it turns, the water is "banked" as though it were following a bobsled course. The surface water flows toward the outside of the bend. Along the bottom the water flows toward the opposite bank. Material eroded from the outside of the bend is deposited on the inside of the bend, forming a **point bar** (Figure 8.24).

8.24 River meanders, showing spiral circulation in meander bends and lack of spiral circulation in straight stretches (known as crossovers) between adjacent meander bends.

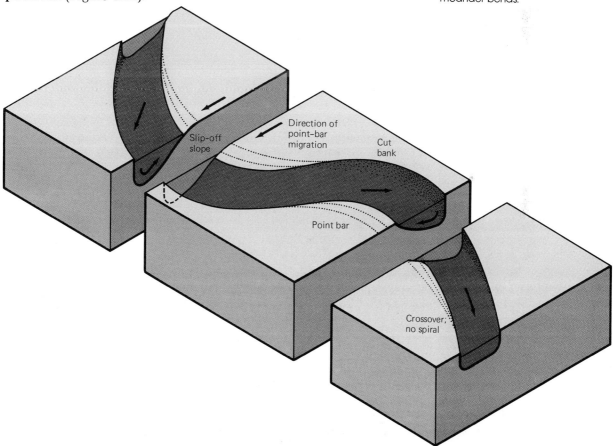

294
Running Water As a Geologic Agent

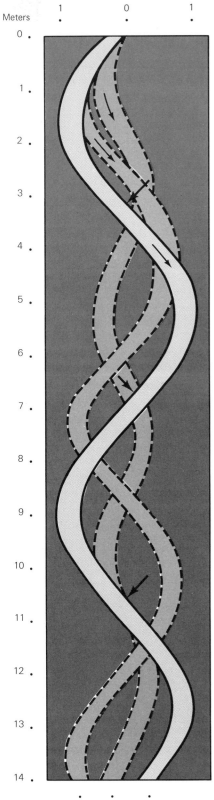

Effects of channel migration The whole system of meanders erodes its way through the valley-bottom sediments like a snake crawling through the grass. As long as all sediments in the path of the meanders are being eroded at a uniform rate, the over-all shape of the meanders will not change, and the meanders all move down the valley at the same rate (Figure 8.25 left). However, typically in valley-bottom sediments are thick bodies of tough clay deposited in flood-plain lakes. Nearly a century ago, Colonel James B. Eads, of the U.S. Army Corps of Engineers, prowled over miles of Mississippi River bottom beneath a diving bell, and observed that the clay river bed resisted erosion so strongly that it might as well have been made of marble.

When a meander encounters one of these clay plugs its forward progress practically stops. Soon the rest of the meanders overtake it. The next meander upstream from the halted one cuts through and isolates a crescent-shaped section which becomes an **ox-bow lake** (Figure 8.25 right). During floods, the radius of curvature of a meander is larger than normal. Larger discharge produces larger meanders. Thus, during floods the channels may cut more or less straight across the point bars.

Another affect of channel migration is to deposit distinctive sediments beneath the point bar. These sediments are as thick as the depth of the channel in a flood. At the bottom of the channel are waterlogged tree trunks, bones of animals, and large clay pebbles. With time, these logs may turn into coal and cause uranium minerals to be deposited in their vicinity. The main uranium mines in the Colorado plateau in the southwestern United States are associated with coalified logs at the bottoms of ancient stream channels. During and after World War II much research was done on meanders as the basis for locating uranium deposits.

Deltas

When a stream flows into a lake or the ocean, its speed is checked, and it deposits its sediment. The body of stream-laid sediment deposited at the mouth of a river is a **delta.**

8.25 Meanders in uniform and nonuniform valley-floor sediments. (Left page) Downvalley migration of meanders, based on constant discharge of experimental stream. (J. F. Friedkin, 1945, U.S. Army Engineers, Waterways Experiment Station) (Right page) Clay plug, deposited in ox-bow lake, prevents meander that encounters the resistant clay from further downvalley migration. Upstream meanders, not thus stopped, overtake the stopped meander and cut off one of its loops, forming a new ox-bow lake. (Map view of stream; not to scale.)

8.26 Nile delta as photographed from Gemini IV spacecraft from altitude of 175–190 km. (NASA)

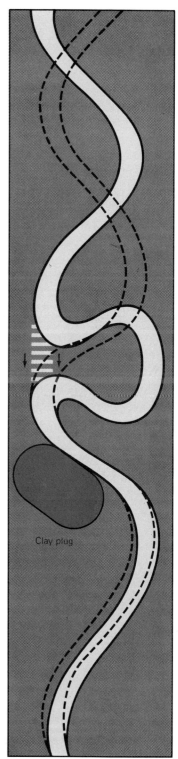

Clay plug

A delta may take the shape of its namesake, the Greek letter delta (Δ), like that of the Nile (Figure 8.26), or it may form radial, arc-shaped, or "bird-foot" patterns. The best-studied delta is that of the Mississippi, near New Orleans.

There are basically two kinds of deltas. These are differentiated by the sediment load and the depth of the channel in relation to the depth of the water in front of the mouth of the river. Shallow channels transporting sand deposit coarse-grained deltas, and deep channels transporting silt and clay deposit fine-grained deltas.

The top surfaces of all deltas are flat. This surface is called the *topset plain*. This plain is underlain by various sediment called *topset beds*. The topset beds of coarse-grained deltas may consist of gravel. The topset beds of fine-grained deltas include fine-grained sands, silts, and clays. The bulk of all deltas is formed by the *foreset beds*. The foreset beds of coarse-grained deltas consist of steeply dipping layers of sand, whereas the foreset beds of a fine-grained delta consist largely of silt in almost horizontal layers (Figure 8.27). In front of the advancing delta are fine-grained *bottomset beds*. As the delta grows forward the bottomset beds are buried by foreset beds, and eventually by topset beds.

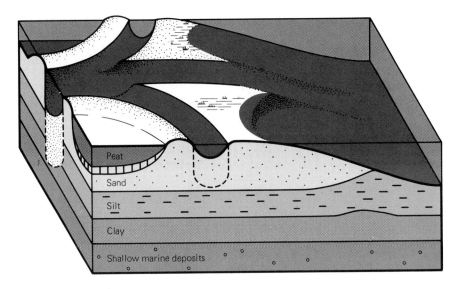

8.27 In fine-grained delta the chief sediments consist of clay (at the base) and silt (which forms the bulk of the delta); fine sand is present in channels at top. In schematic block main channel at left divides into three distributary channels. As distributary channels lengthen, their natural levees extend into the sea, enclosing bays which tend to fill with marsh grasses. The thickness of the foreset silt increases as delta builds into deeper water. Based on Mississippi delta.

Braided Streams

In streams with highly variable discharge and easily erodible banks, the channel may become interspersed with bars and islands in a pattern known as a **braided stream** (Figure 8.28). The process of building up the channel in this way is aggradation; its opposite is degradation, the deepening of a stream channel. A braided stream typically has a broad, shallow channel. A braided stream often forms in response to an excess load of sand-size or coarser sediment during a flood. The stream spreads the excess material widely across the channel bed and, as part of its "self-adjusting" ability, tries to increase its gradient. When the flood waters recede, the stream deposits its load on bars, islands, and in the bed. Channels of braided streams are often seen in arid regions where flash floods may overwhelm a channel with eroded rock waste. A common landform built by braided streams is a **fan,** a low, fan-shaped cone of alluvial sands and gravels ranging from

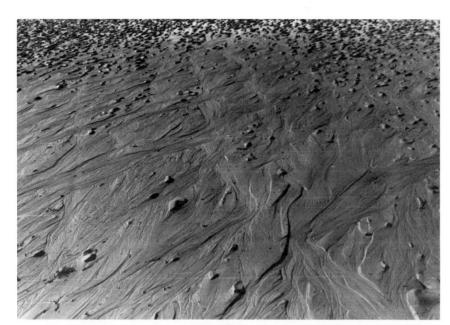

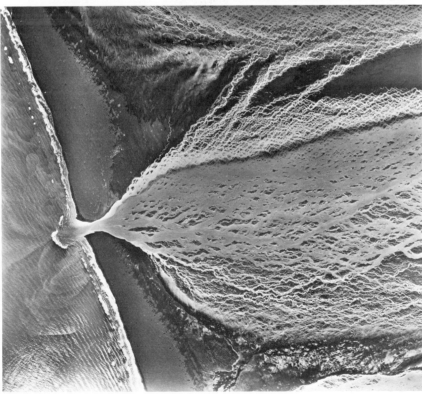

8.28 Braided streams at various scales. (Top) Tiny braided channels formed on sandy intertidal flat by draining away of water during falling tide. Five Islands, Nova Scotia, Canada. (J. E. Sanders) (Bottom) Giant network of braided streams from melting glacier, southeast coast of Iceland, viewed from airplane. (Copyright Icelandic Geodetic Survey)

8.29 Fans, Death Valley, California. (U.S. National Park Service)

a few meters to many kilometers in extent (Figure 8.29). In the Los Angeles basin, the communities of Burbank, Glendale, Montrose, and Pasadena, built along fans from the San Gabriel Mountains, are repeatedly menaced by flash floods like those which built the fans in the first place.

Stream Response to Major Geologic Changes

As we have seen, streams tend to erode in elevated areas, and to deposit in sinking areas. Figure 8.30 summarizes the relationships of the features of these two contrasting kinds of areas. Streams may also change their behavior if the base level changes, if the amount of water changes, or if their valleys are blocked. Therefore, what the streams have done and the sediments they have deposited are useful clues to the geologic history of an area. Two examples of major geologic changes are stream terraces (response to uplift) and the deposition of valley-fill sediments (response to blockage of the channel or to a rise in base level).

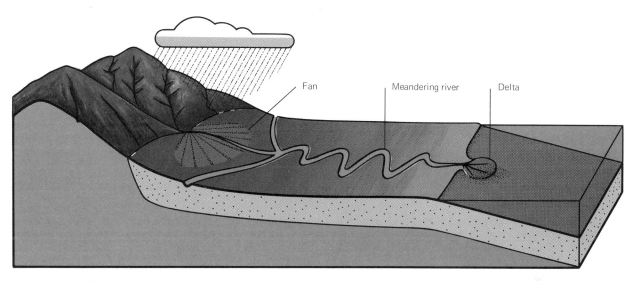

8.30 Streams eroding bedrock compared with streams flowing on and depositing sediments, schematic block diagram. Narrow, V-shaped valleys in highland where bedrock is being eroded, contrast with open country of plains where streams have been depositing sediments. Downstream progression across stream deposits is from fans to meandering flood-plain rivers to delta (where stream enters sea, at right).

Stream Terraces

A **terrace** is a flat, bench-like area formed when a stream has cut downward through its flat valley floor (Figure 8.31). A terrace may be underlain by bedrock or by alluvium.

Often a river may have a series of terraces; along the lower Mississippi, near Forrest City, Arkansas, there are terraces at 12, 30, 60, and 100 meters above the present valley floor. In regions of the world that are tectonically active, such as Japan, terraces are usually the result of uplift. If an area containing a stream is uplifted, the stream cuts downward to reach its old gradient—or a new one of similar slope.

Valley-Fill Sediments

When base level rises or an obstruction appears in its path, an eroding stream may suddenly begin to deposit sediments. These sediments

300
Running Water As a Geologic Agent

fill the valley partially or completely. If a valley is completely filled it disappears as part of the landscape, and its presence may be found only through borings or other testings of the subsurface.

8.31 Sequence of dissected terraces, Rangitikei River, North Island, New Zealand. (S.N. Beatus, New Zealand Geological Survey)

Chapter Review

1. A fundamental concept of hydrology is that all water on the surface of the Earth is participating in the *hydrologic cycle*. The hydrologic cycle includes water in all three of its states. This water is distributed in three reservoir systems: the oceans, the continents, and the atmosphere. We can use *streams* as a general term to include all water flowing in channels, ranging from creeks to rivers.

2. We consume water at enormous rates through household use, industrial use, and agricultural use. In the United States all industrial uses, for example, require 250 billion gallons of water per day, and 200 tons of water are needed to manufacture a ton of steel.

3. In the United States we use only one-quarter of the available water. Yet because of uneven distribution, local problems of water shortage exist.

One of our major problems is finding ways to transport water from areas having abundant rainfall to population centers lacking adequate water supplies.

4   People have always settled close to rivers to use them for transportation. In addition to the uses already mentioned, we use river water to generate electricity, and for the disposal of wastes. Besides the beneficial aspects of rivers we must also be aware of the danger of floods. The low, flat areas next to the stream channel that are occasionally inundated with water are *flood plains*. These flood plains are natural storing places for flood water.

5   There are many interrelated factors about stream mechanics, including energy relationships, discharge, speed, slope, load, and shape of the channel's cross section. The volume of water flowing through a cross section of the stream channel is its *discharge*. The highest natural speeds for running water are about 32 km/hr. The local slope on which a stream flows is its *gradient*. The aggregate of all the local gradients constitutes the *longitudinal profile* of a stream. The level of water in a sea or a lake forms a *base level* for entering streams. The geologic work of a river includes picking up sediment and transporting it downstream.

6   The interaction of all the factors in item 5 above establishes a condition of dynamic balance which characterizes a *graded stream*.

7   Masses of water can move by either laminar flow or turbulent flow. *Laminar flow* can take place only in smooth, straight channels as parallel layers of water shear past one another. Far more common is *turbulent flow*, characterized by a variety of vortices and eddies that are continually forming and disappearing.

8   Flowing water erodes bedrock, carries chemical load in solution, and transports solid particles of sediment. Particles that are supported by water that follows the upward flow paths within turbulent eddies constitute the *suspended load*. Particles that slide, roll, and bounce along the stream bed make up the *bed load*.

9   *Playfair's law* states the principle that streams do not simply occupy valleys, but create them. Rivers create valleys by *headward erosion*, and by *changes in the valley-side slopes*.

10   A river and its tributaries make complex drainage patterns. These patterns depend upon the amount of water and the kinds of bedrock in the drainage basin. The chief drainage patterns, where water is plentiful, are *dendritic, trellis-rectangular, radial,* and *annular*.

11   The rates of erosion are influenced by such factors as the amount of rainfall, kinds and completeness of the vegetation cover, and kinds of geologic materials present. All of these factors are affected by the climate.

12   An idealized *erosion cycle* is thought to begin with the uplift of a large land mass. The cycle ends when this uplifted area has been eroded away and all that is left is an almost flat *peneplain* lying near base level. The cycle can be broken down into three stages: *youth, maturity,* and *old age*. These stages are defined on the basis of the results of stream erosion.

13 All stream deposits are known as *alluvium*. Streams and stream sediments are separated into two groups: (1) where the water is plentiful, channels form great sweeping curves and flow on wide lowlands *(flood-plain rivers)*; (2) where the water supply is discontinuous, streams are intermittent and their channels are braided into short interconnected segments.

14 Some features of flood-plain rivers are *natural levees, meanders,* and *ox-bow lakes*. In streams having highly variable discharges and easily erodible banks, the channel may become interspersed with bars and islands in a pattern known as a *braided stream*.

15 When a stream flows into a lake or the ocean, its speed is checked, and it deposits its sediment. The body of stream-laid sediment deposited at the mouth of a river is a *delta*.

Questions

1   Define *infiltration, runoff,* and *streams*.
2   What is our largest use of water in this country?
3   Explain the problems caused by the uneven distribution of water.
4   Approximately how much of our electricity comes from hydroelectric sources?
5   What is the natural function of a *flood plain*?
6   How do dams affect the natural flow of streams?
7   Define and explain *discharge*.
8   Define *gradient, longitudinal profile,* and *base level*.
9   Explain the three major types of geologic work done by running water.
10 How do rivers create valleys?
11 How is it possible to consider a raindrop as an agent of erosion?
12 Describe the major drainage patterns.
13 Describe the three stages of the *erosion cycle*.
14 Define *natural levees*.
15 How is an *ox-bow lake* formed?

Suggested Readings

Bardach, John, *Downstream: A Natural History of the River.* New York: Harper & Row, Publishers, Inc., 1964.

Janssen, Raymond E., "The History of a River." *Scientific American,* June, 1952. (Offprint No. 826. San Francisco: W. H. Freeman and Company.)

Leopold, Luna B., *Water, a Primer.* San Francisco: W. H. Freeman and Company, 1974.

Leopold, Luna B., and Davis, Kenneth S., *Water.* New York: Time-Life Books, 1966.

Morisawa, M., *Streams: Their Dynamics and Morphology.* New York: McGraw-Hill Book Company, 1968.

Revelle, Roger, "Water." *Scientific American,* September, 1963. (Offprint No. 878. San Francisco: W. H. Freeman and Company.)

Shimer, John A., *This Sculptured Earth: The Landscape of America.* New York: Columbia University Press, 1959.

Starbird, Ethel A., "A River Restored: Oregon's Willamette." *National Geographic,* June, 1972.

U.S. Department of the Interior/Geological Survey, "John Wesley Powell's Exploration of the Colorado River." Washington, D.C.: U.S. Government Printing Office, 1974.

Wyckoff, Jerome, *Rock, Time, and Landforms.* New York: Harper & Row, Publishers, Inc., 1966.

"About one-fifth of the water used in the United States is groundwater."

Chapter Nine
Groundwater

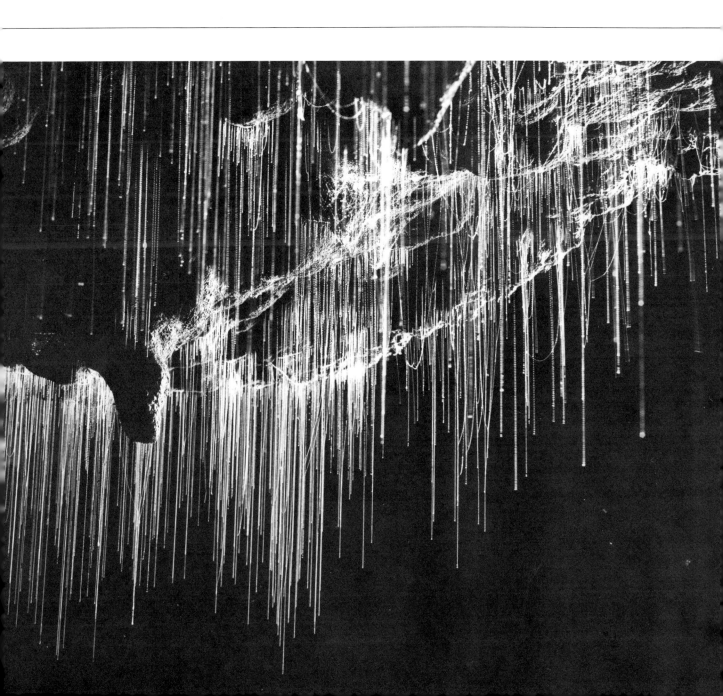

As we have just seen, one needs only to look at a stream of a river to be aware of the effects of surface water. Now we are ready to study the part of the hydrologic cycle that involves water below the surface of the Earth. This water is collectively called **groundwater.** Our objective in studying groundwater will be to learn about the principles that govern its distribution and behavior. We will be particularly interested in how such knowledge affects the daily use of groundwater, and to see the geologic effects of the activities of the water underground. We have already touched on some of these areas in discussing the effect of water on weathering, slopes, and erosion. Now we will look more closely at the effects of water in the lower levels of the Earth.

The Origin and Storage of Groundwater

We have already talked about groundwater moving within the hydrologic cycle, but now we need to discuss specifically where it comes from and how it is stored underground.

Where Does Groundwater Come From?

All shallow groundwater comes from *meteoric water*, which is water that falls out of the atmosphere as rain or snow. Part of this water seeps into the natural openings in regolith or bedrock, where it is stored. It is generally drinkable, and, except near hot springs, has the remarkable property of being at a constant temperature; the water temperature never varies more than a degree from the mean annual air temperature of the area.

Other subsurface water is salty. It is ancient ocean water that is buried with sediments. It is known as *connate water*, or, in the oil fields, *formation water*, because it is the water found in the same rock formations that contain oil or gas.

An unknown amount of water is thought to come to the surface from the mantle. Water which joins the hydrologic cycle for the first time is called *juvenile water*.

How is Groundwater Stored?

Groundwater is stored in openings in the ground, and we will briefly discuss the three major kinds: (1) a framework of rigid openings around the particles of regolith and sedimentary rock, (2) cracks, joints, and fractures in massive bedrock, and (3) huge caverns, caves, and original lava tubes. No matter what kind of openings there are, the water stands in different levels, which we will now discuss. From the top down, these levels are the zone of aeration, the water table, and the zone of saturation.

Zone of aeration In the **zone of aeration** the subsurface pore spaces are occupied partially by water and partially by air. The water occupying the zone of aeration is **vadose** water, from the Latin word for shallow. Vadose water makes up only a small fraction of the total groundwater supply. The thickness of the zone of aeration varies from zero to several hundred meters. The zone of aeration is itself divided into three parts or belts (Figure 9.1)

Zone of saturation Most groundwater is found in the **zone of saturation** (Figure 9.1), where all subsurface pore spaces are filled with water under hydrostatic pressure (pressure exerted evenly from all sides). The saturated zone is bounded at the top either by the zone of aeration or by a layer of impermeable material. Beneath the fresh water in the zone of saturation is a zone of salt water or a water-tight "floor" of strata, such as clay beds, or of bedrock. This floor may be located several kilometers below the surface of the Earth.

Water table The upper boundary of the zone of saturation is the **water table** (Figure 9.1). Usually, the water table is not quite level, as is the surface of a lake, for example. Instead, the water table rises under hills and drops down under valleys. It does so because water moving through tiny pore spaces is impeded by friction, and thus can be piled up. Groundwater escapes at the surface, and flows into streams, swamps, and lakes, lowering the water table.

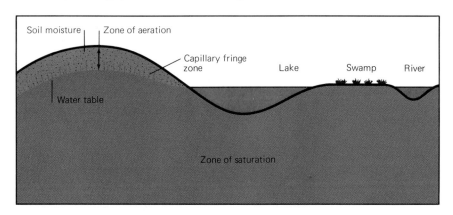

9.1 Zonal arrangement of water underground, schematic profile and section. The upper zone of aeration consists of a belt of soil moisture where plant roots are concentrated, an intermediate belt, and a capillary fringe, which marks the water table. In a moist climate water moves generally downward from the belt of soil moisture to the zone of saturation. However, upward movement takes place much of the time in the belt of soil moisture and in the capillary fringe.

The height of the water table changes easily. It rises when water is added to the zone of saturation, and falls when water is taken away from this zone. Addition of water to the zone of saturation is known as **recharge** of the groundwater supply.

Replenishing the supply of groundwater Rainfall is what supplies the groundwater reservoir. Therefore, the amount of precipitation over a given period governs the input into the groundwater reservoir.

However, the total amount of recharge is subject to other factors besides rainfall. The type and slope of regolith affects the absorption of rainfall. If the pores at the surface of the regolith are open and dry the maximum infiltration is possible. If the pores in the regolith are saturated with water, or if the water is frozen, little infiltration takes place. If the regolith has been built over or paved over, infiltration cannot occur at all. The slope is also important; rainfall tends to flow down steep slopes as surface runoff before much of it can infiltrate.

Another factor is vegetation. During the spring and summer, when plants are growing rapidly, roots soak up most or all of the water that infiltrates the ground. This water is stored by plants or transpired to the atmosphere. In fall and winter, when there is little plant activity, precipitation (except that which falls in frozen form) may travel freely downward past the soil layers to recharge groundwater supplies.

The Mechanics of Groundwater Flow

In this section we will discuss in detail the factors that affect the underground openings in the saturated zone and the flow of groundwater through them.

Characteristics of the Underground Pores

The two important characteristics of openings that affect the flow of water underground are **porosity**, the proportion of pore space to total volume, and **permeability**, the size and degree of connections among the spaces.

Porosity Practically speaking, porosity is a number that indicates how much pore space is present in a volume of material. Therefore, if you want to find out how much water a certain type of material will hold, you must first determine its porosity. Porosity is computed by dividing the volume of the pores in the sample by the total volume of the sample. To convert to a percentage, you multiply the result by 100. Now you can calculate the total space available for fluid. For example, suppose you had a cube of sand, measuring 1 meter on each side, having a porosity of 20 per cent. The volume of pore space is 0.20 × 1 cubic meter, which equals 0.2 cubic meters, or approximately 50 gallons.

Permeability A material's capacity to transmit fluid is known as permeability. In order to determine permeability we need to know the size and the continuity of the openings in the underground sample. If a material has a high permeability it will also have a high porosity. By contrast, a high porosity does not necessarily indicate high permeability. Clay, for example, has a very high porosity, 60 to 70 per cent, but a low permeability because of small pores. Sand, which has only about 25 per cent porosity, has a higher permeability than clay. Thus, although clays contain much water, very little water will pass through; sand, which holds relatively little water, will transmit a great deal of water. Permeability involves not only the amount of fluid transmitted through a material, but also the rate of flow under specific conditions of pressure.

Factors affecting porosity and permeability We will confine this discussion to pores among the particles in sediments and sedimentary rocks. We will not include such irregular large openings as fracture systems, caverns, or lava tubes.

Many factors are working simultaneously. A few simple examples will show how each factor relates to the other.

Consider the two examples in Figure 9.2 (a) and (b), which shows sections of large spheres and small spheres packed the same way. The porosity of each is identical: 46.7 per cent. As long as the packing in each sample remains the same, the porosity will remain the same. This is true because in each sample the proportion of pores to total volume is constant. However, the permeability of the sample having the large spheres is greater because the larger openings between the spheres allows more water to flow faster than through the openings between the smaller spheres.

The spheres in the samples in the previous discussion are considered to be perfectly *sorted* because in each sample, their size is the same.

Up to now we have kept everything constant. Now let us introduce some changes. If we *pack* the spheres closer together the sorting has not changed, but the porosity has been reduced to 26 per cent (Figure 9.2c,d).

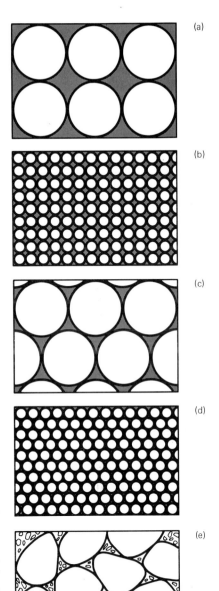

9.2 Porosity and the packing and sorting of spheres.

We can also reduce the porosity by adding cement, because the cement is a solid that occupies former pore space.

If we add other sizes of particles to our sample, the small particles fill up the spaces between the big ones (Figure 9.2e). In this case, not only is the porosity reduced, but so is the permeability.

The fact that clays have such high porosities indicates that this idealized discussion of spheres does not apply to clays. This should be no surprise, because as seen in Figure 5.13, clay particles are flaky and irregular, and even though they are tiny they can be openly packed with large spaces between them. Thus, clay has high porosity and low permeability.

Aquifers

Saturated formations that are porous and permeable, through which groundwater flows readily, are called **aquifers,** from the Latin for water bearer. A saturated clay, which contains water but does not yield it to wells, is not an aquifer. Similarly, a formation with good porosity and permeability is not an aquifer unless it is saturated, even though it may once have held water.

The aquifers having the greatest flow capacities consist of clean, coarse gravel. Next are uniform coarse sands, followed by some mixtures of sand and gravel, sediments of finer grain such as fine sand and loess, mixed alluvial deposits containing minor amounts of silt, and finally the products of rock decay that are still in place. Among consolidated rocks, sandstones form the most valuable aquifers. Next come limestones, which yield good water supplies from solution passages. Extrusive igneous rocks that have fractures, connected vesicles, and other irregularities often yield as well as limestone.

Depending on their kinds of upper boundary, aquifers may be classed as nonconfined or confined.

Nonconfined aquifers In a *nonconfined aquifer* the water table forms the upper surface of the zone of saturation (Figure 9.3). Contour maps and profiles of a water table can be prepared from the elevations of water in wells drilled at different points to the surface of a nonconfined aquifer.

Confined aquifers Groundwater under greater-than-atmospheric pressure that is "held in" by an overlying stratum having low permeability forms a *confined aquifer*. Any well penetrating such an aquifer is an artesian well (Figure 9.4), one in which the water is driven toward, and sometimes to, the surface, by the pressure on the confined aquifer.

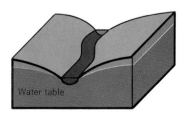

9.3 Nonconfined aquifer, schematic profile and section. In a nonconfined aquifer the shape of the water table is a subdued replica of the shape of the land surface.

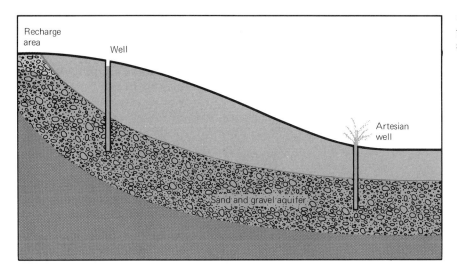

9.4 Confined aquifer has potential for yielding artesian wells, as seen in schematic profile and section.

Size of aquifers Aquifers differ greatly in shape, thickness, and extent. In Texas, the geologic formation known as the Carizzo Sand is an aquifer that ranges in thickness from about 30 to 60 meters and provides potable water to an area 30 to 80 km wide and hundreds of kilometers long. By contrast, some of the permeable glacial deposits in the midwest form tiny aquifers that are only a few meters thick and less than a few square kilometers in extent.

Rate of movement The rate of movement of groundwater depends on both the permeability of the aquifer and the **hydraulic gradient** (the slope of the water table), which causes the water to flow (Figure 9.5). Groundwater flows at widely varying speeds. In an aqui-

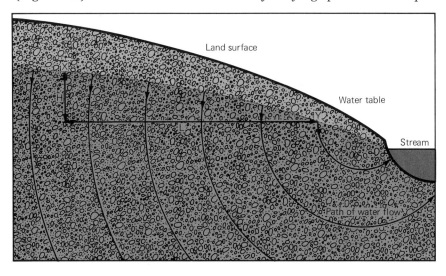

9.5 Hydraulic gradient creates pressure that causes groundwater to flow. The hydraulic gradient is equal to the slope of the water table; it is calculated by finding (a) the difference in height between two points and (b) the horizontal distance between these same points. The horizontal distance divided by the difference in height gives the hydraulic gradient.

fer having a hydraulic gradient of 3 meters per kilometer and low permeability, groundwater may move only 0.00075 meters per day. If the hydraulic gradient is 30 meters per kilometer, and the permeability high, groundwater may move more than 4 meters per day. According to field tests, the normal range of movement is from 1.5 meters per year to 1.5 meters per day, although speeds of 30 meters per day have been reported. The slow rates cited indicate that the flow is laminar.

Darcy's law More than a century ago a French hydraulic engineer named Henry Darcy determined how to measure the rate of groundwater flow. According to **Darcy's law,** the flow rate through porous material is proportional to the pressure driving the water, and inversely proportional to the length of the flow path (Figure 9.5).

Groundwater Consumption and Supply

The two chief reservoirs of fresh water are groundwater and glaciers. The groundwater reservoir is estimated to contain about 30 times as much as all the surface fresh water in lakes, streams, and the atmosphere. Glaciers contain at least three times as much fresh water as is in the ground (Figure 9.6).

At the beginning of Chapter 8 we discussed the tremendous

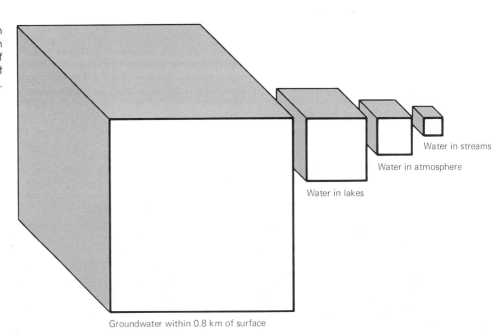

9.6 Volumes of fresh water in various natural reservoirs where fresh water is stored, shown by cubes of various sizes drawn to correct proportion.

thirst of modern society for fresh water, and emphasized the amount of water supplied by streams. Now we can look at the role of groundwater in supplying the world's water supply.

Groundwater Consumption

Throughout the world people are using groundwater in increasing amounts. As supplies of surface water become over-used and polluted, supplies of clean groundwater are even more in demand. Groundwater is easily tapped with wells, comparatively safe from pollution, protected from evaporation, and requires no dams or surface reservoirs.

According to the U.S. Geological Survey, about one-fifth of the water used in the United States is groundwater. Irrigation takes 65 per cent of all groundwater used. Some 91 per cent of this is pumped in the 17 western states where irrigation is extensive and river flow is sparse. Industry uses about 22 per cent of the groundwater, public water supplies, about 10 per cent, and other rural needs, about 3 per cent.

Increasing demands Until the twentieth century, industry and agriculture grew slowly, and so the demand for water was relatively small. With the introduction of irrigation on a large scale, in the first half of this century, and the explosive growth of the petrochemical industry after World War II, water demand skyrocketed in the developed nations. As modern agricultural and industrial techniques spread to the less-developed countries, water needs will continue to grow—even faster than population.

Groundwater Supply

As the demand for supplies of fresh water intensifies throughout the world, so does the search for good aquifers. Geohydrologists urgently press for a better understanding of the movement of known sources of groundwater, the recharging of dwindling reservoirs, and the discovery of new aquifers. Drinkable groundwater comes from springs and wells. It is not always easy to extract it.

Springs A surface stream of flowing water that emerges from the ground is a **spring** (Figure 9.7). The hydraulic gradient forces the water to flow from springs. The late O. E. Meinzer, former head of the groundwater branch of the U.S. Geological Survey, classified gravity springs in eight categories according to the amount of flow. He listed

9.7 Springs as related to various geologic features, schematic diagram.

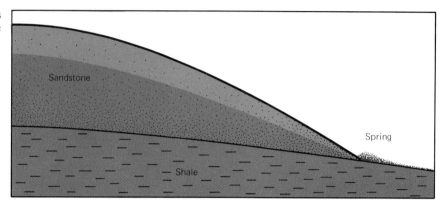

65 "first-magnitude" springs in the United States, discharging more than 27 cubic meters per second. Several hundred second-magnitude springs yield 3 to 27 cubic meters per second, and thousands of third-magnitude springs discharge 1 to 3 cubic meters per second.

Wells A **well** is a hole that is dug or drilled into an aquifer. At one time, farmers and homesteaders could retrieve all the water they needed from shallow cylindrical wells dug with pick and shovel and lined with bricks or stone. Nowadays, most wells are drilled by powerful machinery and equipped with electric pumps. Wells for agriculture and industry have been drilled to depths of 300 meters or more and their diameters are 30 to 40 cm. Such wells may yield millions of gallons a day. Household wells are smaller, typically drawing 30 gallons per hour, or 720 gallons per day.

Whenever water is removed from a well by artesian flow or by pumping, the water table is immediately lowered in the vicinity of the well, an effect known as **drawdown.** The extent of drawdown decreases at greater distances from the well, so that a **cone of depression** forms around the well (Figure 9.8). The cone of depression increases the hydraulic gradient so that groundwater near the well flows faster toward it. At a certain point, according to Darcy's law, the increase ceases. A cone of depression may extend as far as 16 km from a big well; many wells with overlapping cones may lower the entire water table.

Problems of extracting groundwater Anyone who extracts groundwater should be aware that the supply in the aquifer being tapped is limited. Intelligent extraction of water requires knowledge of how rates of extraction compare with rates of recharge and flow. Some aquifers are not now being recharged, and may never be. The water in them has been isolated from the hydrologic cycle. Such an

9.8 A cone of depression is formed when the flow of water into a well lowers water table.

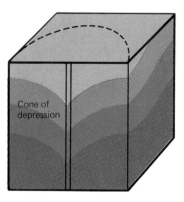

9.9 **Effects of land subsidence** as a result of removal of oil from underground reservoir. A trench is being dug to relocate pipe to keep it from breaking. Long Beach, California. (Wide World Photos)

aquifer can be drained only once, like an oil well.

Since 1911 farmers and ranchers have pumped water from the Ogallala Formation in the arid plains of Texas, which is 60 to 90 meters thick. More than 20 per cent of the water has been used and, at current rates of pumping, the supply will eventually be exhausted.

The worst thing that can happen to an aquifer is to be pumped faster than it is recharged. This lowers the water table, and creates problems of land subsidence (Figure 9.9) and salt-water encroachment (Figure 9.10).

In the San Joaquin Valley in California, groundwater loss has created problems of land subsidence in an area extending nearly 3500 square kilometers. In some places the land has subsided as much as 10 meters. Extracting oil or gas from the ground can also cause land subsidence, and in Baytown, Texas the land has been lowered about three meters since 1920 because oil has been removed.

Damage related to land subsidence can be extensive. Besides decreasing the efficiency of wells and aquifers, irrigation systems can be affected seriously, and flood possibilities may be increased along coastlines.

Artificial recharge The life of an aquifer can be prolonged by artificially increasing the recharge. California, which recharges hundreds of millions of gallons of groundwater per day, operates more than half of the recharging facilities in the country. The two common methods of artificial recharge are water spreading and induced recharge. **Water spreading** refers to the release of water in broad basins, in ditches, or directly on the land. If the regolith is permeable, this water percolates down to the groundwater supply, raising the water table as much as a meter a day. **Induced recharge** involves lowering the water table near a lake or stream so that water will enter the ground from the surface source.

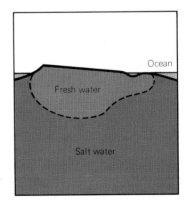

9.10 **Freshwater "floating" on saltwater,** schematic profile and section, Long Island, New York. Pumping in excess of recharge not only lowers water table, but causes fresh water/salt water interface to rise. (U.S. Geological Survey)

Problems of Groundwater Use

Two problems facing users of groundwater are finding new supplies and dealing with pollution.

Seeking new sources New groundwater supplies, hidden from sight, are never easy to find. However, as demand for water is rising quickly, the search for new sources is intensifying. Geohydrologists use a number of techniques in their search, including tools developed by petroleum and mineral prospectors. It is possible to estimate the amount of water in a formation by the behavior of electricity passing through it. Seismic waves may be speeded up or otherwise affected by underground water. Knowledge of the depositional history of an area, the types of rocks to be found, and other basic geologic information are crucial clues in locating aquifers. As detailed satellite photographs have become available in recent years, geologists have learned to use such clues as vegetation, drainage patterns, erosion, and color as indicators of subsurface features. The seven vast basins of groundwater beneath the Sahara were predicted in 1958 on the basis of data obtained from air photographs. Infrared instruments enable geologists to map minute temperature differences at the surface of the Earth. Desert aquifers have been found in Iran and Saudi Arabia during oil-drilling operations. Deep-sea drilling has revealed the presence of aquifers beneath the ocean floor off Florida.

The ancient practice of dowsing, or using a divining rod to locate water, persists even to the present day. Although the technique lacks scientific confirmation, "water witchers" diligently trudge to and fro in many countries, holding a forked stick in both hands (Figure 9.11) until the butt end is drawn downward—supposedly by groundwater. Whether or not there is any scientific merit in what they do, dowsers have at least one condition in their favor: In most parts of the world, almost any hole dug deep enough will yield water.

9.11 **Water dowser** conducting a "survey," using sixteenth-century "tools." (Georg Agricola, 1571, De Re Metallica; Basel)

Groundwater pollution A groundwater supply that is to be used by humans must be guarded carefully against pollution. Just as an aquifer is recharged by seeping rainwater, it may be invaded by polluted seepage from the ground or from higher along in a tilted stratum. The most common pollutant of wells is sewage, which may seep from septic tanks, sewers, cesspools, barnyards, livestock areas, and polluted streams (Figure 9.12). Ordinarily, wells should be placed at least 15 to 30 meters away from such sources. Across greater distances granular material such as sandstone acts as a natural purifier. By contrast, limestone is a very poor purifier and may transmit polluted water great distances. The most common disease communi-

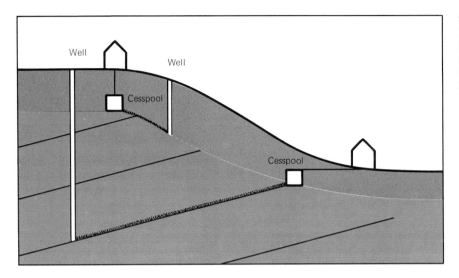

9.12 **Pollution of aquifer** by introduction of household wastes. Knowledge of the geologic relationships underground provides a sound basis for locating wells where they will not be contaminated.

cated by polluted wells is typhoid, which is caused by a bacterium that can survive for a long time in groundwater.

The largest potential sources of groundwater pollution are urban wastes, in the form of either sewage or solid wastes. In arid regions, where other recharge water is not available, the treated water coming out of sewage-treatment plants is potentially valuable for artificially recharging aquifers.

Where solid urban wastes have been dumped as "sanitary landfills," a different kind of groundwater-pollution hazard may appear. Rainwater that seeps into these dumps becomes groundwater. In contact with solid household and other wastes, such groundwater becomes heated and acquires a chemical load by dissolving the garbage. The heated effluent, containing materials dissolved from the garbage, is named *leachate*. In some places, leachate has locally contaminated the groundwater reservoirs near the landfills.

Some industrial wastes do not decompose with time and thus cannot be recycled. Some are toxic to organisms, or even radioactive. As unused land becomes more scarce, disposal of such "hard" wastes becomes difficult and costly. They can sometimes be injected deep into the ground in porous zones or other formations where they cannot be carried by groundwater into aquifers. The most acute waste-disposal problem facing the United States in the coming years is that of radioactive wastes from nuclear electricity-generating plants. Some of these wastes, such as plutonium, will remain "hot" for many thousands of years, so that any deep formations considered for disposal should be earthquake-free, tectonically stable, and dry—a difficult order. Because many energy experts consider that nuclear power is

The Geologic Effects of Groundwater

So far, we have discussed groundwater chiefly as it relates to the hydrologic cycle, and to its use by humans. Groundwater also has much to do with the rock cycle. Groundwater dissolves minerals from bedrock and from regolith and transports the ions in solution, both to other formations in the Earth and to streams and the ocean. In moving great amounts of dissolved material from place to place, groundwater both destroys and creates geologic features on a large scale.

Groundwater and the Landscape

As we saw in Chapter 5, most groundwater contains carbon dioxide in solution, and thus is a weak acid capable of dissolving calcite, the chief mineral of limestone. Such dissolution creates many features at and near the surface of the Earth. These include caves and their deposits, sinkholes, and the deposits around hot springs and geysers.

Caves Percolating groundwater charged with carbon dioxide invades carbonate rocks such as limestones along fractures and bedding planes. By dissolving away the calcite the water slowly widens these openings and builds an intricate drainage network of channels and chambers. When these chambers can be reached from the surface and are large enough for a human to enter, they are called **caves.**

Most of the world's large caves form in gently dipping strata of limestone, just below the water table, where water may move as slowly as 10 meters per year. Because the water follows the principal joints in the limestone, which intersect nearly at right angles, the "floor plan" of a cave often resembles that of an ancient city (Figure 9.13).

Limestone caves are usually decorated with a variety of **speleothems,** elaborate whitish deposits in the shape of icicles, slabs, and mounds. Most of these deposits are calcium carbonate, precipitated when groundwater drips into the cave. Perhaps the most familiar forms are **stalactites,** which grow vertically downward from the ceiling like icicles, and **stalagmites,** which grow upward from the cave floor toward a drip source on the ceiling (Figure 9.14). Stalactites and

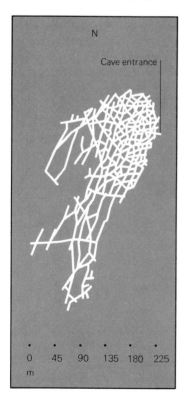

9.13 Map of cave passages underground clearly reflects the work of groundwater in enlarging joints.

9.14 Speleothems formed where previously enlarging cavern is now filling. Cango Cave, Cape Province, South Africa (South African Tourist Corp.)

stalagmites range from slender straw-like structures to the massive columns which are created when the two structures meet. Undulating "drapes" or "curtains" may grow when rivulets of groundwater dribble down an inclined roof or wall.

Our largest caves, with their complex speleothems, huge caverns, and unique fauna, have become major tourist attractions. Many are national parks, such as Carlsbad Caverns in New Mexico, a series of giant chambers in deposits of limestone and evaporites laid down about 210 million years ago beneath a shallow, reef-restricted sea. One chamber, the "Big Room," is 1200 meters long and high enough to hold a 26-story building. Several years ago a connection between Mammoth Cave and Flint Ridge Cave in Kentucky was found, producing the largest cave system in the world, with more than 82 km of charted passageways.

The Mammoth Cave of Kentucky
Frank Blackwell Mayer

In the morning a negro drove me to the Mamoth cave hotel & soon after being provided with a guide I proceeded to visit that great natural curiosity. Proceeding a few hundred yards down a richly wooded ravine you arrive at a large cavity in the hill side from which so cold a draught of air issues that the change startled you by it[s] suddeness & seeking the cause you perceive that the yawning mouth of the cavern is before you. descending several steps & following rapidly the footsteps of the guide you are past the current of air & find yourself in perfect stillness within the cave. For some distance the harmony of the scene is destroyed by the present [presence] of the remains of salt-petre works formerly used during the war of 1812, great quantities of this article being extracted from the dirt of the cave.

Beyond this the subterranean scenery begins & it is probably unequally [unequaled] by any other example known. An immense cavern fifty to sixty feet in width & often more in height extends for four miles from the entrance, & constitutes the main cave. Here perfect stillness reigns & awful silence, never broken but by the hasty flight of some solitary bat & that only near the entrance. Impenetrable darkness fills every cavity & our lamps rather hint that [than] fully reveal the immense masses of jagged, & riven rocks strewned upon the bed of the cave apparently tossed & torn by some tumultous torrent. Rising from this confusion wild the ledges of rocks tower above one another in grand masses to the ceiling or roof which is nearly flat & appears as tho' laid over this chasm chaotic by some cunning architect—beyond is impenetrable darkness and all around are scattered the enormous fragments casting shadows of alarming form & magnitude. It seems a fit residence for gigantic spirits & but little imagination is required to believe their presence, for the form of an immense rock has given it the name of the giant's coffin.

Passing two houses of stone which had been erected for the residence of consumptive invalids who for the sake of a restoration to health were willing to become denizens of this dismal abode for some months, the atmosphere having at one period been supposed beneficial to patients of that class, you arrive at the "Star chamber" a portion of the main cave which has been thus named from the circumstance of certain crystals or white substances reflected from the midst of the dark mass of the roof giving the idea of a starry sky, the deception is remarkable. Beyond the cave increases in rugged & apalling grandness passing th[r]o' many scenes of great picturesqueness. One of these is the cascade hall where a stream dropps from the ceiling a hundred feet to the floor beneath, & the grand temple an immense rotunda near the termination of this portion of the cave. Numerous avenues branch from this one to great distances the greatest distance from the entrance to the extreme end being about 9 miles. These are of lesser size than the first & seem to bear greater evidence of the action of water than other portions. Numerous pits, domes & stalactite & stalagmite formations are found & the general direction is a descent until you reach the "river," a stream some thirty feet wide & often as deep as running as the poet says "thro' caverns measureless to man down to a sunless sea". This region partakes more of the horrible, dismal & dreary & suggests the memory of the unhallow[e]d journey of Dante & Virgil, it is a fit place for the wandering wearied damned & realizes our ideas of Pluto's realmns.

One branch of the river is aptly named the Styx, the other Echo river. On this you embark, the guide acting as oarman. Seldom do you encounter such a situation. In a small flat-boat steered by a guide who by the dim light might easily be imagined an uneasy spirit of the place, the still quiet stream cloudily revealing the sharp angles & points of treacherous rocks lurking just beneath the surface, the low arched roof with the sides rising perpendicularly from the river robbing the drowning wretch of even the sight of a shore on which a handhold could be found—& the wild song of the guide dying away in echo[e]s

From WITH PEN AND PENCIL ON THE FRONTIER IN 1851. The Minnesota Historical Society. Abridged with permission.

Field Trip

which give the effect of an organ accompaniament of some unseen hand—& you have a scene where you feel yourself doubtful of your waking existence. At one point of the pass[a]ge and for a space of thirty yards I was obliged to compress myself to the smallest compass in order to pass under the superincumbent rock. Had the river risen, as it has been known to do, before our return our situation would have been truly awful for our chances of escape would have [been] resting on finding a passage thro' Purgatory[,] a dangerous & intricate ravine where the water does not rise *as* rapidly as in the river.

On landing you find yourself in the "Infernal" regions & they are well named for a place of greater dreariness & utter absence of all that can serve to occupy the mind or offer repose to the body I cannot imagine. An irregularly arched way with an uneven & tortuous bed, composed or rock seemingly formed of mud of the most fetid colour, not offering even the gratification of an angle to the eye for all the forms are rounded into one another in ungraceful curves the floor being of so *uneasy* a surface as to suggest no idea of repose but rather to cause that species of progression seen in landsman on a pitching ship, then all is damp, dreary, desolate, & the mind is irresistably turned upon itself—a fitting spot for the torments of conscience to be administered. Two other places are suited to Dante's ideas of punishment—here is the "Winding way" "or fat-man's-misery" a path which was once the bed of a torrent of about a foot in width & waist deep, here the gluttons might be compelled to walk for ever & until reduced in flesh. After this is the "valley of humility" where the proud must stoop double. From the infernal regions you walk three miles thro' an arched & tortuous rocky avenue over pointed & brocken rock to the "snow ball room" where the crystalization of gypsum on the ceiling give it the effect of an incrustation of snowballs. Gypsum formations are only found beyond the river the first portion of the cave being of darker limestone.

No living creatures are found within this subterranean world except a few bats who seek shelter in the mouth of the cave during winter, & species of rats who frequent the main cave & who as well as the fish found in the river are of a lighter colour than their brothers of the daylight world. The fish found in the river are small, colourless, & without eyes. There are a few spiders & crawfish of a like unhealthy hue.

I returned from our tour of the cave, having been absent 8 hours & having walked about 18 miles, the effect of so much exercise being much less fatiguing than the same amount above ground, owing probably to the equable & agreeable temperature all parts of the cave at all seasons being about 60° of Fahrenheit.

Another cave in the vicinity, ("White's cave") contains more of the stalactite & stalagmite formation—which in its forms suggests many useful ideas of forms & ornaments to the architect. The stone is not brilliant except when broken when the crystalization is apparent. The whole of this region is limestone & contains many similar caves of smaller size. The only remains found in the Mammoth Cave were the bones of some gigantic human being & some of the bones of a mammoth & a wooden bowl found by an early explorer near a spring supposed to be of Indian manufacture. The mummies *said* to have been found here were discovered in a cave some three or four miles distant (Long's.).

1851

9.15. **Sinkhole** in which collapse has ruined a house in Bartow, Florida. (U.S. Geological Survey)

Sinkholes Under the persistent attack of groundwater, a limestone region becomes riddled with underground channels and caves that may drain the region more efficiently than surface streams. The landscape is pockmarked with **sinkholes** (Figure 9.15), formed by the collapse of cave roofs, which funnel more water below ground. Ultimately, an area of **karst** morphology develops (Figure 9.16), characterized by an almost complete lack of surface streams. Karst landscape,

9.16. **Features typical of karst morphology** that are created by groundwater. Dry valleys and numerous sinkholes near Timaru, New Zealand. (S. N. Beatus, New Zealand Geological Survey)

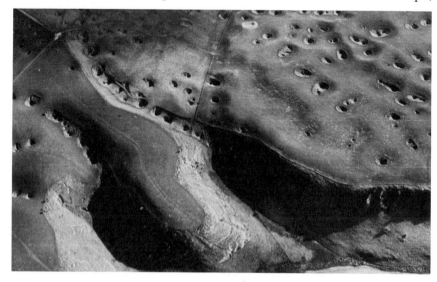

named after the Kars region in Yugoslavia, appears in the United States in Kentucky, northern Florida, and elsewhere in the south.

Deposits around hot springs and geysers Natural vents in the Earth's surface from which hot water issues are called **hot springs.** The water may be juvenile water from a body of magma, or from still-cooling igneous rock. The water may also be groundwater heated by rock at great depths; at a depth of 1.5 km groundwater is 50°C hotter than the average surface temperature.

Water bubbling out of hot springs carries many dissolved salts, chiefly ions of sodium and potassium mixed with carbonates, chlorides, and sulfides. As the water emerges from hot springs it usually deposits siliceous products called **tufa** (TOO-fa) around the vent. The hotter the spring, the greater the amount of siliceous tufa, because hot water can carry more siliceous material in solution than can cooler water. Hot-water deposits around geysers are termed **geyserites** (Figure 9.17). Both hot springs and geysers may construct a variety of mounds, terraces, or bowl-like structures around their vents.

9.17 Geyserites are deposits made where heated groundwater is discharged at the Earth's surface. Minerva terrace mound at Mammoth Hot Springs, Yellowstone National Park, Wyoming. (U.S. National Park Service)

9.18 **Collapse breccia** consisting of particles of dolomite rock that were broken up when underlying evaporites were dissolved. Ordovician, near Comstock, New York. (S. J. Mazzullo)

9.19 **Silicified brachiopod** recovered by dissolving enclosing matrix with hydrochloric acid. Permian Salt Range, Pakistan. (G. Arthur Cooper, U.S. National Museum)

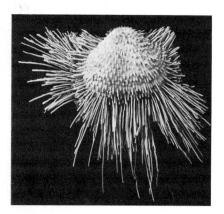

Groundwater in the Rock Cycle

Many of the activities we described as weathering in Chapter 5 occur deeper down than in the top few meters of the bedrock. Groundwater is a persistent underground sculptor, constantly in motion, continually dissolving material from rocks and transporting it elsewhere, depositing it in various forms.

Dissolution In addition to the landscape features resulting from dissolution, which we have just discussed, groundwater tends to dissolve entire formations of rock. When this has happened, the overlying material collapses. The remains of such a decomposition are **collapse breccia** (Figure 9.18). Many of these activities proceed even faster underground than they do above, because both temperatures and pressures are greater beneath the surface.

Cementation of sediment When groundwater dissolves a load of ions from rock the water may move only a short distance before discharging this load. After water has percolated into the zone of saturation it may move at a rate that seems vanishingly small to fast-paced humans; water in fine sand might move only 1.5 km in 135 years. Remaining in contact with rock particles for so long increases the probability that the acidic groundwater will react chemically with rock material. These reactions often precipitate minerals which become the cements which convert sediments into sedimentary rock. The zone where this occurs is known as the **zone of cementation.** Thus groundwater, both under dry land and under the bottom of bodies of water, helps convert sediment into sedimentary rock, a fundamental process in the rock cycle.

Replacement Solution and deposition are often simultaneous: One material may be dissolving as another is deposited in its place. This exchange of one solid element for another is **replacement.** Fossils often show the effects of replacement; the calcite of a seashell may be replaced by silica (Figure 9.19), by pyrite, or by other minerals. A substance is said to be *petrified,* or turned to stone, if organic molecules have been replaced by inorganic minerals. Thus, if a log is buried in a bed of sand which later becomes saturated with groundwater, the wood may slowly be replaced by hydrated silica, or opal. Eventually, an entire tree trunk may be converted to a solid mass of silica and exposed millions of years later after the overlying sediment has washed away. Trees that were buried during the Mesozoic Era 160 million years ago, converted to opal, and later exposed, can be seen today in Arizona at the Petrified Forest National Park (Figure 9.20).

The Geologic Effects of Groundwater

9.20 Petrified logs, Petrified Forest National Park, Arizona. (Fred Mang, Jr., U.S. National Park Service)

9.21 Concretions and geodes. (Left) Large spherical concretions exposed along the shore at Amarie, New Zealand. (G. R. Roberts) (Right) Small geode, seen in section through interior cavity. (E. Dwornik, U.S. Geological Survey)

Concretions and geodes Under certain conditions, chemical precipitation begins around a nucleus such as a bone or a pebble. Deposition of minerals leads to further deposition, forming concentric layers of material in rounded, irregular bodies called **concretions** (Figure 9.21). Concretions, which may consist of calcite, quartz, gypsum, barium sulfate, calcium phosphate, or other crystalline minerals, usually grow in porous sedimentary rock. Similar rounded or egg-shaped deposits called **geodes** sometimes precipitate from groundwater in rock cavities. Whereas concretions grow outward from a central point, geodes grow inward from the walls of a cavity. In some semi-arid regions, capillary action draws lime-bearing groundwater to the surface, where the water evaporates and leaves lime-rich deposits of *caliche*.

Chapter Review

1. That part of the hydrologic cycle involving water below the surface of the Earth is collectively called *groundwater*. The three types of groundwater are *meteoric, connate,* and *juvenile*.

2. Groundwater is stored in three major kinds of openings in the ground: a framework of rigid openings around the particles of regolith and sedimentary rock; cracks, joints, and fractures in massive bedrock; huge caverns, caves, and lava tubes.

3. Groundwater is distributed in two zones. From the top down, these zones are the *zone of aeration* and the *zone of saturation,* separated by the *water table*. Addition of water to the zone of saturation is known as *recharge* of the groundwater supply.

4. The two important characteristics of openings that affect the flow of water underground are *porosity*, the proportion of pore space to total volume, and *permeability*, the size and degree of connections among the spaces. Some factors that affect porosity and permeability are *sorting* and *packing*.

5. Saturated formations that are porous and permeable, through which groundwater flows readily, are called *aquifers*. Depending on their kinds of upper boundary, aquifers may be classified as *nonconfined* or *confined*.

6. The rate of movement of groundwater depends on both the permeability of the aquifer and the *hydraulic gradient*, which causes the water to flow.

7. The two chief reservoirs of fresh water are groundwater and glaciers. The groundwater reservoir contains about 30 times as much as all the surface fresh water. About one-fifth of the water used in the United States is groundwater. Groundwater is extracted mainly from *springs* and *wells*. Serious problems can occur if an aquifer is pumped faster than it is recharged.

8. Although techniques of artificial recharge—*water spreading* and *induced recharge*—are commonly used, there are still problems of finding new supplies of groundwater, and dealing with pollution. The largest potential sources of groundwater pollution are urban wastes.

9. Groundwater has much to do with the rock cycle. Groundwater dissolves minerals from bedrock and from regolith and transports them in solution, both to other formations in the Earth and to streams and the ocean. In

moving great amounts of dissolved material from place to place, groundwater both destroys and creates geologic features on a large scale. Some features created by *dissolution* are *caves, sinkholes,* and the deposits around *hot springs* and *geysers.*

Questions

1. Define *zone of aeration, zone of saturation,* and *water table.*
2. Explain the difference between *porosity* and *permeability.*
3. Define *packing* and *sorting.*
4. Describe the two types of *aquifers.*
5. What is the average rate of movement of groundwater?
6. How much of the water used in the United States is groundwater? Why will our demand for water grow even faster than our population?
7. Explain the principle of *drawdown* and *cone of depression.*
8. Describe two methods of artificially recharging an aquifer.
9. What methods are being used in the search for new sources of groundwater?
10. How are *caves* formed? How large is the largest cave system in the world?
11. What is the difference between a *concretion* and a *geode*?

Suggested Readings

Baldwin, H. L., and McGuinness, C. I., *A Primer on Ground Water.* Washington, D.C.: U.S. Geological Survey, U.S. Government Printing Office, 1963.

Douglas, John Scott, *Caves of Mystery: The Story of Cave Exploration.* New York: Dodd, Mead & Company, 1966.

Halliday, William R., *Depths of the Earth: Caves and Cavers of the United States.* New York: Harper & Row, Publishers, Inc., 1966.

Leopold, Luna B., and Davis, Kenneth S., *Water.* New York: Time-Life Books, 1966.

Moore, George W., and Moore, Nicholas G., *Speleology: The Study of Caves.* Boston: D. C. Heath, 1964.

Sayre, A. N., "Ground Water." *Scientific American,* November, 1950, pp. 14–19. (Offprint No. 818. San Francisco: W. H. Freeman and Company.)

Stenuit, Robert, and Jasinki, Marc, *Caves and the Marvelous World Beneath Us.* New York: A. S. Barnes and Co., 1966.

Waltham, Tony, *Caves.* New York: Crown Publishers, Inc., 1975.

"Today there is hope that the Great Lakes will survive to greet the next glacier."

Chapter Ten
Glaciers and Glaciation

330
Glaciers and Glaciation

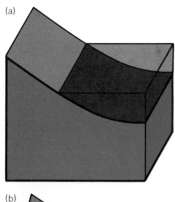

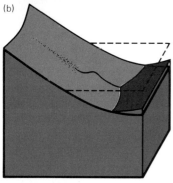

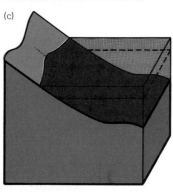

10.1 Changes of sea level that accompanied Quaternary climatic oscillations. (a) High sea level during warm interglacial stage. (b) Low sea level during Ice Age. (c) Higher-than-present sea level (by about 75 m) that would result if all the world's present glaciers melted and their water flowed into the sea.

Glaciers are one of the most powerful agents shaping the surface of the Earth. They carve the bedrock and are the prime creators of sediments. In rasping motions comparable to the massive downslope movements described in Chapter 7, glaciers create "instant regolith." The sediments deposited by glaciers are quite unlike any other kind, and form distinctive shapes and landforms which will be discussed later in this chapter.

Glaciers are closely related to the hydrologic cycle and the world's climate. The water stored on land in glaciers is water that has been temporarily removed from active participation in the hydrologic cycle. Because glacial ice represents such great potential supplies of fresh water, some scientists have suggested that the water problems of coastal cities, such as Los Angeles, might be alleviated by transporting icebergs from the Antarctic to California.

Study of the behavior of the margins of modern glaciers has shown a close relationship between glaciers and climate. When the world's climate warms even slightly, nearly all modern glaciers retreat. When the climate cools, most glaciers advance. Because the water to make glaciers comes ultimately from evaporation of sea water, a great advance of glaciers results in a lowering of sea level. When the glaciers melt, the water returns to the oceans and causes sea level to rise (Figure 10.1). This mutual relationship between glacial behavior and the water in the oceans has been described as one of "robbery and restitution."

Because the geologic effects of modern glaciers on both the bedrock and the regolith are so diagnostic, the effects of ancient glacial activity are easy to recognize. One of the great contributions of geology to our understanding of the history of the Earth has been the demonstration that great oscillations of climate have taken place. During times when glaciers covered millions of square kilometers of the land areas in the Northern Hemisphere and world sea level dropped more than 150 meters, the climate was generally cooler than it now is. During intervening times of warmer climate, the great glaciers melted and sea level rose to where it is now, and even slightly higher.

The deposits of these most recent glaciers still form the regolith in many parts of the world. But in the ancient bedrock, still-older glacial deposits have been lithified; they indicate that glacial episodes took place hundreds of millions of years ago.

The physics and geophysics of glaciers have yielded data useful in understanding the behavior of rocks. Glacial ice flows and creates patterns that are similar to those created by certain processes of rock metamorphism. The weight of a glacier creates a load on the Earth's lithosphere. The downward movements of the lithosphere under the weight of a fully developed glacier contrast with the upward movements that take place after the glacier has melted.

Historical Background

Nearly everyone who lives in the northern United States or in northwestern Europe is familiar with the concept that great glaciers once covered the areas where they live (Figure 10.2). So much has been said and written about the "Ice Ages" and so much speculating has been done about whether or not a new ice age may be at hand, that we may tend to take such subjects for granted. But ideas about the way glaciers work and the geologic proofs that glaciers formerly were much more extensive than they are today are relatively recent developments.

As with most great ideas, the introduction and establishment of modern concepts about glaciers and the Pleistocene Ice Ages created an intellectual revolution. Early in the nineteenth century European scientists were struggling to get accustomed to the idea that some of their lands had been recently covered by red-hot lava from geologically young volcanoes that were no longer active. All of a sudden they began to hear talk about their regions having been recently overspread by a thick sheet of ice.

Although not the first person to propose the idea that glaciers once covered much of Europe, Jean Louis Rodolphe Agassiz was most responsible for winning acceptance of the glacial theory. Originally Agassiz was skeptical about the entire notion. In fact, in 1836 Agassiz made a field trip through the upper Rhône Valley just to disprove the glacial theory—but the exact opposite occurred. Agassiz became an ardent convert and a tireless advocate of the glacial theory. He traveled across Europe and America propagating his

10.2. The last days of the glacial period, as interpreted in this painting of the region of Lucerne, Switzerland about 30,000 years ago. Conceptualization by W. Amrein, Prof. Albert Heim, and Ernst Hodel. (Swiss National Tourist Office)

views. Charles Darwin, among other scientists, quickly accepted Agassiz's arguments, but universal acceptance came only after Agassiz's death in 1873. Today Agassiz's views form the basis of an entire branch of geology known as glacial geology.

Modern Glaciers

A **glacier** is a flowing mass of ice that formed by the recrystallization of snow, is powered by gravity, and has flowed outward beyond the snowline (Figure 10.3). Each year, as a study of detailed precipitation charts will show, at least a little snow falls on each continent. By the end of winter much or all of this snow disappears. Only what remains might eventually become a glacier. A glacier's existence depends on fulfillment of the following three conditions:

1. More fresh snow falls each year than is lost to melting.
2. The compacted ice that forms from repeated melting and refreezing achieves a density greater than 0.84 g/cm^3.
3. A glacier, by definition, always flows internally. If its motion ceases it is still ice, but it is no longer a glacier. But just because glacial ice is always flowing away from the source of accumulation, the forward margin of the ice sheet will not always advance. If melting at the front edge takes place faster than the outward advance of the ice, the forward margin will *retreat*. This does not mean that the glacier will actually move into reverse gear and move backward like a retreating army. Instead, the word *retreat* in this context means that the front edge of the glacier recedes faster than the ice is flowing outward.

Snowfall and the Origin of Glacial Ice

Studies in the Antarctic begun during the International Geophysical Year (1958) have shown that the average annual fresh snowfall at the South Pole is only about 15 cm — an amount of precipitation equivalent to that which falls on the barren, desert Australian "outback." By contrast, New England ski resorts may accumulate nearly 3 meters of snow each winter. Yet in summer New England is green with vegetation. Snow lingers through the year only under certain climatic conditions and above certain altitudes. Such areas are said to be above the **snowline,** and it is here that glaciers can form.

Not all accumulations of snow above the snowline become glaciers. Before it becomes a glacier, a mass of snow must go through a series of changes. These changes result in the formation of snowfields, firn, and finally, glacier ice.

10.3 Small modern glacier on north face of the Bernini Group, Alps, Switzerland. The surface of the upper part of the glacier, which appears smooth, is covered by snow. The part that extends down the valley from the snowline is characterized by deep cracks. Similar cracks are present beneath the snow, but are not visible in this view. (Mario Fantin, Photo Researchers, Inc.)

Snowfields A large area of snow that lasts from one winter to the next is called a **snowfield**. Although the term is not an exact one, it usually means an accumulation that may be coated by a thin layer of ice, but which remains powdery and porous within.

Firn If, however, the snow crystals melt and refreeze into granular particles, the accumulation, after one year, is called **firn**. The density of firn is at least 0.55 grams per cubic centimeter; it is not compact enough to prevent air and water from passing through it.

Glacier ice Finally, if enough firn layers build up year after year, the melting and refreezing cycle is affected by the ever-greater weight of the added layers. This compaction forces out more and more of the air trapped in the particles of ice, and gradually ice crystals grow and join into a single mass (Figure 10.4). When the ice mass attains a density of 0.84 grams per cubic centimeter, becomes impermeable to air, and begins to move under the pressure of its own great weight, it is called **glacier ice**.

10.4 Ice crystals containing air bubbles several centimeters in diameter, on Nenana River, interior Alaska. (Charlie Ott, National Audubon Society)

334
Glaciers and Glaciation

Geologic Significance of Glaciers

Glaciers are important as historical archives of what fell from the ancient atmospheres and of ancient climates. They are also important as creators of sediment, of distinctive deposits, and of characteristic landforms.

Glaciers as historical archives A glacier holds many clues to what was in the atmosphere while the ice was forming, to its own history of flow, and to the climate that prevailed while the ice formed.

The firn layers, and any materials that settle on the surface of the snow-ice, form annual deposits which enable glaciologists to establish the age of the ice by simply counting the layers. A vivid example is the tephra thrown into the air by the 1912 eruption of Mt. Katmai on the Alaskan Peninsula. A film of this tephra has been found in all glaciers in western North America, and even within the Greenland Glacier 3200 km away. Proof indicating the record of the pollution of the atmosphere by lead smelting and the burning of tetraethyl lead in gasoline has been found in cores drilled through glaciers on Greenland (Figure 10.5) and Antarctica (Figure 10.6). In addition, the firn strata of all glaciers may contain anything that can blow onto it between snowfalls, including pollen, spores, and insects.

The proportion of oxygen isotopes in the ice of the world's glaciers provides clues about ancient temperatures. When water falls as precipitation, most of the oxygen atoms are the common isotope oxygen-16. However, depending on the temperature of the air at the time of the precipitation, some ions of the isotope oxygen-18 are always present. The warmer the temperature, the greater the concentration of oxygen-18. A mass spectrometer can determine the ratio of the two isotopes of oxygen. Using this ratio, scientists can infer ancient temperatures.

Grinding bedrock into sediment Glaciers are powerful grinding machines for turning solid bedrock into sediment. Samples of sediment melted out of the Antarctic Ice Sheet form the basis for estimating that each year this vast glacier delivers to the sea 35,000 to 50,000 million tons of sediment. This sediment comes from an area of 14 million square kilometers. The mean rate of sediment production per square kilometer thus is 2500 to 3750 tons. This is 3 to 4 times as much as the aggregate sum brought to the sea by all the world's rivers. The most prolific producer of stream sediment is Asia, which produces 7445 million tons per year from 28.62 million square kilometers. This is an average unit yield of 260 tons per square kilometer. The Antarctic Ice Sheet makes 9.6 to 14.4 times as much sediment. The nearly

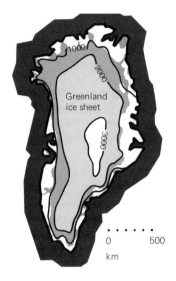

10.5 Map of Greenland, showing altitude of top of the ice sheet. Contours in meters above sea level.

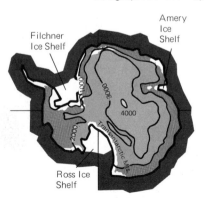

10.6 Map of Antarctica showing altitude of top of ice sheet. Contours in meters above sea level. (American Geographical Society)

continuous blanket of Pleistocene glacial sediments left by now-vanished continental glaciers in North America and in northern Europe is further evidence of how much sediment glaciers create.

Not only do glaciers create huge amounts of sediment, but what they deposit is characteristic. It provides important geologic clues for inferring where ancient glaciers flowed.

Creating distinctive landforms Glaciers have made mountain peaks jagged, enlarged mountain valleys, created fertile farmland, changed the course of mighty rivers, and channeled coastal fjords. We shall examine glacial sediments and glacial landscapes in greater detail in following sections of this chapter.

Kinds of Glaciers

Geologists recognize three chief kinds of glaciers. These are: (1) valley (Alpine) glaciers, (2) piedmont glaciers, and (3) ice sheets (continental glaciers).

10.7 **Valley glaciers,** flowing away from viewer toward terminus at upper right. Snowfield is at upper left. Berner Bay, southeastern Alaska, near Juneau. (Geological Survey of Canada)

Valley or Alpine glaciers The ice masses known as **valley** or **Alpine glaciers** are found in mountain ranges, notably in the Himalayas, the Alps, and the Andes. As their name implies, valley glaciers flow from high altitudes down valley courses (Figure 10.7). A valley glacier can be identified from afar by its distinctive tongue-like pattern where it moves down a mountainside. This "tongue of ice" continues to flow as long as the mountains above it receive adequate snowfall. A valley glacier can be said to resemble a spoonful of syrup flowing partway down the ice cream in a sundae.

Large parts of some valley glaciers often appear not to be nurtured by direct snowfall. That is, most of the glacier may lie well below the snowline, where the glacier grows by avalanches of snow from higher up the mountain (Figure 10.8). Avalanches bring snow from valley-side slopes to a glacier. This movement of snow from the valley-side slopes is analagous to the movement of sediment to streams down valley-side slopes by mass-wasting.

Piedmont glaciers A **piedmont glacier** is a glacier formed by the flowing together of two or more valley glaciers (Figure 10.9). A piedmont glacier is formed by ice that is "third-hand." Snow falls first on the peak, then oozes down into valleys, and finally extrudes on flatlands. Referring back to our analogy of a valley glacier and a spoonful of syrup, we can visualize a piedmont glacier as the result of several spoonfuls. That is, the syrup—or ice—has flowed all the way down the ice cream—or valley—to form a pool below.

Some piedmont glaciers are enormous. In Alaska, the Malaspina Glacier is the product of several separate piedmonts that have joined together to cover an area larger than the state of Rhode Island. It is theoretically possible for a piedmont glacier like the Malaspina to

10.8 Snow avalanches, a major source of materials for forming ice in valley glaciers. (Left) Lobate pattern at margin of snow deposited by an avalanche in 1952. (Photo unit, FAO, Rome) (Right) View downslope along the track left by a wet avalanche. (Swiss Federal Institute for Snow and Avalanche Research)

10.9 A piedmont glacier. Malaspina Glacier, St. Elias Mountains, Alaska. (Austin Post, U.S. Geological Survey)

build upward in height until it eventually surpasses the snowline. Should this happen—as it may have in Antarctica and Greenland—then nearly all the glacier would receive its snow "first-hand," and would grow with considerably more speed.

Ice sheets The greatest glaciers in the world today all fall into the category of **ice sheets.** These include the glaciers of Antarctica (Figure 10.6), Greenland (Figures 10.5 and 10.10), Iceland, Spitsbergen, Novaya Zemlya in the Soviet Union, and the ice caps that cover parts of Canada and Alaska.

To complete the syrup analogy, the formation of ice sheets is like pouring copious quantities of the topping onto a flat surface. The syrup would tend to flow outward in all directions, forming a smooth surface. In sufficient amounts, the flowing mass could even move up small inclines. The center of an ice sheet rests above the snowline. Thus, it is directly nourished by snowfall. As snow accumulates on the surface, it begins to move outward and downward. In addition, the tremendous weight of the ice sheet causes the edges to spread outward. The ice naturally seeks downhill slopes, but when it encounters an incline whose top is lower than the top of the lens-like glacier, the edge will move upward. The center of a sheet always re-

10.10 View of an ice cap, seen from an airplane. "Polaris Glacier" Hall Land, northwest Greenland. (Polar Continental Shelf Project of the Canadian Department of Energy, Mines, and Resources)

mains higher than the sides. The accumulations eventually prove heavy enough to actually depress the land beneath the ice.

World Distribution of Glaciers

Glaciers currently cover about 16 million square kilometers of the Earth's surface, or roughly 10 per cent of the planet's land. In the past, as much as 32 per cent of the land may have been glaciated. Depending on the estimate, these ice masses now contain between 2.5 and 25 million cubic kilometers of water. Should all the glaciers melt, world sea level would rise by an estimated 20 to 60 meters. This is more than enough to inundate most of our great cities. The chief existing ice sheets are on Antarctica and Greenland. By comparison with these two giants, all others are puny.

Antarctic Ice Sheet Of the 16 million square kilometers of area occupied by existing glaciers, some 14 million are located in Antarctica. The giant Antarctic Ice Sheet may be as much as 10 million years old; parts of it are at least 4000 meters thick (Figure 10.6). The sheer weight of the ice has depressed the continental crust more than a kilometer below sea level. Today, however, the land is slowly rising. This is taken as an indication that in the past the sheet was even thicker than now. Scientists from many nations maintain scientific outposts on the Antarctic ice sheet, but the interior of that mighty glacier remains practically unknown.

Greenland Ice Sheet The second-largest glacier is the Greenland Ice Sheet, which covers almost 2 million square kilometers (Figure 10.5). Scientists have now been probing Greenland's ice sheet for three decades. Considerable climatic information has been gathered, but the formation and development of the glacier itself remains virtually as mysterious as the aspects of the Antarctic glacier.

Other major areas Except for the Antarctic and Greenland and other Arctic sheets, the rest of the world's glaciers are located on mountains whose peaks lie above the snowline. In the Western Hemisphere, these include parts of the Alaskan, Cascade, and Rocky Mountain ranges in North America and the Andes chain in South America. In the Eastern Hemisphere, there are glaciers throughout Scandinavia, the Alps, and the Caucasus range; farther east, in the heart of Eurasia, valley glaciers abound in the great Himalayan, Tien Shan, Hindu Kush, Karakoram, and Pamir ranges.

Finally, isolated glaciers exist on the African continent (Mts. Kilimanjaro and Kenya) and on New Guinea in the Pacific Ocean. The only major land mass lacking glaciers is the continent of Australia.

Glacial Movements

Despite their rigid appearances, glaciers are not static ice masses. Rather, as we mentioned previously, they constantly seek equilibrium as snow accumulates and melts, temperature rises and falls, and the Sun appears and disappears with the changing seasons. Because they are in constant flux, glaciers advance and retreat—although gradually—almost daily. We shall examine the measurement and rates of movement, the mechanics of movement, and features created by movement of glacier ice.

Measurement and rates The first attempts to measure glacier movement were simple but effective. Glaciologists drove a line of stakes across the ice, then observed the changing line (Figure 10.11). On the basis of his studies of glacier movement, the British scientist J. D. Forbes, on an Alpine expedition in 1840, was able to make a remarkable forecast. Twenty years earlier, three French mountaineers were killed when they tumbled off Mt. Blanc. Their corpses were buried in a valley glacier. Noting the point where the three bodies had fallen onto the glacier and calculating the mean flow of the glacier, Professor Forbes estimated that the corpses would reach the plain below and would be released from the ice in 1860. The bodies were discovered in 1863.

Early measurements with stakes proved not only that glaciers move, but that flow is fastest in the center and slowest at the sides, as in a river. Modern glaciologists have drilled a series of deep holes through the surface of glaciers and have inserted iron pipes into the holes. With the passage of time, the pipes are bent so that their tops bend forward in the direction of flow. Thus, glaciologists have inferred that the upper parts of a glacier move more rapidly than the lower parts (Figure 10.11).

The measurements on modern glaciers have indicated that the typical rate of advance is 15 to 60 cm per day. Unusual conditions have produced some rapid if short-lived gallops called **surges.** Where large outlet glaciers in Greenland squeeze through narrow mountain passes, their speeds have been clocked at better than 30 meters per day. In the mid-1930s, the Black Rapids Glacier in Alaska suddenly began a six-month surge. The glacier advanced almost 7 km, averaging

10.11 Deflections of stakes driven in a line across a valley glacier prove that the ice flows fastest in the center. Bending over of pipes inserted through holes in ice shows upper levels of ice in valley glacier flow faster than lower levels.

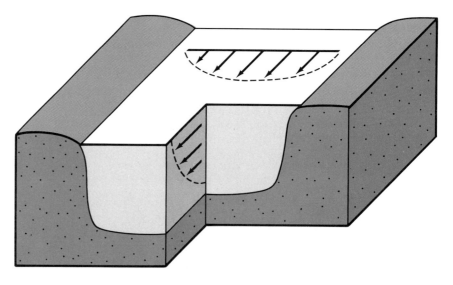

35 meters a day and moving as much as 65 meters in 24 hours. It finally stopped, just short of a hunting lodge and just before it obliterated the only road link between Fairbanks and the "Lower 48."

Mechanics Nonsurging glacial movement is thought to consist of two contrasting modes. These are (1) slippage or sliding along discrete shear planes, and (2) plastic flow from deformation of ice crystals. Both of these mechanisms are related to the fact that the ice in the lower parts of the glacier is under enormous pressure. The factor responsible for making glaciers move is the component of gravity acting along a slope (Figure 7.4). The mechanics of surging are not known.

Sliding. Although the ice tends to slip along many internal shear planes, the weight of overlying ice tends to concentrate most of this motion along a single surface near the base of the glacier. Movement on the bottommost layer is facilitated by the presence of water. This water can come from melting related to air temperature or from the effects of great pressure.

Plastic flow. The secret that enables solid ice to flow is that ice is crystalline (Figure 2.10b) and these crystals can be deformed. The long axes of the ice crystals in the lower parts of the glacier become aligned so that they are roughly parallel to the surface of the glacier. When pressure from above and gravity from below interact, the sheets within the glacier are set in motion.

The zones of fracture and flow Glacial ice under low pressure is a brittle solid that fractures readily. Glacial ice under high pressure

10.12. Deep crevasses in the surface of Blue Glacier, Olympic Park, Washington, seen in close view. This part of the glacier has flowed outward and downward from the snowline. (U.S. Geological Survey)

can flow plastically. The effect of the increase in pressure downward in a glacier is to divide the ice into two zones. These are (1) an upper zone of fracture, and (2) a lower zone of flow.

The most distinctive features of the zone of fracture are **crevasses,** deep fissures in the upper zone of the glacier (Figure 10.12). Crevasses are created by the differing rates and mechanics of flow of the surface and the interior parts of a glacier. The surface is not under much load from above.

Crevasses are much more common in fast-moving valley glaciers of mountain chains than in slow-moving continental ice sheets. Many crevasses are created because of the drag between moving ice and stationary valley walls (Figure 10.13).

In the zone of flow, crevasses become closed, and layers of sediment become intricately contorted (Figure 10.14). Both on a small scale and on a large scale features found in the zone of flow in a glacier resemble the features found in metamorphic rocks in the interiors of mountain chains (Chapter 15).

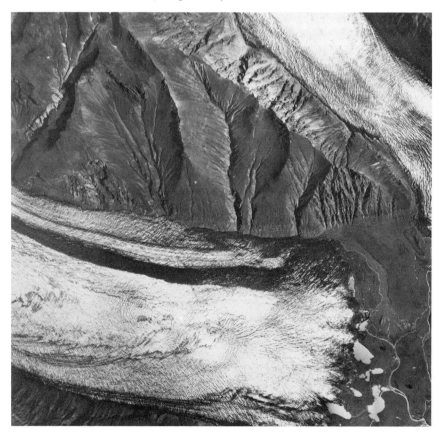

10.13 Patterns of parallel crevasses diagonal to margins of two valley glaciers, Iceland. (Copyright Icelandic Geodetic Survey)

10.14 Deformation patterns at edge of glacier on Baffin Island, Canada, have resulted from complex plastic flow of the ice. Dark streaks are sediment eroded by the glacier. (Stan Wayman, Photo Researchers, Inc.)

The results of **glaciation** (the advance and retreat of a glacier over an area) include erosion and deposition. We shall examine the mechanics of glacial erosion and the landforms eroded in bedrock by glaciers. Then we shall discuss glacial deposits and the landforms created by a glacier shaping its deposits.

Results of Glaciation and Related Processes

Erosion

Nowhere on Earth is there a more potent agent of erosion than an advancing glacier. A glacier that may be 2000 meters thick can sculpt the land almost in a few centuries. To achieve comparable results wind and water would be required to work for millions of years. Glaciers

10.15 Glacier removing blocks of rock defined by intersecting joints, on downflow side of hill. Schematic profile and section.

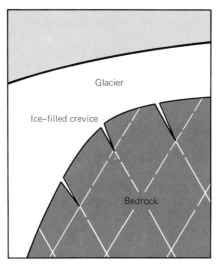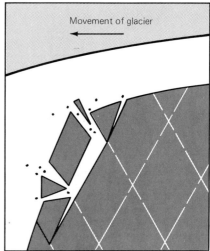

grind solid rocks into fine fragments; pull loose large chunks of bedrock; smooth, shape, and polish surfaces underlain by solid bedrock; and score these smoothed surfaces with scratches and grooves. Glacier-related erosional processes include avalanching and frost-wedging.

Quarrying and plucking In places where the ice is moving slowly a glacier can attach itself to surfaces of bedrock. Water formed from pressure melting or from warm air temperatures fills cracks in bedrock. When this water freezes, it forms a solid jacket of ice around a block of bedrock. As the ice eventually flows away, it pulls out the blocks that have been jacketed by ice (Figure 10.15). Such removal of large blocks of bedrock is named **quarrying,** or **plucking.**

Abrasion Acting chiefly in its basal parts (and along the sides of a valley glacier) a glacier acquires a load of rock particles. It grinds these particles even finer, and in the process uses its load of rock particles to abrade the underlying bedrock.

The fine chips and powder created by abrasion of rocks embedded in the basal and side layers of a glacier against one another and against solid bedrock constitute **rock flour.** Because of their tiny sizes the particles of rock flour are most conspicuous where they leave the glacier in a stream of meltwater. The rock flour makes the water appear milky. Where the milky stream enters a lake the effect is to color the water. The pale blue of the water of Lake Louise, Alberta (Canada), and the cerulean hues of the water of Kashmir's Dal Lake are examples.

The rasping, abrasive effect of glaciers on massive solid bedrock

10.16 Knob of resistant bedrock that was rounded, polished, and striated by flow of Pleistocene continental glacier. Pre-Triassic metamorphic rock, Woodbridge, Connecticut. (Marian Ksiazkiewicz)

is to smooth and and polish the surface and to sculpt it into large-scale rounded forms (Figure 10.16). No other geologic agent creates such features.

Striae and grooves A thick ice sheet tends to flow in a single direction regionally. The ice flows relentlessly in a single direction that crosses hills and valleys as if they did not exist. The direction of flow of a thinner sheet or of a valley glacier is influenced by local morphology. The result of this directed flow is the rasping out of tiny linear grooves, named **striae** (Figure 10.17), or of larger troughs or deep grooves. The striae and grooves are parallel to the direction of

10.17 Glacial grooves at base of glacier, Pumori South Glacier, Himalayas. (Fritz Muller, Copyright Swiss Foundation for Alpine Research, Zurich)

10.18 Smoothly rounded and striated bedrock sculpted by a late Pleistocene glacier that flowed across New York City from the northwest. View facing S20°E, in direction of ice flow. Riverside Park at 168 Street, New York City. (Bruce Caplan)

the glacier's flow (Figure 10.18). When the directions of striae are plotted on a map, they present a picture of the way the ice moved.

Glacial Landforms Eroded in Bedrock

Distinctive landforms are eroded in bedrock by glaciers. These include cirques; glaciated and hanging valleys and fjords; *arêtes*, cols, and horns; and *roches moutonnées*.

    Cirques The most common mountain landscape feature created by glaciation is a **cirque** (seerk), a steeped-walled, bowl-shaped niche in which a valley glacier originated. (See Figure 10.19, opposite page.) A cirque forms in the headwater region of an already existing valley, not just anywhere above the snowline. While a valley glacier is flowing, the cirque is the area above the snowline that provides the downward-flowing ice mass with its nourishment. As a glacier recedes, the cirque is the place where the last traces of ice remain.

    Because the ice stays in a cirque longer than anywhere else, erosion is deepest there. As a result, cirques resemble the top portion of old-fashioned keyholes. They are rimmed on three sides by mountain; the fourth side is the downhill opening through which the glacier descends. Typically, the remaining mountain rim towers high above the bottom of a cirque because centuries of frost wedging enables the glacier to quarry deeply into the rock. The walls of a cirque tend to be

Wind River Mountains, Wyoming. (Austin Post, U.S. Geological Survey, University of Washington)

10.20 U-shaped transverse profile. Now-vanished valley glacier eroded this valley into this characteristic shape. Head of Rock Creek, viewed from summit of the Red Lodge Highway, Beartooth Mountains, Montana. (George A. Grant, U.S. National Park Service)

more or less smooth, because continued abrasion has worn away most of the high spots.

Glaciated and hanging valleys; fjords As a valley glacier creeps out of its cirque, the main body follows the course of least resistance down the mountainside. Nearly always, the ice sheet retraces the bed of an older waterway. In so doing, a valley glacier characteristically alters the V-shaped profile created by running water into a rounder U shape (Figure 10.20). For example, a cross section of the Yellowstone River (Figure 10.21) shows a much sharper cut than a similar view of a glaciated valley.

As the valley glacier creeps down the mountain, it is being fed by tributary glaciers. Each tributary has its own head or source, and finds its own route until it joins the main mass. Because these feeders are by definition smaller, they pack less erosive force. They do not dig as deeply downward, so that they often empty into the valley glacier not at the floor of the main sheet but at a higher level. When the main sheet melts away, its bottom may be several hundred meters lower than the bottoms of its tributaries. These higher tributaries are aptly called **hanging valleys** (Figure 10.22). This landform is characteristic of the European Alps.

Glaciers that deepen valleys near a coastline create U-shaped

10.21 Steep V-shaped valley carved by river in massive bedrock. Yellowstone River, Yellowstone Park, Wyoming. (Larry Kramer)

valleys that can be flooded by the sea after the glacier has melted. Such a glaciated valley occupied by an arm of the sea is a **fjord** (Figure 10.23). Some fjords extend as far as 190 km inland.

The most renowned fjords are those in Norway, especially the Sogne and the Hardanger which have walls almost a kilometer high. The coasts of Alaska and British Columbia, too, are laced with so many fjords that coastal streamers can pick their way through a maze of interconnecting fjords known as the Intercoastal Waterway. The entire coast of Greenland, parts of other North Atlantic islands, New Zealand, and the Chilean coast below the Andes range also display fjords.

Arêtes, cols, horns Rarely does a mountain undergoing glaciation have just one glacier. Commonly, separate glaciers, each fed from snowfall at the peak, will move down different sides of the mountain. Eventually each glacier forms its own cirque. When adjacent cirques grow until they are separated by only a relatively thin vertical remnant of mountain, that remnant is called an **arête** (ah-RET) (from the French word for stop).

Erosion, whether from continued glacial action or from water and wind, will eventually crumble sections of the once-solid arête. The

10.22 Hanging valley (right, at waterfall) formed when main valley was greatly deepened by valley glacier. Such a tributary valley, not deepened, is said to be a hanging valley. Lauterbrunnental, Bernese Oberland, Switzerland. (Swiss National Tourist Office)

10.23 A fjord at the head of Milford Sound, New Zealand. (New Zealand Consulate General, New York)

gaps in this wall, through which it is possible to climb from one cirque to another, are called **cols** (colls).

Finally, as the several arêtes on a mountain are worn down, the remaining pinnacle—if it overlooks three or more cirques—is called a **horn**. Because the erosion process leaves such jagged sides on horns, expert mountaineers prefer the challenge of trying to scale this type of peak. In fact, the world's two best-known peaks are horns—Mt. Everest, which is bordered by four mammoth valley glaciers, and the Matterhorn in the Alps (Figure 10.24).

10.24 Matterhorn, a horn formed by glacial erosion from three sides. View from Hamlet of Findelen, near Zermatt, Valais, Switzerland. (Swiss National Tourist Office)

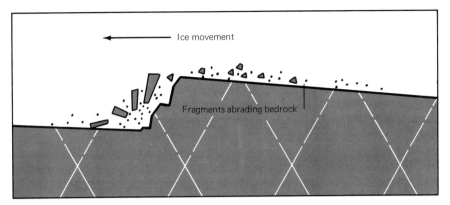

10.25 **Roches moutonnées** seen in schematic profile.

Roches moutonnées The combined action of abrasion, smoothing and rounding on one side (the side up which the ice flowed) and quarrying on the opposite side (downflow side) creates asymmetrical rock hillocks known as **roches moutonnées** (ROASH moo-tahn-AY) (Figure 10.25). The name comes from French and means "rock sheep," in reference to the resemblance of clusters of these hillocks to herds of sheep lying down.

Deposition

In the sense that glaciers are flowing streams of ice that transport sediment, glaciers are comparable to streams. Both flow from high ground to low. They rework the terrain as they travel, and eat away the softest parts. Both gradually polish the bedrock, but, of course, the effects of glacial polishing are much greater than those of stream polishing. And if the glacier was an ice sheet its effects are not confined to valleys.

But unlike rivers and streams, glaciers are not chiefly engaged in transporting the regolith and rock debris they tear from the terrain to a growing accumulation such as a delta. Instead, this matter is strewn in characteristic deposits all along the glacier's path. These signposts from past ice ages allow today's glaciologists to reconstruct the movements of long-vanished glaciers. We shall discuss these subjects under the major headings of glacial sediments and depositional landforms made by glaciers or related to glaciers.

Sediments A general term for any sediment that was in any way connected with a glacier is **drift.** This is short for "northern drift," a term applied in Europe to the material transported southward from Scandinavia. When the term was first applied geologists thought

The Glaciers John Muir

Along the western base of the range a telling series of sedimentary rocks containing the early history of the Sierra are now being studied. But leaving for the present these first chapters, we see that only a very short geological time ago, just before the coming on of that winter of winters called the glacial period, a vast deluge of molten rocks poured from many a chasm and crater on the flanks and summit of the range, filling lake basins and river channel, and obliterating nearly every existing feature on the northern portion. At length these all-destroying floods ceased to flow. But while the great volcanic cones built up along the axis still burned and smoked, the whole Sierra passed under the domain of ice and snow. Then over the bald, featureless, fire-blackened mountains, glaciers began to crawl, covering them from the summits to the sea with a mantle of ice; and then with infinite deliberation the work went on of sculpturing the range anew. These mighty agents of erosion, halting never through unnumbered centuries, crushed and ground the flinty lavas and granites beneath their crystal folds, wasting and building until in the fullness of time the Sierra was born again, brought to light nearly as we behold it today, with glaciers and snowcrushed pines at the top of the range, wheat-fields and orange-groves at the foot of it.

This change from icy darkness and death to life and beauty was slow, as we count time, and is still going on, north and south, over all the world wherever glaciers exist, whether in the form of distinct rivers, as in Switzerland, Norway, the mountains of Asia, and the Pacific Coast; or in continuous mantling folds, as in portions of Alaska, Greenland, Franz-Joseph-Land, Nova Zembla, Spitzbergen, and the lands about the South Pole. But in no country, as far as I know, may these majestic changes be studied to better advantage than in the plains and mountains of California.

Toward the close of the glacial period, when the snowclouds became less fertile and the melting waste of sunshine became greater, the lower folds of the ice-sheet in California, discharging fleets of icebergs into the sea, began to shallow and recede from the lowlands, and then move slowly up the flanks of the Sierra in compliance with the changes of climate. The great white mantle on the mountains broke up in a series of glaciers more or less distinct and river-like, with many tributaries, and these again were melted and divided into still smaller glaciers, until now only a few of the smallest residual topmost branches of the grand system exist on the cool slopes of the summit peaks.

Plants and animals, biding their time, closely followed the retiring ice, bestowing quick and joyous animation on the new-born landscapes. Pine-trees marched up the sun-warmed moraines in long, hopeful files, taking the ground and establishing themselves as soon as it was ready for them; brown-spiked sedges fringed the shores of the new-born lakes; young rivers roared in the abandoned channels of the glaciers; flowers bloomed around the feet of the great burnished domes,—while with quick fertility mellow beds of soil, settling and warming, offered food to multitudes of Nature's waiting children, great and small, animals as well as plants; mice, squirrels, marmots, dear, bears, elephants, etc. The ground burst into bloom with magical rapidity, and the young forests into birdsong: life in every form warming and sweetening and growing richer as the years passed away over the mighty Sierra so lately suggestive of death and consummate desolation only.

It is hard without long and loving study to realize the magnitude of the work done on these mountains during the last glacial periods by glaciers, which are only streams of closely compacted snow-crystals. Careful study of the phenomena presented goes to show that the preglacial condition of the range was comparatively simple: one vast wave of stone in which a thousand mountains, domes, canons, ridges, etc., lay concealed. And in the development of these Nature chose for a tool not the earthquake or lightning to rend and split asunder, not the stormy torrent or eroding rain, but the tender snowflowers noiselessly falling through unnumbered centuries, the offspring of the sun and sea. Laboring harmoniously in united strength they crushed and ground and wore away the rocks in their march, making vast

From THE MOUNTAINS OF CALIFORNIA by John Muir, 1894. Doubleday & Company, Inc

Field Trip

beds of soil, and at the same time developed and fashioned the landscapes into the delightful variety of hill and dale and lordly mountain that mortals call beauty. Perhaps more than a mile in average depth has the range been thus degraded during the last glacial period,—a quantity of mechanical work almost inconceivably great.

I discovered clear sections where the bedded structure was beautifully revealed. The surface snow, though sprinkled with stones shot down from the cliffs, was in some places almost pure, gradually becoming crystalline and changing to whitish porous ice of different shades of color, and this again changing at a depth of 20 or 30 feet to blue ice, some of the ribbon-like bands of which were nearly pure, and blended with the paler bands in the most gradual and delicate manner imaginable. A series of rugged zigzags enabled me to make my way down into the weird underworld of the crevasse. Its chambered hollows were hung with a multitude of clustered icicles, amid which pale, subdued light pulsed and shimmered with indescribable loveliness. Water dripped and tinkled overhead, and from far below came strange, solemn murmurings from currents that were feeling their way through veins and fissures in the dark. The chambers of the glacier are perfectly enchanting, notwithstanding one feels out of place in their frosty beauty. I was soon cold in my shirt-sleeves, and the leaning wall threatened to engulf me; yet it was hard to leave the delicious music of the water and the lovely light. Coming again to the surface, I noticed boulders of every size on their journeys to the terminal moraine—journeys of more than a hundred years, without a single stop, night or day, winter or summer.

The sun gave birth to a network of sweet-voiced rills that ran gracefully down the glacier, curling and swirling in their shining channels, and cutting clear sections through the porous surface-ice into the solid blue, where the structure of the glacier was beautifully illustrated.

The series of small terminal moraines which I had observed in the morning, along the south wall of the amphitheater, correspond in every way with the moraine of that glacier, and their distribution with reference to shadows was now understood. When the climatic changes came on that caused the melting and retreat of the main glacier that filled the amphitheater, a series of residual glaciers were left in the cliff shadows, under the protection of which they lingered, until they formed the moraines we are studying. Then, as the snow became still less abundant, all of them vanished in succession, except the one just described; and the cause of its longer life is sufficiently apparent in the greater area of snowbasin it drains, and its more perfect protection from wasting sunshine. How much longer this little glacier will last depends, of course, on the amount of snow it receives from year to year, as compared with melting waste.

On August 21, I set a series of stakes in the Maclure Glacier, near Mount Lyell, and found its rate of motion to be a little more than an inch a day in the middle, showing a great contrast to the Muir Glacier in Alaska, which, near the front, flows at a rate of from five to ten feet in twenty-four hours.

Mount Shasta has three glaciers, but Mount Whitney, although it is the highest mountain in the range, does not now cherish a single glacier. Small patches of lasting snow and ice occur on its northern slopes, but they are shallow, and present no well marked evidence of glacial motion. Its sides, however, are scored and polished in many places by the action of its ancient glaciers that flowed east and west as tributaries of the great glaciers that once filled the valleys of the Kern and Owen's rivers.

1894

the sediment had been rafted by icebergs. The sediments deposited directly by a glacier include erratics and till. Sediments that are not deposited directly by glaciers but in bodies of water closely related to a glacier include outwash and varved lake deposits.

Erratics Glaciers possess great powers for transporting large rock fragments long distances. In so doing they can deposit debris on bedrock that is unlike the transported fragments. A rock fragment unlike the bedrock underlying it is an **erratic.** If the erratic rests on a polished, striated rock pavement (Figure 10.26), then almost certainly it was deposited by a glacier. The finding of distinctive rocks from the Alps in parts of Switzerland where no such bedrock exists was an important milestone in gaining support for the "glacial revolution."

The distance an erratic may travel depends on its composition. Limestone and other easily weathered rocks can survive only short trips. Granite can move long distances. Plymouth Rock is a granite boulder that may have been swept from New Hampshire to its resting place near Boston where the *Mayflower* Pilgrims alighted on it. In the nineteenth century, the discovery of 11 diamonds scattered in Wisconsin, Indiana, and Ohio pointed to the existence of a rich lode somewhere in Canada. The precise location, however, is not known. Diamonds are the most resistant of minerals, so that these may well have travelled thousands of kilometers. Prospectors are still searching.

10.26 Light-colored erratics resting on striated pavement cut by Pleistocene glacier across dark-colored mafic rock. Rain has washed away all the fine particles that formerly surrounded these erratics. Farmington, Connecticut. (J. E. Sanders)

Results of Glaciation and Related Processes

10.27 Till, consisting of light-colored particles set randomly in dark-colored matrix that is nonstratified and poorly sorted. Montauk Point, Long Island, New York. (Robert LaFleur)

Till The sediment deposited directly by a glacier is **till.** Because the glacier cannot separate its debris by size, till is nonstratified and nonsorted. Till is a heterogeneous collection of debris whose particles range in size from microscopic rock flour up to immense boulders (Figure 10.27). One rock slab that was part of a deposit of till in England was so large that a town was built on it.

Other glacier-controlled sediments In the history of any glacier there are two important stages: (1) the time when over-all outflow exceeds melting and the glacier expands, and (2) the time when overall melting exceeds outflow and the glacier wastes away. The glacier may retreat as a glacier—the ice always flowing actively but the edges melting back—or as large blocks of stagnant ice. In either case water is plentiful. This water may flow away from the glacier in great streams that deposit stratified sediment derived from the glacier. Or it may form lakes in which the melted-out sediments accumulate. Thus arise, respectively, outwash and varved lake deposits.

Outwash As we have said, the body of the glacier possesses no mechanism for sorting its sediment load by size. By contrast, meltwater from the ice mass can organize glacially transported sediment to a certain degree. The sediments, chiefly sand and gravel, deposited by streams that issue from a glacier are called **outwash.** Another name for all glacially derived sediments deposited by streams is **stratified drift.** As a meltwater stream flows away from a glacier, it deposits its larger and heavier particles first, nearer the glacier, and washes the smaller and lighter material farther away.

Varved lake deposits Lakes that receive meltwater from a glacier are named **proglacial lakes.** Such lakes are frozen over during a large part of the year and are free of ice only briefly, for a month or more in the summer. The supply of sediment to a proglacial lake is strictly seasonal. Meltwater streams are active in the summer and nothing at all enters during the winter. In the summer, rock flour becomes widely dispersed in the lake water, and fine sand and silt may flow along the bottom. In the winter all is quiet; the only thing happening is the settling of the fine particles of the suspended rock flour. The particles reach bottom according to size, largest first. They create a **graded layer,** a layer having its particles arranged in a systematic gradient according to size. Because of the long settling period the wintertime deposit is always graded. When the ice melts the new summer material arrives. Summertime sediment is variable. It is laminated, possible even micro-rippled, and may contain several graded layers. It is invariably coarser than the wintertime deposit. Also, the summer sediment always overlies the previous winter's layer along a sharp contact, but grades upward into the following winter's deposit (Figure 10.28).

In 1879, Gerard De Geer, a Swedish geologist, decided that this cycle of sedimentation would leave distinctive layers on the lake bed. His pioneering research confirmed the existence of the seasonal stratification in proglacial lake sediments. He proposed the name **varve** for the sediment deposited in a year. Once varves have been identified they can be counted like tree rings to tell time. One varve—one year.

10.28 Varved silt and clay deposited by proglacial lake of Pleistocene age, Hamden, Connecticut. (J. E. Sanders)

Results of Glaciation and Related Processes

By counting and analyzing varves in Scandinavia, De Geer was able, in the early twentieth century, to date the demise of the last Würm glacier in Northern Europe fairly accurately at about 10,000 B.C.

Depositional Landforms Made by Glaciers or Related to Glaciers

Not only do glaciers erode distinctive landforms in solid bedrock; they also have shaped their sediments into features that still form parts of the landscape in many areas. Flowing ice is responsible for moraine ridges and drumlins. Meltwater, with or without blocks of stagnant ice, deposits eskers and kames. Collapse of sediment over buried blocks of ice creates kettles.

Moraine ridges A large body of drift (consisting of till, stratified drift, or both) that has been shaped into a rounded ridge is a **moraine.** Several varieties are recognized. At the outer margin of a glacier that has reached its maximum extent, the ice pushes up debris into a ridge whose trend follows the edge of the ice. This ridge is known as *terminal moraine* (Figure 10.29). A single glacier deposits

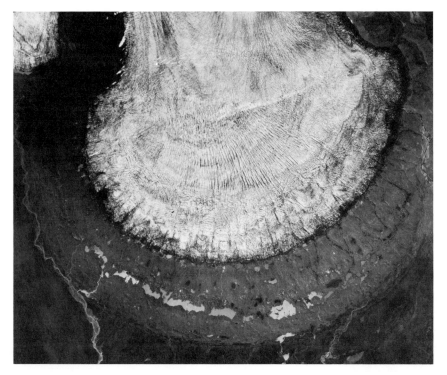

10.29 Terminal moraine at end of valley glacier, Iceland. Radial pattern of cracks in ice resulted from spreading of ice into semicircular outline after it flowed out the end of the valley. The outermost ridge is the terminal moraine; between it and the edge of the glacier are many small lakes. These probably are kettles, formed by collapse of sediment deposited above isolated blocks of ice. (Copyright Icelandic Geodetic Survey)

10.30 Recessional moraines deposited by two retreating glacier lobes. Dark areas are lakes. Western Quebec, Canada, about 80 km east of Hudson Bay. Width of view is 15 km. (Geological Survey of Canada)

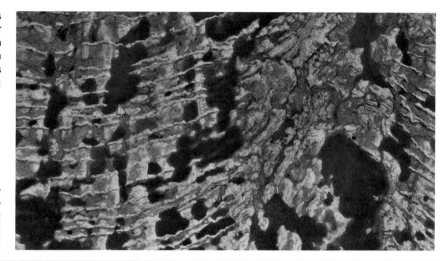

10.31 Medial moraines formed by flowing together of several valley glaciers. Barnard Glacier, Wrangell Mountains, Alaska. (U.S. Geological Survey)

only one terminal moraine—at the point of its greatest extent. But during its retreat it may deposit other morainic ridges along its margins at places where the rate of retreat is temporarily slowed. These ridges are known as *recessional moraines* (Figure 10.30).

A *lateral moraine* is spread along the nonleading edges of a valley glacier, like the banks formed by a V-shaped snowplow. A *medial moraine* occurs atop and within the glacier itself. A medial moraine is the product of converging lateral moraines that flow together where two valley glaciers meet (Figure 10.31).

The late Pleistocene glaciers left huge moraines throughout northern North America and much of Europe.

One complex terminal moraine extends—underwater much of the way—from Nova Scotia in a crescent through Cape Cod, Long Island, northern New Jersey, and into the hills of Pennsylvania (Figure 10.32). In the midwest, the Defiance Moraine begins near Ann Arbor in south-central Michigan and curves through northern Ohio into northwest Pennsylvania. Travellers along the Ohio Turnpike must traverse this deposit, which looms noticeably above the state's generally flat terrain. They cross it through passes cut near Akron and Toledo.

There are equally impressive moraines throughout north-central Europe; in fact, much of Denmark was deposited by successive glaciers that terminated along the eastern edge of the North Sea.

10.32 Limits of Quaternary glacial ice in continental United States, shown by combining outermost points reached by various individual glacial advances.

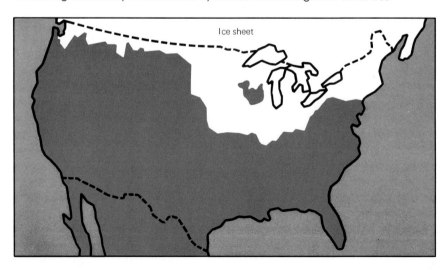

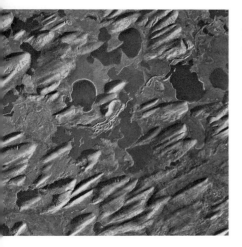

10.33 Swarm of drumlins seen from above, northern Saskatchewan, Canada, deposited by Pleistocene continental glacier that flowed from NE (upper right) to SW. Lakes (dark areas) are about 1 km in diameter. (Geological Survey of Canada)

10.34 An esker, a sinuous deposit formed by a subglacial stream. Width of view is about 2 km. (Austin Post)

Drumlins In contrast to moraines, which can be considered the peripheral leavings of a glacier, a **drumlin** consists of till that a glacier has fashioned into a streamlined shape. Characteristically, drumlins are eroded into an elongated shape with the steepest end facing the direction from which the glacier originated. The lowest end points to the direction in which the ice mass moved (Figure 10.33).

A drumlin forms when drift collects in a depression that is leeward of a protecting bedrock hill, or when till with a large amount of cohesive clay collects into a mass which the glacier can smooth as it passes.

Drumlins range in height from 15 to 60 meters, and may reach almost a kilometer in length. Parts of Canada and the states of Massachusetts, New York, and Wisconsin are dotted with drumlins. In the central Hudson Valley and northeastward into New England, apple orchards have been planted on the north-facing slopes of drumlins.

Eskers An **esker** is a long, winding, continuous embankment that can stretch for more than 60 km; in fact, eskers have been compared in appearance to raised railroad beds (Figure 10.34). An esker is composed of a layer of gravel beneath a low, narrow mound of sand and silt. The esker is a distinctive deposit made by a subglacial stream of meltwater. First the water eroded a tunnel out of the stagnant ice. Then it partially or fully filled this tunnel with stream deposits. When the ice melted away, these deposits were left as an embankment standing above the surrounding lowlands. An esker could

10.35 Origin of kames and kame terraces. Braided stream flowing between valley wall and stagnant block of ice from former glacier deposits sediments (left). After ice has melted the stream-laid sediments are no longer at the bottom of a valley but form a terrace along the side of a large valley (right). Sediments dumped from above into fissure in ice remain as small raised knoll (a kame). Schematic block.

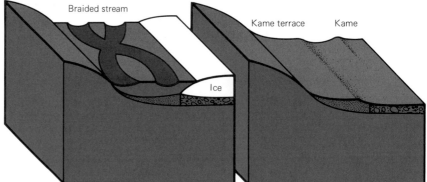

not form where the flow of glacial ice would destroy it. Eskers imply stagnation of large blocks of ice.

Kames and kame terraces Kames and kame terraces are bodies of outwash built against stagnant blocks of ice. These take two forms: (1) short, steep-sided mounds **(kames),** or (2) terrace-like bodies built against the margin of a former glacier **(kame terrace)** (Figure 10.35).

Kettles Scattered within the outwash or the till deposited near the edge of a glacier may be large blocks of ice. These may be buried and slowly melt away. As they disappear the overlying surface is let down and, if no more sediment is added, forms a closed depression, known as a **kettle,** above the former block of ice (Figure 10.36). Many kettles have filled with water and have become small lakes. Some of these have been filled in and have become swamps.

10.36 A kettle, showing a distinct depressed area. (G. R. Roberts)

Former Glaciers

Once geologists became aware of the distinctive geologic work of glaciers they began to study the evidence related to former glacial episodes. They found widespread evidence in the regolith and landscapes for many Pleistocene glaciations (during the last 2.5 million years or so). And in the ancient bedrock they found diagnostic deposits left by glaciers hundreds of millions of years ago.

Pleistocene Glaciations

The most recent episode of widespread glaciation took place less than 20,000 years ago. Indeed, considerable evidence, in the form of radiocarbon dates, shows that in the North American continent as far south as Toronto, Canada, was under ice as recently as 6000 years ago. Let us examine the evidence for Pleistocene glaciations and then consider the extent of these glaciers.

Evidence The evidence consists of till and erratics resting on striated and polished bedrock surfaces and of distinctive landforms created by glaciers in sediments. These include moraine ridges, drumlins, eskers (rare features), kames, and kettles. In addition, varved lake sediments are present locally.

Study of Pleistocene glacial sediments indicates that till and its related glacial sediments are interbedded with nonglacial sediments

362
Glaciers and Glaciation

containing evidence that the climate when they were deposited was comparable to today's climate or even warmer. Along the south shore of Long Island, outwash is interstratified with bay sediments. Both of these examples indicate that glaciers advanced, then retreated and that both the climate and sea level must have fluctuated. Finally, on the deep-sea floor and elsewhere in the sea, the sediments contain indications of Pleistocene changes of climate and sea level.

Extent In North America, Pleistocene glaciers covered all of Canada, the upper tier of the plains states, the midwest as far south as what is now Cincinnati, and all of the east coast down to mid-New Jersey (Figure 10.32). In Europe, glaciers moved as far south as the Mediterranean countries. Large parts of the Soviet Union were covered by ice and major glaciers crept down on all sides of the Himalayan range. In the Southern Hemisphere, the Antarctic ice mass extended north into South America as far as the Patagonian plateau in Argentina (Figure 10.37).

The number of Pleistocene glacial advances and retreats, the directions from which the ice spread, and the over-all effects of the enormous changes of climate outside the areas glaciated are subjects being actively studied by many researchers.

Older Glaciers

The evidence for pre-Pleistocene glaciers consists of distinctive kinds of bedrock. The glacial landscapes have long since been eroded away.

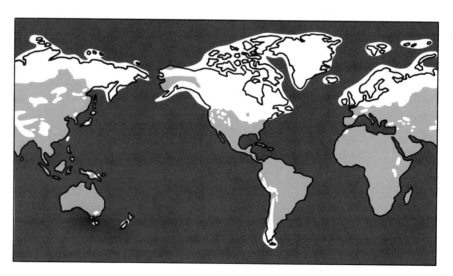

10.37 Generalized extent of Quaternary continental glaciers during ice ages.

Evidence When till has become cemented and forms part of the bedrock of an area, it is a sedimentary rock named **tillite**. Ancient tillites, containing erratics and resting on striated and polished rock surfaces, have been found in rocks of Permian, Devonian, Ordovician, early Cambrian, and older Precambrian ages.

Age and extent Late in the Paleozoic Era, more than 200 million years ago, glaciers again advanced northward and southward from the poles. During this time, the ice reached not only North America and Europe but parts of Africa, Australia, the subcontinent of India, and South America.

Tillite and related glacial deposits of Ordovician age covered what is now the Sahara. Tillites of Precambrian age are known from Australia, Utah, Canada, Scotland, and Scandinavia.

Theories about Climate Change

Widespread interest centers on the question, "Will another Ice Age happen?" Many geologists contend that we live in an interglacial time—not the post-glacial time. In other words, according to them, the Pleistocene Ice Age is not yet finished. In seeking an answer to this question we must examine theories on climate. In so doing we shall look into the subject of the causes of ancient glaciations.

The theories deal with external factors (such as changes in solar radiation or changing orbital relationships in the solar system), or factors on Earth (such as changes in the Earth's magnetic field, variations in volcanic dust in the atmosphere, circulation of the oceans, altitude of continents, and shifts of the Earth's crust).

Theories Based on External Factors

The amount of energy received on Earth is a function of the amount generated in the Sun and the distance from the Earth to the Sun. Both are subject to periodic variations. The variations in the Sun are thought to be related to the sunspot cycle. The distance changes periodically because the Earth's orbit around the Sun is elliptical, and so some points on it are closer to the Sun than others.

The theories based on external factors are known as astronomic theories. They are based on the variations that recur with exact precision and have always happened. They do not explain why, throughout geologic time, continental glaciers have been widespread only now and then. One astronomic theory does, however, predict that short, very warm times should intervene between glacial ages and interglacial ages. The geologic record suggests that exactly such warm times have happened.

Theories Based on Factors on Earth

Factors on Earth that can affect the climate include variations in the Earth's magnetic field, atmospheric conditions (especially the amount of dust), circulation of water in the oceans, altitude of the continents, and shifts of the crust with respect to the Earth's interior.

Variations in the Earth's magnetic field The magnetic field extends far out into space and interacts with incoming solar energy. The magnetic field serves as a filter, trapping some energy waves (and particles) and allowing other energy waves and particles to pass. It has been suggested that a change in the Earth's magnetic field can result in a change in solar energy received, and hence in a change of climate.

Atmospheric conditions; variations in volcanic dust The Earth's atmosphere responds dynamically to solar energy. What is more, the quantity of volcanic dust in the stratosphere is known to vary and also to be effective in blocking incoming solar energy. After major volcanic eruptions—Skaptar Jokull on Iceland in 1683, Tomboro in Indonesia in 1815, Krakatoa in 1883, Mount Pelée in the West Indies in 1902—vast amounts of volcanic ash have entered the atmosphere and drifted with high-altitude winds around the Earth. These particles not only created spectacular red sunsets, but decreased the amount of radiation entering the atmosphere, lowering atmospheric temperatures.

Circulation of the oceans Because solar heating is greatest in a belt near the Equator, and because of the Earth's rotation, a zone of equatorial surface water flows from west to east across all major oceans. If no lands interfered, this equatorial surface current would circle the globe. As it is, however, land carriers divert this flow. For example, the Isthmus of Panama diverts an equatorial Atlantic current into the northern Gulf of Mexico and from there it passes Florida and flows northward in the Atlantic Ocean as the Gulf Stream.

At present, the North Pole lies within the Arctic Ocean basin. The surface of the Arctic Ocean is frozen over. Because of this ice pack, winds blowing toward land from the Arctic Ocean are dry—they do not pick up moisture from the ice. Hence, the surrounding lands are cold, polar deserts. It has been argued that warm water coming in from the North Atlantic could melt the Arctic ice pack. If so, goes the argument, the winds crossing an ice-free Arctic Ocean would pick up moisture. When they blew over land they would drop the moisture as snow and glaciers could grow.

Altitude of continents The present altitude of the continental masses resulted from post-Miocene uplift. Therefore, today the continents stand much higher than they did earlier in the Cenozoic Era, for example. The height of continents affects accumulation of snow. No matter what the climate is worldwide, any mountain having altitude great enough will project above the snowline and thus be able to create glaciers.

Shifts of crust with respect to the Earth's interior The present location of the Earth's axis of rotation is such that the North Pole, as we mentioned before, lies within the Arctic Ocean. As we shall see in Chapter 14, the position of the Earth's magnetic axis generally coincides with that of the axis of rotation. The crust is not fixed above these axes, but has been shown to be in motion. If the crust moves relative to what lies below, then the location of the climate belts would change.

All these factors that affect climate are operating, changing, and interacting. For example, if temperature variations are caused by changing amounts of solar radiation, then atmospheric circulation will be affected. This will happen because the atmospheric circulation depends on temperature. Thus, a long-term temperature differential will move the routes of high-altitude jet streams and disrupt the patterns of weather movements. Changes of this magnitude would necessarily alter the Earth's climate.

Theories of Quaternary Climatic Oscillations

As far as the Quaternary oscillations are concerned, any theory of origin must explain how the climate changed worldwide from warmer than now to much cooler than now, and why this oscillation has been repeated perhaps a dozen times during the last 10 million years or so. Moreover, the cycles seem to be asymmetrical, the times of maximum warmth seeming to be short and to intervene between generally cooler times (glacial climates) and generally warmer times (interglacial climate, similar to today's). Four theories have been proposed to explain Quaternary climatic oscillations. These are: (1) volcanic dust, (2) astronomic, (3) solar-topographic, and (4) a theory based on an alternating graphic, and (4) a theory based on an alternating ice-free and frozen Arctic Ocean.

The variations called for by the first three theories are known to have occurred. The fourth theory proposes a neat mechanism for starting and stopping the glaciers. The predicted association of events is: (a) high sea level, melted pack ice, open Arctic Ocean, expanding glaciers. As glaciers grow sea level drops and the predicted association is: (b) low sea level, refrozen pack ice, iced-over Arctic Ocean, starved (and hence ultimately melted) glaciers.

This theory does not take account of variations of solar radiation received on Earth. Its difficulties concern: (1) lack of correspondence between the predicted ice-free Arctic and evidence in cores of Arctic Ocean sediments, (2) disputed meteorological relationships, and (3) failure to predict the brief warm spells that separate glacial from interglacial times.

There may be partial truth in all four of these theories. The newest evidence suggests that increased volcanic activity needs to be added to the list of established factors.

The Great Lakes and Their Geologic History

In north-central North America there are many large fresh-water lakes that have resulted from the effects of Pleistocene glaciers on the preglacial landscape. The modern lakes are one of a series of lakes that were formed as each glacier retreated from the area. In this section we shall describe the modern Great Lakes, discuss their Pleistocene ancestors, and mention their present difficulties.

The Great Lakes

What are they? The interconnecting bodies of fresh water called the Great Lakes (Figure 10.38) stretch from Minnesota to New York, border eight states and two Canadian provinces, and contain a volume of 23,000 cubic kilometers of water. These lakes constitute far and away the largest such network in the world. Together, they cover 248,000 square kilometers of land, an area larger than 40 of the 50 United States. The five individual lakes, and their worldwide ranks in size, are: Superior (1), Huron (4), Michigan (5), Erie (11), and Ontario (13).

The Great Lakes are fed by rivers, by groundwater, and by precipitation on their immense surfaces. They drain eastward from Lake Superior, whose normal water level lies 180 meters above sea level,

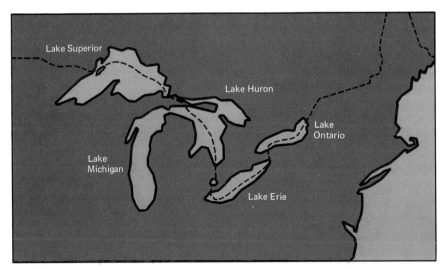

10.38 **The Great Lakes** as they exist today.

through Lake Michigan, Lake Huron, and Lake Erie, then over Niagara Falls into Lake Ontario. This last lake, 74 meters above sea level, empties into the St. Lawrence River, which flows to the Atlantic Ocean.

What is their importance? Although the Great Lakes basin amounts to only 3.5 per cent of the land of the United States, almost 15 per cent of the population lives within the basin, as do almost one-third of all Canadians.

This concentration of population is a testimony to the importance of fresh water. Water from the lakes directly and indirectly nourishes some of the most fertile farmlands in the continent. And what once seemed to be an inexhaustible supply of fresh water fed the heavy industries in lakefront cities ranging from Rochester, New York, to Milwaukee, Wisconsin. The lakes themselves served as the medium for low-cost transportation of shipping goods.

The Ancestral Great Lakes

When a huge glacier melts and retreats, it leaves a terminal moraine and possibly one or many recessional moraines behind as distinctive landmarks. If the plowed depressed ground left by the vanished ice mass then fills with water, the waves rework the till at the shores into beaches (as explained in Chapter 12). In the Great Lakes basin, glaciologists have found a number of such beaches. Through various dating techniques, they have been able to discern that these features were created at different times. Experts now calculate that there have been

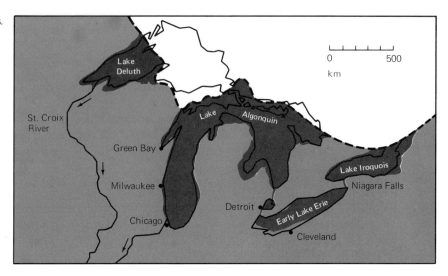

10.39 The Ancestral Great Lakes.

at least seven major lakes that successively covered the area occupied by the present five bodies.

The Wisconsin glacier The last ice mass to cover the Great Lakes basin arrived during the Late Wisconsin Stage. Moving over soft shale bedrock that had already been broken and shaped by previous glaciers, this ice sheet did the final landscaping. The water-filled depressions that are today's five lakes were scooped to their present contours by this last glacier.

Deglaciation When this last glacier retreated north, it left its load, derived from the fine crushing of shale, over much of the upper midwest. That crushed debris became the rich topsoil that now graces the farmlands of the region. Geologists have determined that the bedrock underlying the soil is rugged and hilly, but few traces of these irregularities remain in the flattened landscape. As the glacier retreated, its shifting margin formed a massive dam of ice (Figure 10.39). As meltwater and natural precipitation gathered in the depressions the ice sheet left behind, the water collected and gradually found its way to the present perimeters of the Great Lakes.

The Modern Great Lakes

How new are they? By the carbon-dating of vegetation known to have been killed by the advance of the last glacier, scientists have established that glaciation reached its maximum extent about 11,000 years ago. This ice mass did not retreat above the Great Lakes basin until about 8000 years ago, and still lingered in lower Canada until about 6000 years ago.

Human-made perils The Great Lakes were once vibrant with populations of fish and other aquatic life. Today, however, these populations have been greatly reduced. There are even large "dead" spots in the middle of Lakes Erie and Ontario where there is not enough oxygen to support fish. Much of the shorelines of the various bodies are unfit for human bathing, and only the water of Lake Superior is free enough from bacteria to be drunk, but even it may contain harmful asbestos fibers.

The culprits are humans. The large cities and factories built near the lakes to tap the seemingly endless supply of fresh water have in turn discharged a continual flow of toxic effluents back into the lakes. So heavy has this pollution been at times that the Cuyahoga River has actually caught fire from shore to shore. The results of pollution have been the loss of valuable plant life (except for the oxygen-depleting algae that thrive on pollution) and an ensuing loss of fish species. Lake trout and other fish species have been further reduced by the migration through the St. Lawrence Seaway of parasitic sea lampreys from the Atlantic Ocean.

Since the turn of the century, the Great Lakes have deteriorated markedly. By the late 1960s, conditions were so bad that the public began to apply heavy pressure to industries and cities to filter their wastes before dumping them into the lakes. The clean-up was complicated by interstate and international quarreling, but by now Americans and Canadians alike have recognized that life without the Great Lakes would be neither pleasant nor easy. Today there is hope that the Great Lakes will survive to greet the next glacier, the successor of the former glaciers that created them.

Chapter Review

1. Glaciers are one of the most powerful agents shaping the surface of the Earth. Glaciers are also closely related to the hydrologic cycle and the world's climate.
2. A *glacier* is a flowing mass of ice that formed by the recrystallization of snow, is powered by gravity, and has flowed outward beyond the snowline.
3. Before it becomes a glacier, a mass of snow must go through a series of changes. These changes result in the formation of snowfields, firn, and glacier ice.

4. Glaciers are important as historical archives of what fell from the ancient atmospheres and of ancient climates; and as creators of sediment, distinctive deposits, and characteristic landforms.

5. Geologists recognize three chief kinds of glaciers. These are: *valley* (Alpine) glaciers, *piedmont* glaciers, and *ice sheets* (continental glaciers).

6. Glaciers currently cover about 16 million square kilometers, or roughly 10 per cent, of the Earth's surface. In the past, as much as 32 per cent of the land may have been glaciated. If all the present glaciers melted, the world sea level would rise 20 to 60 meters, enough to inundate most major cities. The chief existing ice sheets are on Antarctica and Greenland.

7. Because they are in constant flux, glaciers advance and retreat almost daily. Modern glaciers advance about 15 to 60 cm per day, excluding short-lived *surges*, which have moved glaciers as much as an average of 35 meters per day. Nonsurging glacial movement is thought to consist of *slippage* or *sliding* along discrete shear planes, and *plastic flow* from deformation of ice crystals. The mechanics of surging are not known.

8. The results of *glaciation* include *erosion* and *deposition*. Nowhere on Earth is there a more potent agent of erosion than an advancing glacier. Processes of erosion include *quarrying* (or *plucking*) and *abrasion*. Distinctive landforms are eroded in bedrock by glaciers. These include *cirques*; *glaciated* and *hanging valleys* and *fjords*; *arêtes*, *cols*, and *horns*; and *roches moutonnées*.

9. Unlike rivers and streams, glaciers do not transport the bulk of the regolith and rock debris they tear from the terrain to a growing delta. Instead, this matter is strewn in characteristic deposits all along the glacier's path. Some glacial sediments and depositional landforms include *drift*, *erratics*, *till*, *outwash*, *varved lake deposits*, *moraine ridges*, *drumlins*, *eskers*, *kames* and *kame terraces*, and *kettles*.

10. The most recent episode of widespread glaciation took place less than 20,000 years ago. In North America, Pleistocene glaciers covered all of Canada, the upper tier of the Plains States, the midwest, and all of the east coast. The evidence for pre-Pleistocene glaciers consists of distinctive kinds of bedrock. The glacial landscapes formed by these very ancient glaciers have long since been eroded away.

11. According to many geologists, the Pleistocene Ice Age is not yet finished. Theories about climate change deal with *external changes* (such as changes in solar radiation or changing orbital relationships in the solar system), or *internal changes* (such as volcanic activity, circulation of the oceans, changes in the Earth's magnetic field, and elevation of continents).

12. The Great Lakes have resulted from the effects of Pleistocene glaciers on the pre-glacial landscape. The modern lakes together are one of a series of lakes that were formed as each glacier retreated from the area.

Questions

1. How do expansions and contractions of glaciers affect sea level?
2. Why is it said that glaciers cause "instant regolith"?
3. In what way did new concepts about glaciers cause an intellectual revolution in the nineteenth century?
4. If a glacier must always move outward to retain its identity as a glacier, how can you explain a glacier's *retreat*?
5. Describe the historical evidence that glaciers have created great amounts of sediment.
6. Describe the three chief kinds of glaciers.
7. Where are most of the modern glaciers located?
8. How have scientists estimated the rate of glacial movement? Define *surge*.
9. Define *quarrying* and *plucking*.
10. List and describe at least four distinctive landforms eroded in bedrock by glaciers.
11. Define *drift, erratic,* and *till*.
12. What evidence exists of Pleistocene glaciation?
13. How can changes in climate bring about another ice age? If changes occur in the Earth's magnetic field or the circulation of the oceans, how might glaciation be affected?
14. Why are the Great Lakes important to the study of glaciers and glaciation?

Suggested Readings

Deevey, Edward S., Jr., "Living Records of the Ice Age." *Scientific American,* May, 1949. (Offprint No. 834. San Francisco: W. H. Freeman and Company.)

Dyson, James L., *The World of Ice.* New York: Alfred A. Knopf, 1963.

Field, W. O., "Glaciers." *Scientific American,* September, 1955. (Offprint No. 809. San Francisco: W. H. Freeman and Company.)

Schultz, Gwen, *Glaciers and the Ice Age: Earth and Its Inhabitants During the Ice Age.* New York: Holt, Rinehart and Winston, Inc., 1963.

Schultz, Gwen, *Icebergs and Their Voyages.* New York: William Morrow and Company, 1975.

Sharp, Robert P., *Glaciers.* Eugene, Oregon: University of Oregon Press, 1960.

Woodbury, David O., *The Great White Mantle: The Story of the Ice Ages and the Coming of Man.* New York: The Viking Press, 1962.

"About one-third of all land area is desert, an impressive amount, considering that two-thirds of the Earth's surface is covered by water."

Chapter Eleven
Deserts and the Wind

The word *desert* derives from the Latin word for abandoned, or forsaken. Thus, formally defined, a desert is an abandoned place. Deserts are indeed among the Earth's most forbidding and least-known lands. Yet for the geologist, deserts hold fascination and an opportunity to observe the surface of the Earth "in the raw," a chance to study Earth materials and processes without an obscuring mantle of vegetation. For human populations, hard-pressed for space and raw materials, deserts offer the potential of new zones of settlement as well as of vast supplies of oil, groundwater, and other resources.

In this chapter we shall study deserts as special effects of extreme climatic conditions. As far as we can tell from examining modern deserts and their products, there have always been deserts. Determining the distribution of ancient deserts is very important in connection with reconstructing the ancient positions of the continents. We shall also deal with how ancient atmospheres have transported sediments. The importance of such transport is emphasized by the fact that many of our present soils have formed by the weathering of ancient windblown dust. Movement of sediment in the atmosphere has given us important clues to the distribution of tephra as it is related to ancient volcanoes and to the fallout patterns of the radioactive products of atmospheric nuclear weapons tests.

Our approach to the material of this chapter will be to review the causes and distribution of deserts, to study desert climates and desert landforms, to look at desert lakes, and finally, to see how the wind transports sediments and shapes desert landscapes.

Causes and Distribution of Deserts

Under this heading we shall define deserts, describe the causes of deserts, and summarize the distribution of deserts in the modern world.

What Is a Desert?

A **desert** is a region characterized by evaporation that greatly exceeds precipitation. Most deserts receive less than 25 cm of rainfall per year; extremely arid deserts receive less than 5 cm of rainfall per year. We will confine our discussion of deserts to the hot, dry regions of low latitudes. The so-called polar deserts of the tundra regions and barren high-altitude plateaus will not be considered here. Although these areas may receive as little precipitation as low-latitude warm

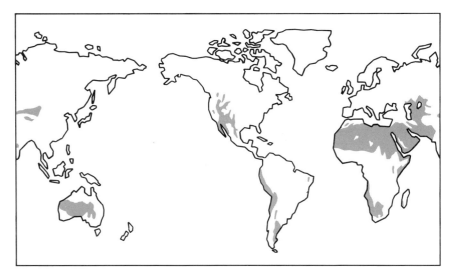

11.1 The distribution of deserts is shown on this world map.

deserts, they are set apart by low evaporation and by such distinctive features caused by low temperatures as permafrost, ice cover, and glaciers.

Most true deserts are surrounded by semi-arid regions called **steppes** (Figure 11.1). In steppes, average rainfall is 25 to 50 cm per year.

We shall look in more detail at deserts in terms of three important factors: (1) precipitation and evaporation, (2) temperature, and (3) vegetation.

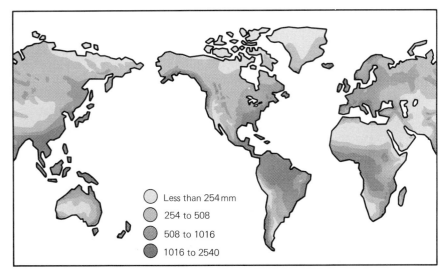

11.2 Annual precipitation throughout the world. In desert regions, evaporation is consistently greater than precipitation.

- Less than 254 mm
- 254 to 508
- 508 to 1016
- 1016 to 2540

Precipitation and evaporation As we have seen, the most significant feature of deserts is their lack of abundant water. The relationship between precipitation and evaporation typically is defined in terms of "potential evaporation," which refers to the amount of water that would evaporate if it were present to do so. Thus, deserts include regions where abundant rain may fall for a few weeks or days, but where it is dry the rest of the year. One weather station in the Kalahari Desert in Africa recorded no rain at all for 15 consecutive years. The lowest mean annual precipitation in the world (0.4 mm) occurs at Dakla, Egypt. In Australia, potential evaporation may exceed precipitation by 10 to 1; in the Sahara, by 100 to 1. In these regions, the amount of water brought to the land in the form of dew and fog is often greater than that brought by precipitation (Figure 11.2).

Temperature Despite the fact that no desert occurs within 15 degrees of the Equator, deserts are the world's hottest places. The reason is that desert air holds little moisture to form clouds and thus to block incoming solar radiation. In humid regions, roughly 40 per cent of incoming solar radiation may reach the Earth's surface; in some deserts, as much as 90 per cent of the Sun's rays may penetrate to the ground. The lack of surface and atmospheric moisture above deserts also allows temperatures to fluctuate widely from day to night and from summer to winter. A peak summer temperature of 58°C has been reported in el Azizia, Libya, the hottest place on Earth (Figure 11.3). The ground surface is up to 10°C hotter. California's Death Valley is almost as hot. With little moisture in the air to retain warmth, the

11.3 Desert landscape in Libya formed of sediment that has been piled into dunes by the wind. Temperatures in this area can vary from the hottest in the world during the day to below zero at night. (FAO photo)

temperature may also drop precipitously as insolation decreases at night or in the winter. One station in Libya recorded a nighttime temperature of −9°C, and the midwinter mean in the Gobi Desert is below −12°C. Daily and annual temperature fluctuations of 60°C are on record for West Africa and Central Australia.

Vegetation Sparse desert vegetation is adapted to make maximum use of scanty rainfall and groundwater. Perennial plants are spaced widely, leaving room for extensive root systems. This thin plant cover offers the surface little protection from winds, heating and cooling, and rapid erosion by the little rain that does fall. *Xerophytes*, from the Greek for dry plants, have evolved ingenious means of surviving in deserts. Foliage is leathery and waxy in order to slow transpiration. Shrubs such as the creosote bush combine shallow roots with deep taproots that seek moisture far underground. Such plants as salt cedar, mesquite, and greasewood may extend roots 6 meters or more below the surface to reach water. By contrast, cactus plants concentrate root systems near the surface to trap rainfall before it can seep beyond reach. Moreover, in their thick trunks, cacti store water which they dole out sparingly between rains.

Types and Distribution of Modern Deserts

In the modern world areas classified as "dry" cover more land area than any other. About one-third of all land area is desert, an impressive amount in relation to the fact that two-thirds of the Earth's surface is covered by water. By one reckoning, 4 per cent of all land is "extremely arid" (12 or more consecutive months without recorded rainfall). Another 15 per cent is classed as "arid," and about 14.6 per cent as "semi-arid."

According to their locations and causes, most modern deserts can be arranged in four groups: (1) subtropical deserts, (2) continental-interior deserts, (3) rainshadow deserts, and (4) coastline deserts.

Subtropical deserts Most of the world's deserts lie in two broad belts of high pressure encircling the globe north and south of the Equator. The location of deserts in these zones is ordained by the planetary weather system. Most solar energy reaches the Earth in the equatorial regions where the Sun's rays strike most directly. Little energy reaches the ground near the poles, where radiation strikes at oblique angles. This establishes a persistent lack of energy balance—an excess at the Equator and a deficit at the poles. To adjust this energy imbalance, the atmosphere and ocean constantly

11.4 Effects of drought in Kalahari. (UPI)

transport heat away from the Equator. Warm, wet air rises from the Equator and moves north and south. As this moisture-laden air cools at higher altitudes, it drops its water as rain. The cooled air, dry and heavy now, descends along the desert belts, drying the land and preventing lighter, moist air from moving into desert regions. The descending air forms stationary high-pressure zones. Such stationary high-pressure zones over North Africa, Asia, and Australia have created the Earth's largest deserts, as well as vast expanses of semi-arid prairie and steppe.

Deserts created primarily by the cool, dry, descending air just outside the tropics are **subtropical deserts.** The greatest deserts on Earth belong to a vast subtropical swath of arid land extending for 8000 km from the west coast of North Africa to Pakistan and western India. Within this swath, the Sahara and the Arabian Desert contain most of the world's extremely arid land. The Sahara alone, covering about 9 million square kilometers, is almost as large as the 50 United States. This belt of deserts is broken at its eastern end by the monsoon region of Asia. Another subtropical desert covering nearly 1.3 million square kilometers is centered at 30 degrees North Latitude in the southwestern United States and northern Mexico. This desert, too, is relieved at its eastern border by humid air moving northward from the Gulf of Mexico. The Kalahari Desert is a subtropical desert occupying most of southwest Africa (Namibia), and a surrounding semi-arid area extends north and east (Figure 11.4). More than 80 per cent of Australia is arid or semi-arid, for it is situated in (and helps to maintain) one of the most persistent zones of dry descending air.

Continental-interior deserts Some deserts form in regions so far from the sea that little moisture ever reaches them. Such continen-

11.5 Canadian prairie is one of the world's outstanding areas for growing wheat, and yet in some ways it is similar to some desert regions. (Canadian National Film Board)

11.6 Local rainshadow desert, Death Valley, California. (American Airlines)

tal-interior deserts, where air is heated and dried in the summer and cooled in the winter, occupy much of central Asia. The Gobi Desert and the Takla Makan Desert of western China and the surrounding Russian steppes are among the harshest environments on Earth. In North America, the Great Plains and Canadian prairie are in many ways similar to the Asian steppe (Figure 11.5). In both cases, stable, dry cells of high-pressure air are protected by adjacent highlands. Only chill, and often equally dry, polar air interrupts the cold monotony—with air that is colder still. Agriculture, though possible, is perennially threatened by drought.

Rainshadow deserts Any mechanism that dries the air can produce a desert—even in cool regions, or areas adjacent to the ocean itself. The most powerful such mechanism is a *landform barrier*, such as a mountain range, which blocks or deflects air currents. In the western United States, the Cascades and Sierra Nevada deflect upward the humid winds blowing inland from the Pacific Ocean. As the air rises, it cools; because cool air cannot hold as much water vapor as warm air, the water vapor condenses and falls as rain or snow. Most of this precipitation falls on the western flanks of the mountain ranges, feeding coastal rivers. By the time the winds reach eastern Oregon, eastern California, Nevada, and Utah, there is little moisture left. The result is a *"rainshadow" desert* (Figure 11.6); see Figure 5.5.

The best examples of rainshadow deserts, those shielded from rain by mountains, are those in the lee of the American cordillera—

the mountain "spine" that extends from northwestern Canada to southernmost Argentina. In North America, as just mentioned, the Sierra Nevada and the Rocky Mountains are responsible for blocking moist Pacific air from Death Valley, the Sonoran Desert and other areas east of the Coast Ranges. In South America, Patagonia (Argentina) is shielded in the same way from Pacific westerlies by the Andes. The deserts of Sonora and Patagonia, 6400 km apart, are so similar that many of the plant species in both areas are the same.

Coastline deserts Some of the world's driest deserts lie along the coasts of Africa, North America, and South America where currents of cold water exert a powerful effect on climate. Winds sweeping over cold water are cooled so that they cannot hold much water vapor. When such winds blow onshore, they dry the coastal land. The cold breeze may carry small amounts of moisture, in the form of mist or fog, but not enough to form rain. This heavy, dry air not only brings no rain of its own, but also blocks the penetration of lighter, moist air from elsewhere.

Cold ocean currents create equally dry "rainshadows" on adjacent land. The Humboldt Current is responsible for the Atacama-Peruvian Desert in Chile and Peru, the world's smallest desert system (Figure 11.7). Although the coast is usually foggy along this desert, it is the world's driest, averaging less than 1.5 cm of rain per year. In Southwest Africa, the Namib Desert, the coastal extension of the Kalahari, lies adjacent to the cold Benguela Current. The cold California Current dries southern California and Baja California. Without rainfall, such regions lack the vegetation to halt the advance of wind-

11.7 Desert scene in Chile, viewed from the air. Fans at upper left lie along a fault that trends North-Northeast. Growth of fans has pushed river (white strip) to right. Irregular light-colored areas are bedrock; dark areas are underlain by sediment. Width of view is 10 km.

blown sand, so that inland conditions resemble great extensions of coastal beaches.

In summary, as we have seen, desert regions are not scattered at random but are distributed in two broad zones that are centered near the Tropics of Cancer and Capricorn, between 15 and 40 degrees North and South Latitude. Most deserts lie in one of five great zones of aridity: North Africa-Eurasia, South Africa, North America, South America, and Australia, each surrounded by semi-arid borders. There are numerous departures from the double belt pattern, caused by such factors as distance from the sea and mountain ranges. One of the largest departures is the monsoon region of southeast Asia, where moist air moves inland for part of each year over land that would otherwise be desert.

The Geologic Cycle in Deserts: Weathering, Sediments, and Landforms

In deserts the geologic cycle still operates, but, as one might expect, the outcome is drastically affected by the climate, particularly by the way water behaves. In addition, the geologic cycle is affected by the lack of vegetation. In this section we examine the major factors in the operation of the geologic cycle in deserts and then explore how these affect weathering in deserts, the sediments of deserts, and the landforms of deserts.

The Effects of Desert Climate on the Geologic Cycle

The chief impact of a desert climate on the geologic cycle concerns temperature, distribution of rainfall and its affect on stream flow, and wind activity.

Temperature Desert temperatures are characterized by great daily extremes, from nearly 60°C to about −5°C. As we read in Chapter 5, such great temperature changes have been cited as a possible mechanism for breaking rocks.

Distribution of rainfall Generally, the driest deserts show no seasonal pattern of rainfall: the less the precipitation, the less predictable its arrival. An unusually extensive or fast-moving storm system elsewhere may dampen only the fringe of a desert, blocked by dense air from further penetration. In less-extreme deserts, which emerge into semi-arid scrub country, rain showers may fall during the same two-week period every year.

382
Deserts and the Wind

11.8 Box canyon, showing typical vertical sides and flat, sediment-covered floor. Dry River, Martinique, the same valley down which the nuées ardentes from Mount Pelée traveled from the vent to the sea. (Press & Inf. Office French Embassy)

11.9 Surface of thin edge of mudflow showing cracks formed when mud dried out. (American Airlines)

Running water When rain does come to the desert, it often comes in brief, high-intensity "cloudbursts" which may dump an entire year's precipitation in an hour. Once water is on the ground, its great transport capacity makes it more important than wind in shaping desert landscapes. On gentle slopes, water from a cloudburst characteristically moves as **sheetwash,** or sheet floods, a broad surface runoff laden with suspended sediment. Sheetwash collects rock debris into broad, thin alluvial deposits. Where the slope is steeper, the surface is quickly scarred by gullies which develop into the *arroyos* and box canyons typical of the American southwest (Figure 11.8), or into the *wadis* of the Sahara and Arabian Deserts. Cloudbursts may cause flash floods which fill an arroyo in a minute or two with a fast-moving mixture of water and sediment 3 meters or more deep. Such muddy waters typically form *mudflows* (Figure 11.9). Flash floods and mudflows may do their erosive work for only a few hours, yet they can move tremendous quantities of material, including large boulders and other large objects (Figure 11.10). Hills in arid regions are often intricately dissected into badlands or intersecting gullies and arroyos. Where running water leaves its channels and spreads over flatlands, the resulting sheet floods form broad and prominent alluvial fans.

Interior drainage Only a few desert streams, generally originating in mountains where melting snow and summer rains provide an

The Geologic Cycle in Deserts: Weathering, Sediments, and Landforms

11.10 Bus overturned by mudflow. (U.S. Geological Survey)

inexhaustible supply of water, flow the year around. The Nile and Colorado are among a handful of rivers large enough to traverse deserts and reach the sea. Far more common is **interior drainage,** the disappearance of streams by evaporation and infiltration into the thirsty soil (Figure 11.11). Groundwater, always deep in deserts, is

11.11 Interior drainage basin at the end of the Amargosa River, Death Valley, California. (Bill Belknap, Rapho/Photo Researchers, Inc.)

nearest the surface under these streams. In the Sahara and several other deserts, deep sandstone beds may bring artesian water to deserts from distant mountains, where their tilted edges catch abundant rainfall. An **oasis** is a spot where the desert surface intersects the water table. Where streams disappear by interior drainage, aquifers afford the only reliable source of water and must be conserved carefully.

Wind The wide temperature ranges within the deserts, and between deserts and adjoining regions, encourage steady winds. Desert winds tend to be stronger and more consistent than winds of more humid regions, and the barren ground of deserts is far more vulnerable to the erosive effects of wind.

Weathering and Mass-Wasting

The key elements of weathering and mass-wasting are high temperatures, which cause the evaporation of salts, low temperatures, which may freeze water, a general upward movement of water through the regolith, and the absence of plant cover.

Mechanisms of desert weathering The effects of weathering are more apparent to the eye because of the absence of a covering of soil and vegetation. Nonetheless, weathering is considerably slower in deserts than in humid regions, and includes as its chief processes salt weathering, thermal effects, and oxidation.

Salt weathering (Chapter 5), a combination of chemical and mechanical weathering, predominates in deserts, producing a coarse, scanty, and usually thin regolith, interspersed with bare bedrock. Salt weathering may disintegrate rock by three mechanisms: (1) stresses exerted by the expansion of confined salts as they are heated, (2) stresses caused by the hydration of salts in confined spaces, and (3) stresses caused by crystal growth from salt solution in confined spaces. These three processes cause granular disintegration which tends to round off sharp edges and corners of coarsely crystalline rocks (Figure 5.10).

Some geologists consider thermal expansion and contraction the most important mechanical weathering agent in deserts. Others argue that thermal effects are negligible without water (Chapter 5). In deserts where temperatures drop below freezing, the expansion of ice crystals within rock pores is probably the most important weathering agent. The water that forms this ice comes either from occasional rainfall or from tiny amounts of groundwater pulled upward by evaporation and capillary action. Even without freezing, repeated cycles of wetting and drying promote flaking and splitting of rock.

Lichens, which grow on many desert rock surfaces, extend tiny finger-like thalli into the cracks and pores of rocks, hastening disintegration and producing acids that dissolve silicates.

Oxidation is an important chemical process in desert regolith. For example, fresh particles of ferromagnesian silicates such as biotite and hornblende become oxidized, and hematite forms. The powdered hematite colors the desert regolith red.

Soils Most desert soils are coarse, thin, and poorly developed. Gray desert soils, typical of Wyoming, Utah, and Nevada, contain little humus and inconspicuous zones. As a consequence of the lack of water and of organic matter, there are few earthworms and burrowing animals to mix and aerate the soil. As water evaporates, salty groundwater is pulled upward by capillary action, leaving deposits of calcium carbonate or hydrous calcium sulfate in the soil. These salts may form crust called *caliche* that resists erosion, or they may cement gravels into conglomerate rock. In Chile, nitrate deposits formed in this way are mined to make fertilizer.

Red desert soils are produced in hotter, more arid deserts where humus content is minimal. Here, soil zones are poorly developed. Partially weathered rock fragments are mixed throughout the regolith.

Mass-wasting in deserts As would be expected from their coarseness and dryness, desert soils tend to form steeper slopes than soils of humid areas. At the same time, mass-wasting is accelerated by the absence of protective vegetation, and slowed by the absence of abundant rain. The bottom of a slope often meets level ground at a sharp angle, unless water smooths the junction with alluvial deposits. In the absence of rivers to carry off erosion products, large accumulations of angular, partially weathered rock debris are common sights in deserts.

Desert Landforms

Deserts display a distinctive variety of landforms quite unlike anything seen in humid climatic regions. Desert landforms we will discuss include fans, bolsons and bajadas, pediments, and playas.

Fans During desert cloudbursts, arroyos or wadis fill quickly with water that flows vigorously enough to sweep sand, pebbles, and even rocks down mountain slopes. As these mountain streams leave the mountains, they spread out rapidly. This spreading out causes deposition of the stream load in **fans** (Figure 11.12). The surfaces of

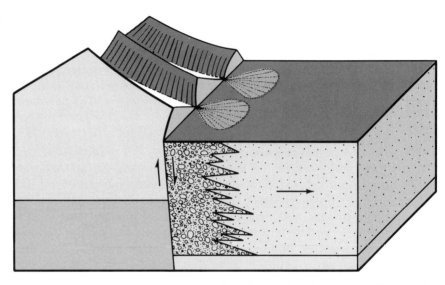

11.12 Fan sediments, greatly thickened by continued accumulation along a fault that dropped one block downward and elevated the adjacent block.

these fans, concave skyward, are ideal forms for the dispersal of peak discharges and sediment loads from mountain streams during flash floods.

Bolsons and bajadas The largest fans accumulate in arid basins surrounded by mountains. As long as nearby mountains are high enough to furnish erosion products, sediment continues to be shed into such basins. In some of the enclosed basins of the southwestern United States, sediment is hundreds of meters thick. A basin that is surrounded by eroding mountains is known as a **bolson,** the Spanish word for purse. In a bolson, many arroyos may contribute their load in such a manner that neighboring fans begin to overlap. Such a wide, coalescing expanse of eroded debris is a **bajada** (ba-HA-da).

Pediments Between the foot of the mountains and the bajada, there is often a gently sloping bedrock surface called a **pediment.** In cross section, extending from either side of a mountain range, a pediment resembles the triangular gable for which it is named. Pediments, bare or thinly covered with sediment, are the dominant erosional features of many desert landscapes (Figure 11.13). When fully developed, they correspond to the peneplains of humid regions.

The evolution of pediments is not completely understood. They serve as broad highways, or transportation surfaces, for sediment that is washed downhill from the highlands. Pediments resemble huge fans, except that pediments are surfaces cut across bare bedrock with only a small coating of sediment. Geologists do not agree on the relative importance of sheetwash, stream flow, and mass-wasting in creat-

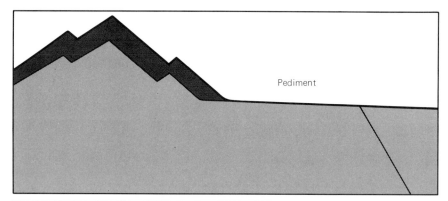

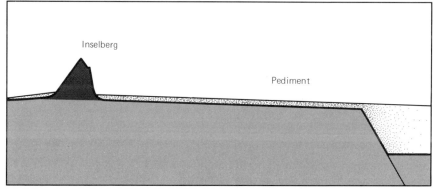

11.13 Retreat of mountain slope creates wide pediment and leaves uneroded remnants named inselbergs.

ing pediments. What has become apparent through mathematical analysis is that a pediment is perfectly graded to move large but intermittent masses of water and rock debris downslope. And, like the bed of a stream, a pediment is continually adjusting its slope to maintain this perfect grade.

Gradually the sediment carried across the pediment rises in the bolson, and the lower portion of the pediment is buried under a bajada. At the same time, the upper edge builds upward and encroaches on the talus slopes or cliff faces, sometimes reaching into major canyons a kilometer or more. Eventually pediments on both sides of a mountain range meet, opening broad passes. At last, erosion reduces the mountains to scattered remnants called **inselbergs** (Figure 11.13). This raising and merging of pediments is a slow process and is usually thought to be interrupted by new mountain building before completion.

Playas Farther from the mountains, beyond pediments and bajadas, the slope of a bolson becomes flatter. A sun-baked flat underlain by clays and salts in a bolson is a *playa* (PLY-ya). Sudden rains may produce enough water to create a desert lake, or *playa lake*, at the

11.14 Playa lake, seen from an airplane. Australia. (G. R. Roberts)

388
Deserts and the Wind

11.15 Large stone that inscribed a track in the dried-out sediment as it traveled across Racetrack Playa, Death Valley, California. It is thought that the stone moves when it has been surrounded by a thin sheet of ice and a strong wind blows on the ice, moving it and the stone together. (D. Hiser, Photo Researchers, Inc.)

center of a bolson (Figure 11.14). Playa lakes may cover large areas with water to depths of a few meters for brief periods. Their beds are normally crusted with sediment, with salts precipitated by evaporation, or with clay (Figure 11.15).

Uniqueness of desert landforms As indicated in Chapter 8, erosive forces in a humid environment work *downward* toward a local base level or toward the ultimate base level, sea level. In many deserts, base level is formed by the surface of sediments in a bolson. A unique feature of desert landforms is that base level is constantly moving *upward* as the bolson fills with sediment. Fans thicken and back up into mountain passes. The fans rise as the mountains dwindle. Eventually, sediments engulf the peaks altogether.

Desert Lakes

Although most surface water in deserts disappears by interior drainage, under certain conditions permanent lakes form as dominant features of the landscape. We shall discuss some modern desert lakes, and then we will examine the evidence of ancient desert lakes which were much larger.

Description of Desert Lakes

When a playa lake is replenished by stream flow faster than the water can drain away, or when interior drainage is blocked by bedrock or other surface features, permanent desert lakes may form. These lakes range in size from modest ponds to Utah's Great Salt Lake, whose surface area is 5.2 million square kilometers.

Water content Almost all desert lakes are brackish, or saline, to varying degrees. The brackishness is largely a result of evaporation: as the lake water continually evaporates in the desert heat, salts dissolved from rocks are left behind. In dry years, Great Salt Lake may contain as much as 27 per cent salt by weight—eight times as much as the ocean.

Sensitivity to climate Desert lakes are extremely sensitive to slight changes in climate, especially those which vary the amount of rainfall. During dry years, the streams that feed desert lakes shrink and fail to keep pace with the persistent evaporation. Lake levels drop steadily, especially during droughts that last as long as several years, leaving an ever-widening belt of salt around the shores. After many cycles of evaporation and replenishment, concluded by disappearance of the lake, the amount of salt deposited may warrant commercial mining. Borax and gypsum are mined in ancient lake beds in the Mojave Desert in California; potash is taken from the salt flats of Utah and the Dead Sea.

Evidence of past existence Greater rainfall and lower temperatures in many desert regions allowed much larger lakes to survive in the past where meager desert lakes exist today. Deltas, beaches, "sea"

11.16 Pluvial lakes in western United States, as reconstructed from studies of ancient lake deposits. (U.S. Geological Survey)

cliffs, and other shoreline features now high above water level provide evidence of the sizes of ancient lakes (Figure 11.16). These larger ancestors of desert lakes are called **pluvial lakes,** because they were fed by abundant rainfall in the past. Geologists believe that the maximum sizes of pluvial lakes coincided with periods of glaciation, because in some places old shorelines are cut on sediments deposited by glaciers (Chapter 10).

In the western United States there are some spectacular remnants of ancient pluvial lakes. Great Salt Lake, and the Bonneville Salt Flats around it, are the remains of Lake Bonneville, a giant pluvial lake which rivaled today's Lake Michigan in size. The water level of Lake Bonneville used to be as much as 300 meters above that of its shrunken descendant (Figure 11.17). To the west lay another huge pluvial lake, Lake Lahontan, covering much of Nevada north of Reno, filling all the low valleys between the Sierra Nevada and the Rockies. Tribes of Indian fishermen lived along its shores; the caves they once inhabited are now high on arid slopes. In California, the proverbially dry Death Valley was once covered to a depth of more than 100 meters by ancient Lake Manley.

Probably the world's largest pluvial lake once spread across the lowlands that now surround the Caspian Sea, connecting it with the

11.17 **Shorelines of Ancient Lake Bonneville,** Utah, appearing as terraces high on the sides of the mountain slopes that enclosed the basin. (American Airlines)

Aral Sea and the Black Sea. Archeologists have found some of the earliest traces of agriculture in this region, known popularly as the Fertile Crescent. Neolithic man first tamed cattle in the Fertile Crescent, and the increase in grazing and browsing by large herds helped to denude the soil and promote erosion. Since then the great lake has shrunk until all that remains is the Caspian Sea, and even that body of water continues to dwindle.

Other desert lakes which were larger during the times of Quaternary glaciations include Lop Nor and Lakes Balkhash and Baikal in Asia, Lake Chad in West Africa, and Lake Eyre in Australia. When the glaciers advance southward and northward from the poles again, which climatic experts think will happen within a few thousand years, such lakes could once again resume their full size and grandeur.

The Geologic Work of Wind

The geologic work of the wind can be discussed conveniently under the sequence that starts with the picking up of loose granular material, the mechanics of transport by the wind, sediments deposited by the wind, including dunes, and the features formed in rock as a result of abrasion by wind-transported particles.

Wind Erosion of Loose Sediment Particles

The effect of wind blowing over loose sediment particles is in many ways comparable to stream flow. Like flowing water, wind is capable of picking up and transporting loose material. The wind moves the sediment chiefly where it is dry and lacks plant cover. If winds from a particular direction are accompanied by rainfall they will pick up very little sediment. Therefore, in order to understand the effectiveness of wind in picking up sediment from a given area, one needs to know not only the speed of the wind and its duration, but also whether or not it is a dry wind or a wet wind.

Deflation and deflation basins The picking up and carrying away of dry, weathered granular material, usually ranging in size downward from coarse sand to the finest dust, is **deflation.** The effects of deflation in deserts are difficult to measure where it lowers broad areas uniformly by equal amounts. Where the process has been locally retarded by moisture, vegetation, or exposed bedrock, the contrast can be striking. In certain conditions, deflation may lift away a

11.18 Abandoned farm, a result of the drought and formation of the "Dust Bowl" in the 1930s. Near Guymon, Oklahoma. (B. C. McLean, U.S. Department of Agriculture)

meter of topsoil in a single year, especially when plowed soil is dried during a drought. Many midwestern American farms were stripped bare of topsoil in the "Dust Bowl" years of the 1930s (Figure 11.18).

When semi-arid grassland becomes desert, the change is accompanied by the deflation of fine materials which were bound by the fine root systems of the grasses. This process is particularly harmful to cropland, because the first materials to blow away are those which alone could encourage new plant growth upon the return of the rains. Extensive deflation is halted only when it approaches the water table (where moisture keeps the surface saturated) or when the remaining pebbles and larger stones ("lag gravel") accumulate so thickly that they protect the sand and dust beneath. Such an accumulation is called **deflation armor** (Figure 11.19); when the stones are close and tightly fitted, it may be given the name of **desert pavement.**

Depressions scooped out of loose materials by deflation are *deflation basins,* or **blowouts.** They may appear as bare, sandy patches a few meters deep and up to a kilometer wide in semi-arid prairie, or as much larger features in more arid deserts. Some geologists believe that the great depressions in the Sahara, a hundred meters deep and thousands of square kilometers in area, are enormous deflation basins. The Qattara Depression, 298 km long, was scoured by the wind down to the water table; its floor is now a forbidding quagmire of salt and shifting sand.

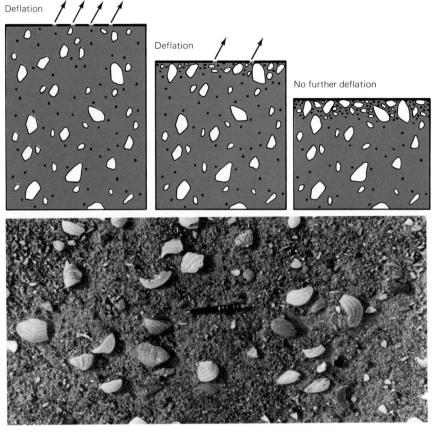

11.19 Deflation armor. (Top) Successive stages in development, starting with pebbly or shelly sand. As wind removes the sand-size particles the coarser materials are left behind and eventually form a continuous layer which protects the sediment below from further erosion. (Bottom) Photograph of deflation lag formed of shells and pebbles on a beach. (J. E. Sanders)

Mechanics of Sediment Transport by the Wind

Like streams, the wind may carry sediment as bed load and as suspended load. The former consists of sand grains, averaging between 0.15 and 0.30 mm in diameter; the latter consists chiefly of silt.

Movement of bed load Most of the particles in the wind's bed load move by **saltation,** a series of long, elastic bounces and impacts that rarely carry them more than 10 cm above the ground (Figure 11.20). The maintenance of any saltation at all requires a wind of at least 16 km/hr. Even much more powerful winds cannot cause the bed load of saltating sand grains to bounce more than a few meters above the surface. Particles within this moving carpet of sand collide with one another, further reducing wind energy and hence the amount of the bed load itself.

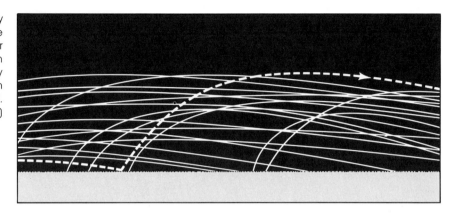

11.20 Saltating particles blown by the wind. The wind pushes the particles forward at the tops of their excursions and gravity pulls them back to the ground, where they strike other particles, spattering them upward and pushing them forward. (Based on R. A. Bagnold, 1941)

A wind moving over a surface lacking loose particles is retarded along the surface by friction. The roughness of the surface is crucial in determining how the wind behaves. There is a very thin layer of air just above the surface where wind speed is zero. This dead-air layer is about 1/30 the diameter of the mean height of the features on the surface. Thus, large rocks and boulders tend to break up the airflow and cause saltating sand to accumulate on the ground around them.

A strong, steady wind may hurl low clouds of saltating sand just above ground level as **sandstorms.** Such sandstorms are strictly limited in altitude; the majority of the blown sand remains within 10 cm of the ground and almost none bounces as high as a man's head. Sandstorms carve low rocks into strange, topheavy shapes, and saw off telegraph poles unless the poles are sheathed in metal protectors.

Movement of suspended load The wind's suspended load is made up of silt particles and clay flakes smaller than 0.06 mm in diameter. Suspended particles are much lighter than sand grains, which usually remain in the bed load. Often, silt and clay particles are harder to lift into the air than sand. Such tiny flakes absorb even small amounts of surface moisture and are therefore more coherent than sand. Also, they are usually packed so closely that they fall into the dead-air layer created by sand grains. An initial disturbance such as a saltation impact or a powerful downdraft in the wind may be required to start them moving. Once lifted, suspended particles can remain aloft for many hours or even weeks, for their settling speeds are less than the average speed of upward-moving eddies in the turbulent updrafts. Windborne dust from Colorado in the 1930s was swept all the way to New England and even over the North Atlantic Ocean.

During the 1930s, **dust storms** sometimes dimmed the midday sun over cities of the eastern seaboard. Dust storms sweep finer material to great heights and over great distances. Although the

11.21 Leading edge of dust storm arriving in Union County, New Mexico on May 21, 1937. The dust cloud moved in on a day having gentle breezes, but 15 minutes later hurricane-force winds blew for 30 minutes. (Al Carter, U.S. Department of Agriculture)

suspended particles are much lighter than those carried in sandstorms, the total weight entrained during a dust storm may be far greater. Winds during the Dust Bowl years raised dust-filled clouds as much as 3 km high, holding many thousands of tons per cubic kilometer (Figure 11.21). Little wonder that vast areas of the Great Plains were denuded and whole houses buried by settling dust. Soil-conservation experts made enormous efforts to limit the damage by planting windbreaks and soil-binding grasses. In the end, only the termination of the sustained drought quenched the dust storms that threatened to convert wheatfields and grazing lands into a desert.

Sediments Deposited by the Wind

Sediments deposited by the wind include fine-grained deposits of dust, and great bodies of sand which typically form dunes.

Loess Evidence of enormous ancient dust storms have been left in the form of extensive deposits of **loess** (luhss) (Figure 11.22), usually defined as nonstratified silt (sometimes mixed with fine sand and clay), that is laid down by the wind during long periods. Loess varies

11.22 **Loess,** forming vertical slope, overlying gravel. (G. R. Roberts)

in color from reddish-brown through yellow to gray, and consists mostly of quartz, feldspars, and carbonates. The grains are angular and pack irregularly, resulting in a high porosity of 60 per cent or more.

Deposits of loess are widespread, forming a mantle over the underlying soil to depths from a few meters to more than 30 meters. Large areas of the central United States in the Mississippi and Missouri basins (including most of Illinois, Iowa, Nebraska, and large parts of South Dakota, Kansas, and Missouri) are coated by loess deposits 1.5 to 24 meters thick. Much of this loess in the central United States was derived from dried-out, fine-grained outwash deposited by meltwater streams draining southward from the retreating Pleistocene glaciers.

An elongated belt of loess reaches from western Europe through the Ukraine and Russian steppes into northern China. The world's thickest deposits cover thousand of kilometers of Shensi Province in China to a depth of 60 meters or more. This thick loess is thought to have blown there from the Gobi Desert. Similarly, the Argentine deposits are believed to have come from the Patagonian Desert.

Like windblown dust today, loess stopped only when it settled in places where existing vegetation provided shelter from the wind. The thickest loess deposits are found where the winds are gentle enough for a stabilizing ground cover to maintain itself. Land morphology was also a factor; prevailing westerly winds blew loess against the eastern slopes of river valleys, such as the Mississippi Valley.

Because silt resists wind erosion, exposed loess faces can remain vertical for long periods (Figure 11.22). Such faces stand out as high bluffs in the Mississippi-Missouri region and in road cuts. Long-used roads in the Chinese loess zone have sunk 10 meters or more below the surrounding ground surface, and yet the banks remain stable. Other loess deposits, which are easy to excavate even with primitive tools, have provided cave dwellings for thousands of years in China and Central Europe. More importantly, loess is an excellent moisture-retaining base for fertile loam. Most of the world's grain is cultivated where wild grasses once trapped windblown dust.

Sand Dunes

Large areas of shifting eolian sand are the most distinctive of desert landforms, covering between one-quarter and one-third of all deserts. Virtually all active or moving eolian sand occurs not in isolated dunes but in vast, migrating **ergs,** or sand seas, greater than 125 square kilometers in area.

Despite the violence and seeming chaos of desert sandstorms, the wind does not drop its sand in random fashion. Rather, the sand gathers in patterns of surprising consistency and even beauty. The leading pioneer student of such patterns, British military engineer Ralph A. Bagnold, observed: "Instead of finding chaos and disorder, the observer never fails to be amazed at a simplicity of form, an exactitude of repetition, and a geometric order unknown in nature on a scale larger than that of crystalline structure. In places, vast accumulations of sand weighing millions of tons, move inexorably, in regular formation, over the surface of the country, retaining their shape, even breeding, in a manner which, by its grotesque imitation of life, is vaguely disturbing to an imaginative mind."

Dune formation Like streams, winds, as their speed decreases, deposit the largest particles first. The dynamics of air flow and turbulence are extremely complicated, but clearly a rock, shrub, or other obstacle that causes deposition can begin an accumulation—a "sand shadow"—that encourages further growth (Figure 11.23). Passing air currents tend to eddy into this shadow and there slow down, adding to the deposit. This irregularity causes further alterations in the windstream, giving rise to repeated zones of deposition that range from ripples a few millimeters high to great **sand dunes** as high as 200 meters. Such deposits soon become independent of the obstacle that caused them originally. Bagnold defined a sand dune as a mobile heap of sand independent of either ground form or fixed wind obstruction.

Coral Dunes
Joseph Wood Krutch

There are several different ways of enjoying scenery and the American Southwest is one of the best places to practice them.

If you are one of those to whom the beauty of a landscape means primarily form and color for their own sakes, then the bright colors and strange but fascinating forms found almost everywhere will be outstandingly rewarding.

Because vegetation is mostly sparse in the dry climate, the earth reveals its structure as it does not in regions where the shape of hills, mountains, and cliffs is smothered in green. Because so many of the formations are sandstone, they are multicolored in vivid reds, yellows, and, occasionally, blues. Because they have been sculptured for centuries by windblown sand, they have assumed all sorts of fantastically beautiful shapes.

These great sandstone buttes are commonly called "monuments," and ancient as they are, there is something curiously modern in their style. They are not fussy and overly ornamented in the Victorian fashion. They are bold, stark, and angled, and a visit to, say, Monument Valley, which straddles the border between Arizona and Utah, is like visiting a gallery of hugely magnified modern sculpture. It is both strange and strangely beautiful.

There is another way of enjoying a landscape. It is to see it as not merely interesting shapes and interesting colors but as a record of our earth's history, a plain tale which tells those who know how to read it how and when these mountains, these valleys, these canyons, and these monuments came to be what they are.

Geology is the science of reading that record with a full understanding of all its details, and most of us are not geologists. But the main outlines of the story are readily understandable by anyone who takes the trouble to look at the large striking features of the landscape with that in mind.

Every year thousands of tourists pass through the little Utah town of Kanab, heading northward toward Zion and Bryce canyons or eastward toward the new Glen Canyon National Recreation Area. Most of them do not even notice a modest wooden arrow a few miles east of the town which points down a dirt road and is marked "Coral Dunes." Still fewer accept its invitation, and that is perhaps just as well, for the undisturbed beauty of the dunes (some dozen miles from the main road) is one of their charms. On an early June morning, my companions and I had them to ourselves in the cool of five thousand feet.

Much of northern Arizona and southern Utah is a land of towering sandstone mesas and buttes, some white, some pink, and some coral-red. They have been sculptured into fantastic shapes and are gradually being eroded away by water, frost, and especially by windblown sand. Some are half buried in their own detritus but still rise sheer above the semidesert plains. At the Coral Dunes, on the other hand, the prevailing wind has heaped and shaped sand into dunes as high as twenty-five feet, which, incidentally, recently furnished a perfect setting for a motion picture, whose action is supposed to take place in the Arabian Desert.

Nothing quite prepares one for the climactic view. The approach is across a semidesert, increasingly sandy but with the sand held in place by fairly abundant sage and juniper together with a few pines. All of them grow less and less abundant, and then one comes upon a true Sahara of drifting coral-colored dunes sloping gently upward on one side, dropping off abruptly on the other; sometimes rippled as though by waves of a seashore; sometimes almost unbelievably smooth and sleek.

At first, one is unaware of any living thing except, perhaps, for ravens calling derisively overhead. No animals are visible. But the most casual inspection reveals the fact that they are only unseen, not absent. There are tracks which can only be those of a bobcat and there are other curious little bipedal marks that proclaim the kangaroo rat. Most beautiful, and at first most puzzling, are long lines of the most delicate tracery, sketched across the smooth surface uphill and down. Each one is perfect despite its obvious fragility and all suggest some secret mysterious workman. What could have made them? Not a small lizard because there is no trailing tail mark. Certainly not a sidewinder rattlesnake, whose strange tracks are much broader and in other ways quite

From THE BEST NATURE WRITING OF JOSEPH WOOD KRUTCH, 1966. William Morrow & Co., Inc. Abridged with permission.

Field Trip

different.

It doesn't take much looking to find the answer. The tiny workmen are everywhere busy, often no more than a few yards from one another. They are little, shiny, quarter-inch, scarablike beetles *(Sphaeriontis muricata)*. From the order to which they belong, it seems a pretty safe guess that their unresisting progress uphill and down is motivated by nothing more spiritual than a search for the droppings of a jack rabbit. But it would not be difficult to imagine that they are artists, endlessly engaged in beautifying the dunes with the perfect but changeless pattern of lace which nature has ordained their six legs to make during many millenniums and which they will continue to make for untold millenniums hence—unless, as seems not improbable, man destroys their environment as he is destroying that of many more conspicuous creatures.

Most of the dunes are shifting, and they are often marching forward in the direction of the prevailing wind. Any sizable plant which manages to get established tends to anchor the sand around it, but more often than not it loses the struggle and is overwhelmed by the slowly advancing waves which may ultimately pass over it, leaving behind a shallower bed of sand in which a new generation of plants may be able to grow.

Evidence of that process is plainly visible in the Coral Dunes. One may see, for example, the skeleton of a pine killed sometime in the past by an advancing wave; beside it are younger pines or junipers which have grown since the crest of the wave passed by.

Given a flat, open surface and hard grains of material too heavy to be blown away by the prevailing wind but not too heavy to be moved by it, the dunes are nearly inevitable. If the grains are too light, they blow away as dust; if they are too heavy (small pebbles, for example), they cannot be piled up by the winds. But if they are just right, you get the similar outlines of the same geomorphic features. And there are minor variations in shapes, depending partly upon the character of the winds. But a dune is a dune is a dune.

Geographers tell us that the wind seldom lifts sand particles more than a foot or two, that the grains are usually simply rolled along, and that they come to rest when the wind's velocity decreases. Once started, a dune itself becomes an obstruction around which the winds swirl and deposit more sand. Because of the variations in wind velocity and sand supply there are different characteristic shapes in different regions, and those most characteristic of the Coral Dunes are what are called "barchans"—that is, hillocks with a sloping side up which the prevailing winds blow the sand and a steeper leeward side where the grains have come to their angle of repose.

Though the Coral Dunes of Utah are in many respects typical, there is one fact about them of additional interest: the material of which they are composed has been twice reduced to sand in the course of many millions of years.

The mesas and buttes which surround them are composed of Navajo sandstone formed during the Jurassic period (say 100 or 150 million years ago). But that sandstone was composed of the detritus from which more ancient mountains long before wore down. Now, this Navajo sandstone has itself been eroded away to make the sand which may (given more millions of years) again solidify into stone.

There are few more striking examples of the restlessness of our earth—always building up, tearing down, and then building up again as the result of the processes which will probably continue until a completely cooled earth can no longer raise mountains and its whole surface is reduced to a featureless plain. 1966

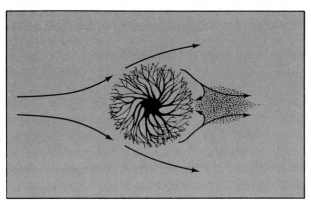
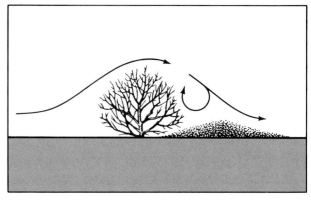

(Top view)   (Side view)

11.23 Sand shadow, an accumulation of sand in the lee of an obstacle.

Dune movement and stabilization   A sand dune has a characteristic profile, featuring a gentle windward face and a steeper lee slope with a maximum angle of repose of approximately 34 degrees. Blown sand migrates up to the crest (whose height may depend on wind speed, size of sand particles, degree of sorting, and other factors) and falls over into the "slip face." There is a net transfer of sand and hence a net forward motion of the dune in the direction of the prevailing wind (Figure 11.23). A migrating dune's advance (some travel as much as 50 meters per year) can be halted and the dune stabilized by the growth of hardy grasses whose windbreaking effect and roots hold the surface sand in place.

Kinds of dunes   The common kinds of dunes are barchan dunes, longitudinal dunes, transverse dunes, and parabolic dunes (Figure 11.24).

Barchan dunes. Dunes having a distinctive crescent shape (in plan view) are **barchan dunes.** The horns of the crescent curve downwind from the main body. Barchans are most common in deserts with steady, moderate winds and limited sand cover. Crosswinds may cause the elongation of one horn, and in deeper sand barchans may coalesce into curved networks called star dunes.

Longitudinal dunes. In deserts where winds blow regularly enough to establish great cylindrical flow cells, longitudinal dunes are common. Such dunes are aligned in the general direction of the wind. Called **seif dunes** by Arabs, longitudinal dunes reach lengths of almost 100 km and heights of 90 meters or more. As the wind speed varies in the cylindrical flow cells, the cross currents build a string (or several strings) of shifting slip faces along the spine of a seif dune.

Transverse dunes. Dunes that are linear, usually short, and aligned at right angles to the prevailing wind are **transverse dunes.** This dune form is most often seen along coastlines which have steady winds; transverse dunes may be irregular in shape if vegetation interrupts their formation.

Parabolic dunes. A dune which resembles an elongated, reversed barchan is a **parabolic dune,** with horns extending upwind instead of downwind. Parabolic dunes are sometimes found among beach dunes migrating inland, and may form as sand drifts behind a gap in an obstructing ridge. Wind funneled through such a gap spreads out and slows enough to drop the sand it carries (just as a stream may deposit its load if the channel volume is suddenly increased). This sand accumulates in a U-shaped, or parabolic, curve just beyond the gap. Later a parabolic dune may detach itself from its site of formation and migrate independently.

Ancient dunes The long, straight cross strata formed on the slip faces of some dunes have been buried and eventually converted into sandstone (see Figure 6.11b). We can make reliable inferences about the winds that blew millenia ago from such preserved dune faces.

The composition of dune sand (mostly quartz, with varying amounts of calcite, gypsum, and other minerals) and its sorting may give clues to its source. The nature of the cement or recrystallization that converts sand to sandstone tells us its history since deposition. Even the shapes of sand grains are informative, because long periods of saltation round off the originally angular fragments of materials.

Features Formed in Rock by Natural Sandblasting

We will now take up abrasion and the various landforms created by it. The landforms include ventifacts, yardangs, and grooves.

Abrasion The process by which sandblown sand and dust wear away rock surfaces is called **abrasion.** The effects of abrasion are not as widespread as was once supposed. Early in this century, a school of "eolianists" contended that wind and windblown sand leveled whole mountain ranges, cutting into them as waves cut into a coastline. Although this vision is now discounted, abrasion doubtless can severely erode both small and large desert structures. **Impact erosion** is a kind of abrasion in which wind-driven sand and tiny pebbles chip shallow irregularities in bedrock.

11.24 Major kinds of dunes, schematic block diagrams.

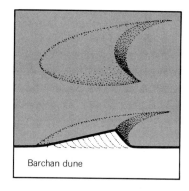
Barchan dune

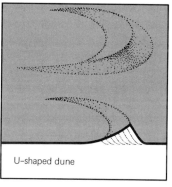
U-shaped dune

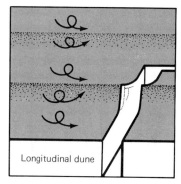
Longitudinal dune

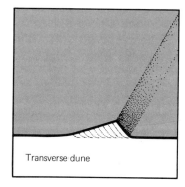
Transverse dune

Wind direction

11.25 Pedestal (top left) formed by sandblasting of bedrock by saltating windblown particles. Death Valley, California. (American airlines) **Wind-formed arch** (top right) in Monument Valley, Arizona. (American Airlines) **Ventifacts** (right). (G. R. Roberts)

Ventifacts, yardangs, and grooves Abrasion grinds away at isolated rock masses, sculpturing slim-legged pedestals (Figure 11.25), hollows, arches (Figure 11.25), and other weird shapes characteristic of desert landscapes. **Ventifacts** (from the Latin for "made by wind") are pebbles and larger rocks with smooth facets cut and polished by abrasion (Figure 11.25). Most ventifacts have two or three facets, although changes in wind direction or shifting of the rock may cause the carving of more facets. **Yardangs,** found in China, West Africa, and Peru, are outcrops of relatively soft rock with rounded upwind faces and slanting, elongated downwind faces. Yardangs range from a few meters to several thousand meters in length and may be several hundred meters high. They and the troughs or grooves between them are clearly aligned with prevailing winds. **Grooves** and **pans** of even larger size, some with orientations which suggest that abrasion and deflation alternated in shaping them, have been photographed from airplanes and satellites.

Geologic History of Deserts

On a geologic time scale, today's deserts are probably shortlived features. The question is hotly disputed, but most geologists would agree that no modern desert has existed as such for more than a few tens of millions of years. In fact, much evidence from history and from archeology indicates that a few thousand years ago many of today's deserts were fertile. This change is thought to be related to Pleistocene climatic variations, which were also responsible for the shrinking of large Pleistocene pluvial lakes to small modern desert lakes.

The climatic processes that determine the expansion, contraction, and migration of deserts are only beginning to be understood. Sandstone strata, inferred to be ancient dunes, provide strong evidence that during millions of years different regions have been alternately in or out of arid zones. Similarly, other kinds of geologic evidence indicate that some present desert areas once lay beneath the sea. For example, marine strata containing Devonian fossils indicate that 350 million years ago the Sahara was covered by ocean. Continental movement has also altered the distribution of deserts by changing the flow of ocean currents. Tectonic activity and erosion, which build and wear away mountains, change atmospheric circulation enough to create or erase deserts. Clearly, the climatic characteristics of any part of the Earth's surface—including deserts—are shortlived and are subject to the vast processes that shape the lithosphere.

Chapter Review

1 A *desert* is a region characterized by evaporation that greatly exceeds precipitation. Most deserts receive less than 25 cm of rainfall per year. Deserts are the world's hottest places because the desert air holds little moisture to form clouds and thus to block solar radiation. Sparse vegetation is adapted to make maximum use of scanty rainfall and groundwater.

2 About one-third of all land area is desert. According to their locations and causes, most modern deserts can be arranged in four groups: *subtropical, continental-interior, rainshadow,* and *coastline* deserts.

3   The chief impact of a desert climate on the geologic cycle concerns *temperature,* *distribution of rainfall* and its effect on stream flow, and *wind activity.* When rain does come to the desert, it often comes in brief, high-intensity "cloudbursts," which may dump an entire year's precipitation in an hour.

4   The key elements of weathering and mass-wasting in the desert are *high temperatures,* which cause the evaporation of salts, *low temperatures,* which may freeze water, a *general upward movement of water* through the regolith, and the *absence of plant cover.*

5   Deserts display a distinctive variety of landforms quite unlike anything seen in humid climatic regions. Desert landforms include *fans, bolsons* and *bajadas, pediments,* and *playas.*

6   Although most surface water in deserts disappears by interior drainage, under certain conditions permanent lakes form as dominant features of the landscape. There is evidence of ancient *pluvial lakes,* which were much larger than the modern desert lakes.

7   The geologic work of the wind can be discussed conveniently under the sequence that starts with the picking up of loose granular material, the mechanics of transport by the wind, including dunes, and the features formed in rock as a result of abrasion by wind-transported particles.

8   Sediments deposited by the wind include fine-grained deposits of dust *(loess),* and great bodies of sand which typically form dunes. The common kinds of dunes are *barchan, longitudinal, transverse,* and *parabolic.*

9   The process by which windblown sand and dust wear away rock surfaces is *abrasion.* The landforms created by abrasion include *ventifacts, yardangs,* and *grooves.*

10 Most geologists agree that no modern desert has existed as such for more than a few tens of millions of years. Evidence indicates that a few thousand years ago many of today's deserts were fertile. This change is thought to have been related to Pleistocene climatic variations.

Questions

1   What climatic conditions must exist to create a *desert?*
2   Name and describe the four main groups of modern deserts.
3   How does weathering take place in a desert?
4   Define *bolson, bajada, pediment,* and *playa.*
5   How is a desert lake formed?
6   Define *pluvial lake,* and describe how it offers evidence of past deserts.
7   Describe *deflation* and the wind erosion of loose sediment particles.
8   Describe the mechanics of sediment transport by the wind. Define *saltation.*

9 Define *loess*. Describe the method of dune formation and name the common kinds of dunes.

10 Are existing deserts considered to be geologically young or old?

Suggested Readings

Leopold, A. S., and the Editors of *Life, The Desert*. Life Nature Library. New York: Time, Inc., 1962.

Shimer, John A., *This Scuptured Earth: The Landscape of America*. New York: Columbia University Press, 1959.

Thompson, Philip D., and O'Brien, Robert, *Weather*. New York: Time-Life Books, 1965.

Wyckoff, Jerome, *Rock, Time, and Landforms*. New York: Harper & Row, Publishers, Inc., 1966.

"It may be many years before nations decide whether to divide the riches of the sea peacefully."

Chapter Twelve
Oceans and Shorelines

Oceans and Shorelines

The Earth is not known as the "water planet" for nothing. Among the planets of our solar system the Earth is unique in possessing a continuous body of salt water, which we call the **oceans.** In some way or other, either as friend or as foe, the oceans affect our daily lives. They are a vast part of the biosphere and are the habitat for hordes of organisms, many of which provide us with significant quantities of food. The interactions between ocean water and the atmosphere affect the weather; both are responsive to the forces that determine the Earth's climate. The oceans are the ultimate reservoir of the water that circulates in the hydrologic cycle. In Chapters 8 and 9 we studied movements of water on the surface of the lands and underground. In Chapter 10 we examined water in its solid form as glacier ice. In Chapter 11 we concentrated on the absence of water in deserts. Now, in this chapter we shall take up the peculiar aspects of salt water.

The Oceans As Part of the Biosphere

Many kinds of organisms live in the oceans. We shall deal with this part of the biosphere by starting with some general information on the distribution and characteristics of the modern oceans. Then we discuss the origin of the oceans and some of their chief habitats. Finally, we shall explore some new developments in using the oceans to grow food supplies.

Distribution and Characteristics of Modern Oceans

The oceans make up a continuous body of salt water that covers 70.8 per cent of the Earth's surface. The distribution of lands and seas over the Earth is asymmetric. The Northern Hemisphere is dominated by continents, and hence has been named the "land hemisphere." The Southern Hemisphere is the "water hemisphere"; it consists chiefly of oceans (Figure 12.1). Ocean water is distributed in four deep-ocean basins: (1) Pacific (as large as all the others put together); (2) Atlantic, (3) Indian, and (4) Arctic. The southern ocean, known as the Antarctic Ocean, is separately named, but occupies a depression that is formed by southward extensions of the Pacific, Atlantic, and Indian ocean basins. Smaller, partially isolated bodies of ocean water are **seas.** A few examples include the Caribbean Sea, the Mediterranean Sea, and the South China Sea. Still smaller arms of the oceans are gulfs, bays, channels, and straits.

The total surface area of the oceans is about 358 million square kilometers. As we have seen in earlier chapters, the average depth of the ocean is about 4 km. The total volume of sea water is 1350 cubic kilometers—18 times the volume of all the land lying above sea level.

Properties of sea water Sea water is described in terms of its proportion of dissolved solids (known as **salinity**), its proportion of dissolved gases, its temperature, and its density. These properties affect the kind of habitat the water provides for organisms and influence the ways in which sea water circulates.

The salinity of sea water includes two distinct subjects: (1) the amount of dissolved solids, which is expressed as the number of parts per thousand by weight, and (2) the chemical composition of the solid residue that forms when a sample of sea water has been evaporated to dryness.

The average salinity of sea water is about 35 parts per thousand. The range is from about 20 parts per thousand, in places where large rivers bring in fresh water, to more than 65 parts per thousand, in the upper ends of embayments in arid climates where evaporation greatly exceeds precipitation. Because salinity affects density and density determines movements of sea water, we shall be having more to say about salinity in following sections.

Despite the variations in the numerical value of salinity from place to place, it is a remarkable fact that if one collects a sample of sea water from any part of the ocean and evaporates it away, the chemical composition of the salt left in the evaporating device will be constant.

The sea possesses astonishing ways of cleansing itself of trace elements, largely heavy metals, which are brought to it by rivers or dumped in by humans. For example, throughout geologic time enough arsenic has been weathered naturally from sulfide minerals

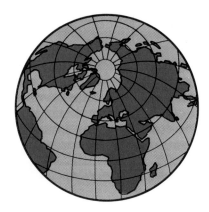 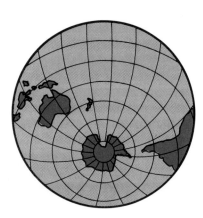

12.1 Uneven distribution of land and water. Land is concentrated in the Northern Hemisphere (left), and water in the Southern Hemisphere (right).

and brought by rivers to the sea to have rendered the oceans toxic. However, as soon as arsenic appears in sea water, organisms incorporate it into their tissues. Thus, the arsenic is removed from sea water and eventually, when the organisms die, it becomes buried in the bottom sediments. Other toxic heavy metals evidently become attached to clay minerals and are removed from sea water when the clay-mineral particles settle to the bottom. The fact that the sea possesses these natural cleansing mechanisms does not mean that we can continue to dump industrial wastes into it without being concerned about the consequences. It is urgent that we try to understand these self-cleansing mechanisms and to know their natural limits so that we do not impose excessive chemical loads.

The *temperature* of sea water varies from 26° to 28°C in the tropics to −4°C at the polar regions, where sea ice forms. The freezing point of sea water varies with its salinity and is lower than that of fresh water. As we shall see, these systematic differences in the temperature of sea water are the basis for an ocean-wide system of circulation.

Depth zones and habitats The depths of the sea are controlled by the underlying major parts of the Earth's crust. The depth of the water determines the penetration of sunlight, which, in turn, controls the distribution of many marine organisms, particularly the plants. Finally, the depths affect the distribution of various habitat groups. We shall examine each of these topics briefly.

As we have seen, the two chief parts of the Earth's crust are the

12.2 Coral reefs, Buck Island. (M. M. Williams, U.S. National Park Service)

continental masses and the ocean floor. Continental masses are rarely submerged by more than a few hundred meters. By contrast, the ocean floor is covered by water averaging 4000 meters, and ranging to about 11,000 meters deep.

Because sunlight penetrates only small distances into sea water, most plants are able to grow only near the surface of the sea. Using the amount of sunlight as a basis, it is possible to recognize three life zones in the oceans: (1) The upper, lighted zone, up to 80 meters thick, where sunlight supports plant growth, is the **photic zone.** (2) Beneath it is a transitional zone of fading light, which extends from 80 to about 600 meters in depth, the **disphotic zone.** (3) Still lower is the lightless bulk of the ocean, the **aphotic zone.** The aphotic zone is not a lifeless zone, as it was once thought to be. Because plants do not grow in the aphotic zone, all animal life there depends on the downward movement of food from above.

The organisms living in the sea display many life styles. Many single-celled organisms, both plants and animals, float freely in the near-surface waters of the photic zone.

Numerous animals and plants live on or in the bottom; these may be attached or mobile. The chief attached plants include turtle grass, kelp, and many kinds of green algae. The most voluminous of the attached animals are the *corals,* many of which build wave-resistant structures in the tropical parts of the ocean, known as *reefs,* shown in Figure 12.2.

In addition to corals, many other marine invertebrate animals live on the bottom. Many kinds of algae spread thin, sticky sheets of tissue over the bottom, forming algal mats. Mat-forming algae create distinctive features known as *stromatolites* found in ancient rocks. As we shall see later, many kinds of marine sediment consist partially or wholly of the remains of marine organisms.

Can We "Farm" the Sea?

Because the surface of the sea absorbs so much solar energy, and floating plants convert this energy to plant tissue, much thought has been given to the possibility that the amount of organic tissue that grows in the sea (that is, its productivity) might be increased artificially. Algae, for example, are known to grow prolifically where supplies of nitrogen and phosphorus are abundant.

In several parts of the world, including the United States and Japan, most of the oysters consumed by humans are grown artificially. The later stages of oyster growth take place in a natural setting, but the reproduction and early stages are carefully controlled in tanks.

In recent years, marine biologists have suggested a number of new oceanic food sources that could easily be exploited. Krill, the shrimp-like crustaceans on which most whales feed, flourish in enormous quantity in the Antarctic Ocean. Other planktonic plants and animals could be grown rapidly on a commercial basis. Huge kelp, and other marine algae, could be "farmed." Menhaden and other so-called "trash" fish, which are now discarded, could be processed into high-protein meal.

One barrier to widespread use of the new sources is not easily surmounted: human taste. It is one thing to raise tons of nourishing micro-organisms or other food in sea water, but quite another to make them appetizing to the human palate.

Movements of Sea Water

The waters of the oceans are constantly moving. They are subjected to uneven heating by the Sun, blown by the winds, evaporated more in some places than in others, pulled by the gravitational attraction of the Moon and the Sun, and affected by the Earth's rotation. All these factors create changes of density or level that cause currents to flow. We need to understand how these main currents work, both to appreciate how the oceans affect the weather and the climate, and to be able to relate the currents to sediments. We shall discuss movements of sea water under four major sub-headings: (1) movement of surface waters, (2) circulation and density currents, (3) waves, and (4) tides.

Movement of Surface Waters

The movement of surface waters results from the interplay of uneven solar heating and the rotation of the Earth. The Sun is responsible for creating the Earth's climate belts and its major wind belts. The rotation of the Earth is responsible for the Coriolis effect. Together, these factors create major surface currents in all oceans.

Relation to climate zones and major wind belts Intense heating by the Sun in the equatorial regions creates uniform masses of warm tropical sea water, whose temperature is one of the world's great constants (about 28°C). As explained in Chapter 11, the heated equatorial air rises, drops its moisture, flows away from the Equator toward both the north and the south, and descends in regions of persistent high pressure. Here the descending air is heated and takes up moisture from the Earth's surface. It creates belts of deserts on the continents and zones of maximum evaporation at sea. Some of the descending air returns to the Equator, and some of it moves toward the poles. The winds blowing over the surface of the oceans drive the water in great surface currents (Figure 12.3) and also create waves.

Any motion on the surface of an object that is rotating, such as the wind or an ocean current on the surface of the Earth, is subjected to a systematic deflection, which is known as the **Coriolis effect.** The effect of moving objects in a rotating frame of reference was first described by a French mathematician, Gaspard Coriolis.

The Coriolis effect The first thing to understand about the Coriolis effect is the direction of the Earth's rotation. As viewed from above the North Pole, the Earth rotates counterclockwise. However, when viewed from above the South Pole, the same rotation of the Earth appears to be in a *clockwise* direction. (Figure 12.4). You can confirm this by spinning a small globe in a counterclockwise direction when viewed from above the globe's north pole. Now, look at the same spinning globe from above its south pole. How would you describe the sense of rotation?

Because of the rotation of the Earth, neither the winds nor the water currents created by them move in simple north-to-south or south-to-north directions. The solid Earth spins independently of the water and air above it, and because this movement is greatest at the Equator and descreases regularly toward the poles, all currents are deflected. The direction of this deflection is determined by the sense of rota-

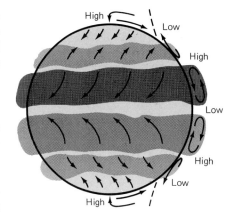

12.3 Large-scale circulation of air, which creates surface currents in oceans. Schematic view of Earth and lower atmosphere.

12.4 Although the Earth rotates in only one direction, the sense of rotation differs depending on one's vantage point. (a) From above North Pole the sense of rotation is counterclockwise. (b) From above the South Pole the sense of rotation is clockwise. The composite satellite photograph in the margin shows an extraordinary view of the world looking directly down at the South Pole. (Department of Commerce/NOAA)

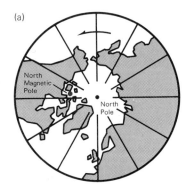
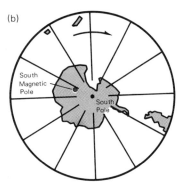

414
Oceans and Shorelines

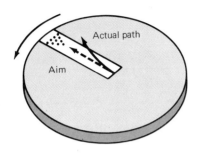

12.5 Coriolis effect shown by the effect of rotating a bowling alley on a giant turntable. The ball is aimed directly at the head pin, but because the turntable rotates the alley counterclockwise, the ball will always roll into the right-hand gutter. If turntable were rotated clockwise, the ball would always roll into the left-hand gutter.

tion of the Earth at a specific place. Wherever the sense of rotation of the Earth is counterclockwise, all objects are deflected to their right (seen in the direction they are moving). In a frame of reference that is rotating clockwise, the deflection is to the left (Figure 12.5).

Major surface currents The major surface currents of the oceans involve four great spirals or **gyres** (JHY-ers), two in the Northern Hemisphere and two in the Southern Hemisphere; four lesser gyres; and a single round-the-world current in the Antarctic Ocean (Figure 12.6).

Circulation and Density Currents

Whereas the surface currents flow largely in response to changes of level, subsurface movements are created by differences of density. Accordingly, subsurface movements of sea water are **density currents.** A density current flows because the Earth's gravity pulls denser water masses downward toward the deepest parts of the oceans. Denser water thus tends to sink and to flow beneath less-dense water.

Turbidity currents A density current that owes its excess density to the effects of suspended sediment is a **turbidity current.**

What are now known as turbidity currents were first discovered in lakes and reservoirs. Geologists interested in the oceans learned about such currents in the 1930s, as a result of studies and observa-

12.6 **Major surface currents of the oceans.** (U.S. Navy Oceanographic Office)

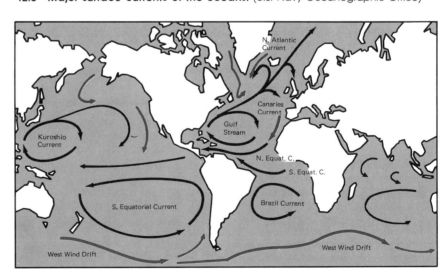

Movements of Sea Water

12.7 Laboratory turbidity current. (California Institute of Technology)

tions made at Lake Mead, which was being formed by the backing up of the Colorado River behind Hoover Dam. Small-scale laboratory experiments showed the essential characteristics of turbidity currents—a dense, sediment-laden current flows along the bottom beneath clear water above (Figure 12.7).

The idea that turbidity currents might be active in the modern oceans was proposed as a mechanism for eroding submarine canyons. In the 1950s layers of sand showing a systematic upward gradation in particle size, named **graded layers,** were found in cores of deep-sea sediment (Figure 12.8). Other indications that turbidity currents are active in the modern oceans include the discovery on the deep-sea floor, far from land, of skeletal debris of organisms that grow only in shallow water, of the remains of land plants where floating out from shore and ice rafting can be excluded, and of great flat-topped bodies of sediment that form vast expanses in the deep parts of the ocean basins.

Waves

We begin our discussion with some fundamental definitions and characteristics of simple waves which cross the water surface without being influenced by the bottom. Such waves are **deep-water waves.** Then we shall describe the waves that are generated by the wind, and shall close this section with a brief account of waves that are created by displacements of the sea floor. In a later section we shall see what happens when waves approach the shore.

12.8 Graded layer of deep-sea sediment raised from Hatteras Abyssal Plain, western Atlantic. Three segments of a single layer more than a meter thick that occupies most of the second meter of the core. Hence, centimeter marks on meter stick at right should read 120 cm below top instead of 20 cm; the 50 should be 150 cm; and the 80, 180 cm. (Lamont-Doherty Geological Observatory of Columbia University, Courtesy E. D. Schneider)

Oceans and Shorelines

Definitions and characteristics of deep-water waves An idealized set of deep-water waves consists of rounded, convex-up crests separated by rounded, concave-up troughs. The horizontal distance between two adjacent crests is **wavelength;** the vertical distance between the top of a crest and the bottom of a trough is **wave height** (Figure 12.9). The time required for a complete wave form to pass a given point is the **wave period.** The speed of advance of the waves is **celerity.** On ocean waves the easiest measurement to make is their period. One simply counts the number of waves that pass a given point in one minute and then divides the number of counted waves into 60 seconds to get the number of seconds per wave (the period).

A single train of waves moves along in a direction that is perpendicular to the lines joining the crests and troughs. Deep-water waves not only raise and lower the water surface but cause the underlying water particles to move in circles. As each wave crest approaches, the water surface is raised and the underlying water particles move forward in the same direction the wave is traveling. Simultaneously, beneath each trough the water surface is depressed and the motion of the underlying particles is backward, in a direction opposite from the advance of the waves (Figure 12.9).

12.9 Waves on water surface passing from deep water (depth more than half the wavelength, at left) to shallow water (depth less than half the wavelength, right). Curved lines show orbits of water particles after passage of each wave. Orbits are circles in deep water and ellipses in shallow water. Schematic block.

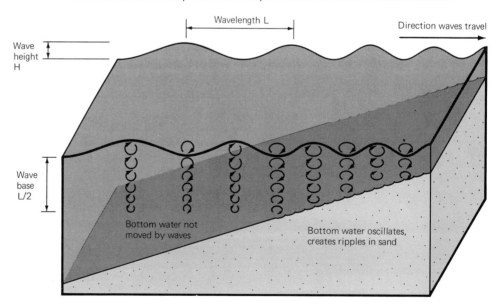

Wind-generated waves Most of the waves on the surface of the sea result from transfer of wind energy into the water. A great storm crossing the open sea churns the water almost as if it were a gigantic egg beater having a diameter of several hundred kilometers (Figure 12.10). Within a low-pressure storm system in the Northern Hemisphere the flow of air is counterclockwise. The wind blows around the circumference of the circular disturbance. Although at a single locality affected by a given storm the wind blows from only a single direction, viewed as a whole storm, winds blow from all possible directions.

The waves being actively blown by winds are steep, choppy, and capped by sheets of spray; they are named **sea waves.** In the open sea the sizes of sea waves depend on the speed and duration of the wind. The periods of sea waves in the open sea range from about 3 to 7 seconds; periods increase with the speed of the wind. Once these waves have traveled beyond the places where the wind actively blows the water, they become transformed into longer, lower, and more regular waves than sea waves. Such transformed waves that have traveled away from the storm center are **swells.** The periods of swells range from about 5 seconds upward to as much as 20 or 22 seconds; the average is 8 to 12 seconds.

12.10 Storm at sea viewed from Gemini X spacecraft. Diameter of storm is approximately 200 km. (NASA)

Tsunami When the sea floor is suddenly displaced, the disturbance is propagated upward to the surface and there becomes a series of long, low waves having periods of 12 to 15 minutes, wavelengths of 100 to 200 km, and wave heights in the open sea of only a meter or so. Such extraordinary waves have been named **tsunami** (Japanese for "harbor wave"). Because of their enormous wavelengths, tsunami are never deep-water waves. Instead, they are a special kind of very-shallow-water wave whose speed is a function of water depth. The celerity of tsunami in the Pacific Ocean ranges from 720 to 800 km/hr.

The kinds of displacements of the bottom which create tsunami are shifting of a body of sediment, faulting, and collapse of a volcanic caldera. Most tsunami in the Pacific originate when earthquakes are recorded beneath the one of marginal deep trenches. The chief cause is thought to be sudden shifting of bottom sediment. The close relationship between earthquakes and tsunami has been developed into a reliable tsunami-warning system. Although tsunami cannot be detected by ordinary observations in the open sea, along certain coasts they do become plainly and menacingly obvious (Figure 12.11). On some shores tsunami have risen to heights of 20 meters, and at the heads of V-shaped coastal indentations, to more than 30 meters. Tsunami cannot be prevented, but loss of life can be avoided by evacuating low-lying coastal areas. Thanks to the tsunami-warning system, this is

12.11 Tsunami breaking at Kawela Bay, north coast of Oahu, Hawaii, April 1, 1946. (F. P. Shepard, Scripps Institution of Oceanography, University of California at San Diego, La Jolla, California)

now possible. After the large Chilean earthquake in 1960, a tsunami was predicted for Hilo, Hawaii, 15 hours after the earthquake. The tsunami arrived at the predicted time.

One of the most lethal tsunami of record was triggered by caldera collapse during the eruption of Krakatoa in 1883. The Krakatoa tsunami killed 36,000 people and was recorded all around the world.

Tides

The **tide** is a rhythmic rise and fall of water level that results from astronomic causes and from the ways in which extremely long waves move into and out of basins. **Tidal amplitude** (or range) is the vertical difference in altitude of the water surface between the level of high water and that of low water. Tidal amplitudes range from half a meter or so on mid-ocean islands such as Tahiti to 15 meters or so in the Minas Basin at the northeast end of the Bay of Fundy, Nova Scotia. Tidal amplitudes larger than normal are **spring tides**. Tidal amplitudes less than normal are **neap tides** (Figure 12.12).

419
Movements of Sea Water

Astronomic aspects of tides The phenomenon of *lunar tides* results from gravity and the motions of the Earth and the Moon. Disregarding the complicating effects of the irregular shape of the ocean basins, we can visualize that the gravitational pull exerted on the Earth by the Moon bulges out the Earth's ocean water. The long axis of this water bulge points toward the Moon. The solid Earth spins on its polar axis once per day independently of this water bulge. Any time a part of the Earth lies within the bulge the tide is high. When a point spins out of the bulge the tide drops.

The Moon travels around the Earth every 29½ days. Strictly speaking, the Moon does not revolve around the center of the Earth. Instead, the centers of these two bodies orbit around their common center of mass, a point known as the Earth-Moon **barycenter** (Figure 12.13). The Earth-Moon barycenter lies about 2700 km beneath the Earth's surface. The orbiting of the center of the Earth and the center of the Moon around the Earth-Moon barycenter creates a centrifugal force, as does the revolution of two figure skaters holding hands. This force, although slight, is sufficient to create a water bulge on the side of the Earth opposite the Moon, just as a skater's skirt or scarf flies backward.

Kinds of tides In many localities the tide rises and falls twice a day with a period of about 12 hours *(semi-daily tides)*. Other places experience only one time of high water and one time of low water

12.12 Record of heights of high tide and low tide for one month and corresponding phases of the Moon. On days when high tides are highest, low tides are lowest. (Data of H. A. Marmer, 1926, replotted)

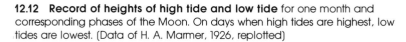

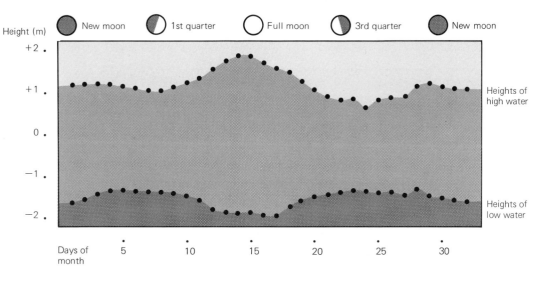

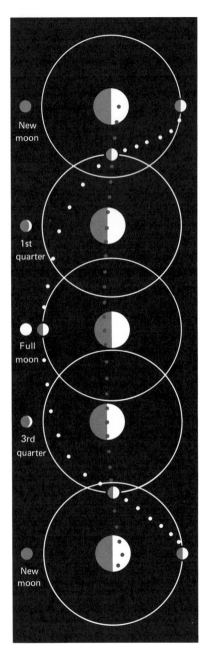

12.13 **Orbits of Earth and Moon** for one month, schematic map with Sun to the right. Brown dots show daily positions of Earth-Moon barycenter; white dots, daily positions of center of Moon. From 1st-quarter phase to Full Moon and to 3rd-quarter phase Moon travels faster than it does in passing from 3rd-quarter phase to New Moon and to 1st-quarter phase. Moon is closest to Earth at Full Moon, hence tidal amplitudes are greatest even though gravitational pull of Moon directly opposes that of the Sun.

each day *(daily tide)*. Semi-daily and most daily tides reach their high and low levels at times that become later each day by about 50 minutes. In addition, spring tides coincide with Full Moon and New Moon, and neap tides with the Moon's quarter phases (Figure 12.12).

Because the Earth spins on its polar axis once per day independently of the traveling lunar tidal bulges, the actual tide at any locality becomes a function of two motions: (1) the Earth's spinning, and (2) the Moon's orbit. It takes 24 hours and 52 minutes for the Earth to rotate once with respect to the Moon (Figure 12.13). Therefore, the time of the tides is displaced each day by 52 minutes. Finally, the theoretical behavior of the tide depends on location of the point on Earth with respect to the shape of the lunar tidal bulge.

Effect of basin shape The foregoing analysis of the tides did not mention that the bulged water is a very long wave that has to move into and out of basins having irregular shapes and limited depths. Viewed as "wave" forms, the tidal bulges have wavelengths equal to half the circumference of the Earth. Therefore, they move as

12.14 **Two-layered arrangement of waters of an estuary,** schematic block. Further explanation in text.

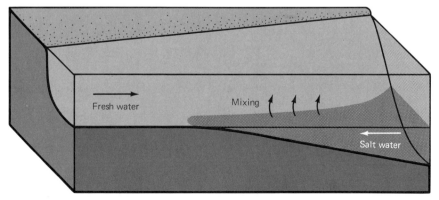

very-shallow-water waves and are greatly influenced by the local depths and shapes of basins.

Tidal currents As the tide rises and falls it creates horizontal tidal currents. These range from barely perceptible currents to rushing streams that make navigation difficult or dangerous. Tides flow into and out of San Francisco Bay at speeds of about 8 km/hr and through the entrance to the Minas Basin, Nova Scotia, at 18 km/hr. A rising tide is known as a *flood tide*; associated currents are flood-tidal currents. A falling tide is known as an *ebb tide*; currents related to a falling tide are ebb-tidal currents. In between are times of slack water, named high-water or flood slack and low-water or ebb slack, respectively.

Coastal Features

A **coast** is a great zone of interaction among the sea, the land, the atmosphere, and some remarkable organisms. The **coastal zone** is a strip of land of varying width that is affected by the sea. The **coastal-ocean zone** is that part of the sea which is influenced by the adjacent land. The **shore zone** extends from the level of lowest tides to the highest point on land that is subjected to wave action. The actual **shoreline** itself is the place where the edge of the sea meets the land at any particular time.

Estuaries and Tide-dominated Coasts

Where many rivers enter the sea the shoreline forms a V-shaped indentation, pointing landward, that is named an *estuary*. The shores of most estuaries are relatively sheltered from vigorous wave action. Therefore, their chief characteristics are determined by the pattern of water circulation established by the entering fresh water from the river and the salt water pushing in from the sea, and by the rise and fall of the tides (Figure 12.14).

As the tide rises and falls it brings suspended sediment toward shore and deposits it on vast intertidal mudflats. In temperate-climate zones species of salt-tolerant grasses colonize these flats. The grasses convert the inner edges of the intertidal mudflats into tidal marshes.

Estuaries and their marginal tidal marshes (popularly named "tidal wetlands") are sites of great biological productivity. So many species of fish and other organisms start their lives in these areas that estuaries and wetlands are aptly named the "nurseries of the sea."

Wave-dominated Coasts: Beaches

Where any kind of sand-size or coarser sediment is available at the shore, wave action arranges and rearranges it into characteristic forms having distinctive profiles and plan views. The sediment forming the overwhelming majority of beaches, however, is of terrigenous origin and consists of sand-size quartz (Figure 12.15).

Definition A **beach** is a body of nonconsolidated sand-size or coarser sediment along the shore of a lake or an ocean. The inner limit of a beach is a boundary on which all agree; it is the upper limit of wave action or the place where the material changes (from beach sediment to bedrock, for example). The outer limit of a beach has been variously defined. Outer boundaries that have been proposed include the low-water line, seaward limit of wave action, and outermost line of breakers. Because we think the distinctive features of beaches result from the effects of breaking waves, we prefer the boundary based on the outermost breakers. In the open sea this boundary lies at a depth that changes with the sizes of the waves, and during great storms may extend down to 10 meters or so.

12.15 **Beach sediment** consisting of sand-size quartz (light-colored layers) and of heavy minerals (black layers). Scarp (about 1 m high) was eroded by sheet flow of water along beach, which created small overhang shown at right. After water creates the overhang, large chunks of sand break off and the scarp retreats. Such scarp retreat is a major process of beach erosion. Jones Beach, Long Island. (Rhoda Galyn)

12.16 **Pointed ripples in sand** created by passage of shallow-water waves, which cause bottom water to move in elliptical orbits. Where back-and-forth movement of water is of equal force it heaps sand into pointed ripples.

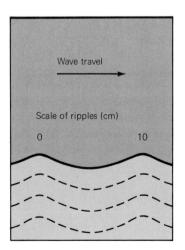
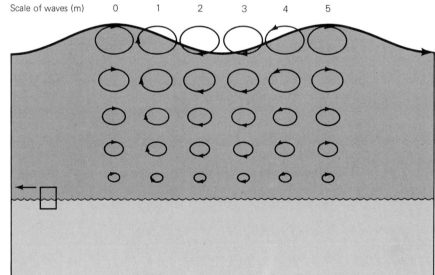

Shallow-water waves and oscillation ripples As deep-water waves approach the shore they eventually reach *wave base*, where they begin to "feel bottom." As a result, the shape of the wave profile changes (crests become narrower and steeper, and troughs wider and flatter) and the orbiting water particles stop moving in circles and begin to travel in ellipses. As the depth of water decreases, the ellipses become flatter and flatter. Waves traveling between wave base (depth equal to half the wavelength) and the point where water depth becomes 1/20 of wavelength are **shallow-water waves.**

Shallow-water waves impart to the bottom water *elliptical-oscillatory* motions that are both *cyclical* and *asymmetric*. The cycle is upward as a crest approaches, landward beneath a crest, downward after a crest has passed, and seaward beneath a trough (Figure 12.16). The upward pulse of the bottom water before the landward surge beneath wave crests can be compared to throwing a tennis ball up into the air before serving it. Because shallow-water waves push the water downward before they cause it to surge seaward, any coarse particles that can be moved at all tend to be driven toward shore.

Shallow-water waves create two kinds of rhythmic features in bottom sediment. These are: (1) small-scale, symmetrical oscillation ripples in fine sand, and (2) large-scale rounded sediment ridges (offshore bars) and intervening rounded troughs (Figure 12.17)

12.17 Offshore bars formed by waves in shallow water seaward of eroding beach, Nantucket Island, Massachusetts. Short stone structures built at right angles to shore are groins. Build up of sand on right side and erosion of sand on left side of each groin shows that dominant direction of sand movement along this shore is from lower right to upper left. (American Airlines)

The Marginal World
Rachel Carson

The edge of the sea is a strange and beautiful place. All through the long history of Earth it has been an area of unrest where waves have broken heavily against the land, where the tides have pressed forward over the continents, receded, and then returned. For no two successive days is the shore line precisely the same. Not only do the tides advance and retreat in their eternal rhythms, but the level of the sea itself is never at rest. It rises or falls as the glaciers melt or grow, as the floor of the deep ocean basins shifts under its increasing load of sediments, or as the earth's crust along the continental margin warps up or down in adjustment to strain and tension. Today a little more land may belong to the sea, tomorrow a little less. Always the edge of the sea remains an elusive and indefinable boundary.

The shore has a dual nature, changing with the swing of the tides, belonging now to the land, now to the sea. On the ebb tide it knows the harsh extremes of the land world, being exposed to heat and cold, to wind, to rain and drying sun. On the flood tide it is a water world, returning briefly to the relative stability of the open sea.

Only the most hardy and adaptable can survive in a region so mutable, yet the area between the tide lines is crowded with plants and animals. In this difficult world of the shore, life displays its enormous toughness and vitality by occupying almost every conceivable niche. Visibly, it carpets the intertidal rocks; or half hidden, it descends into fissures and crevices, or hides under boulders, or lurks in the wet gloom of sea caves. Invisibly, where the casual observer would say there is no life, it lies deep in the sand, in burrows and tubes and passageways. It tunnels into solid rocks and bores into peat and clay. It encrusts weeds or drifting spars or the hard, chitinous shell of a lobster. It exists minutely, as the film of bacteria that spreads over a rock surface or a wharf piling; as spheres of protozoa, small as pinpricks, sparkling at the surface of the sea; and as Lilliputian beings swimming through dark pools that lie between the grains of sand.

The shore at night is a different world, in which the very darkness that hides the distractions of daylight brings into sharper focus the elemental realities. Once, exploring the night beach, I surprised a small ghost crab in the searching beam of my torch. He was lying in a pit he had dug just above the surf, as though watching the sea and waiting. The blackness of the night possessed water, air, and beach. It was the darkness of an older world, before Man. There was no sound but the all-enveloping, primeval sounds of wind blowing over the water and sand, and of waves crashing on the beach. There was no other visible life—just one small crab near the sea. I have seen hundreds of ghost crabs in other settings, but suddenly I was filled with the odd sensation that for the first time I knew the creature in its own world—that I understood, as never before, the essence of its being. In that moment time was suspended; the world to which I belonged did not exist and I might have been an onlooker from outer space. The little crab alone with the sea became a symbol that stood for life itself—for the delicate, destructible, yet incredibly vital force that somehow holds its place amid the harsh realities of the inorganic world.

The sense of creation comes with memories of a southern coast, where the sea and the mangroves, working together, are building a wilderness of thousands of small islands off the south western coast of Florida, separated from each other by a tortuous pattern of bays, lagoons, and narrow waterways. I remember a winter day when the sky was blue and drenched with sunlight; though there was no wind one was conscious of flowing air like cold clear crystal. I had landed on the surf-washed tip of one of those islands, and then worked my way around the sheltered bay side. There I found the tide far out, exposing the broad mud flat of a cove bordered by the mangroves with their twisted branches, their glossy leaves, and their long prop roots reaching down, grasping and holding

From THE EDGE OF THE SEA by Rachel Carson, 1955. Houghton Mifflin Company. Abridged with permission.

Field Trip

the mud, building the land out a little more, then again a little more.

The mud flats were strewn with the shells of that small, exquisitely colored mollusk, the rose tellin, looking like scattered petals of pink roses. There must have been a colony nearby, living buried just under the surface of the mud. At first the only creature visible was a small heron in gray and rusty plumage—a reddish egret that waded across the flat with the stealthy, hesitant movements of its kind. But other land creatures had been there, for a line of fresh tracks wound in and out among the mangrove roots, marking the path of a raccoon feeding on the oysters that gripped the supporting roots with projections from their shells. Soon I found the tracks of a shore bird, prob-

ably a sanderling, and followed them a little; then they turned toward the water and were lost, for the tide had erased them and made them as though they had never been.

Looking out over the cove I felt a strong sense of the interchangeability of land and sea in this marginal world of the shore, and of the links between the life of the two. There was also an awareness of the past and of the continuing flow of time, obliterating much that had gone before, as the sea had that morning washed away the tracks of the bird.

The sequence and meaning of the drift of time were quietly summarized in the existence of hundreds of small snails—the mangrove periwinkles—browsing on the branches and roots of the trees. Once their ancestors had been sea dwellers, bound to the salt waters by every tie of their life processes. Little by little over the thousands and millions of years the ties had been broken, the snails had adjusted themselves to life out of water, and now today they were living many feet above the tide to which they only occasionally returned. And perhaps, who could say how many ages hence, there would be in their descendants not even this gesture of remembrance for the sea.

The spiral shells of other snails—these quite minute—left winding tracks on the mud as they moved about in search of food. They were horn shells, and when I saw them I had a nostalgic moment when I wished I might see what Audubon saw, a century and more ago. For such little horn shells were the food of the flamingo, once so numerous on this coast, and when I half closed my eyes I could almost imagine a flock of these magnificent flame birds feeding in that cove, filling it with their color. It was a mere yesterday in the life of the earth that they were there; in nature, time and space are relative matters, perhaps most truly perceived subjectively in occasional flashes of insight, sparked by such a magical hour and place.

There is a common thread that links these scenes and memories—the spectacle of life in all its varied manifestations as it has appeared, evolved, and sometimes died out. Underlying the beauty of the spectacle there is meaning and significance. It is the elusiveness of that meaning that haunts us, that sends us again and again into the natural world where the key to the riddle is hidden. It sends us back to the edge of the sea, where the drama of life played its first scene on earth and perhaps even its prelude; where the forces of evolution are at work today, as they have been since the appearance of what we know as life; and where the spectacle of living creatures faced by the cosmic realities of their world is crystal clear. 1955

12.18 Refraction of waves in shallow water, viewed obliquely from airplane at low sun angle. Wave crests become narrower and steeper toward upper left, in direction of shallower water. Fire Island, New York. (Bruce Caplan)

Very-shallow-water waves; wave refraction Waves traveling in water ranging in depth from 1/20 of wavelength to about 1.3 times wave height are **very-shallow-water waves.** Waves approaching shore obliquely interact with the bottom and thus experience drag at different times. The parts nearest shore are dragged first. Because of drag on the bottom the crest lines become curved around so that the wave tends to be nearly parallel to the shore when it breaks (Figure 12.18). This change in direction of the crest lines of waves is *refraction*.

Waves that cause the water over which they pass to oscillate, but not to move forward bodily with the wave crests, are known collectively as *waves of oscillation*.

Breakers and surf At a depth of about 1.3 times wave height the water in the wave crests reaches its critical limiting steepness of 1 in 7 and the crests collapse or break, forming **breakers.** The water to build up the heights of incipient breakers rushes toward the crests from both sides, dragging bottom sediment with it and depositing a submerged ridge of coarse material known as a *breaker bar*.

At some point the water from breaking waves may surge landward as a bore or mass of water that moves forward bodily with a wave crest. Such a wave is a *wave of translation*. Waves of translation from breaking waves collectively constitute the **surf.** The shoreward parts of the sea where the surf is active is the **surf zone.**

Nearshore water circulation; rip currents The water that surges landward from the line of breakers into a longshore trough or into the surf zone tends to pile up. Any piled-up water tends to create a current to equalize the level. In this case the easiest direction of flow is parallel to shore as a **longshore current.**

At intervals along the coast, the water flowing as a longshore current in the surf zone (and in troughs lying landward of the breaker zone) piles up to the point where it has enough force to return seaward by cutting through the line of breakers. The narrow, seaward-flowing currents that cut through the breaker zone are **rip currents.** These currents flow fast enough to carry swimmers out from shore. However, as experienced swimmers and surfers know, it is possible to escape from these narrow rip currents by swimming parallel to shore.

Beach profiles; barriers; longshore drift A typical marine beach consists of two chief levels: (1) a lower level, submerged except at low tide; and (2) an upper level, known as a **berm,** which is exposed to the air except during occasional times of high water. In between these two levels is a comparatively even, regular slope, the **beach face** (Figure 12.19).

427
Coastal Features

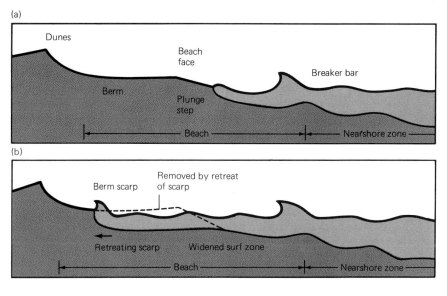

12.19 Profiles at right angles to shore through accreting beach (a) and eroding beach (b). Most waves that approach the shore head-on usually add sand to the beach face, thus widening the berm. When short-period, steep waves approach the beach from extremely oblique directions, however, they create swift currents flowing parallel to shore that cut a scarp. As the water undermines its steep face the scarp retreats and thus the surf zone widens and the berm is eroded. See also Figure 12.15.

Oblique approach of waves creates a parabolic, zig-zag motion of water and sediment on the beach face. This motion consists of oblique swash and straight-down backwash which transports sediment parallel to shore in a process named **beach drifting.** During particularly vigorous beach drifting each wave can shift particles of sediment along the beachface by as much as several meters.

Oblique approach of the waves creates currents that transport sand parallel to shore at many levels. All such transport of sediment parallel to shore is known collectively as **longshore drift** (or longshore transport).

Kinds of beaches Depending on the directions of the waves and the shape of the coast, beaches may be built into distinctive

12.20 Kinds of beaches built by waves and currents along an embayed coast, schematic block.

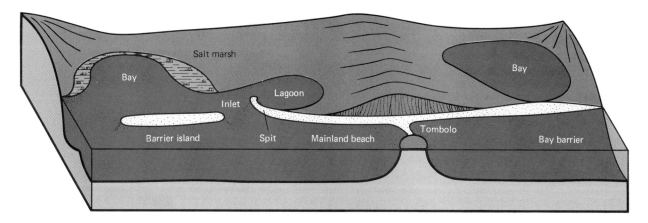

12.21 Stack, sea arch, and cliffs eroded in bedrock by waves. Cliffs of Etretat, France. (Pierre Berger, Photo Researchers, Inc.)

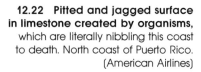

12.22 Pitted and jagged surface in limestone created by organisms, which are literally nibbling this coast to death. North coast of Puerto Rico. (American Airlines)

shapes. Refraction of waves around an island causes sand to accumulate between the island and the mainland or between adjacent islands. Beaches connecting an island to a mainland or to another island are *tombolos*. These and other kinds of features built of sediment along a coast are shown in Figure 12.20.

Bedrock Coasts

Where bedrock and the sea are in contact the dominant geologic process is erosion. The effects of wave action and abundant salt-water spray are added to the usual activities of weathering and mass-wasting described in previous chapters. By removing fallen products of mass-wasting, waves maintain steep slopes and thus encourage further instability of slopes.

The particular effects of the waves involve the impact of water, compression of air in joints as water crashes into the cliffs, and impacts of hurled rocks. Storm waves can drive boulders toward shore with such great intensity that the boulders can richochet off the bottom (or off another boulder) and fly out of the water as if they were mortar shells. One large boulder, weighing about 60 kg, dropped through the roof of the lighthouse keeper's cottage on Tillamook Rock, off the Oregon coast a few kilometers south of the mouth of the Columbia River. How the boulder was propelled is not known, but the result was astonishing. The house was built on top of a cliff about 40 meters above sea level!

Erosion of bedrock by waves creates complex landforms (Figure 12.21). Ultimately these are removed and the result is a wave-cut bench. In tropical regions, where limestone coasts are common, various organisms are effective agents of erosion (Figure 12.22).

Tropical Shorelines and Reefs

Very distinctive features of tropical coasts that are not influenced by large rivers are **reefs**. Modern reefs are built by corals that are hosts to unicellular algae, hence require light for vigorous growth. Surrounding reefs are great bodies of calcium-carbonate skeletal debris originally secreted not only by the organisms that form the reef but also by those that live on and near reefs.

Reefs can grow upward as sea level rises, provided that the rate of rise is not excessive. Kinds of reefs are shown in Figure 12.23.

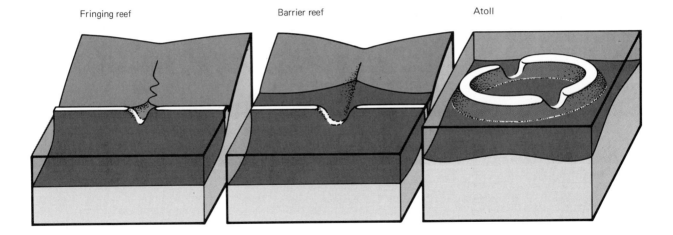

12.23 Kinds of reefs, schematic blocks. (a) Fringing reef, a reef that is attached to a landmass. (b) Barrier reef, a reef built offshore of a landmass and separated from it by a lagoon. (c) Atoll, a circular reef enclosing a lagoon.

Classification and Evolution of Coasts

Coasts can be classified on the basis of three factors: (1) the kinds of material forming them, (2) the kinds of geologic processes operating on them, and (3) the geologic history of the areas involved. Thus it is possible to identify eroding bedrock coasts, barrier coasts, deltaic coasts, faulted coasts, volcanic coasts, and coasts composed of glacial sediments.

If coastal processes were able to act for a long time on a coast that began its history with the submerging of an area that had been shaped by rivers and other processes acting on land, then the sea would alter the morphology. Long-continued operation of coastal activities has been visualized in terms of three stages—youth, maturity, and old age—just as in land sculptured by streams and mass-wasting. In the initial state the coast shows obvious evidence of submergence, including most particularly drowned river valleys. Wave erosion planes off the headlands and fills in the entrances to the bays, establishing a less-indented shoreline. Finally, waves create a wave-cut bench (Figure 12.24). With gradual submergence this bench can widen landward beyond the narrow strip that forms at a fixed sea level.

Marine Sediments

Sediments deposited in the sea, or **marine sediments,** are accumulating in many settings that extend from the coastal areas just described across the continental shelves, along the continental margins, and into the deep-sea basins. In this section we summarize briefly the sediments and morphology of these major parts of the oceans. Understanding of the sediments found in many parts of the modern oceans involves knowledge of what happens when sea level changes.

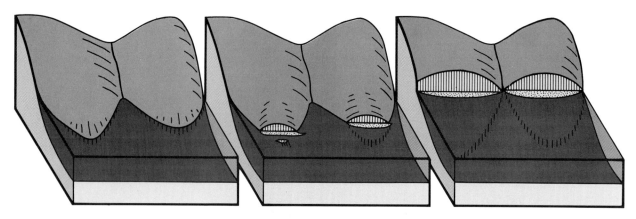

12.24 Stages in wave erosion of submerged coast, schematic blocks. (a) Initial appearance (youthful stage) as water submerges landscape previously dissected by streams and mass-wasting when sea level stood lower. (b) Stage of maturity; by erosion and deposition waves have created a straight shoreline, but shapes of valleys are still prominent. (c) Old age; waves have planed off a nearly horizontal surface beneath the water and have cut so far into the landmass that they have virtually destroyed the shapes of the valleys.

Coastal Sediments

Modern coastal sediments vary according to three chief factors. These are: (1) the kinds and amounts of material being brought from the land to the sea by rivers, by the wind, by glaciers, and by mass-wasting, and eroded from the land by the sea; (2) the kinds of material contributed by the sea, such as sediment brought in by tides and the kinds of skeletal debris secreted by organisms; and (3) the climate, which affects, among other things, precipitation of evaporites, locations of reefs, distribution of glaciers, and quantities of sediment delivered to rivers from inland.

Shelf Sediments

A **continental shelf** is defined as the submerged outer part of a continent that begins at the shoreline and extends to the first prominent change in bottom slope. This change typically takes place at depths of 130 to 200 meters at distances that range from a few kilometers to several hundred kilometers and more from shore.

The sediments on the continental shelves are complex mixtures of materials that were deposited during Pleistocene low stands of sea level, materials that were reworked by the sea as it rose from the outer edge of the shelf to its present level, and sediments that have been deposited since the sea reached its present level.

Sediments of the Continental Margin

The parts of the continental margin we shall consider in this discussion of sediments include continental slopes, submarine canyons, and continental rises. Beyond the break in slope at the outer edge of a continental shelf the gradient of the bottom changes to a steepness of 1:40 (that is depth increases by 1 meter for every 40 meters in a horizontal direction); this is the **continental slope.** The widths of continental slopes vary from 20 to about 100 km; water depths at their lower ends may be several thousand meters.

Cutting into many continental slopes are distinctive, steep-walled **submarine canyons.** Some submarine canyons have walls as high or higher than the Grand Canyon. Some have been cut through rock as hard as granite, and observers using a diving saucer of the French oceanographer Jacques-Yves Cousteau report that they feature overhanging cliffs and irregular walls (Figure 12.25).

The origin of submarine canyons has been a much-debated subject. Now most marine geologists agree that these huge valleys are sculpted by the erosive effects of oscillating tidal currents, submarine avalanches, and turbidity currents. On the steep continental slopes submarine avalanches apparently reach speeds of 10 to 50 km/hr. The sediment load of such avalanches and of turbidity currents is thought to erode the sea floor in much the same way that sand and gravel scour a stream bed. When turbidity currents reach the base of the continental slope, they spread out in the manner of mountain streams in arid regions.

At the foot of the continental slope the gradient of the bottom decreases and becomes as gentle as 1:2000. This gentler slope defines the **continental rise,** a smooth-surfaced accumulation of sediment extending from a continental slope to the deep-ocean floor. The width of a typical continental rise ranges from a few hundred to several thousand kilometers. A comparable feature on land is the Great Plains in the central United States.

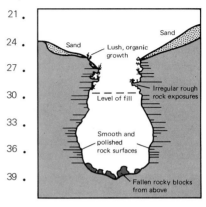

12.25 Steep, locally overhanging bedrock walls of submarine canyon. Jagged profile overgrown by marine organisms (some of which cause erosion) at top contrasts with smoother profile that has been formed beneath level of temporary sediment fill, which periodically moves down the canyon. Sketch based on observations made by geologists working underwater with SCUBA and from research submarines. (Modified from R. S. Dill, 1964)

Deep-sea Sediments

The sediments that are deposited in the deep sea consist of anything that can settle through the water column, flow along the sea floor, or be precipitated out of the bottom water. The chief contributors are the lands, planktonic organisms, volcanoes, the chemicals dissolved in sea water, icebergs (which derive their sediment from land), and the atmosphere (which derives its sediment from the lands, from volcanic explosions, and from falling meteorites). The deep-sea sediments that

432
Oceans and Shorelines

have flowed along the bottom build flat-topped bodies that are named **abyssal plains.** Abyssal plains are widespread on the floors of the major ocean basins as well as in the Caribbean and Mediterranean Seas and in the Gulf of Mexico. The sediments underlying abyssal plains contain numerous layers of *turbidites*, sediments deposited by turbidity currents (see Figure 12.8).

The deep-bottom waters are capable of dissolving skeletal debris composed of calcium carbonate. Hence, in the deepest parts of the ocean calcium carbonate is rare. The chief sediment is brown clay.

Effects of climate The broad-scale distribution of modern deep-sea sediments reflects the Earth's present climate belts (Figure 12.26). The polar regions are characterized by sediments containing abundant cold-water diatoms and considerable quantities of ice-rafted sediment. The tropical regions shallow enough to preserve calcium carbonate contain oozes composed of warm-water planktonic remains.

Effects of sea-level Changes The subject of sea-level changes involves many topics of geologic importance. In Chapter 10 we discussed the changes related to continental glaciers and changes of climate. In Chapter 15 we shall be examining relative changes of sea level in connection with tectonic activity. Here we are concerned with how changes of sea level can affect marine sediments.

12.26 **Major surface water masses** of the sea based on temperature.

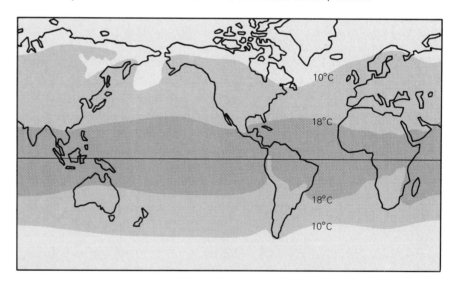

The most obvious changes affecting marine sediments take place when the sea submerges the land. As a result, marine sediments are deposited over a former land surface, and thus bury whatever material underlies that surface. Examples range from bedrock to river-deposited sediments. As the sea submerges a land area the progressively deeper zones migrate landward. Sediments that are closely controlled by depth of water, such as those composed of the remains of certain bottom-dwelling organisms, shift their positions. Sediments from deeper zones overlie those deposited in shallower zones.

The Oceans and You

People have always depended on the sea for food, transportation, recreation, and disposal of wastes. These uses have increased with rising population and more complex industrial activities. In recent years attention has shifted to recovery of natural resources of the sea floor, and this in turn has raised unsolved questions about who should control the sea floor. The critical area of the sea begins at its margin. This fact has been recognized in the United States by the passage of the Coastal-Zone Management Act of 1972.

Coastal-Zone Management

As more and more people want to use coastal lands for more and more purposes, conflicts of usage arise. In recognition of this situation, many states are now preparing studies and plans for use of coastal areas. California has published a detailed analysis and plan for its diverse coastal zone. Texas has published a series of environmental atlases. Other states have comparable projects underway. Outstanding problems of coastal zones include defining exactly where it is, first of all, then describing its characteristics, and ultimately compiling all the present uses and ranking them according to their dependence on the sea and position in the local economic fabric. For example, what can be done about loss of sand from beaches, coastal wetlands, and accommodation for supertankers having drafts of 30 meters, far larger than other ships?

Accommodating supertankers is a necessity that accompanies our dependence on foreign sources of petroleum. Because existing harbors lack the requisite depths and spaces to maneuver, offshore unloading terminals have been built, and others are planned. These require large areas for storage tanks on land. The problem is that no one wants an oil-storage tank for a next-door neighbor, yet all want the petroleum and its products. One plan suggested has been to construct major new petroleum-handling facilities along the shores of Delaware Bay. How the states of New Jersey and Delaware, which border the bay, will settle the conflicts that have arisen over this proposal remains to be determined.

Beach problems   The fundamental difficulty facing most beaches is that sea level is rising on most coasts and the supply of sand has been diminishing. Damming of streams inland has reduced the supply coming from the land, and wave action over the years may have brought ashore all that is readily available from that direction. As a result, large sums of public funds have been spent on placing sand on beaches and on building structures intended to hold the sand in place.

These expenditures have now become very large and the measures have not been as effective as hoped. As a result, a serious debate is in progress over whether such expenditures should be made at all. One philosophy, urged by officials of the National Park Service, is that no structures should be built on beaches, and the public funds should not be spent on restoring structures that have been damaged by storms. A corollary argument is that beaches should be left in their natural conditions and that individuals who have suffered losses should be compensated, but that all further construction should cease. This is a completely different point of view from the traditional "defend-the-coast" responsibility given by Congress to the U. S. Corps of Engineers. These contrasting points of view have not yet been resolved.

Traditionally, communities have sought to block erosion by building *groins*, wall-like stone or concrete structures extending out from shore 30 meters or more and projecting a meter or so above high-tide level. However, sand motion is so complex that groins often achieve unexpected and undesirable results, such as the erosion of other beaches downdrift. A more effective way to replenish eroding beaches is *beach nourishment*, the dumping of sand at a point where natural drifting distributes it to desired locations.

As a result of their perceived needs for growth of urban centers in coastal areas people have greatly altered many estuaries, harbors, and marshes. The commonest kinds of alteration of coastal wetlands have been dredging and filling. The dredging of channels for ships is perhaps the oldest form of coastline modification. These ship channels act as natural traps for sand by slowing longshore currents, so that continuous dredging is needed to keep the channels open. *Filling* is an expression for the dumping of solid waste or other matter onto wetlands to create dry land. About 20 per cent of Manhattan Island is built on "reclaimed" marsh and shallow harbor, as are the principal airports of New York, Newark, Boston, Washington, and San Francisco. Other coastal marshes have been altered for housing developments, small-boat harbors, and waste disposal. In the process, the wetlands have been altered or destroyed. In 1970, it was estimated that 23 per cent of all such areas in the United States had been severely modified, and about 50 per cent had been moderately altered. Only recently have people begun to realize just how important these wetlands are as spawning grounds for many of the fish we eat as well as for other wildlife. With this knowledge, it may not be too late to prevent further damage to our irreplaceable wetlands.

Pollution The sea has been used as a dumping ground for human wastes and for various industrial chemicals, dredge spoil, and debris from torn-down buildings. Ocean dumping finally increased to the point where its undesirable effects became so apparent that the United States Congress passed an Ocean-Dumping Act, prohibiting many previous practices. The amount of dredge spoil and building rubble dumped off shore from New York City is so great that the barges are the chief suppliers of marine sediment in eastern North America.

A major concern about marine pollution centers on oil spills from supertankers and debris from other ships at sea, which traditionally have dumped garbage, bilge wastes, and toilet flushings overboard. Many of these practices are being altered, particularly in coastal areas. Nevertheless, serious concern has been expressed over the extent to which the ocean has already been polluted.

Political and Economic Aspects of the Sea

Serious problems threaten the peaceful sharing of oceanic resources among nations. As these resources become more valuable, countries become more reluctant to share them.

Ownership of the oceans Legally, the sea is a quagmire. Some nations have extended their "ownership" of ocean resources to 320 km from their shoreline; others argue that a country should not be allowed to control such rights as fishing and mineral mining beyond a limit of 20 km, or 5 km. Countries with extensive coasts would generally like to control their own ocean resources. Those with small coasts or none at all protest that an international agency should be created to regulate ownership of ocean resources.

Law of the Sea Conference In 1973, the first Law of the Sea Conference was convened. The intent of the Conference was to draw up a treaty, patterned after that governing uses of Outer Space, that would assure a fair distribution of ocean resources.

A tangle of disagreements crippled the attempt. It may be many years before nations decide whether to divide the riches of the sea peacefully, or to compete for them in traditional warlike fashion.

Chapter Review

1 Among the planets of our solar system the Earth is unique in possessing a continuous body of salt water, which we call the *oceans*. The oceans are a vast part of the biosphere.

2 The oceans cover 70.8 per cent of the Earth's surface. Ocean water is distributed in four deep-ocean basins: *Pacific, Atlantic, Indian,* and *Arctic*. Smaller, partially isolated bodies of ocean water are seas.

3 Some important features of sea water are *salinity, dissolved gases, self-cleansing abilities, temperature,* and *density*.

4 The depth of the water influences the penetration of sunlight, which in turn controls the distribution of many marine organisms, particularly the plants.

5 It is possible that the amount of organic tissue that grows in the sea might be increased artificially. In recent years, marine biologists have suggested a number of new oceanic food sources that could easily be exploited.

6 The movement of water results from the interplay of uneven solar heating and the rotation of the Earth. The Sun is responsible for creating the Earth's climate belts and its major wind belts. The rotation of the Earth is responsible for the *Coriolis effect*. Together, these factors create major surface currents in all oceans.

7 Some important types of waves are *deep-water waves, wind-generated waves,* and *tsunami*.

8 The *tide* is a rhythmic rise and fall of water level that results from astronomic causes and from the ways in which extremely long waves move into and out of basins.

9 A *coast* is a great zone of interaction among the sea, the land, the atmosphere, and some remarkable organisms. A *beach* is a body of nonconsolidated sand-size or coarser sediment along the shore of a lake or an ocean.

10 Sediments deposited in the sea, or *marine sediments*, are today accumulating in many settings that extend from the coastal areas across the *continental shelves,* along the *continental margins,* and into *deep-sea basins*.

Questions

1. In what ways are the oceans important to the geologic cycle?
2. What are the major oceans of the world? How is the ocean area distributed in the Northern and Southern Hemispheres?
3. Define *salinity*.
4. How are sunlight and marine organisms related? Is there life in the lightless zone of the ocean?
5. What are the major causes of the movement of surface water? Define the *Coriolis effect*.
6. What is the difference between wavelength and wave period?
7. What causes tides? Are tides the same throughout the world?
8. In what way can coastal sediments inform us about the geologic record?
9. What kind of sediment normally forms beaches?
10. Describe *submarine canyons* and *abyssal plains*.
11. Explain how the distribution of deep-sea sediments reflects the climate.

Suggested Readings

Carson, Rachel L., *The Sea Around Us*. New York: Oxford University Press, 1961. (Also available in A Mentor Book paperback.)

Engle, Leonard, *The Sea*. New York: Time-Life Books, 1961.

Gross, M. G., *Oceanography*, Second Edition. Columbus, Ohio: Charles E. Merrill Publishing Co., 1971.

Kelley, D. G., *Edge of a Continent: The Pacific Coast from Alaska to Baja*. New York: Galahad Books, 1972.

Laurie, Alec, *The Living Oceans*. New York: Doubleday and Company, Inc., 1973.

Pennington, Howard, *The New Ocean Explorers: Into the Sea in the Space Age*. Boston: Little, Brown and Company, 1972.

Ricker, W., "Food from the Sea," pp. 87–108 from *Resources and Man*. San Francisco: W. H. Freeman and Company, 1969.

Taber, Robert W., and Dubach, Harold W., *1001 Questions Answered About the Oceans and Oceanography*. New York: Dodd, Mead & Company, 1972.

Turekian, K. K., *Oceans*. Englewood Cliffs, N.J.: Prentice-Hall, Inc., 1968.

"The energy released during a great earthquake has been compared to the force of 100,000 atomic bombs."

Chapter Thirteen
Earthquakes and Seismology

440
Earthquakes and Seismology

Like an unseen shark attacking from the depths of the ocean, an earthquake is a fearsome thing because it can strike from below without warning. Today we are learning that earthquakes are not totally unpredictable, and for most of us, scientific explanations about earthquakes have replaced mythology and superstition. But like so many people before us, we must ultimately stand in awe of the furious natural force of an earthquake.

Early Japanese people thought that earthquakes were the result of sudden movements by a giant spider that carried the Earth on its back. Mongolian mythology recorded that the huge, but occasionally unbalanced, supporter of the Earth was a giant mole. Greek and Roman scholars, considerably more scientific, ascribed earthquakes to air escaping from underground caverns or to the collapse of vast subterranean cavities. In the Middle Ages earthquakes were often thought to be divine punishment inflicted upon cities for the sins of the inhabitants.

European scholars began a more scientific treatment of earthquakes after the great tremor that devastated Lisbon, Portugal, in 1755. Seismology in the United States derived more from the disastrous 1906 San Francisco earthquake than from any other single such disaster. That earthquake killed almost 700 people and reduced much of San Francisco to ruins. The fault movement responsible left surface records that were carefully studied by geologists, who produced the first fully documented accounts of the effects of horizontal fault displacement on the Earth's surface.

In the previous three chapters, as we progressed from glaciers to deserts to oceans, our view of the Earth began to grow in scale. We started our study of geology with atoms (which are too small to see), talked about rock specimens we could hold in our hands, and continued to move through topics such as downslope movement and running water. In the next three chapters we shall discuss features that are as unseen as atoms, not because they are so small, but because they are so huge. We can take photographs of the Earth from outer space, but to see it as a whole does not help us enough to understand its interior processes.

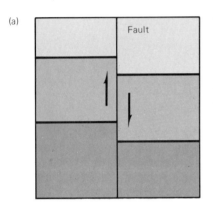
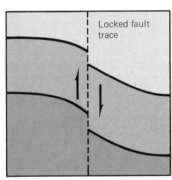
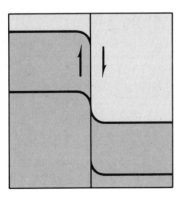

13.1 Sudden slippage on fault creates earthquakes. (a) Fault separating two blocks that have been offset from previous movement, schematic map view with North at top of page. (b) Fault becomes "locked" along dashed line; further movement of right-hand block toward South and of left-hand block toward North bends the blocks, storing elastic energy in bent parts. (c) Fault suddenly becomes "unlocked"; when bent parts snap back into nonbent positions, they release their stored-up energy, which creates an earthquake.

To learn more about the interior of the Earth, scientists depend on increasingly sophisticated instruments and, to some degree, on speculation based on accumulated knowledge. As in our early studies of energy and the atom, we can again employ our basic knowledge of physics and chemistry to discuss waves that radiate deep within the Earth's surface, and to understand some fundamental concepts of energy and motion.

Earthquakes and Seismic Waves

We have said that the Earth is not a static body, but is changing, moving, and full of energy. When parts of the Earth move, the rocks are subjected to large forces which cause **deformation**. When rocks are deformed, their volume, shape, or both shape and volume change. The changes in shape or volume which result from deformation are known as **strain**. When accumulated strain is suddenly released it creates kinetic energy which travels through the Earth as **seismic waves**. Other names for seismic waves include **tremors**, if gentle, and **shocks**, if powerful.

What is an Earthquake?

What we call **earthquakes** are tremors that we can feel. The earthquake itself is the principal event in a sequence of seismic waves. This sequence of waves begins with precursory tremors that become increasingly large, known as **foreshocks**, and ends with waves that begin to fade after the shock of the earthquake itself, known as **aftershocks**. Many earthquakes are related to motion along faults. At some times the opposite sides of a fault become locked together and both bend as further motion takes place (Figure 13.1). Eventually, the faults become unlocked and snap out of their bent positions. Such motion creates an earthquake.

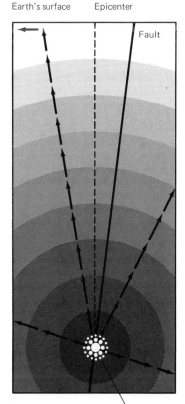

13.2 Energy released at earthquake focus, 20 km beneath the surface, radiates outward in all directions. Paths of waves in four directions shown by short arrows; fronts of waves by arcs of circles. (Where they travel in uniform materials wave fronts in three dimensions define a series of concentric spheres.) Waves that reach the surface change direction and travel along it (indicated by short arrow at top left). Waves traveling within the Earth are body waves; those traveling along the surface are surface waves. The point on the Earth's surface directly above the focus, the epicenter, is where the body waves first reach the surface.

Earthquakes and Seismology

The place where the energy is released is the **focus,** and the place on the surface of the Earth directly above the focus is the **epicenter** (Figure 13.2). The depth of the focus of an earthquake can vary from approximately 5 km to 700 km. Systematic mapping of these depths of focus show a distribution throughout the world that seems to be related to the location of plates. (See Figures 13 and 14 in the Introduction.)

The energy released during a great earthquake (such as the 1906

13.3 Seismograph recording drum. As the drum revolves beneath it, the stylus is driven back and forth by seismic signals, and a thin layer of carbon is scraped away. Other types of seismographs record seismic activity on graph paper. (Hawaiian Volcano Observatory, U.S. Geological Survey)

13.4 Seismogram of a typical high-frequency earthquake recorded in the summit region of Kilauea Volcano. The predominant frequency of this event is about 10 cycles per second and its duration about 1½ minutes. Similar earthquakes related to times of volcanic activity occur in swarms of many hundreds per day. (Hawaiian Volcano Observatory, U.S. Geological Survey)

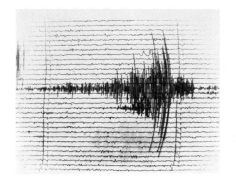

Table 13.1
SCALE OF EARTHQUAKE INTENSITIES WITH CORRESPONDING MAGNITUDES

	Modified Mercalli Intensity Scale	Corresponding Richter Magnitude
I	Detected only by seismographs.	
II	Felt only by a few people, usually on upper floors of buildings.	3.5
III	Vibrations similar to passing truck. Felt by people at rest. Duration long enough to be estimated.	4.2
IV	Shaking noticed by some people outdoors and many indoors. Crockery rattles, standing vehicles rock.	4.3
V	Felt by nearly everyone; many sleepers awakened. Shaking of furniture and beds. Buildings and other tall objects may be disturbed. Pendulum clocks may stop.	4.8
VI	Felt by all. Some windows break, plaster may crack, chandeliers swing. Heavy furniture may be moved. Damage slight.	4.9–5.4
VII	General alarm; everyone runs outdoors. Little damage to well-constructed buildings. Noticed by people driving cars.	5.5–6.1
VIII	Panel walls thrown out of frame structures. Fall of chimneys, factory stacks, monuments, walls. Heavy furniture overturned. Sand and mud ejected in small amounts.	6.2
IX	Buildings shifted off foundations. Ground cracked conspicuously. Underground pipes broken.	6.9
X	Most stone buildings destroyed. Bridges and solid wooden buildings badly damaged. Railway lines bent. Slope failure on steep slopes. Water splashed over banks.	7–7.3
XI	Few buildings remain standing. Bridges destroyed. Broad fissures in ground. Underground pipes completely out of service. Earth slumps and land slips in soft ground.	7.4–8.1
XII	Damage total. Waves seen on ground surfaces. Lines of sight and level distorted. Objects thrown into air.	Greater than 8.1

California earthquake) has been compared to the force of 100,000 atomic bombs. To seismologists, the most meaningful way to describe earthquakes is not in terms of atomic bombs or tons of TNT, but in terms of ground motion.

The first crude instrument for recording the waves passing during an earthquake was used in Italy in 1841. Modern descendants of this instrument, called **seismographs** (Figure 13.3) are still being refined, but the principle remains the same. The recording of a seismograph is a **seismogram** (Figure 13.4). The study of seismic waves and earthquakes is **seismology** (from the Greek *seismos*, for earthquake).

The amount of ground motion is rated by a scale devised in 1935 by Charles F. Richter of the California Institute of Technology. The **Richter Magnitude Scale** (Table 13.1), as it has come to be known, is logarithmic; that is, the amount of energy released in an earthquake of magnitude 6 is 10 times greater than a magnitude 5 earthquake. An earthquake of less than magnitude 2 is usually not felt by humans; any earthquake with a Richter magnitude of 6 or more is considered a major quake.

Early seismologists used quite a different scale, which defines earthquakes by a subjective assessment of their damage and other observable effects. First worked out in the 1880s, it was revised in 1902 by the Italian seismologist Giuseppe Mercalli and then modified again in 1931. This modern version, the **Modified Mercalli Scale,** runs from Roman numeral I through XII, and is still sometimes used along with the Richter scale (Table 13.1).

Some Historic Earthquakes

Most earthquakes do only nuisance harm, such as stopping pendulum clocks, setting off burglar alarms, and rattling tableware. But about once a year a "great" earthquake—greater than magnitude 7.0— happens near enough to a population center to wreak widespread devastation (Table 13.2).

In addition to damaging structures and disrupting people's lives, earthquakes trigger other natural events which themselves may be lethal. Earthquakes can cause slope failures, tsunami, and fires. During the 1964 Alaska earthquake, shock-induced slope failures caused major damage in Anchorage and other areas, and tsunami accounted for more damage in coastal regions than actual "quaking."

An indirect, but severe, effect of earthquakes is fire. Perhaps the most fearsome example of the destructive power of an earthquake fire in the United States followed the 1906 San Francisco earthquake. (See the Field Trip on pages 454–455.) Although the earthquake spent its fury in less than a minute, it started fires, by means of overturned stoves, bared electrical wires, broken gas pipes, and crumbled chimneys, that lasted for days. Movement along the San Andreas Fault broke water mains and levelled pumping stations in San Francisco, crippling fire-fighting efforts.

Our examples of historic earthquakes are Lisbon, 1755; New Madrid, Missouri, 1811-1812; Alaska, 1964; and Managua, Nicaragua, 1972.

Lisbon, Portugal, 1755

The Lisbon earthquake of 1755 aroused tremendous interest because it was the first great European earthquake in recent centuries (Figure 13.5). The shocks lasted six or seven minutes. Within a few minutes all large buildings were destroyed, and half of the houses in the city were ruined. Because the quake occurred on All Saints' Day, November 1, many of the 30,000 fatalities happened when churches tumbled down on the Holy-Day worshippers. (Some reports estimate that one-third of the total population of 235,000 was killed.) Tsunami reached Ireland and even Antigua in the West Indies, 5600 km across the Atlantic.

After the earthquake, Portugese priests documented their observations, and their records, still preserved today, formed the first systematic attempt to describe an earthquake and its effects. The earthquake also prompted the scientific investigation of "elastic waves" by Robert Hooke, British physicist and mathematician.

13.5 Devastation during Lisbon earthquake of November 1, 1755 shown in old engraving (artist not known). (New York Public Library)

New Madrid, Missouri, 1811–1812

Just after two o'clock in the morning of December 16, 1811, an earthquake centered near New Madrid, Missouri, devastated much of the Mississippi River Valley, and shook the Earth's crust with such violence that it awakened people in cities as far away as Pittsburgh, Pennsylvania, and Norfolk, Virginia. In Louisville, Kentucky, about 400 km from the epicenter, chimneys were toppled. The three largest shocks in the series were felt in Quebec, on the Atlantic Coast, and in the Rocky Mountains. The shaking continued on and off through March 1812, and aftershocks were still felt five years later.

In some places the land surface was elevated more than 3 meters. A lake formed by the St. Francis River was turned into dry land when the earthquake heaved out all the water. Large fissures, too wide to be traversed on horseback, opened in the soft alluvium. Some regions sank below the water table, turning good farmland into swamp.

Table 13.2
MAJOR WORLDWIDE EARTHQUAKES

Date	Location	Death Toll	Richter Magnitude
1456	Naples, Italy	30,000	
1556	Shensi, China	830,000	
1716	Algiers, Africa	20,000	
1755	Lisbon, Portugal	30,000+	
1759	Baalbek, Lebanon	20,000	
1783	Calabria, Italy	50,000	
1891	Mino-Owari, Japan	7,300	
1899	Yakutat, Alaska		8.6
1905	Kangra, India	19,000	8.6
1906	Andes of Colombia/Ecuador		8.6
1906	Valparaiso, Chile		8.4
1908	Messina, Italy	100,000	
1911	Tien Shan, China		8.4
1915	Avezzano, Italy	30,000	
1920	Kansu, China	180,000	8.5
1923	Kwanto, Japan	140,000	8.2
1933	Sanriku, Japan	3,000	8.5
1939	Concepción, Chile	25,000	8.3
1950	North Assam, India	1,500	8.6
1960	Chile (three major shocks)	10,000	8.3–8.9
1964	Prince William Sound, Alaska	130	8.6
1972	Managua, Nicaragua	10,000	5.7
1974	Pakistan	5,200	6.3
1975	Ankara, Turkey	2,400	6.8
1975	Hilo, Hawaii	1	7.3

In comparison with many other famous earthquakes, the quake at New Madrid appears to have been neither one of the most dramatic nor the most destructive. It was one of the largest in magnitude, however, and it is important for us to remember that earthquakes can happen, and have happened, in the United States in areas other than the west coast and Alaska.

Prince William Sound, Alaska, 1964

About 7 per cent of all the seismic energy on the globe is released in Alaska. In this century, more than 60 Alaskan earthquakes have exceeded magnitude 7, and seven earthquakes have exceeded magnitude 8. The greatest of these jolted the state on March 27, 1964, killing 114 people in Alaska and causing $350 million worth of damage. Twelve more persons were killed in Crescent City, California, and four people in Oregon perished in a tsunami triggered by the earthquake.

Severe slope failures and slumps destroyed business and residential areas in the principal city of Anchorage (Figure 13.6). Railroad lines twisted like string, highways crumpled, gas lines ruptured, and the airport control tower collapsed. Soft, watery clay beneath the Turnagain Heights section of Anchorage failed, and carried houses partway down a mudflat toward Cook Inlet (see Figure 7.22). The main tremor lasted three to four minutes, causing significant damage to ground and structures throughout an area of 130,000 square kilometers.

13.6 Damaged buildings on Fourth Avenue near C Street, Anchorage, Alaska, during Prince William Sound earthquake of March 27, 1964. Most of the damage shown here resulted from slope failure as clay underlying the center of the city moved out from under the depressed area in the center of the view. (U.S. Geological Survey)

The magnitude of the quake on the Richter scale was 8.6, probably the largest of this century, and one of the largest in history. The main jolt was followed by 10 or more tremors during the next 24 hours. In the ensuing 69 days, about 12,000 additional aftershocks greater than magnitude 3.5 were recorded. Most of the seismic activity took place along a linear zone aligned in a northeast-southwest direction centered about 120 km east-southeast of Anchorage. To the west of this zone of activity, the surface within an area of about 78,000 square kilometers was lowered as much as 2 meters; marine shellfish are now growing on submerged spruce trees. To the east, an area of as much as 128,000 square kilometers was raised, in some places as much as 10 meters. Docks were elevated beyond the reach, except at the highest tides, of the boats that formerly used them.

Coastal towns and villages, smashed by the tsunami, suffered the most. The waves wiped out whole fishing fleets, harbor facilities, and canneries. Tsunami destroyed or damaged the waterfronts of Seward, Whittier, Homer, Cordova, and Kodiak (Figure 13.7).

The earthquake spread destruction even beyond Alaska. The sudden displacement of the sea floor generated tsunami that struck along the coasts of British Columbia, Washington, Oregon, and California (Figure 13.8). Small waves were recorded even as far away as Hawaii, Japan, and Antarctica.

13.7 Damage along shorefront at Kodiak, Alaska, created by tsunami generated by sudden displacement of ocean floor during Prince William Sound earthquake. (Official U.S. Navy photograph)

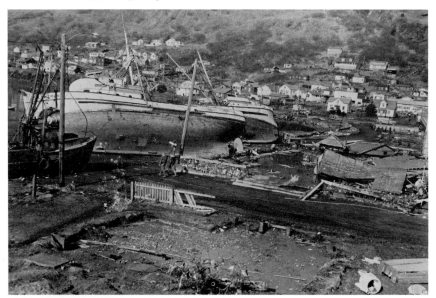

13.8 Travel-time map for tsunami generated by Prince William Sound earthquake. Wavelengths of tsunami are so large that even in the deep-ocean basins they travel as very shallow-water waves. Therefore, speeds of tsunami are controlled by the square root of the Earth's gravitational acceleration times depth of water.

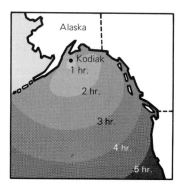

13.9 Rubble of collapsed buildings in Managua, Nicaragua, photographed on December 27, 1972, four days after the earthquake. (UPI)

Managua, Nicaragua, 1972

Nicaragua, the largest country in Central America, is primarily an agricultural region, whose farmers understandably cluster along stretches of the fertile volcanic soil near Lake Managua. At the southern end of the lake lies the capital city of Managua, where more than 300,000 people share a balmy climate, and picture-postcard views of the lake and the surrounding volcanoes.

Two days before Christmas in 1972, the gala nighttime street dancing and holiday parties were interrupted by a series of six seismic tremors that rattled and bounced the city for two hours, destroyed or damaged 75 per cent of the city's buildings, and killed some 10,000 people. All three city jails collapsed, crushing 80 prisoners and allowing another 400 to slip away into the night.

The quake came as no surprise to seismologists—or to Managuans who remembered earlier seismic disasters. The city shook violently in 1885, and again in 1931, when 1450 died. Along with Alaska, California, and some other border regions around the Pacific Ocean, Nicaragua is located in one of the most seismically active areas in the world. Those who decided to rebuild the capital on the ruins left by previous quakes did so more on the basis of faith than of geologic wisdom.

Striking photos taken recently by satellites show why faith was not enough. The city lies along a line of threatening volcanoes. At its very heart is the volcanic Lake Tiscapa, and beneath it lies a porous foundation of pyroclastic debris, lapilli, and fine tephra. When this foundation was shaken by the moderate (magnitude 5.7) 1972 quake, buildings collapsed like jackstraws (Figure 13.9). "Construction," reported a geologist who investigated the damage, "was infested with bad detailing and materials." As a result of such poor building practices, property damages exceeded $280 million.

Four years after the 1972 earthquake Managua was still not close to being rebuilt, and the original estimate of $7\frac{1}{2}$ to 10 years had been revised to 10 to 15 years. But many people are worried more about the long-term effects of rebuilding rather than current delays. Although the new Managua is being built farther from the heart of the most dangerous fault area, it will still be close enough to the fault line to suffer major damage if another earthquake strikes. And so the $280 million rebuilding program continues amidst the same fears of future catastrophe.

Seismology and Seismic Waves

Now that we have looked at some examples of earthquakes, we should discuss seismology and seismic waves in order to understand the details of the waves and their effects on the surface of the Earth. Later we will see how seismic waves can help us predict earthquakes.

Types of Seismic Waves

The seismic waves that race through and around the Earth during an earthquake (a typical speed is 25,000 km/hr) are complex, and may fall into half a dozen or more classes. The most common types have been labelled *P*, *S*, and *L* waves (Figure 13.10).

P, or **compressional waves,** is the seismologists' name for sound waves. Such waves alternately push and pull particles of rock or other material through which they travel. For this reason, *P* waves are sometimes called "push-pull" waves. The back-and-forth motion is along the travel path of the waves. The *P* is an abbreviation for "pri-

450
Earthquakes and Seismology

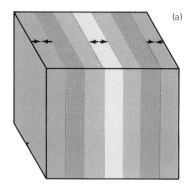

13.10 Particle movements created by passage of various seismic waves traveling from left to right. Schematic blocks. (a) In P waves, particles oscillate back and forth in direction of travel. (b) In S waves, particles oscillate back and forth along lines at right angles to direction of travel of waves. (c) In one kind of surface waves particles move in ellipses that lie in a vertical plane parallel to direction of wave travel. Motion around ellipses is in opposite direction of wave travel beneath crests and in same direction of wave travel in troughs.

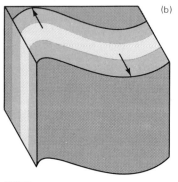

mary." These are the fastest waves, and thus are the first to be recorded on a seismogram.

S, or **shear waves,** tend to displace particles at right angles to the direction in which the waves travel. In crude diagram form, S waves resemble a rope that is fixed at one end and shaken at the other. Because they travel through the main body of the Earth, P and S waves are also known as body waves.

L, or **long waves,** travel along the surface of the Earth. L waves cause the strongest ground motion during an earthquake and accomplish most of the damage.

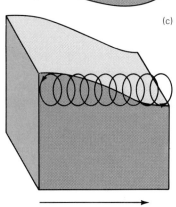

Direction of waves

The Scientific Use of Seismic Waves

If earthquakes generated only one kind of seismic wave, or if the different kinds of seismic waves all traveled with the same speeds, knowledge of them would not be of much value to seismologists. Fortunately, the behaviors of these waves differ so much that seismologists can use these variations to locate earthquake epicenters, to distinguish liquids and solids underground, and to infer the densities of rocks. In later sections we will look at other useful applications of seismic waves.

Locating earthquake epicenters We have mentioned earthquake epicenters, but we have not discussed how they can be located. The way seismologists locate epicenters is very much like the way one estimates how far away a bolt of lightning has struck. The lightning creates two kinds of waves, light and sound (thunder). These waves begin at one point and travel outward at different speeds. No matter what the distance, the light reaches us almost instantly. The sound waves travel slower than light waves, and there is a time lag between the flash of lightning and the sound of thunder. The farther away the storm, the greater the gap between lightning and thunder. (The time lag is about 3 sec/km.)

13.11 Graph of time versus distance for P waves and S waves. (K. E. Bullen, 1954, Seismology, London, Methuen and Company, p. 35)

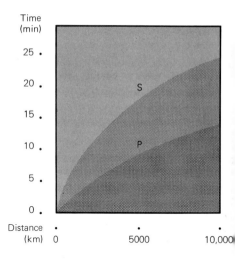

The difference in speeds between *P* and *S* waves is the most useful way to locate an earthquake epicenter (Figure 13.11). *P* waves travel about 1.75 times as fast as *S* waves. The time interval that elapses while only *P* waves are being recorded until *S* waves arrive is a direct function of distance from station to epicenter. If *P* waves arrive 2 minutes, 41 seconds before *S* waves, for example, seismologists know that the distance from the station to the epicenter is 1600 km; if the difference is 6 minutes, 27 seconds, the station-epicenter distance is 4800 km.

The station-epicenter distance defines the radius of a circle which can be drawn with the station at the center; the epicenter lies along the outer rim of the circle. When the circles are drawn from two stations they intersect at two points, one of which is the epicenter. We can tell which of these two points is the epicenter by drawing a circle from a third station, which will pass through just one point, the epicenter (Figure 13.12). This method of locating earthquake epicenters has made it possible to determine the distribution of earthquakes, which will be discussed below.

Seismic waves in liquids and solids Another contrast in wave behavior is of great usefulness to geologists. *P* waves pass through matter in all three of its states: solid, liquid, and gas. By contrast, *S* waves travel only in solids; they do not pass through fluids. If *P* waves from an earthquake, but not *S* waves, are recorded at a given station, we know that a gas-filled or liquid-filled space lies between the focus and the seismograph.

Speeds of waves and densities of materials The speeds of both *P* and *S* waves are determined by the kinds of material through which they travel. In general, the speeds of body waves increase with the density of the material. If seismographs at two separate stations

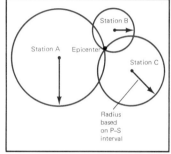

13.12 Locating earthquake epicenter by finding point of intersection of three circles drawn around three seismic recording stations, using as radius for each circle the distance based on the time elapsed between the first arrival of P waves and that of S waves.

receive P waves which have traveled at different speeds from a common focus, then we can conclude that the density of the material along the two routes must be different.

The Distribution of Earthquake Epicenters

We have shown how earthquake epicenters can be located. Now we will discuss the formal organization of a worldwide network of stations known as the World Wide Standardized Seismograph Network (WWSSN), which was established in 1961 by the U.S. Coast and Geodetic Survey. (All United States governmental involvement in the study of earthquakes has now been transferred to the U.S. Geological Survey.) One of the prime reasons for the development of the WWSSN was to monitor underground nuclear blasts throughout the world. Presently the WWSSN has 116 stations in 61 countries.

The first worldwide network of epicenter locations was mapped more than 75 years ago by an enterprising Scot named John Milne, who took advantage of the global extent of the British Empire by placing simple measuring devices at strategic locations throughout the colonies.

Mapping Earthquake Epicenters

In the early days of seismology, scientists were lucky if they could plot an earthquake epicenter to within 150 km. Nowadays, it is possible to locate earthquake epicenters within 8 km. The plotting of epicenters has shown concentrations of long, linear belts, with relatively quiet areas in between. The three worldwide belts are the Circum-Pacific belt, the Alpine-Himalayan belt, and the mid-oceanic-ridge belt (Figure 13.13).

Prediction of Earthquakes

In trying to predict earthquakes, seismologists are concentrating on the killer quakes and the moment of "snapping" of the faults discussed earlier. Seismologists are trying to find out if the effects of this tremendous accumulation of strain on rocks can be measured. If a series of measurements through time shows when a critical point has been reached, then it may be possible to predict when the snapping point will occur.

Prediction of Earthquakes

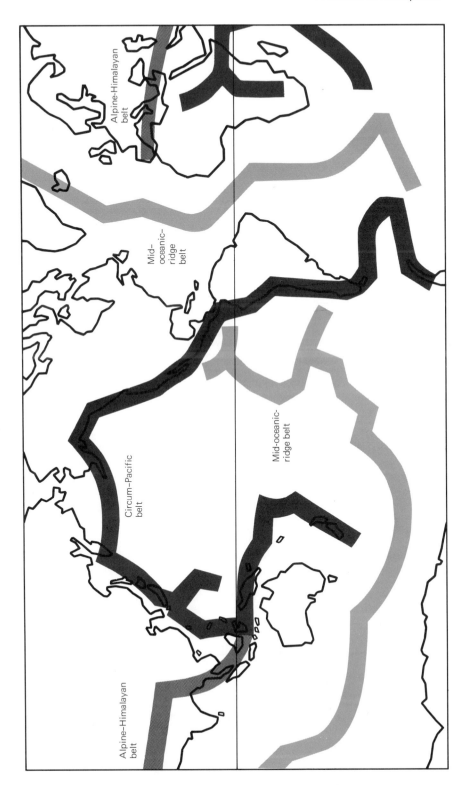

13.13 World map showing belts where most earthquake epicenters are clustered.

Earthquake and Fire: San Francisco in Ruins

The Call-Chronicle-Examiner April 19, 1906

Death and destruction have been the fate of San Francisco. Shaken by a tremblor at 5:13 o'clock yesterday morning, the shock lasting 48 seconds, and scourged by flames that raged diametrically in all directions, the city is a mass of smouldering ruins. At six o'clock last evening the flames seemingly playing with increased vigor, threatened to destroy such sections as their fury had spared during the earlier portion of the day. Building their path in a triangular circuit from the start in the early morning, they jockeyed as the day waned, left the business section, which they had entirely devastated, and skipped in a dozen directions to the residence portions. As night fell they had made their way over into the North Beach section and springing anew to the south they reached out along the shipping section down the bay shore, over the hills and across toward Third and Townsend Streets. Warehouses, wholesale houses and manufacturing concerns fell in their path. This completed the destruction of the entire district known as the "South of Market Street." How far they are reaching to the south across the channel cannot be told as this part of the city is shut off from San Francisco papers.

After darkness, thousands of the homeless were making their way with their blankets and scant provisions to Golden Gate Park and the beach to find shelter. Those in the homes on the hills just north of the Hayes Valley wrecked section piled their belongings in the streets and express wagons and automobiles were hauling the things away to the sparsely settled regions. Everybody in San Francisco is prepared to leave the city, for the belief is firm that San Francisco will be totally destroyed.

Downtown everything is ruined. Not a business house stands. Theatres are crumbled into heaps. Factories and commission houses lie smouldering on their former sites. All of the newspaper plants have been rendered useless. The "Call" and the "Examiner" buildings, excluding the "Call's" editorial rooms on Stevenson Street being entirely destroyed.

It is estimated that the loss in San Francisco will reach from $150,000,000 to $200,000,000. These figures are in the rough and nothing can be told until partial accounting is taken.

On every side there was death and suffering yesterday. Hundreds were injured, either burned, crushed or struck by falling pieces from the buildings, and one of ten died while on the operating table at Mechanics' Pavilion, improvised as a hospital for the comfort and care of 300 of the injured. The number of dead is not known but it is estimated that at least 500 met their death in the horror.

At nine o'clock, under a special message from President Roosevelt, the city was placed under martial law. Hundreds of troops patrolled the streets, and drove the crowds back, while hundreds more were set at work assisting the Fire and Police Departments. The strictest orders were issued, and in true military spirit the soldiers obeyed. During the afternoon three thieves met their death by rifle bullets while at work in the ruins. The curious were driven back at the breasts of the horses that the cavalrymen rode and all the crowds were forced from the level district to the hilly section beyond to the north.

The water supply was entirely cut off, and may be it was just as well, for the lines of Fire Department would have been absolutely useless at any stage. Assistant Chief Dougherty supervised the work of his men and early in the morning it was seen that the only possible chance to save the city lay in effort to check the flames by the use of dynamite. During the day a blast could be heard in any section at intervals of only a few minutes, and buildings not destroyed by fire were blown to atoms. But through the gaps made the flames jumped and although the failures of the heroic efforts of the police, firemen and soldiers were at times sickening, the work was continued with a desperation that will live as one of the features of the terrible disaster. Men worked like fiends to combat the laughing, roaring, onrushing fire demon.

No Hope Left for Safety of Any Buildings

San Francisco seems doomed to entire destruction. With a lapse in the raging of the flames just before dark, the hope was raised that with the use

Field Trip

of the tons of dynamite the course of the fire might be checked and confined to the triangular sections it had cut out for its path. But on the Barbary Coast the fire broke out anew and as night closed in the flames were eating their way into parts untouched in their ravages during the day. To the south and the north they spread; down to the docks and out into the resident section, in and to the north of Hayes Valley. By six o'clock practically all of St. Ignatius' great buildings were no more. They had been leveled to the fiery heap that marked what was once the metropolis of the West.

The City Hall is a complete wreck. The entire part of the building from Larkin street down City Hall avenue to Leavenworth, down from top of dome to the steps is ruined. The colossal pillars supporting the arches at the entrance fell into the avenue far out across the car tracks and the thousands of tons of bricks and debris that followed them piled into a mountainous heap. The west wing sagged and crumpled, caving into a shapeless mass. At the last every vestige of stone was swept away by the shock and the building laid bare nearly to its McAllister street side. Only a shell remained to the north, and the huge steel frame stood gaping until the fire that swept from the Hayes Valley set the debris ablaze and hid the structure in a cloud of smoke. Every document of the City government is destroyed. Nothing remains but a ghastly past of the once beautiful structure. It will be necessary to entirely rebuild the Hall.

Mechanics' Pavilion, covering an entire block, went before the flames in a quarter of an hour. The big wooden structure burned like tinder and in less time than it takes to write it was flat upon the ground.

Confusion reigned. Women fainted and men fought their way into the adjoining apartment houses to rescue something from destruction—anything, if only enough to cover their wives and babies when the cold of the night came on. There was a scene that made big, brave men cry. There were the weeping tots in their mothers' arms wailing with fear of the awful calamity; salesmen and soldiers fighting to get the women out of harm's way through the crowd; heroic dashes in the ambulances and the patrol wagons after the sick and injured and willing men, powerless as the mouse in the clutch of the lion, ready to fight the destroyer, but driven back step by step while their homes went down before them.

It was when the terrible shock of the first big rumbler was passing off, that San Franciscans, sent scurrying into the streets in their nightclothes, turned to the east and south and first saw the pillars of flame that have bred such wicked destruction.

San Jose is Ruined

Passengers arriving on trains from other cities in California bring tales of death and disaster from nearly all of them. The loss of life and property in San Jose was great, it being estimated that nearly 50 people were killed and many more injured. The Vendome Hotel Annex was badly wrecked, between 10 and 15 people being killed there. The St. Francis Hotel there was badly damaged, one aged woman being killed. Hiram Bailey sustained internal injuries. Dr. DeCrow was killed and his wife badly injured. Every business building in the city was demolished to such an extent that nearly all will have to be torn down. The postoffice building was half demolished, the front of the new Court house fell into the street and the entire building is a wreck.

The State Insane Asylum at Agnew is reported demolished, the superintendent and his wife being killed and seventeen nurses injured. Two hundred inmates of the asylum escaped and are roaming over the countryside.

1906

The Search for Predictable Behavior

In their attempts to predict when an earthquake will take place, seismologists have searched for some predictable relationship between strained rocks and some measureable property. These indicators have included changes of speeds of seismic waves, changes in the tilt of the ground, and various other properties.

Changes in speeds of seismic waves The most consistently observed precursors evidently are changes in speeds of seismic waves (Figure 13.14). The first observation of these changes came from Russian seismologists recently working in the Garm region of the Soviet Republic of Tadzhik in central Asia. In examining the seismograms of medium-sized tremors coming from a known source, the Russians noticed that for several months before an earthquake, the speeds of the P waves consistently dropped. They were able to notice this because the $P-S$ interval was changing. Ordinarily, this would not occur, but in this case the strain on the rocks had built up enough to slow down the P waves. This drop was followed by a return to normal shortly before a severe earthquake took place. American scientists soon found comparable peculiarities in their own data, and proposed explanations to account for them.

What do we know about rocks being strained that might account for this remarkable change in the speeds of P waves? It has been seen in laboratory tests that when a rock is strained to the point of breaking it will expand, or dilate, just before it breaks. What is happening is that a network of micro-fractures has developed, and this has increased the volume of the specimen. It is thought that this dilatancy can explain the changes in P waves.

According to the **rock-dilatancy theory,** proposed by Yash Aggarwal, Christopher Scholz, and Lynn Sykes at Columbia University's Lamont-Doherty Geological Observatory, dilation affects the speed of P waves in the following way: Most crustal rock is saturated with groundwater, as discussed in Chapter 9. Any expansion of the rock tends to make it more porous, or less dense, than saturated rock. This is because the air (or possibly steam) occupying the new pores is less dense than water. Eventually, water from adjacent regions seeps into the rock and P waves return to their normal speeds. This new water raises pore pressure beyond its original value, decreasing the frictional resistance along the fault and helping to trigger an earthquake.

Several successful predictions of small quakes have already been made by monitoring the speeds of P waves. The first successful prediction was made in 1973 in New York State. On August 1 a prediction was made that a quake of magnitude 2.5 or more would strike the

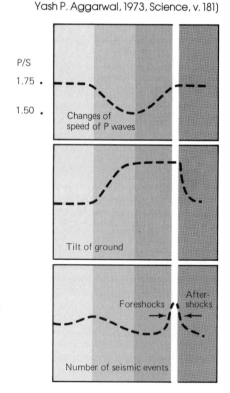

13.14 Graphs of measurement used for predicting earthquakes. Top curve records changing speeds of P and S waves from a known source. The ratio of the speeds of the P waves and S waves decreases by 10 to 20 per cent and then returns to normal just before an earthquake takes place. (C. H. Scholz, L. R. Sykes, and Yash P. Aggarwal, 1973, Science, v. 181)

Blue Mountain Lake region "in a couple of days." Two days later the quake came, at magnitude 2.5.

Changes in the tilt of the ground It is still too early to judge whether rock dilatancy is a sufficiently widespread phenomenon to be used as an earthquake predictor. It does not, for example, seem to apply to deep-focus tremors. But for shallow-focus quakes, dilatancy does seem to explain a number of other precursors. In addition to causing the speeds of P waves to change, the dilatancy of a body of rock would cause uplift and ground tilt. Tilting of the ground has been studied in detail in California and Japan, and has been found to be a reliable indicator of a forthcoming quake. Just before the quake occurs, the tiltmeters record rapid changes. In the part of the San Andreas Fault near San Francisco that has been locked for many years, a closely controlled network of triangulation stations exists (Figure 13.15). These points are surveyed by instruments that use laser beams so that accuracy within millimeters is possible. Early in 1970, many of the points of the triangular network seemed to be moving away from one another. How this change may be related to a forthcoming earthquake is yet to be determined, but it seems that enough strain has accumulated so that the breaking point is not far away.

Pangaea 200 million years ago

Can We Learn to Live with Earthquakes?

Since time immemorial, people have been experiencing earthquakes. In a practical way, what have they done about it? In the remotest parts of the Himalayas the natives will not allow geologists and paleontologists to hammer on rocks for fear of arousing the gods who control earthquakes. In some other areas of the world, supposedly more advanced, attitudes exist about earthquakes that do not differ greatly from those of the Himalayan natives.

People have known for some time how to construct buildings that would withstand the vibrations of an earthquake, and yet some recent United States earthquakes have caused millions (or even billions) of dollars' worth of damage to structures.

In 1921 the great American architect Frank Lloyd Wright designed Tokyo's Imperial Hotel. The sprawling hotel was located in an earthquake-prone area, but Wright boasted that his hotel, designed with a flexible foundation, could survive any earthquake. The test came quickly. In September, 1923, a great earthquake leveled both Yokohama and Tokyo, reducing most parts of both cities to heaps of rubble (Figure 13.16). Initial reports to Wright in Los Angeles declared that his building had been destroyed also, but he laughed at them confidently. A week later, the accurate report finally arrived in a cable from Tokyo:

458
Earthquakes and Seismology

13.15 San Andreas fault and other major faults in California. (Left) Locations of faults and stations of triangular network. Inset shows selected lines near San Francisco where points of stations have been moving farther apart. (Right) Airplane view in 1965 looking northward along the San Andreas fault, which offsets stream at lower center. Carrizo Plains at left, Elkhorn Plains at right. (R. E. Wallace, U.S. Geological Survey)

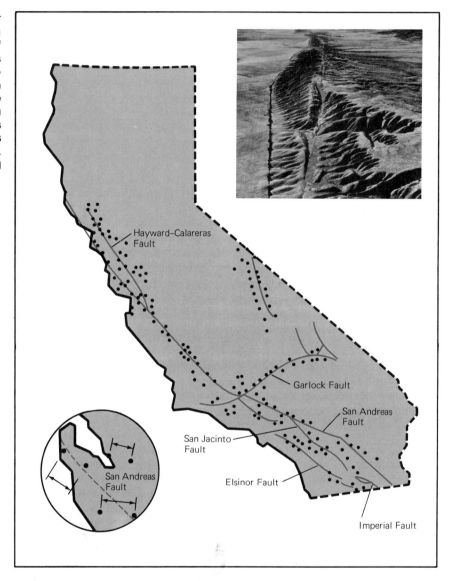

HOTEL STANDS UNDAMAGED AS MONUMENT TO YOUR GENIUS. HUNDREDS OF HOMELESS PROVIDED BY PERFECTLY MAINTAINED SERVICE. CONGRATULATIONS. OKURA.

Obviously, there are lessons to be learned from the past. We can also heed the scientific knowledge of the present. We can try to coexist with earthquakes or we can ignore them. Let us consider briefly the repercussions of these alternatives.

13.16 Ruins of Tokyo, Japan (Byobashiku, one of the business quarters) as a result of the 1923 earthquake and fire. (UPI)

The Effects of Earthquakes

From a human standpoint, the most important effect of earthquakes is destruction of structures and loss of life. With respect to loss of life, the United States has been relatively lucky. Although earthquakes during the past century have caused some $2 billion worth of property damage, only about 1500 Americans have died. (Compare American fatalities with some of the figures listed in Table 13.2.) The great San Francisco earthquake of 1906 occurred early in the morning when few people were exposed to the danger of falling buildings; fewer than 700 persons died. When a severe earthquake struck Long Beach, California, on March 10, 1933, and destroyed many school buildings, school was not in session. At the time of the 1964 Alaska earthquake, a festival was scheduled to have been in progress along the Seward waterfront, which was completely destroyed by tsunami. At the last minute, the festival had been cancelled. Other countries, by contrast, have suffered enormous losses of life, as well as property damage, and seismologists warn that the luck of the United States cannot continue indefinitely (Table 13.3).

460
Earthquakes and Seismology

Table 13.3
DESTRUCTIVE CALIFORNIA EARTHQUAKES

Date	Location	Richter Magnitude	Death Toll	Damage
1812	San Juan Capistrano	7–8	50+	
1857	Fort Tejon	8+	1	
1868	Hayward	7?	30	$350,000
1872	Owens Valley	8+	27	$250,000
1899	San Jacinto	7?	6	
1906	San Francisco	8.3	700	$1 billion
1915	Imperial Valley	6–7	9	$1 million
1925	Santa Barbara	6.3	13	$8 million
1933	Long Beach	6.3	115	$40 million
1940	Imperial Valley (El Centro)	7.1	9	$6 million
1952	Tehachapi (Kern County)	7.7	14	$60 million
1954	Eureka	6.5	1	$2.1 million
1971	San Fernando Valley	6.6	65	$550 million

Source: 1972 Britannica Yearbook of Science and the Future. (Copyright 1971 by Encyclopaedia Britannica, Inc.)

Based on records of previous earthquakes, government seismologists have prepared earthquake-hazard maps of the United States and Canada. These maps have been combined as Figure 13.17.

Why buildings fall down What an earthquake does to a building is to give it a lot of energy it does not need, and is not usually built to accommodate. A building is first hit by the arrival of P waves; this sudden blow is followed seconds or minutes later by the S waves, leading some people to think there have been two earthquakes. Finally, the L waves arrive, often widely spaced, throwing the ground from side to side. The most damaging seismic waves are "choppy" L waves—steep and close together, like storm waves on the sea. Gentle, long-period waves may do little or no damage.

If a building cannot absorb the energy of these arriving waves, it disintegrates. Earthquake engineers attempt to design buildings that can absorb excess energy without falling down, as a tree absorbs energy from the wind without toppling. Steel beams are the best reinforcement for buildings, because they provide a tree-like capacity to sway, stretch, or vibrate instead of breaking. Brittle materials, such as brick, concrete, glass, and adobe, are inflexible and tend to crack easily during earthquakes.

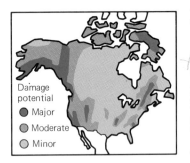

13.17 Earthquake hazard map of North America, showing the areas with the greatest damage potential as a result of an earthquake.

Building regulations and common sense If we know how to construct buildings that can resist earthquake damage, why do buildings keep falling down? We know that the closer a structure is to a

461
Can We Learn to Live with Earthquakes?

fault, the more likely it is to be damaged during an earthquake. Unfortunately, building regulations have been so lax that this warning is commonly ignored. Near San Francisco Bay, for example, many structures, including a Mormon temple; several freeway overpasses; the new, automated BART subway line; and the football stadium of the University of California at Berkeley have been built on or

Table 13.4
WHAT TO DO WHEN AN EARTHQUAKE STRIKES

When an earthquake strikes, for a minute or two the solid Earth may pitch and roll like the deck of a ship. The motion is frightening, but unless it shakes something down on you, it is harmless. Keep calm and ride it out. Your chances or survival are good if you know how to act.

During the Shaking:

1. If indoors, stay indoors. Hide under sturdy furniture. Stay near the center of the building. Stay away from glass.
2. Don't use candles, matches, or other open flames.
3. Don't run through or near buildings where there is danger of falling debris.
4. If outside, stay in the open away from buildings and utility wires.
5. If in a moving car, stop but stay inside.

After the Shaking:

1. Check utilities. If water pipes are damaged or electrical wires are shorting, turn off at primary control point. If gas leakage is detected, shut off at main valve, open windows, leave house, report to authorities, and stay away until utility officials say it is safe.
2. Turn on transistor radio or television for emergency bulletins.
3. Stay out of damaged buildings; aftershocks can shake them down.

Tsunami: The Earthquake's Deadly Companion

Tsunami are a danger on our western and Alaskan coastlines and on Pacific islands. Generated by undersea earthquakes, these great seawaves can be a quake's worst killer. Sooner or later, tsunami visit every coastline in the Pacific, so these cautions apply to you if you live in any Pacific coastal area.

1. If there is an earthquake in your area, leave low-lying coastal sections for high ground.
2. A tsunami is a series of waves, and the first wave may not be the largest.
3. Never go to the beach to watch for a tsunami.
4. Stay out of all danger areas until an all-clear is issued by competent authority.
5. Stay tuned to radio or television during a tsunami emergency. Bulletins issued through local public safety agencies and National Oceanic and Atmospheric Administration (NOAA) offices can help you save your life.

Source: U.S. Department of Commerce, National Oceanic and Atmospheric Administration. (Copies of this table in a convenient card format are available for 10 cents, or $7.50/hundred, from the Superintendent of Documents, U.S. Government Printing Office, Washington, D.C., 20402.)

adjacent to a major active fault zone, the Hayward Fault, which is parallel to the San Andreas Fault. As many as 50 new housing developments are built each year across known fault zones. According to official surveys, there are about 50 schools and hospitals on sites subject to severe earthquake damage. In the summer of 1975, city officials in Los Angeles approved the idea for construction of an underground mass-transit subway system, possibly extending all the way from the San Fernando Valley to downtown Los Angeles. Considering the severity of the 1971 San Fernando Valley earthquake (Figure 13.18), this new underground construction will probably be observed with great interest by scientists and city residents alike.

In the August 22, 1975, issue of *Science*, Robert A. Page, John A. Blume, and William B. Joyner published an article entitled "Earthquake Shaking and Damage to Buildings," adapted from papers presented at the 140th meeting of the American Association for the Advancement of Science in 1974. The following statement is taken from that article: "We consider what the effects of a repetition of the 1906 San Francisco earthquake would be on present-day structures in the San Francisco Bay area. We conclude that . . . many structures, in-

13.18 The San Fernando Valley earthquake in 1971 caused $550 million damage, including extensive damage to public structures and freeways such as this collapsed overpass connecting Foothill Boulevard and the Golden State Freeway. (U.S. Geological Survey)

cluding some modern ones designed to meet earthquake code requirements [which are often outdated] cannot withstand the severe shaking that can occur close to a fault. . . . A repetition of the 1906 San Francisco earthquake today could cause tens of thousands of deaths and billions of dollars of damage. Such potential losses could be reduced through improved engineering and construction practices and through more judicious land utilization."

The 4000-year-old Code of Hammurabi states: "If a builder build a house for a man and do not make its construction firm, and the house which he has built collapses and causes the death of the owner of the house, that builder should be put to death." Few people today would advocate such a powerful judgment. But one way or another, we continue to kill people when we build without regard for the lessons of the past. It can happen here. Indeed it has.

Chapter Review

1. *Seismic waves* are special forms by which kinetic energy is transmitted inside the Earth and along the rocky parts of its surface. Seismic waves are created when strain is released by abrupt movement along *faults*. Parts of some faults occasionally become locked. Further movement bends rocks out of shape, and they eventually snap out of their bent positions. This snapping causes seismic waves and tremors. When we can feel these tremors we call them *earthquakes*.

2. The place inside the Earth where the energy is released is the *focus*, and the place on the surface of the Earth directly above the focus is the *epicenter*. Instruments called *seismographs* record the waves passing during an earthquake. The *Richter Magnitude Scale* and the *Modified Mercalli Scale* are two methods used to express earthquake intensity.

3. About once a year a "great" earthquake—greater than Richter magnitude 7.0—happens near enough to a population center to wreak widespread devastation. Earthquakes can also cause other, equally harmful events such as slope failures, tsunami, and fire. About 7 per cent of all the seismic energy on the globe is released in Alaska. In 1964 the Prince William Sound earthquake in Alaska registered a magnitude of 8.6 on the Richter scale—probably the largest of the century, and one of the largest in history.

4. The seismic waves that race through and around the Earth during an earthquake are complex. The most common types have been labelled *P waves* (compressional waves), *S waves* (shear waves), and *L waves* (long waves). *L waves* cause the strongest ground motion during an earthquake and accomplish most of the damage.

5   The behaviors of different kinds of seismic waves vary so much that seismologists can use these variations to locate earthquake epicenters, to distinguish liquids and solids underground, and to infer densities of rocks.

6   Currently, it is possible to locate earthquake epicenters within 8 km. The plotting of epicenters has shown some concentrations of long, linear belts, with relatively quiet areas in between. The three worldwide belts are the *Circum-Pacific belt,* the *Alpine-Himalayan belt,* and the *mid-oceanic-ridge belt.*

7   In trying to *predict earthquakes,* seismologists are concentrating on the killer quakes and the moment of "snapping" of the faults. Seismologists are trying to find out if the effects of this tremendous accumulation of strain on rocks can be measured. If a series of measurements through time shows when a critical point has been reached, then it may be possible to predict when the snapping point will occur.

Questions

1   Define *earthquake, focus,* and *epicenter.*

2   What methods are used to measure seismic waves?

3   Why was the Lisbon earthquake of 1755 scientifically important?

4   What factors caused the most damage during the 1964 Alaska earthquake? Are these factors typically destructive when they accompany an earthquake?

5   Earthquakes generate different kinds of seismic waves, and these waves all travel at different speeds. How do these differences help seismologists locate earthquake epicenters? What else can seismologists infer from the information they receive from seismic wave differences?

6   Name and locate the three worldwide belts of earthquake epicenters.

7   Has there been any indication that seismologists can predict earthquakes? What methods are being studied in the search for predictable behavior? Describe the *rock-dilatancy theory.*

8   Will an earthquake be more lethal in an urban area than in a rural area of comparable population? If so, why?

9   What sections of the United States are most vulnerable to earthquakes? Explain why certain parts of the country are more susceptible to earthquakes than others.

Suggested Readings   Anderson, D. L., "The San Andreas Fault." *Scientific American,* v. 225, no. 5, pp. 52–66. (Offprint No. 896. San Francisco: W. H. Freeman and Company, 1971.)

Suggested Readings

Brown, B. W., and Brown, W. R., *Historical Catastrophes: Earthquakes*. Reading, Mass.: Addison-Wesley Publishing Co., Inc., 1974.

Clark, S. P., Jr., *Structure of the Earth*. Englewood Cliffs, N.J.: Prentice-Hall, Inc., 1971.

Coffman, Jerry L., and von Hake, Carl A. (Editors), *Earthquake History of the United States*. Washington, D.C.: U.S. Department of Commerce, U.S. Government Printing Office, 1973.

Iacopi, Robert, *Earthquake Country*. Menlo Park, Calif.: Lane Book Company, 1964.

Roberts, Elliot, *Our Quaking Earth*. Boston: Little, Brown and Company, 1963.

Tazieff, Haroun, *When the Earth Trembles*. New York: Harcourt, Brace & World, Inc., 1964.

Thomas, Gordon, and Watts, M. M., *The San Francisco Earthquake*. New York: Stein and Day, Publishers, 1971.

U.S. Department of the Interior/Geological Survey, "Active Faults of California." Washington, D.C.: U.S. Government Printing Office, 1975.

U.S. Department of the Interior/Geological Survey, "Earthquakes." Washington, D.C.: U.S. Government Printing Office, 1974.

U.S. Department of the Interior/Geological Survey, "The San Andreas Fault." Washington, D.C.: U.S. Government Printing Office, 1975.

"The 1950s drew to a close in some confusion, with a great number of new facts in search of a theory."

Chapter Fourteen
Plate Tectonics and the Interior of the Earth

Plate Tectonics and the Interior of the Earth

Before we can talk intelligently about the theories of drifting continents, moving plates, spreading sea floors, and other dynamic activities of the Earth, we must know the basic information about its architecture. In this chapter we will be discussing the internal structure of the Earth and how we discover it. Our purpose is to try to connect theories about dynamic activities with observed behavior of the structure of the Earth. In the next chapter we will examine the large-scale fabric of the Earth's crust.

Because we cannot actually see the interior of the Earth for ourselves, we must depend on theory supported by observation and experimentation. In this sense, we are dealing with a situation comparable to our study of atoms, where we could not see the atoms but could infer their properties—from the way they react to X-rays, for example. Similarly, we have learned about the interior of the Earth from our observations of seismic waves. Until just after the turn of the century the accepted theory was that the interior of the Earth was molten, and that liquid magma escaped through cracks in the solid "crust" of this hot interior (volcanoes).

The Interior of the Earth

The basic technique for X-raying the Earth is to measure the time elapsed between an earthquake and the arrival of seismic waves at a seismograph station.

The first important discovery made through the use of seismic wave behavior was that the Earth beneath the crust was not totally molten, but consists of shells having different properties.

The Crust-Mantle Boundary

The **Mohorovicic discontinuity** is a seismic interface between the crust and the mantle. This interface was discovered in 1909 by an obscure Yugoslav geophysicist named Andrija Mohorovicic. He was studying earthquakes in the Balkan Peninsula when he found that below depths of a few tens of kilometers both P and S waves traveled faster than at shallower depths. The speeds of P waves increased from about 7 km/sec to 8.2 km/sec (Figure 14.1). This discontinuity was later found to be a worldwide phenomenon, and was named in honor of Mohorovicic.

The mantle The same methods used by Mohorovicic could be used to examine deeper parts of the Earth. Other methods for learning

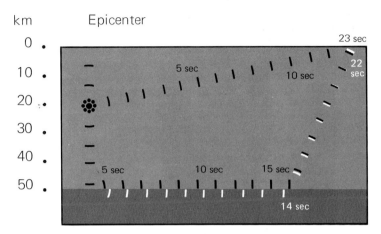

about the mantle include study of the depth distribution of earthquake foci, and examination of exotic blocks thrown out of volcanoes.

In order to look deeper into the Earth, the network of seismic stations must be spread farther and farther. In addition, precise information must be available about the travel times of the various seismic waves. The objective is to find out if any changes in the speeds of the seismic waves take place that can be related to changes in physical properties that occur with changes in depth. The following account is not strictly chronological, but rather is a sequence arranged by depth.

When variations in the properties of the upper mantle were first reported, almost no one accepted the results. A zone was found in which the speeds of the S and P waves did not increase regularly with depth, as they should if the properties of the mantle are changing gradually, but decrease before they increase again. This zone lies in the depth range of about 75 to 200 km (Figure 14.1). Together, the mantle and crust above this zone where the speed of seismic waves reverses constitute the *lithosphere*. The reversal of speeds of seismic waves is taken to indicate the existence of a zone of low strength and easy flow, which has been named the *asthenosphere*.

Another boundary within the upper mantle lies at a depth of 700 km. This marks the lower limit of deep-focus earthquakes.

No pronounced seismic irregularities deeper than 700 km have been found until a remarkable interface is encountered where the S waves vanish and the speeds and directions of the P waves change notably. The effect of this disappearance of the S waves is to create a "shadow zone" for each earthquake, within which no S waves and no direct arrivals of the P waves are recorded (Figure 14.2).

14.1 Seismic behavior of the Earth's mantle. (Upper left) Conditions under which the Mohorovicic discontinuity was discovered. Seismic station at right receives P waves coming directly from earthquake focus 78.4 km distant. These P waves traveled through the Earth's crust at a speed of 7 km/sec and thus reached the station in 14 sec. P waves that traveled vertically downward came to a seismic interface in about 4.2 sec, and then traveled at 8.2 km/sec in material below the interface reaching the station in about 23 sec, approximately 1 sec sooner than the waves that had traveled along the interface in the material of the upper layer. (Below) Variations of the speeds of seismic waves with depth in the Earth.

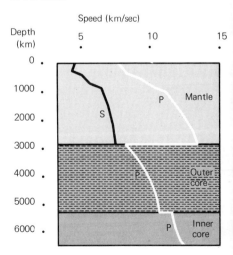

470
Plate Tectonics and the Interior of the Earth

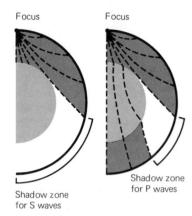

14.2 Shadow zone created by effect of Earth's core on paths of seismic waves from distant sources, schematic sections showing half of the Earth. Dashed lines show paths of waves, which are curved because the properties of the mantle change with depth. The S waves (left) that encounter the Earth's core are stopped, an indication that the outer core is composed of nonsolid material. P waves (right) can pass through core, but their paths are bent.

These changes in seismic-wave behavior indicate that the shadow zone is caused by a pronounced seismic interface at a depth of 2900 km. Above this interface, the material is solid; it transmits both S and P waves. Below this interface, the material does not transmit S waves. The only materials we know at the surface of the Earth that do not transmit S waves are fluids (liquids or gases). We are not certain if material under the great pressures at a depth of 2900 km is exactly like the fluids at the Earth's surface. Whatever its state, this material resembles fluids in not transmitting S waves. Accordingly, we will discuss the material that does not transmit S waves as if it were a fluid. The **mantle** is then defined as the mostly solid portion of the Earth between the crust and a depth of 2900 km.

The seismic interface at 2900 km is known as the **mantle-core boundary**.

The Core of the Earth

The **core** is the smooth, dense sphere that forms the innermost portion of the Earth (Figure 14.3). Because S waves are not propagated through the core, at least part of it must be molten. By the speeds of P waves, and other evidence, the chemical composition is thought to be largely iron, with some nickel. The temperature is probably about 4500°C, and the density of the material about 13.5 times that of water.

Movements of the liquid outer core are thought to generate the Earth's magnetic field, which is inferred to result from the flow of electrons. It seems probable that the liquid in the core moves because of the Earth's rotation, convection of heat, and possibly other sources of energy, but it moves in complex ways that are not understood. The liquid, which is thought to be molten iron, is an electrical conductor. The flow of electrons in an electrical conductor creates a magnetic field.

Challenges to the Established Views of the Earth

Early in the twentieth century geologists were pursuing the implications based on fixed continents, and working within the theoretical framework that had been established. On the local scale within which they were working, they found nothing to challenge the existing theories. On a global scale, however, Eduard Suess, the leading tectonic thinker of the early twentieth century, noticed geological similarities in parts of Africa and Brazil, South America, and speculated that

Challenges to the Established Views of the Earth

the two continents had once been united. Suess did not pursue this theory, and therefore saw no reason to challenge the idea of static continents.

A young German meteorologist named Alfred Wegener (Figure 14.4) became interested in the idea of moving continents on a global scale. Beginning with the fundamental notion that Africa and South America may once have been joined, he proceeded to challenge the established school of thought. According to the then-existing Earth model, continents could not move. Wegener was unable to develop an acceptable new Earth model, so he could never establish a theoretical basis in support of his view of mobile continents.

Alfred Wegener and Continental Drift

Wegener published his first essays on continental drift in 1912. In 1924 an English translation of his now-famous book, *The Origin of the Continents and Oceans*, was published. This book brought Wegener's theory to the attention of the English-speaking world, and initiated a long series of debates and criticisms.

In Wegener's own words, the exact fit of the Atlantic coastlines of South America and Africa "was the starting point of a new conception of the nature of the earth's crust and of the movements occurring therein; this new idea is called the theory of the displacement of continents, or, more shortly, the displacement theory, since its most prominent component is the assumption of great horizontal drifting movements which the continental blocks underwent in the course of geological time and which presumably continue even to-day."

Wegener asserted that many surprising and apparently unrelated facts from geology, climatology, and paleontology could be explained by only one idea: that the continents had, until about 200 million years ago, been united in a single land mass which has since broken apart. He called his supercontinent **Pangaea** (Figure 14.5), or "all world."

Whereas Wegener's compilation of the evidence for continental drift was impressive, it was largely circumstantial. Wegener believed that the continents moved, but he was unable to propose a satisfactory mechanism which could move them. He could not propose a new theory, and according to the existing Earth model the continents could not

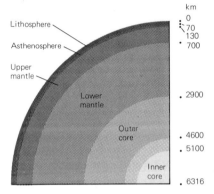

14.3 Cutaway slice of the Earth, showing major boundaries based on behavior of seismic waves. (After U.S. Geological Survey)

14.4 Alfred Wegener, (1880–1930) the German meteorologist and geophysicist who devoted his life to what was a nearly futile attempt to convince geologists that continental drift had taken place. (Wide World Photos)

472
Plate Tectonics and the Interior of the Earth

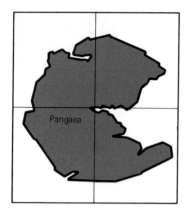

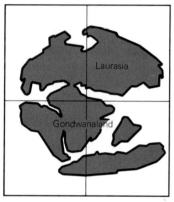

14.5 Pangaea, (top) as reconstructed by Wegener. (Bottom) Continental reconstruction according to Alexander DuToit, involving two major continental masses, a northern Laurasia, and a southern Gondwanaland.

move. The final impasse resulted when Wegener suggested that the ocean floor would have to be younger than the continents, and the opposing scientists argued that the ocean floor was as old as the continents. Wegener's evidence was left intact, his theoretical arguments were at a standstill, and the issues remained unresolved for the most part.

Laurasia and Gondwanaland Among the very few persistent supporters of continental drift was a South African geologist named Alexander DuToit. Whereas Wegener had thought in terms of one supercontinent called Pangaea, DuToit favored a scheme with two large continents, which he called **Laurasia** and **Gondwanaland.** Laurasia stood for the union of the Laurentian rock formations of eastern Canada with what is now Europe and Asia. Gondwanaland signified all the southern continents—Africa, South America, Australia, Antarctica, and India. Gondwana is the name of an ancient kingdom in the Deccan Plateau region of India. The Gondwana sedimentary rocks, like the fossil plants and animals, bore close resemblances to those of other Gondwanaland continents.

Paleomagnetism and the New Look at Continental Drift

For 30 years, continental drift was an inconclusive subject which few scientists took seriously. Suddenly, with the advent of new evidence centered around **paleomagnetism** (the study of ancient magnetic fields) the theory became newly respectable.

Paleomagnetism How do rocks preserve ancient magnetic fields? There are two ways this can happen, and in both cases the key is magnetite. One method involves sediments, and depends on the fact that some particles are magnetic. The second method involves igneous rocks, and depends on the fact that magnetite becomes magnetic only after it has cooled.

As magnetic sedimentary particles of magnetite fall to the bottom of a lake or sea, they become aligned as if they were little compass needles. Assuming that the sedimentary particles do not move after they have settled, they retain a magnetic alignment which is identical to the magnetic field at the time when they settled.

As the magnetite in an igneous rock cools and solidifies, it passes through a critical temperature known as the **Curie point.** For iron at atmospheric pressure the Curie point is about 770°C; for nickel it is

330°C, and for magnetite, 578°C. An increase in pressure lowers the Curie point. Above its Curie point, a substance has no magnetism. As magnetite cools below 578°C the particles become magnetized in a direction parallel to the magnetic field of the time. Rocks preserving evidence of ancient magnetic fields are said to possess **remanent magnetism.**

When beds of Triassic red sandstones were mapped during 1953 in Great Britain, they showed that the remanent magnetism in these ancient rocks defined a Triassic North Magnetic Pole that was 30 degrees from the present pole. The data from the remanent magnetism were thought to be best explained by the idea that since the Triassic Period, about 200 million years ago, England had rotated clockwise by 30 degrees and moved northward thousands of kilometers (Figure 14.6).

Studies of the remanent magnetism of rocks of many ages from many locations indicated that the Magnetic North Pole seemed to have "wandered" from place to place through time, a motion called **polar wandering.** Computations based on polar wanderings reinforced Wegener's idea that the continents had drifted. This unexpected evidence from geophysics removed continental drift from the fringes of science and brought it to the forefront of scientific discussion.

The sea floor and continental drift New advocates of continental drift had little more than Wegener did to offer in the way of a driving mechanism. They still imagined that the continents plowed through the sea-floor sediments. If the continents did plow through the deep-sea floor as suggested, then the deep-sea sediments should have been disturbed, like snow in front of a snow plow. But deep-ocean research found instead undisturbed sediments and jagged bottoms. The mid-oceanic ridge was found to be a center of shallow-focus earthquakes, and had many fractures, suggesting it had been formed by the pulling apart of its opposite sides. Also, if the continents had moved through the oceans, the implication was that they would slide across the ocean bottom and leave a smooth surface behind. Instead of this, the deep ocean floor was found to be jagged and irregularly covered with sediment. According to ocean

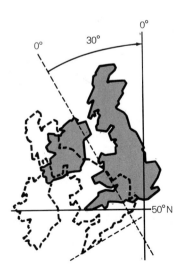

14.6 Clockwise rotation of Great Britain and shift northward from the Triassic Period to the present, based on paleomagnetic measurements. (Smithsonian, January 1975)

Plate Tectonics and the Interior of the Earth

experts, the conditions on the bottom of the ocean did not match the predictions made by the new advocates of continental drift. A theoretical impasse still existed, and only a new theory could break it.

Sea-Floor Spreading

The 1950s drew to a close in some confusion, with a great number of new facts in search of a theory. With most of the old drifters dead and the few young ones afraid to discuss it, the theory itself drifted without a champion of sufficient energy and reputation to lead the fight. But, strong leaders began to emerge in the 1960s.

Harry Hess and the Moving Sea Floor

In a paper published in 1962, Harry H. Hess (Figure 14.7) argued that the ocean floor has been, and still is, moving. Furthermore, Hess proposed that the ocean floor was inexorably locked to the mantle (Figure 14.8). This theory would allow the ocean floor to move along with the mantle without disturbing the ocean sediments. In other words, Hess was saying that instead of the continents plowing through the ocean floor, the ocean floor was moving at a deeper level, taking the continents with it. Hess wrote that along with the vast mid-oceanic ridge system mantle rock is converted into oceanic crustal rock. He thought that this conversion took place at a temperature of 500°C.

Because this new oceanic crust is not piling up anywhere, Hess reasoned that it must be moving away from the ridge and disappearing somewhere. The most likely site of disappearance, said Hess, was the great ocean trenches.

Another possibility to accommodate the continuous addition of new mantle material was that the Earth was expanding. The theory of an expanding Earth is still a viable explanation for drifting continents and, although accepted by only a few scientists, has not been disproved. At the rate of movement that Hess proposed, the ocean floor at any point on the Earth could not be older than about 260 million years—brand new in comparison with land rocks more than three billion years old. The figure also coincided with oceanographic experience: all marine sediments were found to be younger than about 200 million years.

Hess also rejuvenated the idea that convection currents moved the continents—the same idea Arthur Holmes had proposed three decades earlier. Normally, convection currents occur in fluids such as air or water when their molecules are excited by a heat source. A radia-

14.7 Harry H. Hess, (1903–1969), originator of the hypothesis of sea-floor spreading by movement of thick plates along a zone of weakness in the upper mantle. (Orren Jack Turner)

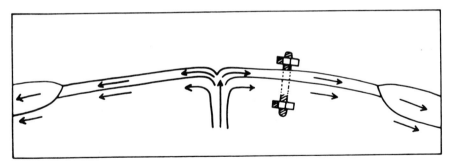

14.8 The Earth's crust firmly attached to the immediately underlying upper mantle; schematic profile and section drawn by Harry H. Hess and published in 1965.

tor produces a convection current, or cell, in a room, as does a flame in a pot of water. The heated fluid rises from the heat source. As it cools, the current flows downward to be reheated, and then rises again (Figure 14.9). In 1928, when Holmes published his idea, convection was thought to be restricted to fluids. The idea of solid mantle rock flowing in currents was considered to be fantastic by most scientists when Holmes first suggested it, but in 1962 it found more support because other evidence favoring continental motion was becoming stronger.

In 1961, Robert S. Dietz proposed the name **sea-floor spreading** to describe the outward movement of new crust away from a mid-oceanic ridge.

14.9 Convection currents as visualized by Arthur Holmes, schematic profile and section through outer part of Earth showing continental block above rising convection cell and ocean basin above descending convection cell. (Redrawn from A. Holmes, 1928.)

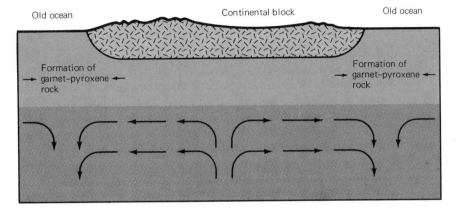

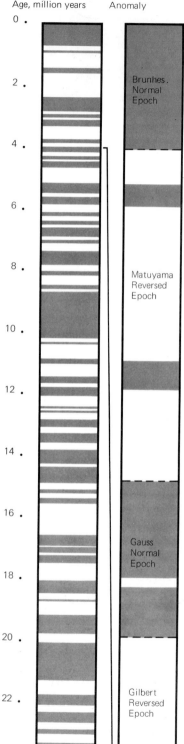

476
Plate Tectonics and the Interior of the Earth

The "Magnetic Tape Recorder"

As we have seen, the Earth's magnetic field reverses itself every half million years or so. This phenomenon of **polar reversals** was first noticed in 1906 by a Frenchman named Bernard Brunhes, who thought at first that his instruments were playing tricks on him. However, in the 1960s researchers on remanent magnetism not only confirmed the information that polar reversals had taken place, but began to establish time scales for the magnetic reversals (Figure 14.10).

Magnetic characteristics of the ocean floor Studies in the 1960s with a sensitive magnetometer towed behind a ship recorded a bafflingly consistent pattern of magnetic stripes on many parts of the deep-sea floor. Half the stripes were **positive anomalies**—that is, zones of sea floor with magnetic strength higher than the Earth's average value. The other half were **negative anomalies,** with readings lower than the adjacent stripes. Drawn on a map, the stripes made a pattern as distinctive as that of the flank of a zebra (Figure 14.11).

The Vine-Matthews hypothesis It was inferred that the positive anomalies were produced because the polarity of the magnetite which had been magnetized in the rocks as the temperature dropped past the Curie point was the same as the Earth's present field. By contrast, negative anomalies were inferred to be present over rocks having magnetite that in the same process had been magnetized so that its polarity was opposite to the Earth's.

In 1962, Frederick J. Vine and Drummond Matthews proposed that the magnetic anomalies around the mid-oceanic ridge were symmetrical. In 1963, they further proposed that if the sea floor had been spreading continuously and magnetic polarity reversals had occurred now and then, a magnetometer survey directly over a ridge crest should reveal a pattern of anomaly stripes extending symmetrically away from the ridge in both directions. The width of the stripes should equal the rate of sea-floor spreading multiplied by the time between reversals. This proposal has since become known as the **Vine-Matthews hypothesis.**

14.10 Magnetic-polarity time scale based on synthesis of paleomagnetic information from deep-sea sediments and continental volcanic rocks and on various fossils. (W. F. B. Ryan)

Other Implications of Sea-Floor Spreading

The two important implications of the hypothesis of sea-floor spreading are (1) the predicted ages of the sea-floor crust and sediments, and (2) the concept of fracture zones and transform faults on the deep-sea floor.

Predicted ages of sea-floor crust and sediment One of the attractive features of the Vine-Matthews hypothesis for the explanation of the magnetic anomalies of the deep-sea floor is that it offers a means of predicting the age of the crust underlying the sea floor at any point. The belts of magnetic anomalies are simply counted outward, and assuming that each anomaly represented a corresponding time of magnetic-polarity reversal, the ages of the sea-floor crust can be assigned by matching the number of the anomaly with the time-scale of magnetic polarity reversals. It is important to emphasize that this procedure absolutely depends on the notion that spreading is continuous, and that no anomalies are skipped. There is no way to decide which anomaly is which; they all look more or less alike. Therefore, they can be used for certain dating only if all are present and simple counting suffices to convert anomaly to geologic age. When this was done, maps were made of the predicted age of the crust underlying various parts of the ocean floors (Figure 14.12).

14.11 Map of magnetic anomalies in Reykjanes Ridge, SW of Iceland, where pattern of symmetry is best displayed. Black areas represent positive anomalies, thought to mark places where the magnetic polarity of the magnetic particles in the oceanic crustal rocks is the same as the Earth's present polarity. White areas designate negative anomalies, inferred to result from magnetic polarity in oceanic crustal rocks that is opposite to Earth's present polarity. (J. R. Heirtzler, X. LePichon, and J. G. Baron, 1966)

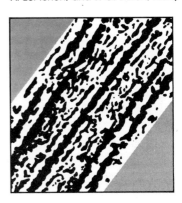

14.12 Ages of deep-sea floor. Rate of over-all spreading may be seen by examining several checkpoints. The center of the ridge (lightest portion) below Australia indicates a spreading of 40 million years; from California to the middle of the Pacific represents 80 million years; to the center of the mid-Atlantic ridge indicates a spreading of 70 million years. (Geological Society of America)

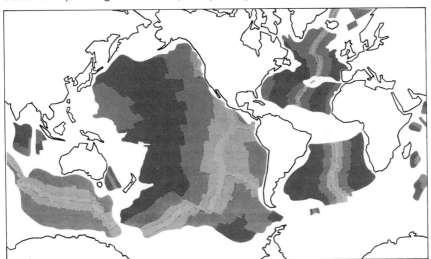

14.13 Deep-sea drilling ship, Glomar Challenger. Ship's length is 125 m; top of drilling derrick is about 60 m above the waterline. Rack on forward part of ship holds 7270 m of drilling pipe that is 12.5 cm in diameter. (Deep Sea Drilling Project, Scripps Institution of Oceanography)

14.14 Transform fault, a special kind of fault along which movement has been horizontal and next to which new crustal material has been formed while movement was in progress. All other kinds of faults cut crustal blocks whose volumes do not change significantly during faulting. (Based on J. T. Wilson)

The Glomar Challenger From such a map as described above, we can make two predictions: (1) the age and magnetic polarity of the magnetite in the mafic crust underlying the deep-sea floor, and (2) the age of the oldest sediments that overlie the rocky crust. One of the chief means of attempting to verify these predictions has been a program of scientific drilling of holes in the sea floor. This program, known as the Joint Oceanographic Institutions Deep Earth Sampling (JOIDES) program, is conducted from **The Glomar Challenger,** a special drilling vessel, 125 meters long, that resembles a floating oil derrick (Figure 14.13).

In the last few years it has become possible to drill through chert and other sedimentary rocks and to sample igneous rocks. Igneous rocks sampled include basalt sills that intrude the deep-sea sediments, basalt breccias, massive basalts, and coarse-grained mafic rocks including gabbro and serpentinized peridotite. It is not certain that the oceanic crust has been penetrated. The igneous rocks drilled so far display complicated magnetic properties and yield radioactive ages that do not agree with existing models of sea-floor spreading. The drilling results are still being evaluated.

In the first two months of 1976, Glomar Challenger drilled approximately 2000 meters into the deep-sea floor near the mid-Atlantic ridge in an attempt to discover more information about the basic mechanism of plate tectonics.

Fracture zones and transform faults One of the most remarkable of the many astonishing discoveries made during the 1950s and 1960s about the deep-sea floor was the delineation of the numerous series of parallel fracture zones. These cut across and displace the mid-oceanic ridges. The trends of the fracture zones are more or less parallel to the Equator in the Atlantic and Pacific Oceans, but are aligned nearly north-south in the northern part of the Indian Ocean.

The concept of transform faults was proposed to account for the geometric arrangement of offsets of a spreading ridge. In fact, the definition of a **transform fault** is that it is a transcurrent fault that ends at a spreading center (Figure 14.14). The difference between a transform fault on the deep-sea floor and a great transcurrent fault on a continental block is that the transform faults are offsets between places where new material is being added to the volume of rock being deformed. The faults which displace continental rocks involve bodies of rock to which nothing new is added.

Present status of the concept of sea-floor spreading In the 15 years since it was proposed, the concept of sea-floor spreading has literally revolutionized our thinking about the ocean floor. It has been the basis for making very specific predictions about rates of movement

479
Plate Tectonics

of the sea floor, and about the ages and magnetic polarities of sea-floor crustal rocks.

Perhaps the most noteworthy thing to come from the concept of sea-floor spreading is a new and worldwide perspective on the whole Earth. This new concept is known as global tectonics, or plate tectonics.

Plate Tectonics

We shall deal with plate tectonics by first summarizing its origins. Then we shall present a statement of the modern version of this idea, list some of the important predictions made by plate-tectonic theory, and close by comparing some of these predictions with the geologic evidence.

Current Concepts of Plate Tectonics: Predictions and Tests

The fundamental concept of **plate tectonics** is that the major dynamic activity of the outer part of the Earth involves great slabs (plates) of lithosphere, and that these plates move laterally around the Earth, interacting with one another in one of three ways (Figure 14.15, overleaf).

Plates and plate boundaries Three kinds of plate boundaries have been defined: (1) divergent, or spreading boundaries, (2) convergent boundaries, and (3) transcurrent boundaries.

A typical example of a modern **divergent** or **spreading boundary** between two plates is a mid-oceanic ridge. Here new crustal material is thought to be manufactured from changes of the underlying mantle. The basis for this kind of plate boundary is sea-floor spreading, as we have previously discussed. According to one version, proposed by John F. Dewey and Kevin Burke, initial spreading begins at isolated domal uplifts (Figure 14.16a). Such domes are thought to lie above cylindrical parts of the mantle having high heat flow, known as **mantle plumes.** The uplift of the dome causes it to fracture along three cracks arranged at angles of 120 degrees (Figure 14.16b). If spreading begins from these fractures, new oceans may open along all of them, or on only one or two (Figure 14.16c). An example in the modern world where this activity is inferred to have taken place is the junction of the Red Sea, the Gulf of Aden, and the African rift at Afar, Ethiopia (Figure 14.16d; also see the chapter-opening photograph for this chapter).

The opposite of a divergent boundary is a **convergent boundary,** where two plates collide (Figure 14.17). If one of the plates contains

14.15 Kinds of Plate Movements

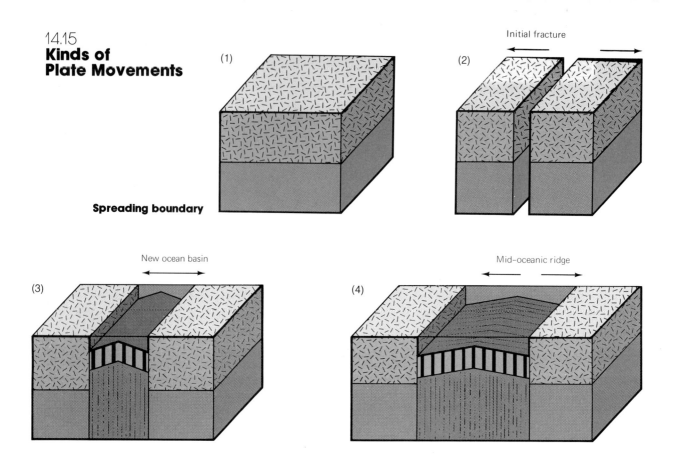

Spreading boundary

Relief drawing of a strip of the bottom of the Atlantic Ocean from Florida at left to northwest Africa at right (below). Mid-Atlantic ridge displays two dominant sets of fractures: (1) prominent linear transform faults which trend across ridge from upper left to lower right, and (2) fractures parallel to the crest of the ridge. Numbers show depths in feet at points marked by dots. Compare with profile of mid-Atlantic ridge on pages 16-17. (B. C. Heezen and Marie Tharp. Copyright © 1973 National Geographic Society)

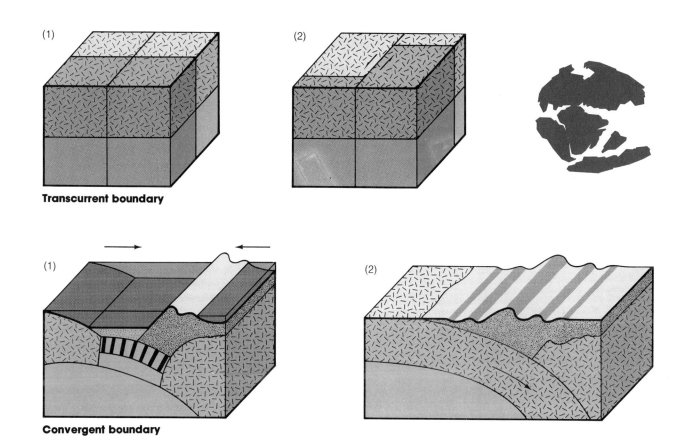

Transcurrent boundary

Convergent boundary

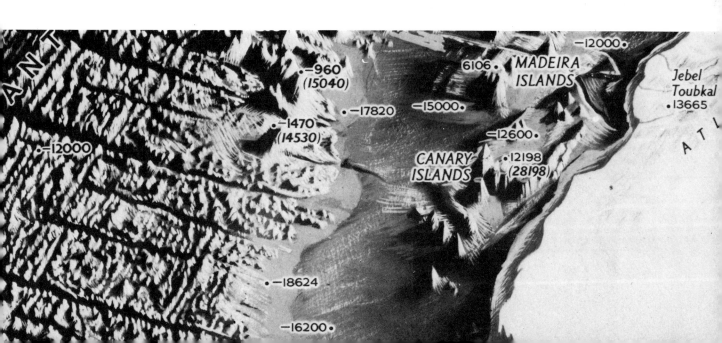

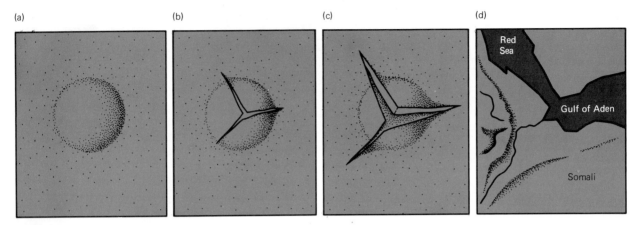

14.16 Domal uplift and fracturing along three cracks at 120-degree angle. (a) Initial domal uplift before fracturing. Diameter of dome is 100 to 200 km or so. (b) Cracks form, at first only within circumference of dome. (c) Spreading begins along two of the three fractures. (d) Junction of the Red Sea, Gulf of Aden, and African Rift system; photo at left shows the same view as seen from Gemini XI on September 15, 1966, with camera aimed southeast along Red Sea toward Indian Ocean (curvature of Earth visible in upper right corner; antenna of spacecraft is seen at lower left) (NASA)

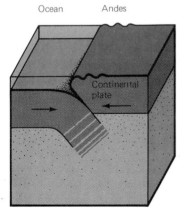

14.17 Convergent boundary between two crustal plates; schematic block. Deep-focus earthquakes are thought to be confined to such boundaries.

oceanic rocks, then the proposition is that the oceanic plate plunges beneath the other plate, in a great underthrust fault, which has been named a *subduction zone*. Because subduction zones have been related to deep-sea trenches, we shall defer our discussion of them until a following section. If two continental plates collide, no subduction is thought to occur. Instead, mountain ranges are supposed to be crumpled and elevated. An example is the present-day Himalayas, which are thought to be the result of collision between the Indo-Australian Plate and the Asian Plate.

Along a **transcurrent boundary,** two plates are in contact along a great transform fault or strike-slip fault (Figure 14.18). An example of such a plate boundary is thought to be the San Andreas Fault, in California, which is considered to be the transcurrent boundary between the Pacific Plate and the North American Plate.

Trenches and subduction Perhaps one of the most dramatic and attention-getting aspects of the whole idea of global tectonics is that the creation of new crustal material at mid-oceanic spreading ridges is exactly matched by corresponding destruction of crustal material at **subduction zones,** the convergent boundaries between plates, whose identifying feature in the modern world is supposed to be a deep-sea trench. According to plate-tectonic theory, the solid crust of the deep sea progresses steadily and relentlessly toward its

destruction at a deep-sea trench. Whatever has accumulated on top of the sea-floor crust, pelagic sediments or even entire volcanoes, is presumed to disappear down the trenches. This version of subduction implies that the trenches should be filled with offscraped sediments that may first have been deposited thousands of kilometers away, perhaps in some greatly different climate zone from the trench where they are presumed to disappear.

Subduction is thought to be indicated by the great inclined **Benioff zones** of earthquake foci falling in the intermediate and deep-focus categories. New detailed seismic studies of the first-motion direction on earthquakes originating in Benioff zones indicate that the upper plate is moving upward over the lower block, in accordance with the theory of subduction. However, very little else predicted by the theory seems to be matched by observation and some observations are opposite to those predicted. In the first place, no trench has yet been found in which vast quantities of pelagic sediments have been offscraped. Many trenches rimming the central Pacific are nearly devoid of sediment. In others the filling sediment consists of gently inclined pelagic sediments underlying horizontal strata of Quaternary sediments. What is more, the JOIDES borings in the North Pacific have not found indications that sediments deposited beneath the Equator have been transported all the way to the Aleutian Trench, as some plate-tectonic theorists predicted.

Clearly, the predictions from the theory of subduction do not match observations in modern trenches. The discrepancy is great enough that some new explanation seems required.

Mantle hot spots and rows of volcanic islands. Within ocean basins are many rows of volcanic islands that become systematically younger in one direction. An example is the Hawaiian chain, which begins on the northwest with the atoll of Midway, and ends on the southeast with the modern volcano Kilauea, on the island of Hawaii (Figure 14.19).

According to the plate-tectonic explanation, the Pacific Plate is moving northwestward over a fixed plume of heat rising through the underlying mantle. Where the plate is over the plume, a volcano is active. As the plate moves away, the volcanic cone is cut off from the rising heat, and becomes inactive. Ultimately, it becomes eroded. Moreover, the island may be submerged as it moves down the slope of the dome created by the hot spot. If this view is correct, then the row of islands marks the track of the Pacific Plate for the past few tens of millions of years.

If the Pacific Plate is moving across a fixed hot spot in the mantle, then movements of the mantle are not driving the plate. This concept

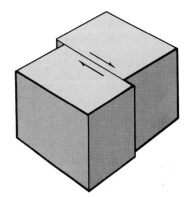

14.18 Transcurrent boundary between two crustal plates; schematic block. Along transcurrent plate boundaries only shallow-focus earthquakes are generated.

Plate Tectonics and the Interior of the Earth

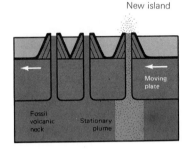

14.19 Concept of row of volcanic islands created by movement of Pacific plate northwestward over fixed, chimney-like "hot spot" in Earth's mantle. According to this concept, only a single volcano at the end of the chain, is active at any one time. Inactive volcanic cones move farther and farther away from the "hot spot," in the process becoming eroded or submerged. (Based on J. T. Wilson)

14.20 Comparison of tectonic trends and ages of ancient rocks in Africa and South America. The close fit of these features is considered by many to constitute a strong argument in favor of the idea that these two now-separated continents formerly were joined. (P. M. Hurley)

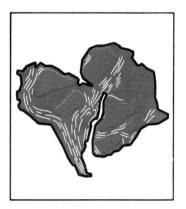

that the plate is moving independently of the mantle is the exact opposite of the Hess view that the sea-floor crust is bolted to the underlying mantle (see Figure 14.8). Both ideas might be wrong, but only one of them can be correct. If the island-row concept is correct, then some mechanism other than large-scale mantle convection must be sought for driving the plates. If the plates are "locked" to the underlying mantle, then the island-row theory needs to be modified. The answer will come only when the mechanism of plate motion has been fully understood.

Plate Tectonics and Continental Drift

According to the theory of plate tectonics, continental drift is a normal part of the Earth's behavior; the great plates that compose the outermost shell of the Earth are always moving. Some continents are split apart as new oceans open. Other continents collide as ocean basins disappear. As an illustration of how continental drift and plate-tectonic theory are related, let us return to Wegener's example of Pangaea and examine it in light of modern concepts.

The break-up of Pangaea From the dates of the oldest sediments and fossils in the Atlantic Ocean, geologists have inferred that Pangaea began to break apart during late Jurassic time (about 180 million years ago). This is the same break-up date assigned by Wegener. By rapidly flipping the upper right-hand corners of the pages of this section of the book the reader can see an animated reconstruction of the inferred break-up of Pangaea.

As Africa slowly disengaged itself from the eastern seaboard of North America, Africa was also beginning to sever its ties with South America (Figure 14.20). The South Atlantic began to open about 135 million years ago—much later than North Atlantic.

At about the same time that the primeval North Atlantic began to open, a seam is thought to have formed between Africa and the rest of Pangaea. Another oceanic ridge cut between the tip of South America and Antarctica. A short distance to the northeast, it branches and divides India from Antarctica and Africa. India began its long northward journey, which was to end in collision with Asia 80 million years and 7000 km later, deforming the crust to form the Himalayan mountain complex.

The last of the great land masses to split apart is thought to have been the combined Australia and Antarctica. These two continents remained joined until perhaps 40 million years ago. Antarctica is inferred to have moved southward. Australia set off northward at high

speed, traveling more than its own length in only a few tens of millions of years.

Meanwhile, in the Pacific, the huge East Pacific Ridge changed its direction of spreading from east-west to roughly northwest-southeast; such changes in direction of inferred spreading are not yet well explained.

Predictions for future plate movements The continued northwesterly movement of the Pacific Plate would bring Los Angeles abreast of San Francisco in about 10 million years. In about 60 million years, at present rates of spreading, Los Angeles will reach the Aleutian Trench, where the Pacific Plate is presumed to be consumed by being re-cycled downward into the mantle.

Some geologists predict that even greater changes than those just listed will affect Africa and the Mediterranean. They expect that as Africa rotates counterclockwise the Red Sea and the Gulf of Aden will widen. At the same time they foresee that the western Mediterranean will swing shut, closing off the Strait of Gibraltar. Finally, they predict that the African Rift Valleys will widen and become new oceans, thus sundering Africa lengthwise.

Some dissenting opinions about plate tectonics As we have said, the current evidence in favor of the hypothesis of plate tectonics seems to be overwhelming, but many basic questions are still not answered. Several dissenting ideas have come forward, and some of the more serious objections should be considered.

It has been suggested by A. A. Meyerhoff and Howard A. Meyerhoff that the Earth would have had to expand greatly to accommodate the shifting of North and South America away from Europe and Africa; such an expansion has not taken place. The Meyerhoffs further submit that strata in areas where subduction has supposedly taken place have not been deformed. Moreover, the strata of other parts of the sea floor, near Iceland, for example, exhibit folds and other evidence that they have been pushed together, instead of indications that they have been pulled apart, as predicted by the idea of sea-floor spreading.

Questions about sea-floor spreading have been raised by the Soviet academician V. V. Beloussov. These include: How can the Red Sea be spreading obliquely to the local magnetic anomalies? How does the hypothesis of plate tectonics explain vertical movements, especially in the interiors of continents?

David W. Scholl and Michael S. Marlow have difficulty finding evidence that deep-sea deposits exist where they are supposed to have been scraped off against trenches or subducted with oceanic plates. Also, Scholl and Marlow agree with some geologists that field mapping does not indicate as much plate motion as would be necessary to confirm the plate-tectonic concept.

Other questions need to be considered, and probably the most critical issue involves the driving mechanism needed for extensive plate movement. Even if it *did* happen, no one is really sure *how*. And the arguments will continue until an acceptable reconciliation is reached. In the meantime, the hypothesis of plate tectonics will continue to fascinate and inspire scientists throughout the world.

Plate Tectonics in Relation to Crustal Deformation and Mountains

Most of the ideas about sea-floor spreading and plate tectonics have been devoted to explaining features of the sea floor. Continents have been involved in the sense that the new global view of the Earth's dynamics offers a much more plausible mechanism for explaining continental drift than has ever been thought of previously. However, plate-tectonic theory also has much more to say about the origin of geologic structures and about how mountains form. These subjects bring us back to the land again and form the topic of Chapter 15.

Chapter Review

1 The basic technique for "X-raying" the interior of the Earth is to measure the time elapsed between an earthquake and the arrival of seismic waves at a seismograph station. As early as the turn of the century it had been thought that the differences in wave behavior indicated that the interior of the Earth is not homogeneous. The first important discovery through the use of seismic-wave behavior was that the Earth beneath the crust is not totally molten.

2 The *Mohorovicic discontinuity* is a seismic interface between the crust and the mantle. The *mantle* is the mostly-solid portion of the Earth between the crust and a depth of 2900 km. The seismic interface at 2900 km is the *mantle-core boundary*.

3 The *core* of the Earth is the smooth, dense sphere that forms the innermost portion of the Earth. Because S waves are not propagated through the core, at least part of it must be molten. By the speeds of P waves, and other evidence, the chemical composition of the core is thought to be largely iron, with some nickel.

180 million years ago

4   By studying waves that have passed through the core, it has been possible to determine that the core consists of two parts. The inner core is solid, and the outer core is molten. Movements of the liquid outer core are thought to generate the Earth's magnetic field.

5   In the early twentieth century, the major tectonic ideas stated that the continents were fixed in position. The Earth's crust could be raised, lowered, and crumpled, but could not be shifted sideways any significant distances. These principles were soon to be challenged.

6   Alfred Wegener and others became interested in the idea of moving continents on a global scale, based on the notion that Africa and South America had once been joined, Wegener challenged the established school of thought which favored the idea of fixed continents. He was not able to develop any acceptable new Earth model, and he could never establish a theoretical basis in support of his view of *continental drift*.

7   For 30 years, continental drift was an inconclusive subject which few scientists took seriously. With the advent of new evidence in the 1950s centered around *paleomagnetism*, continental drift took on new respectability. Studies showed that the Earth's magnetic field had varied in intensity, and that the position of the Earth's magnetic pole had shifted. Computations based on polar wanderings reinforced Wegener's idea that the continents had drifted.

8   In 1962 Harry H. Hess proposed that the ocean floor has been, and still is, moving. Furthermore, Hess said that the ocean floor was locked to the mantle. This theory would allow the ocean floor to move at a deep level, taking the continents with it. Hess also rejuvenated the idea that convection currents move the continents.

9   In 1961, Robert S. Dietz proposed the name *sea-floor spreading* to describe the outward movement of new crust away from a mid-oceanic ridge.

10 In the 1960s, researchers on remanent magnetism not only confirmed the information that polar reversals had taken place, but began to establish time scales for magnetic reversals.

11 The two important implications of the hypothesis of sea-floor spreading are (1) the predicted ages of sea-floor crust, and (2) the concept of fracture zones and transform faults on the deep-sea floor.

12 The fundamental concept of *plate tectonics* is that the major dynamic activity of the outer part of the Earth involves great slabs (plates) of the lithosphere, and that these plates move laterally around the Earth, pulling apart or colliding with one another in the process.

13 Three kinds of plate boundaries have been defined: *divergent*, or spreading boundaries, *convergent* boundaries, and *transcurrent* boundaries.

14 According to the theory of plate tectonics, continental drift is a normal part of the Earth's behavior; the great plates that compose the outermost shell of the Earth are always moving. The plate-tectonic theory rein-

forces Wegener's idea of a supercontinent, *Pangaea*, which is thought to have started breaking apart about 180 million years ago.

Questions

1. How are seismic waves used to help determine the interior structure of the Earth?
2. Define *mantle, mantle-core boundary,* and *core.*
3. Why do scientists think that the core consists of two parts, one solid and one molten?
4. How is the core of the Earth thought to be related to the Earth's magnetic field?
5. Why was Alfred Wegener unable to prove his theory of drifting continents?
6. Define *Pangaea* and explain how it fits into Wegener's theory.
7. Define *paleomagnetism* and explain its role in reinforcing the idea of continental drift.
8. Explain *sea-floor spreading.* Why were Harry Hess' ideas so important?
9. Describe the *Vine-Matthews hypothesis* and the magnetic characteristics of the ocean floor.
10. Why is it possible that the sea floor may be younger than the continents?
11. Explain the concept of *plate tectonics.* Describe the three kinds of plate boundaries.
12. Define *subduction zone.*
13. Is the concept of plate tectonics generally accepted, or are scientists raising questions? What sort of objections would you expect?

Suggested Readings

Anderson, Alan H., Jr., *Drifting Continents.* New York: G. P. Putnam's Sons, 1971.

Calder, Nigel, *The Restless Earth.* New York: The Viking Press, 1972.

Clark, S. P., Jr., *Structure of the Earth.* Englewood Cliffs, N.J.: Prentice-Hall, Inc., 1971.

Dietz, R. S., and Holden, J. C., "The Breakup of Pangaea." *Scientific American,* October 1970. (Offprint No. 892. San Francisco: W. H. Freeman and Company.)

Sullivan, Walter, *Continents in Motion.* New York: McGraw-Hill Book Company, 1974.

Suggested Readings

Takeuchi, H., Uyeda, S., and Kanamori, H., *Debate About the Earth*, Revised Edition. San Francisco: W. H. Freeman and Company, 1970.

Wilson, J. Tuzo (Editor), *Continents Adrift*. A Scientific American Book. San Francisco: W. H. Freeman and Company, 1972.

"Paradoxically, the highest points in the world begin their lives at the lowest points in the world."

Chapter Fifteen
Mountains, Crustal Deformation, and Recycling of Continents

What are mountains? How did they get there? What are they made of? How are they related to the behavior of the lithosphere and to the concepts of plate tectonics? How do mountains relate to the geologic cycle in general? In trying to answer these questions we shall apply the geology we have studied so far toward the understanding of **orogeny** (or-AH-juh-nee). The word orogeny, seemingly technical, is really quite simple, and is basic to our total view of this chapter. Orogeny comes from the Greek words *oros* (mountain) and *genesis* (birth). So, orogeny simply means the birth of mountains—but not just any mountains, only those related to orogenic belts. Orogeny includes all of the processes that affect **orogenic belts.** In these linear belts, the strata of sedimentary rocks are especially thick, have been deformed by folding and faulting, and in some locations may have been invaded by granitic batholiths. Two types of mountains are not related to orogenic belts: volcanic cones and stream-dissected plateaus. The remaining kinds of mountains illustrated on pages 494–495 in Figure 15.2—*large domal uplifts* exposing "basement" rocks, *fault-block mountains, fold mountains,* and *complex mountains*—are related to orogenic belts, and will be discussed in this chapter.

For most purposes, the word **mountain** is appropriate for any part of the Earth's land areas having relief of more than 700 meters.

Orogenic Belts and Their Mountains

In terms of geologic time, mountains exist for a very short time. They are built up, eroded away, and recycled. In the process, they help to keep the continents alive by providing solid mass faster than erosion can erase it. But as mountains come and go, the rocks in a mountain's orogenic belt remain basically untouched, much the way your chromosomes do not change because your appendix has been removed, or your nose has been reshaped. In this chapter we try to treat mountains (no matter *how* they were formed, by plate movement or not) in their proper perspective—as the surface relief of orogenic belts.

Orogenic belts and related mountains may be discussed most simply by means of an analogy with human lives. If, for the purposes of our analogy, we are permitted to include the idea of human reincarnation, the comparative stages are shown in the table on the following page.

Without the existence of a **geosyncline,** a place where the Earth's crust subsides to form a basin or trough, the process of "gestation" and orogeny could not happen (Figure 15.1). Geosynclines are likely

Orogenic Belts and Their Mountains

Orogenic Belts and Related Mountains	Human Lives
Geosyncline (initial sinking of a part of the Earth's crust beneath sea level)	Conception
Accumulation of thick sediments	Gestation
Orogeny	Birth
Initial elevation	Youth
Long-continued erosion	Maturity
Peneplain	Old age/Death
Recycling	Reincarnation

to appear on the sea floor, so paradoxically, the highest points in the world (mountains) begin their lives at the lowest points in the world (the bottom of the ocean), where sediments can pile up and begin the process of forming an orogenic belt.

Figure 15.3 emphasizes how the fabric of an orogenic belt remains the same, whereas the surface expression of the belt may change with time. The sequence shown begins with mountains that are eroded to a peneplain. By later subsidence the peneplained orogenic belt is covered with younger strata. Now the orogenic belt is completely invisible on the Earth's surface; nonetheless it retains its original fabric. Later, the buried orogenic belt comes to the surface and forms mountains once again, this time along fault blocks or in the center of a great domal uplift exposing the "basement rocks," a name given to ancient metamorphic and igneous rocks, of old orogenic belts, which underlie sedimentary strata.

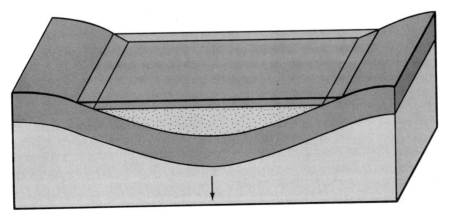

15.1 Geosyncline according to original concept. The great down-bent part of the Earth's surface is several hundred kilometers wide. Sediments (stippled area) are deposited in shallow seaway. With further submergence the elevated areas at the sides of the seaway may be submerged. Schematic block.

15.2 Kinds of Mountains

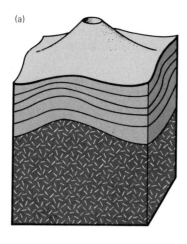

Volcanic cone, an accumulation of volcanic products around a vent. Mountains that are volcanic cones include some of the world's most majestic peaks on land, many volcanic islands, and thousands of seamounts on the deep-sea floor, having summits that do not reach the water surface.

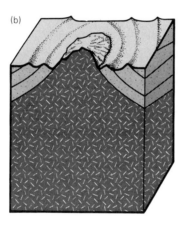

Giant domes exposing "basement rocks," old metamorphic and/or igneous rocks which underlie the sedimentary strata of continents, and around which sedimentary strata dip away in all directions. After extensive uneven erosion the central area of more-resistant basement rock stands high as a mountainous area.

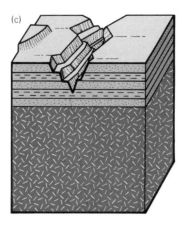

Plateaus, deeply dissected regions of horizontal strata, in which mountainous relief has resulted from regional uplift and deep dissection by stream valleys.

Fault-block mountains, elevated tracts bounded on one or more sides by faults, fractures along which the opposing sides have been displaced.

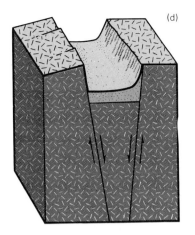

Fold mountains, elevated areas formed by uplift of folded strata, either on individual up-folded strata or by differential erosion of a wide tract including both upfolds and downfolds.

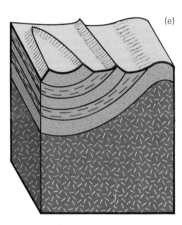

Complex mountains, elevated areas formed by uplift of regions having complex structural fabric, including folds, faults, and batholiths.

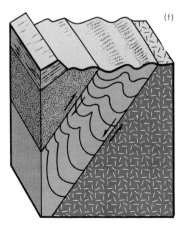

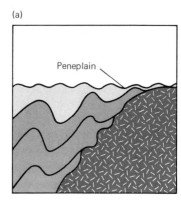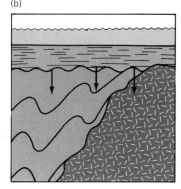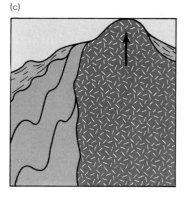

15.3 Rock fabric of an orogenic belt and various surface expressions. (a) Former mountain chain eroded to a peneplain. (b) Peneplain subsides, is submerged by sea, and buried by sediments deposited on the sea floor. (c) Uplift brings rocks or orogenic belt back to the Earth's surface, forming domal mountain.

With minor exceptions, the major orogenic belts that have been uplifted to form mountains within the last 100 million years are arranged in a simple pattern. If the surface of the Earth were flooded with enough additional water so that sea level could rise by 1000 meters, most of the land remaining above sea level would belong to the Circum-Pacific belt or the Alpine-Himalayan belt (Figure 15.4).

In this chapter we will relate orogenic belts to the general topics of crustal deformation and the recycling of continents. We will also study the kinds of rocks found in orogenic belts, their geologic structural features, their dynamic histories, and the mechanics of orogeny. We will conclude with a discussion of theories of orogeny and their relation to plate tectonics and the history of continents.

The Anatomy of Orogenic Belts

Some rocks and structures found in orogenic belts are peculiar to orogenic belts; others could be found elsewhere. It should be emphasized, therefore, that not all the rocks and structures we describe here are found only in orogenic belts.

The Rocks of Orogenic Belts

All three major kinds of rocks can be detected in orogenic belts. A remarkable feature of the sedimentary rock strata of orogenic belts is their great thickness as compared to strata of the same age that lie outside orogenic belts.

135 million years ago

15.4 The world's mountains represented as the only remaining land areas if the sea were to rise by 1000 meters. (Left) The North American and South American Cordillera, Western Hemisphere. (Right) The Alpine-Himalayan belt and island peaks of the Eastern Hemisphere.

Many orogenic belts are typified by their vast tracts of metamorphic rocks. Indeed, some of these zones of metamorphic rocks in orogenic belts are so large that the process of creating them has been named *regional metamorphism*. Many metamorphic rocks of orogenic belts are products of *dynamic metamorphism*. *Dynamic metamorphic rocks* display evidence that internal movement accompanied metamorphic changes. Such evidence includes the great stretching and distortion of familiar shapes such as pillows or boulders to form gneiss (Figure 15.5).

Not all orogenic belts contain igneous rocks. Where present, however, these may be either volcanic or plutonic (intrusive). The volcanic rocks of orogenic belts include numerous examples of those that form strata, particularly pillowed basalts, indicating submarine eruptions. Other volcanic rocks may be products of both submarine and subaerial eruptions. The plutonic rocks or orogenic belts include great batholiths of granitic rocks or, by contrast, pods and slices of mafic and ultramafic rocks. These ultramafic rocks include coarse gabbros, peridotites, and serpentine, which are the kinds of rocks thought to underlie the deep-sea floor.

Geologic Structures

The features that can be delineated by describing the three-dimensional arrangement of rock bodies are named *geologic structures*. The study and mapping of geologic structures and the analysis of crustal deformation is **structural geology.**

15.5 Effects of dynamic metamorphism. (Top) Stretched pillows in Precambrian rocks, southwestern Finland. (Bottom) Stretched boulders in Carboniferous conglomerate, East Greenwich, Rhode Island. (J. E. Sanders)

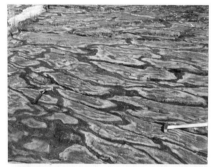

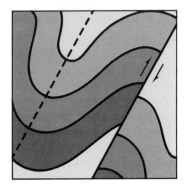

15.7 Thrust fault (solid line) formed by breaking along the crest of an asymmetric anticline. Dashed line marks axial plane of adjacent asymmetric syncline.

Folds When strata are folded they typically form elongated upfolds or *anticlines* and intervening downfolds or *synclines*. Alternatively, they may form rather circular or oval features known as *domes* and *basins*. The various kinds of folds are shown in Figure 15.6, below, across both pages.

Thrusts A structural feature found only in orogenic belts is a **thrust,** a great fault along which strata have been piled one on top of another, thus rearranging the normal order of the layers and increasing the thickness of the pile (Figure 15.7).

Other faults In mountains, as well as in other locations, many kinds of faults other than thrusts are common. Most of these are usually steeper than thrusts. The terms employed to describe faults are based on the relationship of direction of dip of the fault and relative sense of movement of the blocks on each side. The common kinds of faults are defined and illustrated in Figure 15.8.

Faults are usually recognized by the displacements of strata (Figure 15.9 right) or by large offsets of entire mountain ranges (Figure 15.9 left). In fact, the term *fault* and many of the ideas about faults came from displacements of coal beds noticed long ago by British miners.

15.6 Various fold structures, schematic blocks having vertical front sides and flat tops formed after folding. Dot-dash line in blocks at right marks traces of axial plane of folds. In simple upright anticlines and synclines, the axial plane is vertical and the dips of the strata forming the limbs are equal. In an anticline the oldest formation is exposed at the crest and the younger strata on the limbs dip away from it. In a syncline the youngest formation is present along the trough and progressively older strata forming the limbs dip beneath it. In an asymmetric fold the dip of the strata on one limb is steeper than that on the other limb. In an overturned fold the strata of one limb have been rotated more than 90 degrees from their formerly horizontal positions.

Dome

Basin

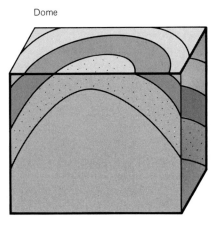
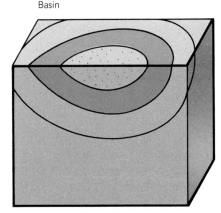

When the layer of coal they were mining came to an abrupt end they called the place a "fault."

Blocks bounded by steep faults that are nearly parallel have been given special names. A block that has dropped between two relatively higher-standing blocks is a **graben** (GRAH-ben). A block that has been elevated above two adjacent lower-standing blocks is a **horst.** (See Figure 15.8c.)

Examples of Kinds of Mountains Based on Geologic Structures

In Figure 15.2 we showed six kinds of mountains. Four of these are defined on the basis of their geologic structure: (1) large domal uplifts exposing "basement" rocks, (2) fault-block mountains, (3) fold mountains, and (4) complex mountains.

Large Domal Uplifts Exposing "Basement" Rocks

Mountains formed by **large domal uplifts** exposing "basement" rocks resulted from *great vertical uplift* of a cylindrical segment of the lithosphere, and from later *differential erosion*. The "basement" rocks that become elevated in the cores of these great domes generally are much more resistant to erosion than the sedimentary strata which originally covered the ancient rocks. After much erosion the resistant cores remain as mountains, whereas the land surface underlain by the covering strata is lowered considerably. Examples include the Black Hills, in South Dakota; the Ozarks, in Missouri; and the Adirondacks, in New York State.

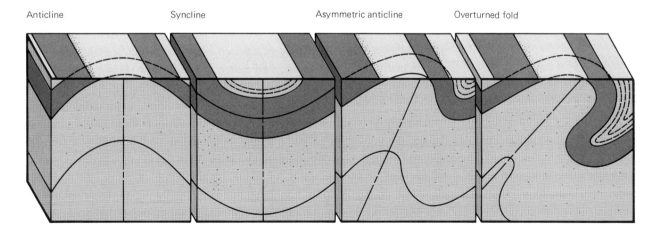

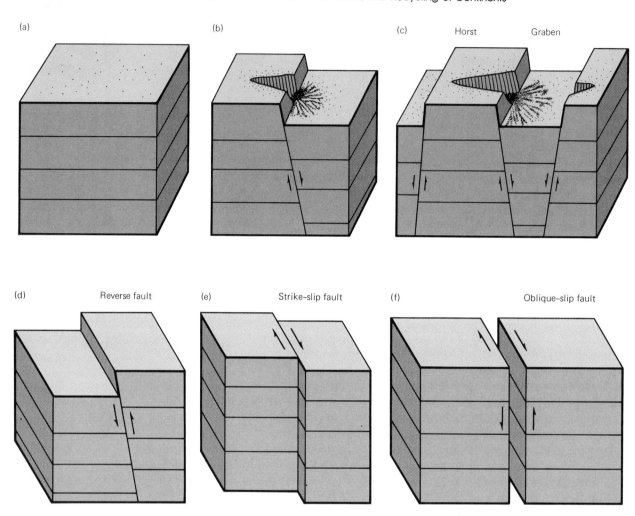

15.8 Common kinds of faults, schematic blocks. (a) Block before movement on any fault. (b) A normal fault dips beneath the relatively downdropped block. (c) Blocks defined by three normal faults whose traces on the Earth's surface (top) are parallel. A horst is a relatively upthrown block between two faults. A graben is a relatively downthrown block between two faults. (d) A reverse fault dips beneath the relatively upthrown block. (e) On a strike-slip fault the relative displacement is horizontal. The sense of displacement is determined by looking across the fault and noting which way the opposite block moved. Right-lateral offset is shown here. (f) Along an oblique-slip fault relative motion contains both horizontal and vertical components.

Examples of Kinds of Mountains Based on Geologic Structure

Fault-Block Mountains

Faults may raise and lower any parts of the Earth's surface. Commonly, however, orogenic belts become broken by networks of faults. A typical sequence would show a newly deformed orogenic belt becoming a fold-mountain chain or a complex-mountain chain. Then the mountain chain may be eroded to a peneplain. Shortly afterward, the peneplain is elevated and the rocks broken into a series of fault blocks, some of which may become fault-block mountains. The mountainous relief of **fault-block mountains** results from vertical elevation and possibly also from differential erosion, as in domal uplifts. (See Figure 15.2d.)

Fault-block mountains may be tilted because they were elevated much more on one fault than on an adjacent fault. The great Sierra Nevada of eastern California is an example of an enormous tilted fault-block range. The steep east-facing escarpment contrasts with the gentle slope toward the west. Numerous other tilted fault-block ranges are found in southern Nevada and western Utah, in the region known as the Basin and Range Province. This region is part of an extensive orogenic belt whose age spans the Palezoic and Mesozoic eras. The

15.9 Offsets along faults. (Left) Mountain range offset, possibly as much as 40 km, along left-lateral strike-slip fault that trends E-W, western China; view from ERT-1 satellite. (NASA) (Right) Thin layers of distinctly foliated gneiss offset up to 10 cm along curved faults. Precambrian, northern Manitoba, Canada. (Geological Survey of Canada, Ottawa)

502
Mountains, Crustal Deformation, and Recycling of Continents

15.10 **Folds exposed in coastal cliff.** Lulworth, Dorset, England. (G. R. Roberts)

belt was subjected to several orogenies, the terminal one about 60 to 70 million years ago. The fault-block ranges are only a few million years old; some of them are still being actively elevated.

Fold Mountains

By **fold mountains** we refer to mountain belts whose major relief is related to the eroded parts of extensive groups of folds.

The earliest geologic observers noticed small folds that could be seen in a coastal cliff, for example (Figure 15.10). However, the idea that the strata of an entire mountain chain had been folded on a large scale did not occur to them. That notion was first proposed about the middle of the nineteenth century, when the Appalachian chain in Pennsylvania was being mapped. In eastern Pennsylvania eroded edges of inclined strata form parts of enormous parallel folds that are hundreds of kilometers long, and up to tens of kilometers wide. These folds have elevated and depressed the strata by thousands of meters. The present-day surface of the Earth is the result of great erosion that

15.11 **The Swiss Alps, complex mountains.** The three ridges in the foreground are underlain by Mesozoic marine strata of great thrust sheets. The jagged peaks on the skyline are composed of Paleozoic granitic gneisses. Peak at extreme left is the Jungfrau. (Swiss National Tourist Office)

took place after folding. This part of the Appalachians has been named the Valley and Ridge Province.

Other fold mountains besides the Valley and Ridge Province include the Zagros chain, in Iran, and the Jura, in France and Switzerland. In young mountain ranges many anticlines form elongated hills and the synclines, valleys. In old fold chains this relationship between folds and the landscape may be reversed. After much differential erosion, relief is controlled by the resistance of the strata. In such mountains the axes of anticlines may underlie valleys, whereas the rocks in the troughs of synclines cap high ridges.

Complex Mountains

Mountainous relief that results from the elevation and differential erosion of orogenic belts having complex internal structures, including folds, great thrusts, batholiths, and belts of regionally metamorphosed rocks, are **complex mountains.** Nearly all orogenic belts contain at least some parts that would be classed as complex mountains. These parts usually are in the interior of the chain. Examples include the Caledonian chain of Wales, Scotland, and Norway; the interior parts of the Alps (Figure 15.11); parts of the Appalachians in eastern Canada, New England, and the Great Smokies in Tennessee and North Carolina.

Dynamic History of Orogenic Belts

The dynamic history of an orogenic belt involves three stages: (1) geosynclinal stage of accumulation of strata, (2) terminal-orogenic stage, and (3) post-orogenic stage of block faulting and vertical movements.

Geosynclinal Stage

As we have mentioned, an orogenic belt begins its history when a geosyncline starts to subside. The region that subsides may be at the margin of a continent. If so, then the strata accumulate at the continental margin, with the continental mass on one side and a deep-ocean basin on the other (Figure 15.12). Alternatively, the subsided area may lie within a continental mass. In this case, the geosyncline is a true trough with land bordering it on both sides. (See Figure 15.1.)

15.12 Strata accumulating on subsiding margin of continental block (schematic). (Left) If water above subsiding block in tropical climate remains shallow carbonate sediments accumulate; they eventually become limestones. Mud accumulates in deeper water (right); such mud may contain blocks of limestone that were transported downslope from shallow-water area at left. (Right) Deep-water trough formed by great subsidence, schematic block. Areas next to trough accumulated shallow-water carbonate sediments as fast as area subsided. In between, the shallow-water carbonate sediments did not accumulate, so water deepened. The tongue of ocean in the Bahamas formed in this way.

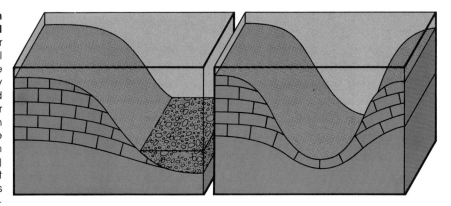

The detailed history of the crust during the geosynclinal stage can be read from the kinds of strata present, their thicknesses, and the sizes of their particles. In some geosynclines, subsidence was matched exactly by accumulation of sediment. The water started shallow, and even though the bottom may have continued to subside and sink through thousands of meters, sediment kept accumulating so that the water remained shallow. For example, this kind of subsidence and accumulation of limestones and carbonate sediments has been going on in the Bahamas for the last hundred million years. Exploratory borings indicate that shallow-water limestones and dolostones and evaporites deposited on intertidal flats are 5500 meters thick.

If that much sinking had taken place without a corresponding accumulation of strata, then the water would have become 5500 meters deep. In some geosynclines such deep water evidently did form (Figure 15.12, right). Later, the deep area filled with sediments. The oldest sediments were deposited in deep water and the younger ones in shallower and shallower depths.

In some geosynclines, folds and faults were active while the strata were accumulating. The growing folds affected the thickness of the strata, which became thinned over crests of folds and thickened in troughs of synclines. Similar changes may have taken place on fault blocks.

Terminal-Orogeny Stage

While the various sedimentary strata of a mountain chain are accumulating some of them may be deformed locally. However, such local deformation usually is overshadowed by what happens during the grand climax, the *terminal orogeny*, which involves deformation, metamorphism, and plutonism of the mountain belt. Great thrusting takes place. The usual arrangement is that deep-water strata are thrust over the shallow-water strata (Figure 15.13). Orogeny typically includes great thermal events—regional metamorphism, formation of granitic batholiths (see Figure 4.8), and the concentration and deposition of many metal deposits. While the strata are being subjected to high temperatures and pressures they may recrystallize as they are being deformed.

The long-distance thrusting of the deep-water strata over the shallow-water strata and thus from the ocean side toward the interior of the continent, is an event of great significance in the history of an orogenic belt. An explanation for this activity is one of the great mysteries of geology, although examples are known from all parts of the world. In North America during the Paleozoic Era such thrusting took place at different times in the three marginal geosynclines: the Appalachian on the east, the Ouachita on the south and the Cordilleran on the west.

During the Ordovician Period in the Appalachian geosyncline a body of dark-colored, fine-grained deep-water strata of Cambrian and Ordovician ages measuring several thousand kilometers long, hundreds of kilometers wide, and thousands of meters thick, traveled as much as 150 km westward across the sea floor and came to rest above shallow-water carbonate rocks, also of Cambrian and Ordovician ages.

15.13 Thrusting of deep-water strata over shallow-water strata at continental margin, schematic profiles. (a) Carbonate sediments collect on shallow shelf; at outer edge of shelf, reef grows and sheds debris into deep water (at right). (b) Shallows shelf subsides greatly; carbonates become overlain by deep-water shales. (c) Thrusting takes place with movement from ocean side toward continent. Debris shed from reef is now located on shelf side of reef.

(a)
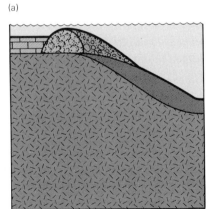

(b)
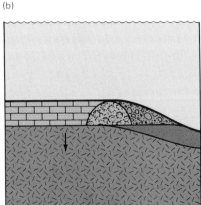

(c)
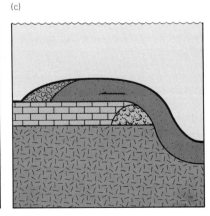

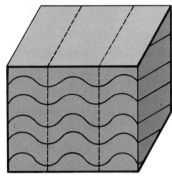

15.14 Fault-bounded basins in recently folded orogenic belt, schematic blocks. (a) Newly folded and elevated strata of orogenic belt. Dashed lines mark sites of future faults. (b) Graben forms between two steep faults.

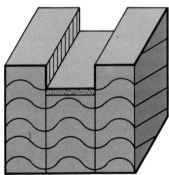

In New York state the transport was from east of what is now the Green Mountains to the Taconic Upland.

During the Pennsylvanian Period comparable activity took place in what is now southwestern Arkansas and southeastern Oklahoma. Once again, dark-colored shales and siltstones, deposited in deep water, were thrust over shallow-water strata. The direction of displacement was northward and westward. During the Mississippian Period deep-water Paleozoic shales were displaced eastward from what is now western Nevada onto a shallow-water platform.

Post-orogenic Stage of Block Faulting and Epeirogenic Movements

Shortly after all the deformation and heating that accompany the terminal orogeny have taken place, the orogenic belt typically becomes greatly elevated and a mountain range is born. Vertical subsidence or elevation that takes place without crumpling the strata is known as **epeirogeny.** While all the vertical movements are in progress local fault-bounded basins may form (Figure 15.14). In these basins thick bodies of sediments and some volcanic rocks may accumulate. Rapid uplift and subsidence may create thick bodies of nonmarine fan deposits along the margins of the basins. (See Figure 11.12.)

Eventually these faulting movements cease and the orogenic belt becomes eroded to a peneplain. It may later subside and start accumulating sediment in a new major cycle of activity.

How does all this happen? What is responsible for the great vertical movements of epeirogeny and the deformation and thermal activity of orogeny? We begin with eperiogeny.

Causes of Epeirogeny: Isostasy

The vertical movements of the lithosphere clearly involve questions about its strength and about what lies beneath it. These are such large-scale subjects that we cannot deal with them directly. Instead, we have to rely on geophysical measurements, theories, and calcula-

Causes of Epeirogeny: Isostasy

tions. The most useful approach to vertical movements and the strength of the lithosphere come from detailed measurements of the value of the Earth's gravitational acceleration and from calculations about densities of rocks that may be affecting these measurements.

Strength of the Lithosphere

Using gravity measurements as a basis for comparing assumptions with reality, it is possible to infer the strength of the Earth's lithosphere. This is done by computing the masses of various surface irregularities. Surface features can be divided into two groups that are considered "small" and "large" loads, respectively. The small loads include such things as deltas and individual volcanic cones. The large loads include big lakes, continental glaciers, and mountain ranges. From the gravity measurements geophysicists have concluded that the lithosphere is strong enough to bear the weights of deltas and of volcanic cones, but is not strong enough to withstand the weights of big lakes, continental glaciers and mountain chains (Figure 15.15).

When a "large" load is placed on it the lithosphere sinks. When the load has been removed the lithosphere "bounces back" to its old level. This has been taken to mean that the lithosphere maintains a condition of flotational balance with the *asthenosphere* the underlying zone of negligible strength. Such a condition of flotational balance maintained by the Earth's gravity among segments of the lithosphere has been named **isostasy** (eye-SAHSS-tuh-see) (from the Greek for "equal standing").

In the early days of the twentieth century, isostasy was explained in terms of two layers, the sial and the sima, separated by the Mohorovicic discontinuity. According to this interpretation, the sial was thought to be a comparatively rigid layer, the crust, which "floated" on a denser, yielding layer, the sima. The concept of plate tectonics now indicates that the balancing movements of isostasy take place at the base of the lithosphere, which lies far below the Mohorovicic discontinuity. Under the new idea, vertical changes are thought to take place because the lithosphere not only floats on the asthenosphere, but also undergoes changes in density.

In the plate-tectonics view of isostasy, the actual weights of columns of material extending from the Earth's surface to some depth in the asthenosphere are equal (Figure 15.16).

Isostatic balance responds to two kinds of processes: (1) flow of the asthenosphere for reasons other than sagging of the lithosphere beneath a load, and (2) changes in the thickness or the density of the lithosphere itself.

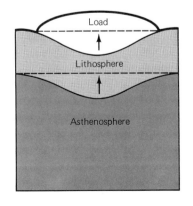

15.15 Large load exerted by a continental glacier on the lithosphere causes it to sag down into the asthenosphere, schematic block. Dashed lines mark pre-load levels; arrows show direction of movement that will take place when load has been removed. Thickness of lithosphere is 70 to 200 km; thickness of large continental glacier may be 4 km.

15.16 Isostasy. Metal blocks floating in mercury. (1) The tops of two blocks having same density but differing in thickness not only project upward to different levels but also extend downward to different depths. To reduce height of right block to that of left block requires erosion equal to both "extra" extension upward as well as downward. As material is removed from top of right block the block rises until its bottom lines up with left block. (2) Same blocks as in (1), but with load at top added to simulate weight of continental glacier. Surface of right block has been depressed, but when load has been removed will return to height shown in (1). (3, 4) Composite blocks composed of three layers of differing densities, least dense at top and most dense at base. In (3), force of unknown kind raises left-hand block. After top layer has been removed and elevating force no longer acts, top of left block sinks to a level lower than that of right block (4).

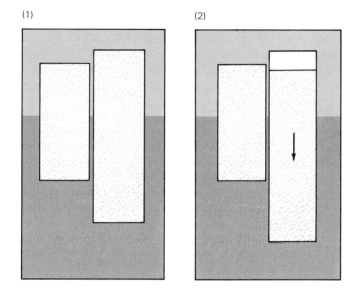

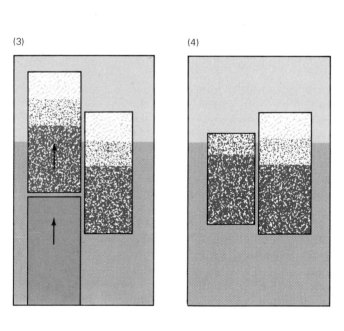

65 million years ago

Flow of the Asthenosphere

The asthenosphere evidently flows laterally when a heavy load causes the lithosphere to sag, and then when the load has been removed it flows back again. Suppose, however, that some process, confined to the asthenosphere, causes the material forming the asthenosphere to flow. A possible way in which such flow might take place is a slow-moving convection current. Over places where the asthenosphere has flowed laterally the lithosphere would sink downward, thus causing the Earth's surface to subside. Similarly, if moving material can pile up anywhere, it would cause the overlying lithosphere to be elevated.

Changes in the Lithosphere

The lithosphere beneath a continental mass consists of three major layers. From the top down these are: (1) sial, (2) sima, and (3) upper mantle, composed of mafic material. The thickness, density or both thickness and density of these layers may change. The top of the sial may be eroded. All the layers may expand or contract with temperature changes or experience changes in density because of increase in metamorphic grade or or other reactions.

Effects of erosion Plate movements may force part of the lithosphere upward for some reason other than isostasy. If this happens, the sialic capping may be thinned or removed altogether by erosion. As the sial is being eroded, its place is being taken by denser material from below. If all the sial were to be eroded away, then all that is left is a block of dense material being held up from below. When the plate movements change, and the dense block is no longer being pushed upward, isostasy will cause the block to sink, possibly below sea level. In the Bahamas, for example, the surface of the lithosphere has subsided about 5 km in the last 125 million years. If it were not for the fantastic ability of marine organisms to create limestones, which replaces the subsided material, the Bahama Islands would have disappeared long ago.

Thermal expansion or contraction If rocks of the lithosphere are heated they expand at first and eventually may undergo metamorphic changes. Because the metamorphic changes involving silicate minerals take place so slowly, however, considerable volume expansion because of heating is thought to be possible before the metamorphism reactions are completed. In the other direction, when heated rocks cool, they contract.

Mountains, Crustal Deformation, and Recycling of Continents

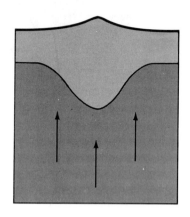

15.17 **Upper surface of thickened lithosphere standing high** because it is underlain by a thick "root." Schematic profile.

Effects of orogeny One of the conspicuous effects of orogeny is a thickening of the sialic part of the lithosphere. Where this part of the lithosphere has been greatly thickened its surface tends to be elevated. The piling up of additional sialic material creates a large load on the lithosphere. The base of the thickened lithosphere tends to be depressed; the top tends to be raised and to become a mountain chain (Figure 15.17). The isostatic elevation of a mountain chain where the sial has been newly thickened is thought to be based on the same principle that causes a thick part of an iceberg floating in the sea to stand higher than a thin part of the same iceberg. The same principle applies to the metal blocks of differing thicknesses floating in mercury. (See (1), Figure 15.16.)

The thickened lithosphere beneath the newly elevated mountain chain is known as the "root" of the mountains. As the elevated parts of the thickened sial are eroded, isostasy takes place. The protrusion on the base of the lithosphere gradually disappears and the lithosphere is slowly "floated" upward. After its "root" has disappeared a former mountain range has become a peneplain. After this review of epeirogeny we now turn to some of the hypotheses of orogeny.

Mechanisms of Orogeny

Orogeny may involve one or all of the following: large-scale folding; great thrusting, especially along enormous low-angle overthrusts; regional metamorphism; and emplacement of granitic batholiths. Two major processes are involved: (1) some kind of lateral movement parallel to the Earth's surface, which takes place to create folds and overthrusts, and (2) great additions of heat, which metamorphoses the rocks on a regional scale and can melt the materials in the orogenic belt to create granitic batholiths.

Mechanisms that have been advanced to account for orogeny include: (1) cooling and shrinking of the Earth on a global scale, (2) gravity sliding of strata down the flanks of an elongated dome of the lithosphere, and (3) lateral movements of great lithosphere plates according to the concepts of plate tectonics.

Global Cooling and Shrinking

One of the most psychologically satisfying ideas, and one that was believed by nearly all geologists during the nineteenth century, stated that the Earth's crust became shortened and folded (and possibly also

thrust) during orogeny because the interior of the Earth was cooling off and thus the crust became wrinkled as it was forced to accommodate itself to a smaller and smaller interior. This idea has now been discarded.

Gravity Sliding of Strata

Some geologists argue that the folds and overthrusts of orogenic belts do not involve the main body of the lithosphere. These geologists visualize that mountain structures are created chiefly by vertical uplift and secondarily by strata sliding down the sides of large domes (Figure 15.18). Fold mountains may be formed as layers of the original, larger mountain slide down and buckle into a corrugated pattern of folded strata. In order for gravity sliding to create forms large enough to be considered mountains, the original dome would have had to be extremely massive, both in height and girth.

The structures formed by such downslope movement do not necessarily imply great shortening of the lithosphere nor the existence of compression within the main body of the lithosphere itself.

The most fashionable explanation for orogeny and the development of orogenic belts at present is that orogenic belts are related to sea-floor spreading and the accompanying lateral movements of lithosphere plates.

Orogenic Belts and Plate Tectonics

According to plate tectonics, orogenic belts are born at continental margins during three great cycles of sea-floor spreading. The early phases of development of some orogenic belts are thought to take place at the margin of a continental mass that is located along the trailing edge of a lithosphere plate. Such a passive margin subsides and traps the sediment eroded from the interior of the continent to build a continental terrace. An example is the great continental terrace along the present east coast of North America (Figure 15.19). Development of a continental terrace will continue as long as the adjacent ocean is spreading.

According to the concept of plate tectonics the formation of some existing mountain chains might be explained this way: the Himalayas were formed when the free-moving India slammed northward into the Asian mainland and pushed the mountains upward as India crunched beneath the surface (Figure 15.20); Africa and Europe revolved in a motion that pushed Italy into central Europe and created the Alps, also building up the Matterhorn from a former piece of Italy; the colli-

15.18 Gravity sliding as a cause of folds. (Top) Strata horizontal. (Bottom) Uplift at left creates slope toward right; strata slip downslope, becoming crumpled as a result.

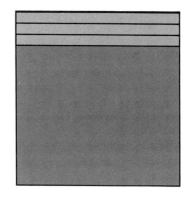

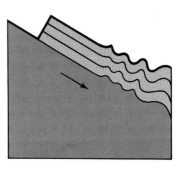

A Near View of the High Sierra
John Muir

Mount Ritter is king of the mountains of the middle portion of the High Sierra, as Shasta of the north and Whitney of the south sections. Moreover, as far as I know, it had never been climbed. I had explored the adjacent wilderness summer after summer, but my studies thus far had never drawn me to the top of it. Its height above sea-level is about 13,300 feet, and it is fenced round by steeply inclined glaciers, and cañons of tremendous depth and ruggedness, which render it almost inaccessible. But difficulties of this kind only exhilarate the mountaineer.

All my first day was pure pleasure; simply mountaineering indulgence, crossing the dry pathways of the ancient glaciers, tracing happy streams, and learning the habits of the birds and marmots in the groves and rocks. Before I had gone a mile from camp, I came to the foot of a white cascade that beats its way down a rugged gorge in the cañon wall, from a height of about nine hundred feet, and pours its throbbing waters into the Tuolumne. I was acquainted with its fountains, which, fortunately, lay in my course. What a fine traveling companion it proved to be, what songs it sang, and how passionately it told the mountain's own joy! Gladly I climbed along its dashing border, absorbing its divine music, and bathing from time to time in waftings of irised spray. Climbing higher, higher, new beauty came streaming on the sight: painted meadows, late-blooming gardens, peaks of rare architecture, lakes here and there, shining like silver, and glimpses of the forested middle region and the yellow lowlands far in the west.

Passing a little way down over the summit until I had reached an elevation of about 10,000 feet, I pushed on southward toward a group of savage peaks that stand guard about Ritter on the north and west, groping my way, and dealing instinctively with every obstacle as it presented itself. Here a huge gorge would be found cutting across my path, along the dizzy edge of which I scrambled until some less precipitous point was discovered where I might safely venture to the bottom and then, selecting some feasible portion of the opposite wall, reascend with the same slow caution. Massive, flat-topped spurs alternate with the gorges, plunging abruptly from the shoulders of the snowy peaks, and planting their feet in the warm desert. These were everywhere marked and adorned with characteristic sculptures of the ancient glaciers that swept over this entire region like one vast ice-wind, and the polished surfaces produced by the ponderous flood are still so perfectly preserved that in many places the sunlight reflected from them is about as trying to the eyes as sheets of snow.

Looking southward along the axis of the range, the eye is first caught by a row of exceedingly sharp and slender spires, which rise openly to a height of about a thousand feet, above a series of short, residual glaciers that lean back against their bases; their fantastic sculpture and the unrelieved sharpness with which they spring out of the ice rendering them peculiarly wild and striking. These are "The Minarets." Beyond them you behold a sublime wilderness of mountains, their snowy summits towering together in crowded abundance, peak beyond peak, swelling higher, higher as they sweep on southward, until the culminating point of the range is reached on Mount Whitney, near the head of the Kern River, at an elevation of nearly 14,700 feet above the level of the sea.

Westward, the general flank of the range is seen flowing sublimely away from the sharp summits, in smooth undulations; a sea of huge gray granite waves dotted with lakes and meadows, and fluted with stupendous cañons that grow steadily deeper as they recede in the distance. Below this gray region lies the dark forest zone, broken here and there by upswelling ridges and domes; and yet beyond lies a yellow, hazy belt, marking the broad plain of the San Joaquin, bounded on its farther side by the blue mountains of the coast.

Turning now to the northward, there in the immediate foreground is the glorious Sierra Crown, with Cathedral Peak, a temple of marvelous architecture, a few degrees to the left of it; the gray, massive form of Mammoth Mountain to the right; while Mounts Ord, Gibbs, Dana, Conness, Tower Peak, Castle Peak, Silver Mountain, and a host of noble companions, as yet nameless, make a sublime show

From THE MOUNTAINS OF CALIFORNIA by John Muir, 1894. Doubleday & Company, Inc.

along the axis of the range.

Eastward, the whole region seems a land of desolation covered with beautiful light. The torrid volcanic basin of Mono, with its one bare lake fourteen miles long; Owen's Valley and the broad lava table-land at its head, dotted with craters, and the massive Inyo Range, rivaling even the Sierra in height; these are spread, map-like, beneath you, with countless ranges beyond, passing and overlapping one another and fading on the glowing horizon.

Lakes are seen gleaming in all sorts of places,—round, or oval, or square, like very mirrors; others narrow and sinuous, drawn close around the peaks like silver zones, the highest reflecting only rocks, snow, and the sky. The eye, rejoicing in its freedom, roves about the vast expanse, yet returns again and again to the fountain peaks. Perhaps some one of the multitude excites special attention, some gigantic castle with turret and battlement, or some Gothic cathedral more abundantly spired than Milan's. But, generally, when looking for the first time from an all-embracing standpoint like this, the inexperienced observer is oppressed by the incomprehensible grandeur, variety, and abundance of the mountains rising shoulder to shoulder beyond the reach of vision; and it is only after they have been studied one by one, long and lovingly, that their far-reaching harmonies become manifest. Then, penetrate the wilderness where you may, the main telling features, to which all the surrounding topography is subordinate, are quickly perceived, and the most complicated clusters of peaks stand revealed harmoniously correlated and fashioned like works of art—eloquent monuments of the ancient ice-rivers that brought them into relief from the general mass of the range. The cañons, too, some of them a mile deep, mazing wildly through the mighty host of mountains, however lawless and ungovernable at first sight they appear, are at length recognized as the necessary effects of causes which followed each other in harmonious sequence—Nature's poems carved on tables of stone—the simplest and most emphatic of her glacial compositions.

Could we have been here to observe during the glacial period, we should have overlooked a wrinkled ocean of ice as continuous as that now covering the landscapes of Greenland; filling every valley and cañon with only the tops of the fountain peaks rising darkly above the rock-encumbered ice-waves like islets in a stormy sea—those islets the only hints of the glorious landscapes now smiling in the sun. Standing here in the deep, brooding silence all the wilderness seems motionless, as if the work of creation were done. But in the midst of this outer steadfastness we know there is incessant motion and change. Ever and anon, avalanches are falling from yonder peaks. These cliff-bound glaciers, seemingly wedged and immovable, are flowing like water and grinding the rocks beneath them. The lakes are lapping their granite shores and wearing them away, and every one of these rills and young rivers is fretting the air into music, and carrying the mountains to the plains. Here are the roots of all the life of the valleys, and here more simply than elsewhere is the eternal flux of nature manifested. Ice changing to water, lakes to meadows, and mountains to plains. And while we thus contemplate Nature's methods of landscape creation, and, reading the records she has carved on the rocks, reconstruct, however imperfectly, the landscapes of the past, we also learn that as these we now behold have succeeded those of the pre-glacial age, so they in turn are withering and vanishing to be succeeded by others yet unborn.

1894

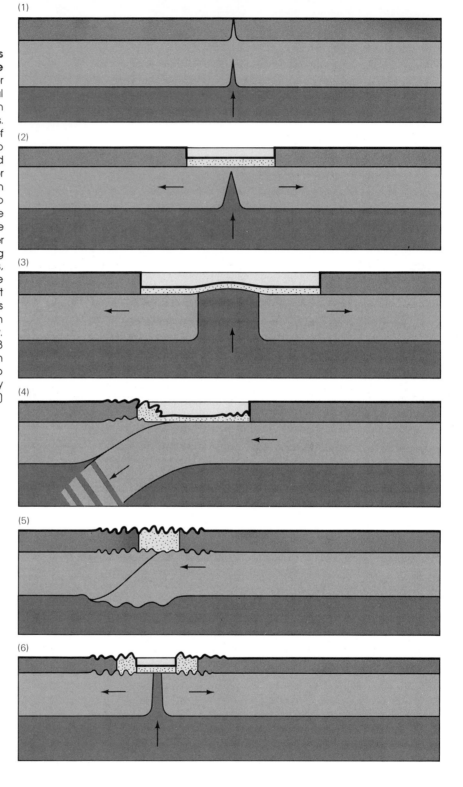

15.19 Origin of mountains according to concepts of plate tectonics. (1-3) In sea-floor spreading cycle No. 1, continental mass at top is split, new ocean grows, and progressively widens. On subsided trailing edges of continents sediments accumulate to form continental terrace and continental rise. (4) In sea-floor spreading cycle No. 2, the ocean created in cycle No. 1 begins to vanish down a subduction zone formed beneath former trailing edge of continent. (Spreading center for cycle No. 2 is out of drawing to right.) (5) As ocean disappears, the continent on its right side approaches the continent on its left side. Collision of two continents crumples the continental-margin strata and causes terminal orogeny. (6) Finally, spreading cycle No. 3 begins; mountain range is split in half and the opposite sides begin to drift away. (Based on J. F. Dewey and J. Bird)

sion of Arabia and Eurasia moving in opposite circular paths forced the uplifting of the Zagros Mountains in Iran.

Unsolved Problems about Mountains

Despite the great new insights and stimulation of effort on the topic of large-scale tectonics that have appeared in the first decade of the "new age of plate tectonics," not all the problems of orogeny have been solved. We still do not know the details of the effects of temperature changes on the rocks, no do we really know why the subsurface

15.20 The Himalayan Mountains may have been forced upward violently if India crashed into Asia according to the concept of plate tectonics. (United Nations)

temperature changes. Lastly, all geologists do not agree on the driving force behind great thrusts along which the "basement" rocks have traveled over sedimentary strata for many tens or even hundreds of kilometers. Despite these difficulties about mechanisms, however, geologists tend to agree on the relationships between orogenic belts and continents, which we summarize next.

Orogenic Belts and Continents

Geologic observations in all parts of the world have established the evidence for recognizing ancient orogenic belts. In this concluding section we examine the relationships between orogenic belts and continental masses and then discuss the question of recognizing in the modern world the early stages of active orogenic belts.

Ancient Orogenic Belts

Once they have experienced a terminal orogeny, the rocks of orogenic belts retain permanent distinctive features. Thus it is possible to trace ancient orogenic belts by their characteristic rock strata and by the uniform trends of their structures, even after their mountainous relief has been eroded away. The nuclei of all continents are the so-called *continental shields,* areas underlain by ancient "basement" rock

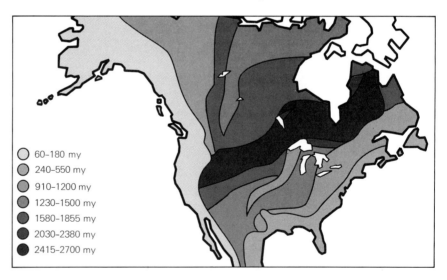

15.21 Orogenic belts of North America. (Geological Survey of Canada)

which during the last 600 million years or so have been subjected only to epeirogenic movements. In North America, the Canadian shield consists of many ancient orogenic belts whose ages become systematically younger outward from a central zone (Figure 15.21). Similar relationships have been found to apply within other continental-shield areas.

Whether continental areas have increased through time or not, despite the ravages of incessant erosion, continental areas have managed to survive. Terminal orogenies thus have kept the continents going, possibly by continually recycling sediment eroded from continents and possibly by additions of new material from the mantle.

Can We Identify Active Modern Orogenic Belts?

Now that we have examined the generalized history of and have summarized the relationship between old orogenic belts and continental masses, we can ask the important question: Can we identify in the modern world any orogenic belts in their geosynclinal stages—places where future fold chains will arise?

At least four likely sites have been suggested. These are: (1) continental terraces and adjoining continental rises, as are found around the eastern and western sides of the Atlantic Ocean; (2) narrow ocean basins bounded on both sides by continental masses, such as the Mediterranean Sea; (3) areas of island arcs and related basins, as off the east coast of Asia and the Aleutians; and (4) the Gulf of Mexico-Caribbean area, including the Bahama Banks. All these areas have subsided and are accumulating thick sequences of sediments. Hence, they all conceivably could be going through initial stages of geosynclinal subsidence comparable with those that ancient mountain chains experienced.

Orogenic Belts and the Geologic Cycle

Geosynclines, orogenic belts, orogeny, mountains, erosion, recycling—all are typical features of the natural cycles we have been describing throughout this book. In order to study the whole process of the deformation of the Earth's crust we have made use of many of the concepts introduced earlier. Rocks, water, geologic time, the interior of the Earth, tectonics, erosion, slopes, and other topics culminate in the study of orogenic belts and their visible products, mountains. Nature itself has provided a fitting grand climax to the study of geology, and nature has shown its pattern of pure and incessant cycles once again.

Chapter Review

1. *Orogeny* defines the birth of mountains from orogenic belts.

2. *Orogenic belts* originate where the strata of sedimentary rocks are especially thick, have been deformed by folding and faulting, and locally may have been invaded by granitic batholiths. The basic fabric of an orogenic belt is permanently imprinted, even though its mountain may have been eroded away.

3. Without the existence of a *geosyncline*, a place where the Earth's crust subsides to form a basin or trough, orogeny could not happen. Geosynclines are likely to appear below sea level, where marine sediments begin to accumulate.

4. All three major kinds of rocks can be detected in orogenic belts. The rock strata of orogenic belts are much thicker than strata of the same age that lie outside orogenic belts.

5. The main geologic structures of orogenic belts are *folds*, *thrusts*, and other kinds of *faults*.

6. Some examples of mountains based on their geologic structure are the Black Hills, South Dakota (large domal uplifts), Sierra Nevada (fault-block), Appalachians (fold), and the interior parts of the Alps (complex).

7. The dynamic history of an orogenic belt involves two contrasting kinds of movements: *epeirogeny* and *orogeny*. These movements are typically arranged in three stages: (1) *geosynclinal*, (2) *terminal-orogenic*, and (3) *post-orogenic*.

8. A condition of flotational balance maintained by the Earth's gravity among segments of the lithosphere is called *isostasy*.

9. The theories that have been advanced to account for orogeny include: (1) cooling and shrinking of the Earth on a global scale, (2) gravity sliding of strata down the flanks of an elongated dome of the lithosphere, and (3) lateral movements of great lithosphere plates according to the concepts of plate tectonics.

10. According to plate tectonics, orogenic belts are born at continental margins during three great cycles of sea-floor spreading.

11. It is possible to trace ancient orogenic belts by their characteristic rock strata and by the uniform trends of the structures even after their moun-

Chapter Review

tainous relief has been eroded away. The nuclei of all continents are the so-called *continental shields*.

12 Based on the relationship between old orogenic belts and continental masses we might be able to identify the sites of future fold chains.

Questions

1 What is the importance of a *geosyncline* in the process of orogeny?
2 Name the four kinds of mountains related to orogenic belts. Why are volcanic cones and stream-dissected plateaus not considered to be related to orogenic belts?
3 Explain the paradox that the process of mountain building begins with subsidence instead of with uplift.
4 Viewed from the perspective of geologic time, why are orogenic belts more important than mountains?
5 What kinds of rocks are usually found in orogenic belts?
6 Define (a) *normal fault*, (b) *reverse fault*, (c) *thrust*.
7 Describe the difference between the movements involved in *epeirogeny* and *orogeny*.
8 What is the effect of a *terminal orogeny*?
9 Explain the concept of *isostasy*, and the relationship of loads to the *asthenosphere* and the *lithosphere*.
10 Describe at least three hypotheses of orogeny.
11 According to the hypothesis of plate tectonics, how were the Himalayas formed?
12 How have orogenic belts helped to maintain the continents?
13 Where in the modern world would one expect to find the indications that an orogenic belt may be in the making and hence that a future mountain chain might develop?

Suggested Readings

Anderson, Don L., "The Plastic Layers of the Earth's Mantle." *Scientific American*, July, 1962. (Offprint No. 855. San Francisco: W. H. Freeman and Company.)

Milne, L. J., et al., *The Mountains*. New York: Time-Life Books, 1962.

Wyckoff, Jerome, *Rock, Time, and Landforms*. New York: Harper & Row, Publishers, Inc., 1966.

"The power of nations has always been closely linked to the abundance of Earth resources"

Chapter Sixteen
The Earth's Resources

The power of nations has always been closely linked to the abundance of Earth resources. Global leaders traditionally have had good supplies of metals and fuels, and the technological insight to put them to use. At the battle of Marathon in 490 B.C., the Persians, many of whom carried only leather shields and stone weapons, were soundly beaten by the Greeks, who had learned to make metal swords and shields. Throughout the nineteenth century, the tiny island of Great Britain ruled much of the world largely because of her vast metal deposits: the rock and regolith of Britain held greater mineral wealth per acre than any other nation advanced enough to utilize it. She was the largest producer of iron, coal, lead, copper, and tin, and her scientists and builders knew how to fashion these riches into machines, mills, railroads, ships, and cities. Today, likewise, the two most powerful nations in the world—the United States and the USSR—are also richly endowed with natural resources. The United States, containing about six per cent of the world population, consumes annually about a third of the world's resources.

As described in Chapter 1, the Earth's resources include energy sources and materials. These can be divided into two groups: resources that can be replenished and resources that can be used one time only. In this chapter, we shall describe where these resources come from, how fast we are using them, and, in some cases, how little there is left.

Renewable Energy Sources

What we shall call *renewable energy sources* either are available in almost infinite amounts or are not permanently altered through use. They include solar energy, plant products (such as wood), and the Earth's internal heat (geothermal energy).

Solar Energy

Solar energy, our most important renewable energy source, consists of a complex barrage of energy waves and particles. We are most familiar with the fraction of solar energy that we can see—visible light. Photovoltaic cells convert solar energy into electricity.

Panels for trapping solar heat have been mounted on houses to provide heat and hot water. Another way to "trap" solar energy is by focusing sunlight on a small area by using mirrors (Figure 16.1). This localized heat may then be used to produce steam, which can drive a turbine to make electricity.

Plant Products

Plants create two basic kinds of material: (1) the woody tissue or **cellulose,** which we can burn directly as wood—this same material is what ultimately makes coal; (2) soft tissues, many of which are **hydrocarbons,** which are used only in their altered fossil forms, oil and natural gas.

Active research is being carried on to find ways by which enzymes can convert solid cellulose into liquid hydrocarbon fuels. This conversion of cellulose (woody plant tissue) into hydrocarbons for fuel (alcohol) may be our most useful future source of energy. History has shown that we will probably use up as much energy as is available. Without careful regulation we could easily deplete our seemingly endless supply of trees. In areas of Asia, Africa, and Latin America, where wood is the main fuel source, there is already a firewood shortage, and many people are burning cow dung instead of wood.

Geothermal Energy

The heat being conducted outward from the interior of the Earth is **geothermal energy.** Geothermal heat escapes into space almost everywhere at a rate that is steady, but is so tiny that it is unnoticeable. However, at volcanoes, hot springs, or perhaps one of the handful of power stations that generate electricity from geothermal heat, so much more heat escapes that its existence becomes spectacularly apparent (Figure 16.2).

Geothermal heat is a large and potentially useful source of energy. In many locations in the United States and elsewhere, hot magma chambers lie close beneath the surface. Engineers are experimenting with technologies to tap this huge heat source and bring its energy into the service of humanity. Even where no bodies of magma are present, deep, warm salt water can be pumped to the surface for heating water and homes.

16.1 Parabolic mirror for concentrating the Sun's rays. Solar furnace, Odeillofont Romeu, Pyrenees, France; Solar energy laboratory. (Centre National Recherche Scientific, France)

Nonrenewable Energy Sources

FINITE

What we shall call *nonrenewable energy sources* are both limited in their total amounts available and permanently altered through use by being converted into other substances in the process of releasing their energy. Nonrenewable energy sources include the so-called fossil fuels (coal and petroleum) which are derived from ancient living organisms, and uranium.

16.2 Commercial geothermal plant. "The Geysers," Sonoma County, about 1200 kilometers north of San Francisco, California. (United States Geological Survey.)

Fossil Fuels

Formerly living organisms that have been converted into forms that can supply energy are known as **fossil fuels.** Fossil fuels now account for more than 95 per cent of all the energy generated in the United States. Fossil fuels are divided into two groups: (1) coals, which are solids, and (2) petroleum, which is a liquid, and includes crude oil and natural gas.

16.3 Impressions of ancient leaves and twigs from black shale immediately overlying a coal layer. (United States Bureau of Mines)

Coal The formation of coal is fairly straightforward. Coal begins as the remains of fresh-water plants, including dead limbs, trunks, branches and leaves (Figure 16.3). The ideal environment for the accumulation of such plant material is thought to be a warm, densely-vegetated swamp located in a basin that is slowly subsiding. The raw plant material first forms peat. As the peat subsides it is compressed by the weight of overburden and is geothermally heated. Slow chemical and physical processes transform peat into coal. Oxygen, hydrogen, water, and other plant elements are released, and carbon is left behind. The greater the proportion of fixed, or elemental, carbon to impurities, the higher the *rank* of a coal. *Lignite*, the lowest-rank coal contains much moisture and burns poorly. *Bituminous* coal contains more fixed carbon than peat and therefore burns hotter. *Anthracite* is the highest-rank coal; it consists of nearly pure carbon, and burns with almost no smoke. The thickness of beds of coal, which are always associated with shale, sandstone, clay, or other sedimentary rocks, ranges from only a few centimeters to 30 meters or more (Figure 16.4). Only those beds more than a meter or so thick are mined.

525
Nonrenewable Energy Sources

If clean, inexpensive ways to burn coal can be found, our abundant coal resources will go far toward filling energy needs. Measured reserves in North America alone amount to some 1560 billion tons (Figure 16.5). The United States has between one-fifth and one-half of the world's coal deposits, concentrated mainly in 17 states. Even though underground mining methods can remove only about half of our coal, there is still enough accessible and usable coal to last for centuries. The U.S. Geological Survey estimates total inferred reserves for the world at 8415 billion tons. Unfortunately for the less-developed nations, 97 per cent of this coal lies in Asia, North America, and Europe (Figure 16.6).

Petroleum Hydrocarbons, compounds of hydrogen and carbon, are the main components of petroleum. There are many kinds of hydro-

16.4 Coal and associated sedimentary strata. (Left) Layer of sub-bituminous coal about 21 meters thick in the Belle Ayr strip mine near Gillette, Wyoming (Amax Inc.) (Right) Schematic column of sedimentary strata, about 50 m thick, showing typical patterned arrangement commonly associated with coal.

16.5 **Locations of principal coal and lignite deposits** in United States and southern Canada. (Compiled from maps published by United States Geological Survey and Geological Survey of Canada)

carbons, and petroleum is a complicated mixture of them.

In addition to hydrocarbons, petroleum contains a very minor amount of complex compounds called *porphyrins* (not related to porphyries of igneous rocks). Porphyrins are significant for two reasons. First, they are formed only in the chlorophyll of green plants and in the hemoglobin of blood. Thus, porphyrins imply an organic origin for petroleum. Second, porphyrins break down at a temperature of 200°C. Therefore, oil containing porphyrins has never been heated above 200°C.

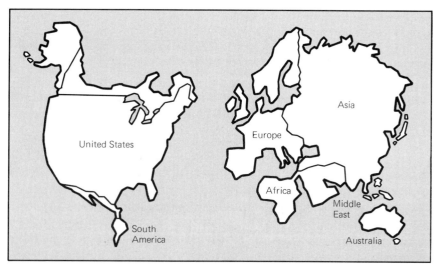

16.6 **World map of coal resources,** with sizes of areas drawn in proportion to their coal resources. (Courtesy Exxon Corporation)

³ Oil typically occurs as tiny droplets widely dispersed through the pore spaces of rock. Any oil deposit require that the appropriate hydrocarbons from some *source* be concentrated in a *reservoir* that has become part of a *trap*. For a petroleum deposit to be *commercial,* the quantity of oil or gas that can be produced *must be large enough* to yield more than the high cost of lease acquisition, exploration, and of drilling and completing the wells. Only about 1 out of 10 holes drilled on land is completed and only about 1 out of 50 is commercial.

One of the great mysteries of petroleum geology centers on the following dilemma. Modern "raw" organic matter accumulates in fine-grained sediments, such as silts and clays, that become shales. The average content of organic carbon of shales is 2.0 to 2.5 per cent by weight. By contrast, the average content of organic carbon in modern sands is less than 0.5 per cent. The *source beds* for petroleum are thus thought to be shales, yet the oil in detrital sedimentary rocks is found in sandstones. How does organic matter get from the source beds into the reservoir sands? If raw organic matter in the muds is deeply buried it is converted first to kerogen and from kerogen to crude oil. After becoming oil, the tiny droplets are thought to migrate out of the shales and move into the nearest sand, a process named *primary migration*. The difficulty with the concept of primary migration is that organic matter is converted to crude oil at depths where the porosity of shales is only about 10 per cent and permeability has vanished.

Another possibility is that the organic matter becomes kerogen during its first cycle of sedimentation, but does not get transferred out of the shale underground. Instead the transfer takes place after the shale has been elevated and eroded. During the second cycle of erosion and deposition tiny solid particles of kerogen can be recycled and deposited with sands. After a second burial these kerogen particles become oil within the sand. If this possibility is correct, then the appropriate sands serve as both source beds and reservoirs.

Various underground arrangements of strata can serve as petroleum traps. For large amounts of oil to accumulate the dip of the sandstone must continue for a great distance without any reversals of direction or without any interruptions. This gives a large subsurface drainage area. The oil migrating up the dip of such layers eventually comes either to the land surface, where it is lost, or to a place where the dip reverses or the sand is cut off by an impermeable material (Figure 16.7).

An important factor in the history of a potential petroleum concerns time of migration and trap formation. If the oil migrates *before* the trap has formed, then all is lost. Oil fills only those traps which formed prior to migration.

528
The Earth's Resources

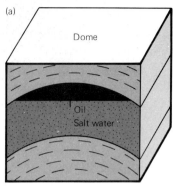
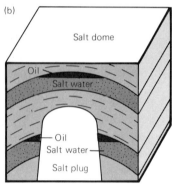
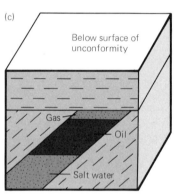
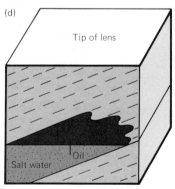

The United States, where the first well was sunk in 1859, has been the world's leading oil producer, but its reserves have dwindled. Currently the Soviet Union is the world's leading oil producer, followed by the United States and Saudi Arabia. Up to 1948, the United States had produced more than one-half of the world's oil. By 1970, it was producing only 21 per cent. Rate of consumption has passed rate of production, and the difference is made up by large imports from oil-rich foreign sources, such as Venezuela and the Middle East (Figure 16.8).

The United States has run through its supplies of natural gas with equal speed. The peak production rate is coming between 1975 and 1980, and the effect of dwindling supplies can already be felt in the form of sporadic shortages and rising prices.

Uranium

Crude oil and natural gas now provide fully 78 per cent of the energy used in the United States (Figure 16.9). Yet these sources represent only 3.5 per cent of our total energy reserves. Thus, many people have come to place their hopes for the future upon the clean burning of coal and upon nuclear energy. Coal resources are about 10 times the combined resources of oil, natural gas, and oil shale, and high-grade uranium resources are perhaps four times greater.

In 1975, nuclear reactors produced 8 per cent of the electricity in the United States. The chief attraction of uranium as fuel is its tremendous efficiency. To fuel one 1000-megawatt power plant for a year requires about 2.3 million tons of coal, 10 million barrels of crude oil—or only 30 tons of uranium. In different terms, one pound of fissionable uranium produces as much energy as 6000 barrels of fuel oil. Thus, the use of uranium as fuel is highly desirable to most electrical utility operators.

16.7 Configurations of strata beneath the surface in four common kinds of oil pools. (a) Dome, with oil confined between saltwater in sandstone and impermeable shale overlying the sandstone. (b) Salt plug, which has created a closed dome in the overlying strata (with oil accumulating as in Figure 16.7a), and has dragged up sandy layers where it has penetrated through them. Oil in these flanking sands is trapped partly by the overlying shale and partly by the salt itself. (c) Unconformity trap, in which oil has migrated up the dip of the tilted sand in the lower sequence but has been trapped by the sealing effect of the unconformably overlying impermeable shale. In fields such as the one shown the oil was not present in the reservoir sand until after the upper, truncated edge of the sand had been sealed by the overlying shale. Many of the world's giant oil fields have resulted from unconformity traps. (d) Stratigraphic trap formed by lateral change from sand to shale and subsequent tilting of all layers. After tilting the oil migrated up the dip of the tilted sand and stopped where the sand pinched out into shale.

529
Nonrenewable Energy Sources

But the reserves of uranium-235 are, like those of fossil fuels, nonrenewable. With the great expansion of uranium-fueled reactors

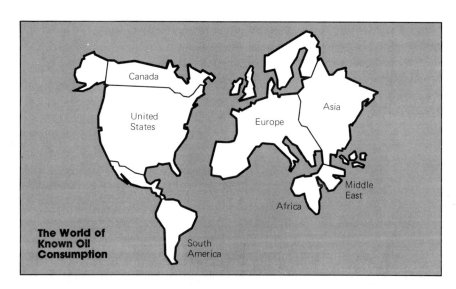

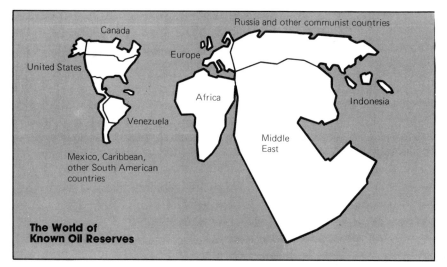

16.8 World oil consumption compared with world oil reserves, with sizes of countries shown in proportion to their rates of use or reserve supply of oil. (Top) Consumption map. (Bottom) Reserves map. Although the United States is currently the world's second-leading oil producer, it appears dwarfed in comparison with the Middle East because we are using up our limited reserves faster than other countries are. Notice in the top map how small the Middle East appears in comparison with the United States. (Courtesy Exxon Corporation)

16.9 Sources of energy consumed in the United States 1850 to 1970, with projection to the year 2000. Projections for the future based on H. H. Landsberg, Resources for the Future, Inc.

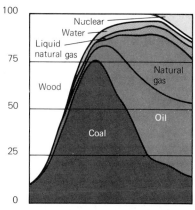

530
The Earth's Resources

planned for the next two decades, most of the high-grade ore will be used up by the 1990s. Fortunately, the so-called breeder reactor, which produces more fuel than it consumes, could help prolong our supplies of nuclear fuel indefinitely. If engineering and safety problems can be solved satisfactorily, breeder reactors will probably be producing significant amounts of power by the next century. Scientists are now experimenting with *nuclear fusion*, which uses abundant isotopes of hydrogen as fuel.

Renewable Resources

As with energy sources, we can divide materials (or, as we shall refer to them, resources) into the categories of renewable and nonrenewable. Again our same generalized definitions apply. That is, renewable resources are those that are not permanently altered through use—for example, water—or those that can be replaced because they grow in response to solar energy—for example, plant products.

Water

Because no life can exist without it, water may be our most important renewable resource. Although 99.35 per cent of the Earth's water is not conveniently available for our use because it exists either in the oceans or in glaciers, we are assured a continous supply by the hydrologic cycle.

There is, however, a limited amount of fresh water falling unevenly over the continents. The fact that it is continuously renewed does not guarantee that there is always enough for all places. More important, perhaps, is the rate of water use, which has increased sharply along with population and the Industrial Revolution (Figure 16.10). As we discussed in relation to running water and groundwater, industry and agriculture use enormous quantities, and they are already running short in the more arid sections of the United States. According to a water specialist with the U.S. Department of the Interior, ". . . all parts of the [United States] either have or will have water problems. The well-watered Eastern and Southern States are beginning to share a concern about water that has been felt since the first settlement of the arid West. As industrial development and

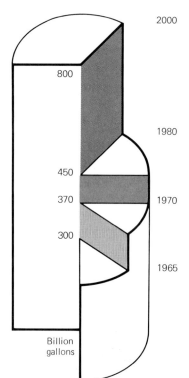

16.10 Rates of water use in the United States 1965 through 1970, with projections to the year 2000. (United States Geological Survey)

urbanization expand in the East, it is becoming more apparent each year that lack of water may deter growth unless action is taken to assure a continued supply."

Plant Products

Before the Industrial Revolution, our dependence upon plant products and the animal products derived from them was nearly total. We used wood for fuel and building, cotton and wool for clothing, and fruits, grain, vegetables, and meat for food. Now we use primarily fossil fuels for energy, rock products for structures, and synthetic fabrics, mostly derived from petroleum, for clothing.

Nonetheless, we still need lumber, cotton, and other renewable plant products and must take care not to use them faster than they can be replaced. Food, for example, is infinitely renewable, but the rate of renewal is finite. The proportion of the worldwide food supply that comes from various sources is shown in Figure 16.11, page 532.

In recent years, the Earth's human population has been increasing faster than the supply of plant food. Although agriculture, with techniques of fertilization and irrigation, is more efficient than ever before, there are finite limits to expansion of the food supply—limits we are fast approaching.

Nonrenewable Resources

Most of the resources familiar to us, and which we have been consuming in ever-increasing quantities, are nonrenewable. These can be divided into two categories: (1) metallic and (2) nonmetallic. Although some metals, such as tin, aluminum, copper, and iron, can be recycled as scrap, they are only partially recovered in re-use and thus we consider them nonrenewable.

The Concept of Ore

A natural deposit from which some material can be extracted at a profit is an **ore.** Because the concept of profitability enters into this definition, the decision about whether a given deposit is an ore must take into account the concentration and mineral form of the desired material, how much money must be spent to remove it and concen-

532
The Earth's Resources

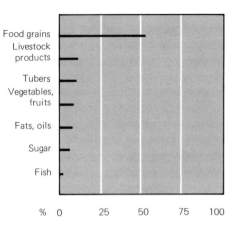

16.11 **Proportions of world's food supply,** indicated according to the various sources. (From Lester B. Brown, "Human Food Production as a Process in the Biosphere," p. 100, The Biosphere, A Scientific American Book, 1970.

trate it (if necessary), costs of transportation, and prices that prevail when the material is sold.

A mineral deposit situated far from a suitable processing site may not be an ore even though its concentration is high. By contrast, new demands for a previously worthless metal may make an ore out of what was previously ordinary rock. Likewise, new processing technologies may "create" ores. Aluminum was first separated in pure form in 1827, but was not used much until the twentieth century when methods of processing large amounts of the metal had been developed. These processes, in turn, depended upon an abundance of electricity and water. The large-scale processing of aluminum did not begin until the 1890s. Price fluctuations and the political situation also determine whether a particular deposit is an ore.

Resources and Reserves

The term *resource* is a general one; it implies little about quantity or quality. Geologists must know how much of a certain resource is in a given place, and whether an ore is rich or lean. Only then is good economic planning possible. The branch of geology that deals with estimating the value of resources, the need for them, and their impact upon the economy is **economic geology**.

Reserves are known, identified deposits from which some material can be extracted profitably with existing technology and under present economic conditions. Resources include reserves plus other mineral deposits that may become available in the future, known deposits that cannot be mined profitably with existing technology or under present economic conditions but that might become profitable with new technology or under different economic circumstances. A second kind of resource is a deposit, rich or lean, that may be inferred to exist, but that has not yet been discovered. A good example of a potential resource has been the large amounts of kerogen in some shales of the western United States. The conversion of shale kerogen to oil is so costly, in terms of dollars, energy, and water, that only pilot projects have been initiated so far.

Metallic Resources

Modern civilization exists because of the sophisticated ways in which we have learned to use metals. The basic metal of modern industrial civilization is steel, which is made from iron that is refined and combined with various other metallic elements to make alloys. Metals

useful for making steel alloys are known as ferro-alloys. These include manganese, nickel, chromium, molybdenum, tungsten, and vanadium. Other metals are used for plating; they are applied as resistant coatings to a steel base. These include chiefly zinc (used in galvanizing) and tin (for food containers). Aluminum, magnesium, and titanium are important lightweight structural materials used in aircraft and other machines. Copper is a basic metal for electrical conductors and pipes. Lead has many uses in storage batteries and as chemically resistant coatings. Gold and silver are used in jewelry, in industrial processes, as the bases for coinage, and as media of exchange.

The Formation of Metallic Ores

Metallic ore deposits are so unusual that they have been called "freaks of nature": they are found only where geological processes have concentrated them. Far less than 1 per cent of crustal rock is in the form of ore. For this reason, a single body of ore may represent a large proportion of the total available metal. A single mine at Climax, Colorado, has furnished more than 60 per cent of the world production of molybdenum (Figure 16.12).

Ore bodies form by processes already described in earlier chapters, including igneous activity, volcanism and plutonism, metamorphism, weathering, sedimentation, and subsurface flow of solutions.

16.12 Climax molybdenum mine, at Fremont Pass, Lake County, Colorado showing scar formed by collapse above underground mine. (AMAX Inc.)

Metallic ores formed by igneous and metamorphic activity
The Earth's heat may concentrate metals during events associated with volcanism, forming ores during the cooling phase of a body of magma. Heavier minerals tend to sink to the bottom of a magma chamber. An illustration of the end product of this process, although it involves a nonmetallic mineral that is not of economic value now, is the layer of olivine toward the bottom of the Palisades sill along the Hudson River. This sinking of minerals in a magma chamber is a kind of *magmatic segregation*. Examples of ores formed by magmatic segregation are the chromite and platinum deposits of the Bushveldt, South Africa, and the chromite near Stillwater, Montana.

As a magma cools and the heavier minerals settle out, the less-dense, more volatile substances begin to escape into the surrounding rock. As this hot material comes in contact with country rock local metamorphism may take place, forming *contact-metamorphic deposits*. The magnetite deposits of Iron Springs, Utah, formed in this way.

Escaping volatiles, which are usually hot solutions rich in dissolved minerals, may be deposited along the walls of channels through country rock, to form *hydrothermal ore deposits*.

Many hydrothermal ore deposits, such as the lead and zinc deposits of the upper Mississippi Valley, the Ozark region, and Tennessee are found far from any bodies of igneous rocks. It is thought that these deposits formed by the action of hot brines on sea-floor sediments. The source of the heat may have been a local "hot spot." The brine may have been sea water or very saline groundwater.

Residual ore deposits As weathering leaches soluble materials from surface rock or regolith, the remaining insoluble portion becomes more concentrated. If this portion contains metals of economic importance, they are *residual ore deposits*.

Sedimentary ores More than 95 per cent of all the ore mined is iron ore, and the most abundant iron ores are *sedimentary* in origin. The mechanism of this origin is not completely understood, partly because the ores were deposited so long ago: most are between 1.7 and 3.2 billion years old. The iron in nearly all is thought to have been transported in solution and deposited chemically. In ancient sedimentary ore, layers of iron commonly alternate with layers of chert-rich siliceous material. The chert-rich layers may represent seasonal episodes of explosive growth of silica-secreting organisms, such as diatoms. These *banded iron formations* contain the world's oldest fossils, and are found in all continental shields.

Banded iron formations, such as those around Lake Superior,

Nonrenewable Resources

may be locally enriched if they have been elevated and exposed to weathering. The siliceous and carbonaceous minerals are leached out, leaving behind *secondarily enriched* ore.

Other important sedimentary deposits are *placers*, which are particles which have been eroded, transported, and concentrated because of superior hardness density. Gold placers are found in Australia, California, Siberia, and Alaska; titanium placers are mined in India and Australia; and tin placers occur in Southeast Asia.

Present

Shields and metallic ore deposits The great Precambrian shields, found on every continent, are the world's richest treasure troves of mineral deposits (Figure 16.13). The shields contain volcanic, residual, and sedimentary ores of all kinds. Even though some of the richest surface deposits have been depleted, it is a safe inference that vast amounts of additional ore underlie the surface.

Nonmetallic Resources

Nonmetallic resources include a long list of materials ranging from asbestos to zeolites. Nonmetallic resources are vital in many industrial processes, in the making of steel, for fertilizers, in construction uses, as fillers and filters, in the making of bricks and pottery, as powders to mix with water to make dense drilling mud, as gemstones, for concrete, and for countless other applications.

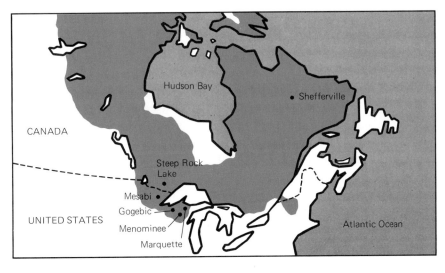

16.13 Canadian shield, north-central United States and Canada, showing locations of chief iron-ore mines. (Geological Survey of Canada)

16.14 Undergound deposit of laminated rock salt (halite), Detroit, Michigan, being extracted at depth of 350 m by room-and-pillar method. The rooms are 15 m wide, 6.5 m high, and up to 1.5 km long. Crosscuts are made at intervals of 25 m leaving pillars to support the roof. (International Salt Company, a part of Akzona Inc.)

Nonmetallic resources of sedimentary origin include strata of limestone and dolostone, sand and gravel, concentrations of diatom skeletons (diatomite), zeolites, and evaporite minerals, gypsum and salt (Figure 16.14). Some sedimentary rocks, notably certain limestones and sandstones, can be cut or broken into pieces that can be used for construction stone (Figure 16.15). Many placers include nonmetallic materials such as diamonds and garnet. Clays may be transported strata or residual accumulations weathered from parent rocks containing feldspars. Sulfur is a nonmetallic resource that is found associated with evaporites; the sulfur probably formed by bacterial reduction of the sulfate minerals. Nonmetallic resources associated with metamorphic rocks include graphite, kyanite, asbestos, talc, and garnet. A few phyllites and schists break into flat slabs that can be used as sidewalks or

16.15 Massive limestone being quarried for building stone at Oolitic, Indiana. The Salem Limestone (Mississippian) is removed by making long cuts with a channeling machine (on tracks in background) and wire saws. Cross cuts are made by using wedges in rows of holes made with a pneumatic drill. (Indiana Limestone Company's PM&B quarry, courtesy Indiana Geological Survey)

patios. Nonmetallic resources from igneous rocks include granite building stones and monumental materials, dolerite for crushed stone in ballast and concrete aggregate (Figure 16.16), feldspar for use in ceramics, mica for use in electrical insulators, and pumice for use in lightweight aggregate.

Resource-Use Curves and Human Activities

After carefully compiling the history of the use of many nonrenewable materials and energy resources, M. King Hubbert has come to the conclusion that the time-*versus*-consumption graphs all follow a simple pattern (Figure 16.17). This graph is controlled by the equation that biologists have derived from studies of growth during the life histories of individual organisms. The remarkable feature of the graph is that the area between the curve and the horizontal axis is equal to the total quantity of the particular resource. Therefore, if one knows a number for this total and has on record the production statistics for drawing the left-hand part of the curve, then one can make surprisingly accurate predictions about how future production will take place. Using such an analysis in the mid-1950's Hubbert accurately forecast that the production of oil in the United States would reach a peak in the early 1970's and would thereafter decline (See Figure 16.9.)

Although such graphs are powerful tools, one's ability to use them obviously depends on the accuracy of the information about the total quantity of the resource that is available. Because of changing economic situations and the invention of new technological processes, it may be extremely difficult or even impossible to decide on a realistic number for the area under the curve. Despite this obvious complication, some aspects of these production-history curves are of the utmost significance to economic practices. As an example, consider the differences between the situation that prevails during the first 50 per cent of consumption and that prevailing during the second 50 per cent.

From the start of production to the 50-per cent line, production increases every year. The rate of increase may be dramatic; whatever the rate is, if it stays the same year after year, when it is plotted on a logarithmic scale the production graph becomes a straight line (Figure 16.18). If the rate of increase changes, then the straight-line segment ends and a new line may begin (marking a new period of a different constant rate of increase). For example, the annual rate of increase in the production of coal and iron ore, the basic ingredients of steel, held steady at about 7 per cent for about one hundred years, from early in

16.16 Dolerite from an ancient lava flow of Jurassic age being quarried at New Britain, Connecticut. Rock blasted from vertical faces, about 18 m high, is trucked to crusher (lower left), and then placed in piles according to size. (New Haven Trap Rock-Tomasso Division of Ashland Oil Company, courtesy Victor Tomasso)

538
The Earth's Resources

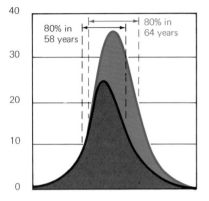

16.17 Production through time for petroleum produced in the United States (lower curve) compared with that produced in the entire world (upper curve). The curve for United States production is based on proved reserves of 200 billion barrels, and the world curve, on proved reserves of 600 billion barrels (M. King Hubbert)

the nineteenth century until 1908. In 1908, the rate curve changed from a 7-per-cent-per-year increase to about a 2-per-cent-per-year increase. As the upper curve in Figure 16.18 shows, this change in the rate of annual increase in the production of coal and steel coincided with the start of an unending increase in the general level of prices.

Whether this coincidence of the two curves is a matter of chance or reflects some underlying economic principle is not for geologists to decide. But geologists, or anyone else, for that matter, are entitled to inquire into the consequences of a continuing difference between the fundamental rates of addition of new raw materials (one of the two kinds of real new wealth) into the system each year and the rates of increase in numbers printed on balance sheets (as established by the interest rates). Under a system that is based on compound interest, is there anything to look forward to but continuing inflation once one approaches the 50-per-cent point on the use curves? And if that is true, can such a system survive the reality of the part of the curve stretching between 50 and 100 per cent? In this part of the curve, the amounts of material produced each year become *less* than they were during the preceding year. With the interest rate demanding larger numbers on the balance sheets and the Earth yielding less and less of the natural resources, what else can happen but inflation, as entrepreneurs are forced to squeeze more out of less?

Geology and Resources

It now appears that the rapid population and industrial growth that has prevailed for the last few centuries cannot be considered the norm in terms of human history. This tumultuous period is, in fact, one of the least-normal phases of human history, a temporary spurt of growth accompanied by new ways to tap energy resources. Population experts now think that this spurt must end in the decades ahead if disaster is to be avoided. It is virtually impossible for the nearly vertical curve of population growth and energy production to continue in a finite world. We can probably expect a flattening of this curve in the near future—if not in our own lifetime, at least in the lifetime of our children. But a sobering question remains: Can we plan rationally

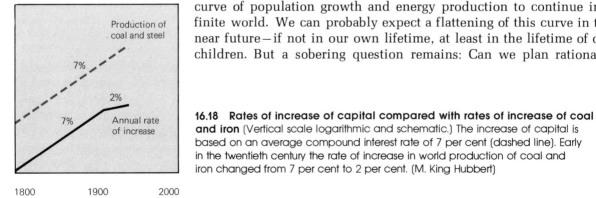

16.18 Rates of increase of capital compared with rates of increase of coal and iron (Vertical scale logarithmic and schematic.) The increase of capital is based on an average compound interest rate of 7 per cent (dashed line). Early in the twentieth century the rate of increase in world production of coal and iron changed from 7 per cent to 2 per cent. (M. King Hubbert)

and act quickly enough to flatten the curve peacefully, without widespread conflict and starvation, or will a struggle for the precious remaining resources undo most of what civilization has accomplished?

Future Resources: Will They Come from the Oceans?

Where will the future supplies of mineral resources and of energy come from? Several answers are possible, but the plain truth is that right now we simply do not know. What we do know is that unless additional supplies are found, and found soon, our present age of affluence and highly mechanized and industrialized activities will end.

Vast areas of the continents have not been thoroughly explored. New methods for studying these areas and for re-evaluating the mineral possibilities of areas near known ore deposits are being applied. Many future discoveries can be expected to follow from study of images of the Earth's surface that are sent back daily from orbiting satellites. Extensive and expensive drilling programs may be required to provide direct information and samples from basement rocks covered by various sedimentary strata.

Many hopes have been expressed about finding vast mineral resources on the sea bed. At the present time, and probably for the immediate future, the chief resource from the sea bed is petroleum that is extracted from wells drilled from offshore platforms or from special drilling ships (Figure 16.19). Oil produced from water-covered areas has come mostly from areas where oil fields are present on the adjacent land.

Other mineral resources now being recovered from the sea bed include placer tin from the shelf sediments off Indonesia and placer diamonds from the shelf sands off southwest Africa. Also, sea water itself is being exploited to recover salt and to extract magnesium.

Future possibilities of mining mineral resources from the sea bed include manganese and associated elements from the manganese nodules that are widespread on many parts of the deep-sea floor, and various metals from the sediments underlying the pools of hot brine in the Red Sea. In these sediments from localized areas in the Red Sea metals have been deposited as a result of the reactions between sea water and rising hot waters that acquired their loads of metals by dissolving the underlying marine evaporites of Miocene age. Other metals may have been concentrated by volcanic activity at mid-oceanic ridge crests.

All future mineral exploration and possibly exploitation will be expensive. The costs may be so great that only governments will dare to undertake such activities.

16.19 Offshore drilling platform used in drilling exploratory borings (this one is off Nova Scotia, Canada) into parts of sea floor covered with water to depths of several hundred meters. Elevated work platform, built well above the level of storm waves, rests on three thick vertical columns which extend downward to large submerged floats located deeper than zone of orbital motion caused by wave action. Such rigs are semisubmersibles. The supports of other drilling platforms extend to the sea floor. (Mobil Oil Corporation)

The Earth's Resources

Chapter Review

1 Global leaders traditionally have had good supplies of metals and fuels, and the technological insight to put them to use.

2 *Renewable energy sources* are either available in almost infinite amounts or not permanently altered through use.

3 Scientists are learning to tap some solar energy for human use by converting it to electricity.

4 Plants create *cellulose,* which we can burn directly as wood, and *hydrocarbons,* which are used in their altered fossil forms, oil and natural gas.

5 The heat being conducted outward from the interior of the Earth is *geothermal energy.*

6 *Nonrenewable energy sources* are both limited in their total amounts available and permanently altered through use.

7 *Fossil fuels* are formerly living organisms that have been converted into coals and petroleum. Fossil fuels now account for more than 95 per cent of all the energy generated in the United States.

8 Even though underground mining methods can remove only about half of our coal, there is still enough accessible and usable coal to last for centuries.

9 Any oil deposit requires that the appropriate hydrocarbons from some *source* be concentrated in a *reservoir* that has become part of a *trap.* For a deposit to be *commercial,* the quality of oil or gas that can be produced must be large enough to yield more than the high costs involved with the development of wells.

10 It is estimated that the total world supply of petroleum is about 2000 billion barrels, and that 80 per cent of this total will be consumed by the year 2025.

11 Nuclear reactors, using uranium, produced 8 per cent of the electricity of the United States in 1975; this process is nuclear *fission.* Scientists are now experimenting with nuclear *fusion,* which uses abundant isotopes of hydrogen as fuel.

12 *Renewable resources* are those that are not permanently altered through use, like water; or those that can be replaced because they grow in response to solar energy, like plant products.

13 *Nonrenewable resources* can be metallic or nonmetallic. The basic metal of modern civilization is steel, which is made from iron that is refined and combined with other metallic elements to make alloys. Nonmetallic resources include a long list ranging from asbestos to zeolite. A natural deposit from which some material can be extracted at a profit is an *ore.*

14 Ore bodies form by processes including igneous activity, volcanism and plutonism, metamorphism, sedimentation, and flow of solutions below the surface.

15 The world population is now growing so fast that if present rates continue it will double every 25 to 30 years. However, population experts predict a flattening of this curve in the near future.

Questions

1. What is the difference between *renewable energy sources* and *nonrenewable energy sources*? What are our main renewable energy sources?
2. Define *fossil fuels*.
3. How does the supply of coal in the United States compare with the supply of oil?
4. Explain the differences between *anticlinal folds*, *salt dome fields*, and *stratigraphic traps* as types of oil reservoirs.
5. Explain the difference between nuclear *fission* and nuclear *fusion*. Why might the breeder reactor be our most important potential source of energy by the next century?
6. Describe the concept of the extraction of *ore* and how it relates to *economic geology*.
7. Define *contact-metamorphic deposits*, *hydrothermal ore deposits*, and *residual ore deposits*.
8. How has it been shown that population and consumption relate to future supplies of resources?

Suggested Readings

Cloud, P., "Mineral Resources from the Sea," pp. 135–158 from *Resources and Man*, San Francisco: W. H. Freeman and Company, 1969.

Cook, E., "The Flow of Energy In An Industrial Society." *Scientific American*, 1971. (Offprint No. 667. San Francisco: W. H. Freeman and Company.)

Hubbert, M. King, *Resources and Man*. San Francisco: W. H. Freeman and Company, 1969.

Hubbert, M. King, "The Energy Resources of the Earth." *Scientific American*, 1971. (Offprint No. 663. San Francisco: W. H. Freeman and Company.)

Laporte, Léo F., *Encounter with the Earth*. San Francisco: Canfield Press, 1975.

McDivitt, James F., and Manners, Gerald, *Minerals and Men*, Revised and Enlarged Edition. Baltimore: The Johns Hopkins University Press for Resources for the Future, Inc., 1974.

Park, Charles F., Jr., "Man and the Earth." *Exxon USA*, fourth quarter, 1975. (Chapter One of *Earthbound: Minerals, Energy, and Man's Future*. San Francisco: Freeman, Cooper & Company, 1975.)

Skinner, Brian J., *Earth Resources*. Englewood Cliffs, N.J.: Prentice-Hall, Inc., 1969.

U.S. Geological Survey, "United States Mineral Resources." U.S. Geological Survey Professional Paper 820. Washington, D.C., 1973.

U.S. Department of the Interior, "United States Energy: A Summary Review." Washington, D.C., 1972.

Appendices

Appendix A **The Origin of the Earth**

Appendix B **Identification of Minerals**

Appendix C **Identification of Rocks**

Appendix D **Metric Equivalents and Powers of Ten**

Appendix E **Understanding Topographic Maps**

Appendix A
The Origin of the Earth

In geology, as in many sciences, the most obvious question is sometimes the most difficult to answer: How was the Earth formed? It might appear that the Principle of Uniformitarianism is all we need: Simply trace present Earth processes backward until we reach the beginning. If you attempt this exercise, you will realize the limits of this otherwise-useful principle. We can trace back into the past as far as the oldest rocks, 3.8 billion years. We think the Earth's age is 4.6 billion years. Therefore, there is a gap in the record of 0.8 billion years. This gap is known as **pre-geologic time.** During this period we think the Earth was a very different place than it is now—there probably was no erosion, no transport of sediment, no ocean, and a very different atmosphere.

One of the reasons geologists are interested in exploring the Moon and the planets of the solar system is that present-day conditions on some of these bodies may correspond to conditions on the Earth during its early years. In addition, data about the interiors of other planets may help geologists understand the relationships among rates of rotation, origin of magnetic fields, and the ways in which internal features are related to tectonic activity at the surface.

Because the Earth's Moon lacks an atmosphere, the Moon's cratered surface has not been modified by weathering, by wind action, by streams, or by glaciers. Mercury possesses a dipole magnetic field about 1 per cent as strong and having the same orientation as the Earth's. Because Mercury's spin axis is perpendicular to its orbital plane, Mercury does not experience seasons, as on Earth. Mercury shows no signs of tectonic activity. Mercury's surface resembles the Moon's; both are cratered and both lack atmospheres.

Mars evidently has a dense core and displays many volcanic features on its surface, yet lacks a magnetic field. The only signs of tectonic activity are fractures that may have resulted from surface swelling near volcanoes and later collapse. The atmosphere of Mars is thin, and contains much carbon dioxide, very little water, and no oxygen. Mars is subject to gigantic yearly dust storms, which begin near the end of the spring in the Southern Hemisphere, at the planet's closest approach to the Sun. Wind speeds in excess of 180 km/hr are required to raise the dust into the atmosphere. Braided channels near the Martian equatorial region are thought to be products of erosion and deposition by running water which took place when the climate was warmer and wetter.

The atmosphere of Venus is about 90 times as dense as the Earth's and consists mostly of carbon dioxide. The rapid movement of the upper atmosphere and its content of acids formed the basis for the idea that the surface of Venus should show effects of great erosion. But when the Soviet spacecraft Venera 9 penetrated the thick yellow veil of Venus in October 1975, it sent back surface photographs which showed instead flat, angular rock fragments up to 50 cm long; the effects of the expected extensive erosion were not evident.

The origin of the Earth is still one of the most actively debated topics in geology, and it must be looked at in terms of the origin of the solar system. We will review two classic theories and also discuss some of the newer concepts.

The Origin of the Solar System

The Earth evolved as part of a larger system of planets and a star, collectively called the solar system. The **solar system** is a name given to our star, the Sun, and the group of nine major planets which orbit the Sun. Also included are satellites, asteroids, meteoroids, and comets. Continuing discoveries of satellites indicate that our knowledge of the solar system is incomplete. However, at the present time we have no evidence that more than nine major planets are present.

Collision concept Two centuries ago, the Comte de Buffon proposed the first comprehensive idea about the origin of the solar system. He thought that the Sun had nearly collided with another star, and that this near-collision pulled material out of both stars. Later, this pulled-out material condensed into the planets. Buffon's idea is now thought to be impossible for two reasons: (1) collisions between stars are extremely rare, and (2) any matter pulled out would be dispersed if it were not condensed.

Buffon also tried to estimate the age of the Earth by heating metals and observing their cooling rates, thus simulating a planet cooling off. His result was 74,832 years.

Nebular-cloud ideas Another approach began in the mid-1800s. The German philosopher Immanuel Kant, the French mathematician Pierre Simon de LaPlace, and the German scientist Hermann Ludwig von Helmholtz all proposed that the solar system began as a large, wispy rotating cloud of gas which, because of its own gravitational attraction, contracted into our Sun and planets. Helmholtz and the British physicist Lord Kelvin also proposed that even today the Sun was obtaining its energy by slow contraction. Calculations showed that if this were so, the age of the solar system would be about 100 million years. Even nineteenth-century geologists knew that this is too brief a time for the rock formation and fossil record found on Earth.

Other factors There are other factors that must be taken into account in any idea of the origin of the solar system. One factor is **angular momentum.** The Sun possesses 99.9 per cent of the mass of the solar system, but only about 2 per cent of its angular momentum, or momentum of rotation. The early theorists thought their contracting cloud would take care of this problem, but later investigations showed that this is not so. Today we believe that some of the angular momentum of the Sun was transferred to the surrounding gases by magnetic coupling. We can demonstrate magnetic coupling by rotating a magnet under a sheet of paper sprinkled with iron filings. The filings will line up with the magnetic field. Also, much of the angular momentum may have been contained in the material which failed to form into either the Sun or planets. Later, this material is thought to have been "blown away" by the solar "wind." The **solar wind** is the name applied to the streams of high-energy particles and energy waves that convey solar radiation outward from the Sun into interplanetary space.

Other questions, such as why all the planets but two rotate the same way, why there is an asteroid belt, and how comets were formed, are still being debated by modern scientists based on the ideas of Kant, LaPlace, and Helmholtz. It is now believed that the nebular-cloud model is basically correct, but needs to be considerably expanded and refined.

From Nebular Cloud to Solid Planet

Much of the material of the primordial nebular cloud went to make up the Sun. Material left behind gradually condensed into small particles, then larger particles, and finally into objects large enough to capture an atmosphere around them and to attract more small objects.

The inner planets (those closest to the Sun) never became large enough to capture much of the hydrogen and helium of the cloud. Therefore, today these planets lack atmospheres, or possess only thin atmospheres. The outer planets, sometimes called the "gas giants," are thought to consist chiefly of the remains of that primordial nebular cloud.

The Earth is thought to have begun as a large accumulation of smaller particles. As more and more particles hit the Earth, they converted their energy of motion into heat, and the proto-Earth heated up. Further, the compression of the interior of the Earth by gravity produced more heat. By far the most important source of heat, however, was the decay of radioactive elements within the Earth, which is still in progress. Within a billion years after it had formed out of the cloud, the temperature of our planet reached the melting point of iron, a very common element which forms about one-third of the Earth. All throughout the globe, iron was melted into enormous "blobs". Because iron is denser than other common elements, it tended to sink toward the center of the Earth, displacing such lighter elements as silicon and aluminum. This melting and sinking of iron is thought to have been a catastrophic event. The sinking iron released large amounts of energy which heated the Earth to a temperature of perhaps 2000°C. This "iron catastrophe" was probably responsible for melting much of the Earth.

After they had melted, the various elements forming the Earth began to differentiate. As iron sank to the core, the less-dense substances rose to create the mantle and crust. A few elements which are very dense in their pure forms, such as uranium, were also carried upward because they tend to form light compounds. The easily melted substances, especially feldspars, rose to the surface. Substances having slightly higher melting points, including the mafic minerals, became the Earth's mantle.

The tremendous heat flowing outward from the interior of the Earth set up huge convection currents, and the crust cooled rapidly. Once the crust had solidified, loss of heat was slowed. At the same time, gases were released from the rocks, forming an early atmosphere that lacked oxygen, and may have consisted of hydrogen, water vapor, methane, and ammonia. Solar energy and chemical reactions broke up some of the gases, particularly water vapor, and the hydrogen escaped into space. After green plants appeared and large-scale photosynthesis took place, oxygen began to accumulate in the atmosphere. Water condensed to fill the ocean basins.

The melting and differentiation is thought to have been completed by about 4 billion years ago. By then, the continents as we know them began to form. The oldest-known rocks are about 3.8 billion years old. If we were transported back in time to that distant age, many of the geologic features would be familiar to us. Living organisms, of course, still would not have begun their arduous and fascinating journey through time to the present.

If you are interested in further reading about the solar system and the origin of the Earth, see the September 1975 issue of **Scientific American**, which devoted the entire issue to the solar system.

Appendix B
Identification of Minerals

Luster

1. Shiny luster (limonite with shiny black surface).
2. Metallic luster (galena, steel gray).
3. Metallic luster (shiny, specular hematite)
4. Dull luster (magnetite; notice tiny iron filings adhering to specimen).
5. Earthy luster (oölitic hematite).
6. Pearly luster (feldspar cleavage fragment).
7. Vitreous luster (glassy quartz crystal).
8. Silky luster (fibrous gypsum).
9. Vitreous luster (massive quartz).
10. Vitreous luster (halite cube).
11. Metallic luster (pyrite).

10	Diamond
9	Corundum
8	Topaz
7	Quartz
6	Orthoclase
5	Apatite
4	Fluorite
3	Calcite
2	Gypsum
1	Talc

Mohs scale of relative hardness. Minerals having high numbers will scratch all minerals having lower numbers. Difference in hardness between diamond (10) and corundum (9) is greater than difference between corundum (9) and talc (1).

Identification of Minerals

Effects of Breaking (Fracture and Cleavage)

Fracture, shown by irregular surfaces of garnet (**3**) and two kinds of quartz (**1** and **2**). Broken surfaces of specimen 1 show both irregular fractures and some conchoidal fractures.

Fibrous mineral, asbestos (**5**), breaks on irregular surfaces parallel to long axes of mineral fibers.

Calcite rhomb (**4**) results from breakage along three cleavages not at right angles.

Cubes of halite (**6**) and of pyrite (**7**) result from three cleavages at right angles.

Two cleavages at right angles shown by plagioclase (**8**) and pyroxene (**9**).

Muscovite (**10**), a mineral having one perfect cleavage direction.

Striae, closely spaced thin parallel lines formed by intersection of internal planes with surface of mineral. (**1**) Plagioclase, distinguished from orthoclase by striae visible on cleavage surfaces; (**2**) Pyrite, with striae on faces of cube.

Table B.1
MINERALS AND PROPERTIES

Note: Boxes contain dots only if the property represented by that box is important in identifying the mineral. For example, the box for streak, which is not used to identify silicate minerals, is empty in the rows for those minerals.

Mineral	Color	Luster	Form and Internal Features	Breakage Fracture	Breakage Cleavage	Streak	Hardness	Magnetism	Chemical Tests	Remarks
Amphibole	● Dark green to black	● Vitreous			● 2, not at right angles		● 5 to 6			Distinguished from pyroxene by two cleavages not at right angles.
Biotite	● Black				● 1 perfect; cleavage flakes are flexible, elastic, and tough.					Color and flexibility plus elasticity of cleavage flakes separate biotite from chlorite.
Calcite					● Three, not at right angles		● 3		● Effervesces with cold dilute acid	Three cleavages not at right angles, hardness of 3, and effervescence with cold dilute hydrochloric acid are diagnostic
Chlorite	● Green	● Silky			● 1 perfect; cleavage flakes are flexible, soft, but not elastic					Green mica with cleavage flakes that are flexible but not elastic.
Fluorite		● Vitreous			● Four directions		● 4			Resembles calcite; distinguished from calcite by hardness of 4 and lack of reaction with acid.
Galena	● Dark gray	● Metallic; shiny	● Cubes		● 3, at right angles	● Black				Specific gravity is 7.6.
Garnet	● Mostly dark burgundy red	● Vitreous		● Irregular			● 6.5 to 7.5			Red color, great hardness, and lack of cleavage, and color typify garnet
Gypsum		● If silky, then fibrous			● 1 prominent cleavage		● 2			Hardness of 2 is diagnostic; no salty taste.
Halite		● Vitreous to greasy	● Cubes		● 3, at right angles		● 2.5		● Salty taste	Salty taste is diagnostic

Identification of Minerals

Mineral	Color	Luster	Form and Internal Features	Breakage Fracture	Breakage Cleavage	Streak	Hardness	Magnetism	Chemical Tests	Remarks
Hematite						● Reddish brown				Many varieties; streak test is diagnostic for all.
Limonite						● Yellow-brown				Streak test is diagnostic.
Magnetite	● Black	● Metallic; dull				● Black		● Only mineral that is strongly magnetic		Use a magnet.
Muscovite	● Silvery	● Vitreous			● 1 perfect; cleavage flakes are tough, elastic					Colorless ("white") mica.
Olivine	● Green	● Vitreous		● Conchoidal fracture			● 6.5 to 7			Green color, conchoidal fracture, vitreous luster, and hardness characterize olivine.
Orthoclase feldspar (including microcline)	● Creamy, pink, flesh-colored, green	● Pearly			● 2 excellent cleavages at right angles		● 5			Separated from plagioclase by color and lack of striae.
Plagioclase feldspar	● Dead white, gray, blue, or transparent	● Pearly, iridescent blue, or glassy	● Striae on cleavage faces		● 2 excellent cleavages at right angles		● 5			Feldspars are generally recognized by pearly luster (exception: vitreous luster in some plagioclase), and two good cleavages at right angles. Striae distinguish plagioclase.
Pyrite	● Brass yellow	● Metallic	● Chiefly cubes		● 3, at right angles	● Black	● 6 to 6.5			Brassy yellow metallic luster and black streak are characteristics.
Pyroxene		● Vitreous			● 2, at right angles		● 5 to 6			Distinguished from amphibole by angles of cleavages; from feldspars by dark color.
Quartz		● Vitreous		● Irregular and conchoidal			● 7			Identified by hardness, luster, and fracture.
Talc		● Greasy					● 1			Softness and greasy feel identify talc.

Appendix C
Identification of Rocks

Igneous Rocks

Equigranular varieties arranged in three lines with felsic (**1, 2, 3**) at left, intermediate (**4, 5, 6**) in center, and mafic (**7, 8, 9,** and **10**) on the right, and grading from coarse in front to fine in the rear.

Felsic Rocks (Granite-Rhyolite Group)

Granite (**1**, coarse grained; **2**, fine grained), easily recognized by its light color, abundant feldspars, quartz, and minor dark-colored ferromagnesian minerals (biotite in specimen **1**).

Rhyolite, **3**, so fine grained that individual minerals cannot be seen with the naked eye.

Intermediate Rocks (Diorite-Andesite Group)

Diorite (**4**, coarse grained; **5**, fine grained), a gray rock dominated by plagioclase feldspar and one or more ferromagnesian silicates, such as biotite or hornblende, (less commonly, pyroxene). Quartz may be present. Orthoclase, if present, is less abundant than plagioclase.

Andesite, **6**, commonly is porphyritic, phenocrysts are plagioclase, biotite, or hornblende.

Mafic Rocks (Gabbro-Dolerite-Basalt Group)

This group is characterized by dark gray to black color and about equal amounts of plagioclase and pyroxene. Individual rocks are recognized by sizes of particles.

Gabbro, **7**, a coarse-grained mafic rock.

Dolerite, a mafic rock of medium particle size. Specimen **8** is coarse-grained dolerite; specimen **9** is fine-grained dolerite.

Basalt, **10**, the fine-grained mafic rock.

Special Features and Ultramafic Varieties

Porphyritic rocks and porphyries, **1** to **3** (distinguished by abundance of phenocrysts; porphyritic rocks contain less than 25 per cent phenocrysts; porphyries contain more than 25 per cent phenocrysts).

1 Coarse diorite porphyry, featuring numerous large phenocrysts of plagioclase feldspar and a few of quartz. Most phenocrysts are about 10 mm long; groundmass is aphanitic.

2 Medium-grained diorite porphyry, with plagioclase phenocrysts about 4 mm long in aphanitic groundmass.

3 Porphyritic andesite, with needle-like phenocrysts of hornblende, up to 3 mm long, in fine-grained groundmass.

4 Anorthosite, a coarse-grained ultramafic rock consisting of bluish plagioclase crystals.

5 Vesicular mafic rock (scoria), characterized by open, sponge-like appearance.

6 Amygdaloidal basalt, formed by deposition of minerals in vesicles.

Sedimentary Rocks: Clastic Varieties

1. Conglomerate, showing characteristic rounded pebbles.
2, 6 Arkose, with feldspars showing as light-colored areas. Both a polished face of a pebbly arkose, **6**, and the rough surface of a coarse-grained variety, **2**, are shown.
3. Claystone. Reddish specimen photographs as gray.
4. Breccia, illustrated by a rounded pebble within which are angular particles.
5. Sandstone, laminated.

555
Identification of Rocks

Sedimentary Rocks: Nonclastic Varieties

The various grades of materials formed by accumulations of plant material, which vary systematically in contents of moisture and carbon, also differ in appearance.

1. Peat. A dried specimen is brownish, crumbly, porous, and contains much recognizable fibrous plant debris.
2. Lignite, known also as "brown coal," is identified by its brown color. Lignite is less porous than peat and tends to break into small chips; recognizable plant debris is less conspicuous than in peat.
3. Sub-bituminous coal is black and blocky, but its luster is dull and its appearance is sooty; it leaves black dust on whatever it touches.
4. Bituminous coal displays shiny, but not an iridescent, luster and breaks in a distinctly blocky pattern.
5. Anthracite (considered by some as a metamorphic rock, but shown here with the other coals) displays a shiny, iridescent luster and breaks with conchoidal fracture.
6, 7 Limestones, coarse-grained kind with fossils visible, **6**, and fine-grained kind without visible fossils, **7**.
8, 9 Dolomite rocks (also called dolostones), medium-grained, **8**, and coarse-grained, **9**, examples.
10. Halite rock, easily identified by its softness, greasy feel, and salty taste.
11. Chert, illustrated by small piece of black flint, can be recognized by its conchoidal fracture, great hardness, and lack of reaction with dilute hydrochloric acid.

Metamorphic Rocks: Common Foliates and Nonfoliates

Foliates, **1–6**:

1 Slate. A thin, flat specimen broken along smooth slaty-cleavage surface; lack of parallelism between slaty cleavage and stratification is not visible in this specimen. Thin specimens of slate can be distinguished from thin specimens of shale (a sedimentary rock) by tapping with a nail. The slate sound is a metallic ring; the shale sound is a dull thud.

2 Phyllite, shiny (or silky) luster; micas not clearly visible to the naked eye.

3, 4 Schist is easily identified by its conspicuously aligned (and sometimes crumpled) micas. Other minerals include quartz and feldspar (usually visible only on surface broken across the layers of aligned micas) and scattered special metamorphic minerals such as muscovite, shown in **3**, and many small garnets in **4**. Schists and gneisses are named according to their chief minerals.

5, 6 Gneiss contains conspicuous amounts of light-colored felsic minerals that form distinct layers, **6**, or pods, **5**, that alternate with dark-colored ferromagnesian minerals. A gneiss having the minerals of granite is a granitic gneiss.

Nonfoliates, **7–9**:

7 Marble, typically glistening white and composed of calcite (will effervesce with cold, dilute hydrochloric acid) or of dolomite (effervesces only if the dilute hydrochloric acid is placed on mineral powder or if the acid is warm).

8 Quartzite, recognized by its crystalline texture, great hardness, lack of reaction with dilute hydrochloric acid, and breakage across quartz particles.

9 Hornfels, a hard-to-break, dark-colored spotted rock featuring scattered crystals of special metamorphic minerals. The conspicuous mineral in specimen **10** is biotite.

Appendix D
Metric Equivalents and Powers of Ten

Metric Conversion Table

English to Metric			Metric to English		
When You Know	Multiply by	To Find	When You Know	Multiply by	To Find
inches	2.54	centimeters	centimeters	0.39	inches
feet	0.30	meters	meters	3.28	feet
yards	0.91	meters	meters	1.09	yards
miles	1.61	kilometers	kilometers	0.62	miles
square inches	6.45	square centimeters	square centimeters	0.15	square inches
square feet	0.09	square meters	square meters	11.0	square feet
square yards	0.84	square meters	square meters	1.20	square yards
acres	0.40	hectares	hectares	2.47	acres
square miles	2.6	square kilometers	square kilometers	0.38	square miles
cubic inches	16.4	cubic centimeters	cubic centimeters	0.06	cubic inches
cubic feet	0.27	cubic meters	cubic meters	0.37	cubic feet
cubic yards	0.76	cubic meters	cubic meters	0.13	cubic yards
cubic miles	4.19	cubic kilometers	cubic kilometers	0.24	cubic miles
ounces	28.3	grams	grams	0.04	ounces
pounds	0.45	kilograms	kilograms	2.20	pounds
tons	0.9	tons	tons	1.1	tons
fluid ounces	30.0	milliliters	milliliters	0.033	ounces
quarts	0.95	liters	liters	1.06	quarts
gallons	3.8	liters	liters	0.26	gallons

Some Useful Metric Conversion Tables

Inches	Inches or Centimeters	Centimeters	Feet	Feet or Meters	Meters
0.39	1	2.54	3.28	1	0.30
0.78	2	5.08	6.56	2	0.60
1.17	3	7.62	9.84	3	0.90
1.56	4	10.16	13.12	4	1.20
1.95	5	12.70	16.40	5	1.50
2.34	6	15.24	19.68	6	1.80
2.73	7	17.78	22.96	7	2.10
3.12	8	20.32	26.24	8	2.40
3.51	9	22.86	29.52	9	2.70
3.90	10	25.40	32.80	10	3.00
7.80	20	50.80	65.60	20	6.00
9.75	25	63.50	82.00	25	7.75

Note: For example, 1 inch equals 2.54 centimeters, and 1 centimeter equals 0.39 inch.

Yards	Yards or Meters	Meters	Miles	Miles or Kilometers	Kilometers
1.09	1	0.91	0.62	1	1.61
2.18	2	1.82	1.24	2	3.22
3.27	3	2.73	3.10	5	8.05
4.36	4	3.64	6.20	10	16.10
5.45	5	4.55	9.30	15	24.15
6.54	6	5.46	12.40	20	32.20
7.63	7	6.37	15.50	25	40.25
8.72	8	7.28	31.00	50	80.50
9.81	9	8.19	62.00	100	161.00
10.90	10	9.10	124.00	200	322.00
21.80	20	18.20	310.00	500	805.00
27.25	25	22.75	620.00	1000	1610.00

Pounds	Pounds or Kilograms	Kilograms
2.20	1	0.45
4.40	2	0.90
8.80	4	1.80
13.20	6	2.70
17.60	8	3.60
22.00	10	4.50
44.00	20	9.00
55.00	25	11.25
110.00	50	22.50
220.00	100	45.00

°C		°F
100	Water boils	212
90		194
80		176
70		158
60		140
50		122
40		104
37	Normal body temperature	98.6
30		86
20		68
10		50
0	Water freezes	32

Use the following formulas to convert Celsius and Fahrenheit temperatures:

$$°C = \frac{(°F - 32)}{1.8} \qquad °F = (1.8 \times °C) + 32$$

Multiples (Powers of 10) and Prefixes

Multiples and Submultiples	Prefix	Symbol
$1{,}000{,}000{,}000{,}000 = 10^{12}$	tera	T
$1{,}000{,}000{,}000 = 10^{9}$	giga	G
$1{,}000{,}000 = 10^{6}$	mega	M
$1{,}000 = 10^{3}$	kilo	k
$100 = 10^{2}$	hecto	h
$10 = 10^{1}$	deka	da
Base Unit $1 = 10^{0}$		
$0.1 = 10^{-1}$	deci	d
$0.01 = 10^{-2}$	centi	c
$0.001 = 10^{-3}$	milli	m
$0.000001 = 10^{-6}$	micro	μ
$0.000000001 = 10^{-9}$	nano	n
$0.000000000001 = 10^{-12}$	pico	p
$0.000000000000001 = 10^{-15}$	femto	f
$0.000000000000000001 = 10^{-18}$	atto	a

Appendix E
Understanding Topographic Maps

Topographic maps are scale models of portions of the Earth's surface which show three dimensions of space in two dimensions. A map view of the ground is from vertically above. Such maps show the configuration of the Earth's surface and aid in making geologic interpretations without actually visiting the region shown. With some basic geologic knowledge and the ability to read such maps, it is often possible to determine general or specific rock structures, the agents which have modified them at the surface, and the extent of the modification.

In addition to their use for geologic interpretation, topographic maps are invaluable for getting around in the field, and serve as bases for plotting the results of geologic observations. The following information is intended to help the student read and understand topographic maps.

Map Symbols

Map symbols are numerous, and would be difficult to reproduce here, but an excellent pamphlet, "Topographic Maps," is available without charge from the Map Information Office, U.S. Geological Survey, Washington, D.C. 20242. This pamphlet contains a comprehensive color representation of map symbols.

As many as five colors may be used with map symbols, and the four most important are listed below:

Red and black: Human-made features.

Blue: Water forms.

Brown: Contours, other relief symbols. (Contours indicating depth of water or elevation on the surface of glaciers are shown in blue.)

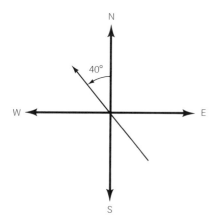

Figure E.1

Scale and Measurement of Distance

A **map scale** is the ratio between a distance shown on the map to the actual corresponding distance on the ground. To be a valid ratio, both distances must be expressed in the same linear units. For example, if 1 cm on the map represents 5000 cm in the field, the ratio is 1:5000.

For practical map use, the scale can be expressed in other ways, all derived from the ratio. In the simplest cases, map scales may be shown as 1 cm = 1 km, for example, thereby varying the units. Most maps show a scale directly in what is known as a **graphic**

scale. A line is marked off with any convenient lengths, such as meters, kilometers, or feet, yards, or miles. A graphic scale can be used as a special ruler to measure directly the distance between points on the map.

Orientation and the Expression of Direction

Orienting a map is the process of placing it horizontally so that map lines are parallel to Earth lines radiating from the point of observation. Two general methods of orientation are:

1. **Alignment by compass.** The compass is placed in a horizontal position and rotated until the north-seeking end of the compass needle points to the proper angle of magnetic declination on the compass dial. A meridian on the map is then aligned parallel to the N-S direction on the compass dial (true N-S direction), making certain that the north ends of both lines coincide.
2. **Visual alignment.** A well-defined linear feature, such as a known stretch of road on the map, is aligned with the actual road in the field. This method is more practical near buildings, power lines, and railroads, where the compass may be affected by local magnetic disturbance resulting in a deviation of the compass needle from the magnetic N-S line.

When the map has been oriented, geographic features appear on it in proper angular relationship to the corresponding features in the field.

Expression of direction A direction is an angle, measured in a horizontal plane, which any given line makes with a standard reference line. If both ends of the standard reference line are used, the directions are called **bearings**. A given line is referred either to the north or south end of the reference line, whichever makes an acute angle with the given line. The acute angle may lie either east or west of the reference line. All bearings are expressed by three parts in the order indicated: (1) North or South, (2) the acute angle, (3) East or West. Figure E.1 shows a line bearing North 40° West (N40°W).

Expression of location Points on the surface of the Earth and on maps are located by using various grid systems. One standard of reference involves two angles. One angle **(longitude)** measures distance east or west of Greenwich, England. The other angle **(latitude)** expresses distance north or south of the Equator. These angles are easily visualized by a schematic sketch in which the Earth has been cut into various slices. On the longitude slice, two hemispheres result, one for the eastern half of the world, and the other for the western half (Figure E.2). The zero line passes through Greenwich Observatory and the 180° point is the International Date Line. Positions are determined by angles from 0° to 180° E or W of Greenwich. Successive longitude lines divide the Earth into wedge-shaped slides, as when a cantaloupe melon is cut.

The latitude slice cuts the Earth in half at the Equator. Angles are measured from 0° at the Equator to 90° N or S, at one of the poles. Successive latitude slices through the Earth define a series of circular slices, as when a series of parallel wires are drawn through a hard-boiled egg (Figure E.3).

In the United States, the most widely used grid system was adopted for land surveys, and makes use of North-South Range lines and East-West Township lines. The **Township-Range grid system** is based on a unit known as a township, which is a square measuring 6 miles on each size, an area of 36 square miles. (In this case the metric system is not yet used.) The Township-Range designation refers to the center of the town-

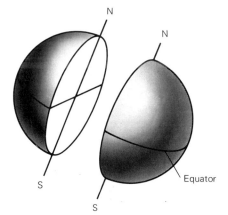

Figure E.2

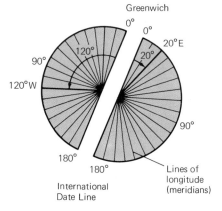

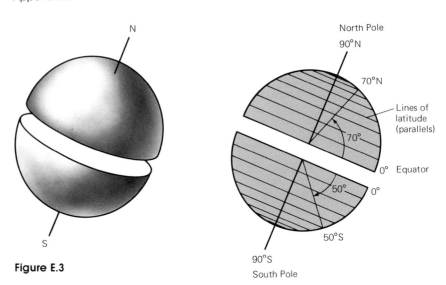

Figure E.3

ship. Each square mile is known as a section; these are numbered in rows, beginning in the northeast corner, and zig-zagging back and forth as shown in Figure E.4. Each square-mile section can be subdivided into quarter sections, and these in turn into quarters, and so forth. Divisions can also be made by N or S and E or W halves.

Contour Maps

To determine a quantitative picture of heights, slopes, and the shapes of land features from a map, a device for showing relative and absolute elevations of all points on the map must be used. The device most commonly employed is the **contour line**. A contour line traces the shape that would be formed by the intersection of a horizontal plane with the land surface at a stated altitude above a reference surface. Successive lines are separated by a uniform vertical distance, called the **contour interval.**

Constructing topographic profiles For many purposes, it is necessary to construct topographic profiles from a contour map. The points at the ends of the profile are selected, a line drawn between them, and the edge of a piece of paper is laid along the line. The intersections of each contour line with the profile line are marked on the edge of the paper. The paper is now moved to a piece of graph paper, on which the vertical scale has been selected. Each point where a contour line crosses the profile line is now drawn at the appropriate height on the graph paper. When the points are connected in a smooth line the profile is finished. In many profiles, the vertical and horizontal scales are not the same. Therefore, it is necessary to label both scales on every profile. In Figure E.5, the length of 1 km on the horizontal scale is equal to only 40 meters on the vertical scale. Therefore, the ratio between the two scales is 1000/40 = 25. This ratio is known as the vertical exaggeration.

In drawing and reading contour maps the following conventions are applied:

1. All points on a given contour line lie at the same elevation.
2. All contours are closed lines but may run off any particular map.

Figure E.4

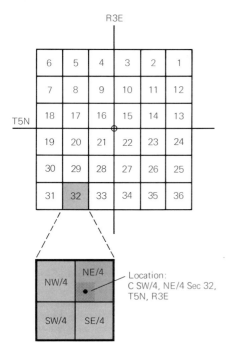

563
Understanding Topographic Maps

3. Two contours of different elevations cannot cross each other, except at overhangs, which are rarely mapped.
4. Contour lines of different elevations may meet or merge at a cliff.
5. The spacing of contours indicates the degree and kind of slope as follows:
   (a) The closer the spacing, the steeper the slope.
   (b) Even spacing indicates a uniform slope.
   (c) Uneven spacing indicates an irregular slope which may be either concave or convex.
6. When contours cross a stream, they always form a V whose apex points **up the valley**. The apex of a V-shaped contour crossing a ridge line points **down the ridge**.

Indexes showing topographic maps published for each state, Puerto Rico, the Virgin Islands, Guam, and American Samoa are available free on request from the U.S. Geological Survey, Washington, D.C. 20242, or Federal Center, Denver, Colorado 80225.

Figure E.5

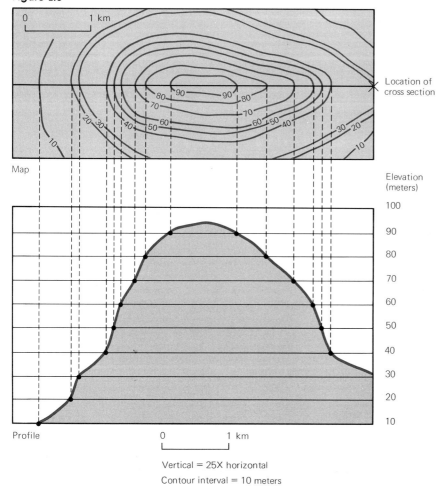

Vertical = 25X horizontal
Contour interval = 10 meters

Glossary

Aa (ah-ah) The jagged, uneven surface on viscous mafic lava.

Abrasion The wearing away of rock surfaces by friction with particles borne by wind, water, or glaciers.

Abyssal plain A flat area of the deep-ocean floor underlain by gravity-transported sediment.

Aftershock Seismic waves that begin to fade after the main shock of the earthquake itself.

Agglomerate Coarse-grained rock composed of fragments of volcanic material of bomb or block size.

Alluvium Sediment deposited by a stream.

Alpine glacier See **valley glacier.**

Altitude The vertical height of a point above sea level.

Amorphous Without form; applied to noncrystalline solids lacking an orderly internal arrangement of ions.

Angle of repose The steepest slope angle at which a body of loose, dry particles remains stable.

Angular momentum The momentum of rotation of the solar system.

Anthracite coal The highest-ranking coal, consisting of nearly pure carbon and burning with almost no smoke.

Anticline An upfold of rock strata.

Aphanitic A textural term applied to igneous rocks in which the particles are smaller than 1 mm in length.

Aphotic zone The lightless portion of the ocean lying deeper than 600 meters.

Aquifer A saturated formation that is porous and permeable, through which groundwater can flow readily.

Arête (ah-RET) A thin, sharp mountain ridge between adjacent cirques.

Ash Volcanic ejecta smaller than 4 mm; also called *volcanic dust.*

Ash flow A volcanic avalanche produced by the eruption of fine particles of frothy magma in a stream of gases.

Asthenosphere A zone of weakness, underlying the lithosphere, where rock flows easily, making isostasy possible.

Atmosphere The gaseous envelope surrounding the Earth.

Atomic energy See **nuclear energy.**

Atomic number The number of protons in the nucleus of an atom or an ion.

Badlands A landscape of closely spaced V-shaped valleys, created by stream erosion.

Bajada (ba-HA-da) A wide, coalescing expanse of eroded debris, characteristic of desert regions.

Barchan dune A crescent-shaped sand dune, with the horns of the crescent curving downwind from the main body.

Basalt Dark-colored, fine-grained dense mafic igneous rock consisting of plagioclase and pyroxene.

Basalt glass A glassy mafic rock resembling obsidian, distinguished by its opaque thin edges.

Basement rocks Ancient metamorphic and igneous rocks, of former orogenic belts, that underlie sedimentary strata.

Batholith A massive discordant pluton having an exposed area of more than 100 square kilometers.

Bauxite Regolith rich in aluminum oxides.

Beach A body of nonconsolidated sediment, sand-size or coarser, along the shore of a lake or an ocean, and extending to the upper limit of wave action or to a change of material on the landward side to the outer line of breakers.

Beach drifting A process by which sediment on a beach-face is transported parallel to the shore by the parabolic flow pattern of the swash and backwash.

Beachface The sloping portion between the berm and the submerged part of a beach.

Bed A sedimentary layer thicker than 1 cm.

Bedload Particles that slide, roll, and bounce along the base of a fluid current, as on a stream bed or in the wind.

Bedrock Continuous solid rock that everywhere underlies the regolith and in some places forms the Earth's surface.

Benioff zones Inclined zones of earthquake foci.

Berm The upper portion of the beach that is predominantly exposed to the air.

Biosphere The part of the Earth in which life exists, limited to regions providing sufficient liquid water and energy from the Sun.

Bituminous coal Middle-ranking coal, rich in carbon and burning with a smokey flame.

Block lava Viscous lava, usually rich in silica that breaks in angular blocks.

Blocks Angular particles larger than 32 mm that were ejected from a volcano in solid form.

Blowout See **deflation basin**.

Bog soil Saturated soil consisting of partly decomposed plant matter that becomes peat.

Bolson A desert basin surrounded by mountains.

Bombs Smoothly curved particles larger than 32 mm that were ejected from a volcano while molten.

Boss A steep-sided hill formed when the rocks around a stock have worn away.

Boulder field An accumulation of angular blocks caused by repeated cycles of freezing and thawing.

Bowen's Reaction Principle A principle that demonstrates the relationships between crystallizing minerals and liquid magma during the formation of igneous rock. See also **continuous reaction series** and **discontinuous reaction series**.

Braided stream A stream interspersed with bars and islands, forming a braid-like pattern.

Breccia (BREH-cha) Any rock consisting of angular coarse fragments. Volcanic breccia is an accumulation of blocks ejected from a volcano, either loose or cemented into a solid mass. Sedimentary breccia consists of a rock composed of angular gravel-size particles.

Broken-glass rock A volcanic product created when lava flows into water, consisting of masses of almost completely glassy particles that become cemented.

Caldera (cahl-DAY-ruh) A large, circular depression having a diameter of more than 1 km created by the explosion or collapse of a volcano.

Caliche (kuh-LEE-chee) Calcite deposited as nodules within soil or as crusts at its surface.

Carbonation A chemical reaction between carbonic acid and minerals or rocks.

Catastrophism See **doctrine of catastrophism**.

Cave A natural underground chamber that can be reached from the surface and that is large enough for a human to enter.

Celerity The speed of advance of a sea wave or a seismic wave.

Cementation A process of lithification of sediments resulting from the crystallization of minerals from solution in the pore spaces among the framework particles.

Chemical energy Energy released or absorbed when ions form bonds with other ions or when matter changes state.

Chemical weathering Chemical processes which change the composition of rock materials. Also called decomposition.

Chert A nondetrital sedimentary rock composed of silica.

Cinder Tephra measuring from 0.06 to 4 mm.

Cinder cone A cone formed by small, uniform-sized tephra that were ejected from a volcanic vent and then fell back nearby.

Cirque (seerk) A steep-walled, bowl-shaped niche in which a valley glacier originated.

Clastic Refers to sediment particles that have been broken.

Cleavage The tendency of certain minerals, crystals, and rocks to break along smooth planes that are parallel to zones of weakness in their internal structure.

Coastal zone A strip of land of varying width that is affected by the sea.

Col (coll) A gap in an arête, caused by erosion.

Collapse breccia Rock material that remains after underlying formations have been dissolved by groundwater.

Collapse crater A crater produced by the collapse of the center of a volcano.

Competence An attribute of streams, measured by the largest particle a stream can move.

Complex mountains Mountainous relief resulting from the elevation and differential erosion of orogenic belts having complex internal structures such as folds, great thrusts, batholiths, and belts of regionally metamorphosed rocks.

Complex weathering Weathering processes which include the crystallization of salts, the combined effects of physical and chemical weathering, and the effects of organisms.

Composite cone A conical volcanic edifice consisting of a steep-sided cone built of alternating layers of tephra and igneous rock lying atop a volcanic shield.

Compromise boundary An irregular boundary of a mineral particle formed when enlarging crystal lattices interfere with one another so that no particles display their characteristic crystal faces.

Concordant A term describing plutons having boundaries that are parallel to the layers of surrounding rock.

Concretion A feature formed in sedimentary strata where circulating groundwater deposits minerals in concentric layers outward from a nucleus, commonly a whole or broken fossil.

Cone (volcanic) A conical structure formed by material ejected from a volcanic vent.

Cone of depression A depression in the level of the groundwater table formed around a well as a result of drawdown.

Connate water Ancient ocean water that is buried with sediments.

Contact The boundary between two different bodies of rock.

Contact-metamorphic aureole The zone surrounding a pluton in which metamorphic changes have taken place.

Continental drift The movement of land areas as a result of tectonic activity.

Continental shelf The submerged outer part of a continent, beginning at the shoreline and extending to the first prominent change in the bottom slope.

Continuous reaction series In Bowen's reaction series, the series in which reactions between individual minerals and magma take place continuously.

Contour line An imaginary line that traces the shape that would be formed by the intersection of a horizontal plane with the land surface at a stated altitude above a reference surface.

Core The central spherical part of the Earth, presumably metallic, and whose inner part is solid and whose outer part is fluid.

Coriolis effect The systematic deflection of bodies or currents moving on the surface of a rotating object.

Corrosion A reaction in which a fluid dissolves a solid, as stream water and its rocky bed, for example.

Country rock A general term for rock penetrated by and surrounding an igneous intrusive.

Covalent bond A bond in which the shared electrons orbit the nucleus of more than one ion.

Creep The imperceptible downslope movement of regolith.

Crevasse A deep fissure in the upper zone of a glacier.

Crust The thin, rocky outermost layer of the Earth.

Crystal A solid body bounded by natural plane surfaces which reflect a regular internal ionic structure.

Crystal form The natural shape of a mineral which reflects its regular, symmetrical internal structure.

Curie point The temperature beyond which a magnetic substance loses its magnetism.

Darcy's law The relationship which states that the rate of flow of groundwater through porous material is proportional to the pressure driving the water, and inversely proportional to the length of the flow path.

Debris flow The rapid downslope movement of a viscous mass of water-saturated regolith.

Debris slide The rapid downslope movement of regolith.

Decomposition See **chemical weathering**.

Deflation basin A depression scooped out of loose materials by deflation. Also called *blowout*.

Deformation Any changes in the form of rocks produced by folding, faulting, and other internal movements.

Delta The body of stream-laid sediment deposited at the mouth of a river.

Dendritic pattern Drainage pattern in which streams dissect material that erodes equally in all directions.

Density Mass per unit volume.

Density current A subsurface current flowing because of differences in density between water masses.

Desert A region characterized by evaporation that greatly exceeds precipitation.

Detritus (deh-TRY-tus) A collective term for particles, derived from the breaking up of bedrock, that were transported and deposited as solids of varying sizes.

Differential weathering The uneven wearing away of rock material.

Dike A discordant tabular pluton that fills a fracture in older, previously formed rock.

Discharge The volume of water flowing through the cross section of a stream channel during a given time.

Discontinuous reaction series The name given to the reaction between a mineral and a melt in which new mineral phases appear and earlier phases disappear when the temperature decreases.

Discordant A term applied to a pluton having contacts that are not parallel to the layers of the country rock.

Disintegration See **physical weathering**.

Disphotic zone The transitional zone of fading light between the photic and aphotic zones of the ocean.

Dissolution A form of chemical weathering in which solid rock materials are dissolved by water.

Doctrine of catastrophism The belief that the geologic record and the Earth's landscape were shaped by sudden, supernatural, and catastrophic events.

Doctrine of uniformitarianism The concept that the geologic record and the Earth's landscape have resulted from the interaction between processes and materials that are subject to established scientific principles. Also called uniformity of process.

Domal mountains Mountainous relief formed by large domal uplifts exposing basement rocks.

Dome In volcanology, a rounded structure formed by magma pushing up from below. In orogeny, a circular structure formed by uplift.

Drawdown The lowering of the water table in the vicinity of a well from which water has been removed. (See **cone of depression**.)

Drift A general term for any sediment related to glaciers, such as till, outwash, and varved lake deposits. See also **continental drift**.

Drumlin An elongated, streamlined hill shaped by a glacier, and consisting of till or of bedrock or of both.

Dune A hill or ridge of sand formed by the wind.

Dust storm A windstorm that sweeps fine particles to great heights and distances.

Dynamic metamorphism Metamorphism resulting from rock deformation.

Earth flow Slow downslope movement in which saturated regolith sags downward in a series of irregular terraces.

Economic geology The branch of geology that deals with estimating the value of resources, the need for them, and their impact on the economy.

Effusive eruption A quiet eruption of fluid, fast-moving mafic lava.

Ejecta Materials ejected explosively by the sudden release of gas from a volcano. (See **tephra**.)

Energy The capacity to do work or to create motion or change.

Energy-level shells Regions into which electrons are organized around the nucleus of an atom or ion.

Environment of weathering The subaerial environment.

Epeirogeny Vertical subsidence or elevation of the Earth's crust that takes place without crumpling the strata.

Epicenter The place on the surface of the Earth directly above the focus of an earthquake.

Equigranular A term describing the texture of igneous rocks in which the crystals are all about the same size.

Erosion A general term for the changes that take place in rock materials at the Earth's surface as a result of the movement of water, wind, or ice.

Erratic A transported rock fragment that is unlike the bedrock underlying it.

Esker A long, winding, continuous ridge of drift created by a subglacial stream of meltwater.

Estuary A V-shaped arm of the sea, pointing landward, formed by the submergence of the mouth of a river.

Evaporite Sedimentary rock that crystallizes when sea water or groundwater in arid regions evaporates.

Exfoliation The spalling off of bodies of rock along sheeting joints.

Explosion crater A crater formed by a violent explosion which blows off the top of a volcanic cone.

Explosive eruption The violent discharge of highly viscous felsic lava.

Extrusive A term applied to igneous rocks formed by the cooling of lava at the Earth's surface.

Fan A low, fan-shaped cone of sands and gravels deposited by a stream.

Fault A fracture along which the opposite sides have been displaced relative to each other.

Fault-block mountains Mountainous relief resulting from vertical elevation along faults and possibly also from differential erosion.

Feldspar A class of silicate minerals containing abundant silicon, oxygen, and aluminum.

Felsic Rich in feldspars.

Felsite Any fine-grained felsic igneous rock.

Ferromagnesian A silicate mineral rich in iron and magnesium.

Firn An accumulation of snow that has melted and refrozen into granular particles.

Fission The splitting of an atomic nucleus.

Fissure eruption A volcanic eruption that emerges through a crack on land or at sea.

Fjord A glaciated U-shaped valley that has been flooded by the sea after the melting of a glacier.

Flank eruption An eruption in which lava breaks through the sides of a volcanic cone.

Flood plains Low, flat areas next to a stream channel that are occasionally inundated with water.

Focus The place within the Earth where the energy of an earthquake is released.

Folding The bending of rock strata.

Fold mountains Mountain belts whose major relief is related to the eroded parts of extensive groups of folds.

Foliation A feature of many metamorphic rocks, characterized by alternating layers of different minerals or parallel arrangements of the same mineral.

Foreshock Seismic waves that begin before the main shock of the earthquake with precursory tremors that become increasingly large.

Fossil The naturally preserved remains of an ancient organism.

Fossil fuel Fuel such as coal or petroleum, formed from the altered remains of organisms.

Framework particle A particle in a sediment or a sedimentary rock, that is of sand size or coarser.

Frost heaving The lifting up of a surface by the growth of ice crystals.

Frost wedging The prying apart of solid bedrock by the pressure of freezing water.

Fumarole A vent in or near a volcano that issues hot gases or fumes.

Fusion The joining of atomic nuclei.

Gabbro A granular igneous rock composed mainly of ferromagnesian minerals, a smaller or equal amount of plagioclase, and negligible quartz.

Gel A solid formed from a colloid.

Geode A deposit of minerals precipitated by groundwater, that grew inward from the walls of a spherical or elliptical cavity in bedrock.

Geologic cycle All of the interactions between materials of the Earth and its sources of energy.

Geologic structures Features that can be delineated by describing the three-dimensional arrangement of rock bodies.

Geologic time Time conceived in large units for the purposes of geologic chronology.

Geologic time scale A scale for classifying the broad units of geologic time, consisting of eras, periods, and epochs.

Geomorphology The study of the development of landforms.

Geosyncline A place where the Earth's crust subsides and sediment accumulates in the early stages of the history of an orogenic belt.

Geothermal energy Energy produced by the heat of the Earth's interior, such as that of hot springs.

Geothermal gradient The regular increase of temperature downward within the Earth.

Geyser A periodically spouting hot spring.

Geyserite A hot-water deposit around a geyser.

Glaciation The advance and retreat of a glacier over an area.

Glacier A flowing mass of ice that was formed by the recrystallization of snow, is powered by gravity, and has flowed outward beyond the snowline.

Glacier ice An ice mass, developed from snow, that has attained a density of 0.84 grams per cubic centimeter, has become impermeable to air, and has begun to move under the pressure of its own weight.

Glass A naturally occurring amorphous solid formed during volcanic activity when molten material cools too quickly to form crystals.

Glowing avalanche See **nuée ardente.**

Graben (GRAH-ben) A block of the Earth's crust that has dropped along faults between two relatively higher-standing blocks.

Graded layer A bed having a systematic gradation in size of particles.

Graded stream A stream in a condition of balance established by the interaction of factors such as energy relationships, discharge, speed, slope, load, and the shape of the channel's cross section.

Gradient The local slope on which a stream flows.

Granite An equigranular, coarse-grained igneous rock consisting mainly of light-colored minerals, chiefly feldspar and quartz.

Granitization The transformation, by metamorphic processes, of sedimentary rocks into granite.

Granular A term applied to rocks in which the crystalline particles are larger than 1 mm.

Granular disintegration A process of physical weathering in which the individual mineral particles of a rock separate from one another.

Graphic granite A granite in which the cleavage faces of feldspar show intergrown crystals of quartz that resemble the characters of Arabic writing.

Gravity The inward-acting force that causes the Earth to attract all objects to its center.

Greenhouse effect The temperature-regulating effect produced by the tendency of carbon dioxide in the atmosphere to trap incoming solar heat.

Groove A desert landform created by abrasion and/or deflation.

Groundwater Water below the surface of the Earth.

Gyre (JHY-er) The spiralling motion of surface currents in the oceans.

Half-life The time required for one-half of the atoms of a radioactive substance to disintegrate into a daughter isotope.

Hanging valley A tributary valley whose floor is higher than the floor of the main valley at the point of juncture.

Hardness The resistance to scratching of a mineral.

Hawaiian A term applied to eruptions involving fluid lava, usually of great quantity, in which gases are liberated continuously and relatively quietly.

Heat energy Energy produced by the constant motion of atoms.

Historical geology The study of the history of the Earth and of life on Earth.

Horst A block of the Earth's crust that has been elevated along faults so that it stands above adjacent lower-standing blocks.

Hot spring A spring which emits liquids that are hotter than the average air temperature of the surrounding region.

Humus Dark-colored, partially decomposed soil material formed when dead plant matter is oxidized slowly and is attacked, but not destroyed, by bacteria.

Hydraulic gradient The slope of the water table.

Hydrocarbon A compound consisting mainly of hydrogen and carbon, created by plants and used for fuel in its altered fossil forms, oil and natural gas.

Hydrologic cycle The natural sequence of changes that water undergoes as it passes into the atmosphere by evaporation from the ocean, precipitates back to land as rain or snow, and returns to the oceans as flow over or below the ground.

Hydrology The study of water.

Hydrolysis A chemical process in which a mineral combines with water.

Hydrosphere The water portion of the Earth, which includes oceans, rivers, lakes, underground water, and glacial ice.

Hydrostatic pressure The pressure exerted by the weight of overlying water.

Ice sheet A type of glacier which, nourished by snowfall, spreads out uniformly over a land surface.

Igneous activity The making of hot liquid material from the mantle, its rising into the crust or onto the surface of the Earth, and its cooling, forming solid igneous rocks.

Igneous rock Rock formed by the solidification of molten material.

Impact erosion A kind of abrasion in which wind-driven sand and tiny pebbles chip shallow irregularities into bedrock.

Induced recharge A method of artificial recharge that involves lowering the water table near a lake or stream so that water will enter the ground from the surface source.

Infiltration The seepage of rainwater into the regolith.

Inselberg Scattered remnants of desert mountains that have been eroded.

Insolation The solar energy received continuously by the Earth.

Interior drainage Streams that disappear by evaporation and infiltration into desert soil and do not drain into the ocean.

Intrusive A term applied to igneous rocks formed by magma that has penetrated older rock within the Earth's crust and has solidified without reaching the surface.

Ion An atom or group of atoms that have gained or lost electrons and thus have acquired electric charges.

Ionic bond A bond formed between two ions having opposite electrical charges.

Island arc A chain of islands that form an arc-like pattern.

Isostasy (eye-SAHSS-tuh-see) The condition of flotational balance maintained by the Earth's gravity among segments of the lithosphere.

Isotope A form of a chemical element having the same number of protons in the nucleus but a different number of neutrons.

Joint A crack in rock along which the opposite sides have not shifted relative to each other.

Juvenile water Water originating from the mantle and joining the hydrologic cycle for the first time.

Kame A short, steep-sided mound of outwash that was deposited in contact with a block of ice left by a glacier.

Kame terrace A terrace-like body of outwash built against the margin of a former glacier.

Karst A type of land surface, found in limestone regions, characterized by sinkholes, underground streams, and caves.

Kerogen A complex solid hydrocarbon, insoluble in organic solvents, that yields petroleum on heating.

Kettle A depression in glacial drift formed after the burial or melting of a large block of ice.

Kinetic energy Released or actualized energy; energy of motion.

Krakatoan A term applied to particularly violent volcanic eruptions.

Laccolith A concordant lenticular pluton that has a flat floor and an arched roof.

Lamina (plural: laminae) A layer of sediment with a thickness of less than 1 cm.

Laminar flow A type of flow in which parallel layers of water shear past one another as smooth planes.

Lapilli (laa-PILL-ee) Small stony fragments or cinder ejected from a volcano.

Laterite Regolith rich in iron oxides.

Latitude An angle expressing distance north or south of the Equator; used in conjunction with *longitude* to determine location on the Earth's surface or on maps.

Latosol A reddish soil of the humid tropics, typically lacking humus and rich in iron oxides and aluminum oxides.

Lattice The regular arrangement of ions in a crystal.

Lava Molten silicate material that has been extruded onto the surface of the Earth.

Lava plain A low-lying region of small relief, underlain by flat-lying sheets of extrusive igneous rock.

Lava plateau An area, underlain by horizontal sheets of extrusive igneous rock, whose top stands at a high altitude.

Law of Constancy of Angles The law which states that the angles between corresponding faces on different crystals of one substance are constant.

Leachate Heated groundwater contaminated with chemical materials dissolved from garbage and other wastes.

Leaching The removal from rock of materials in water solution.

Lenticular Shaped like a lens; thicker in the middle than at the ends.

Lignite The lowest-ranking coal, containing much moisture and burning poorly.

Lithification The conversion of loose sediments into sedimentary rocks by cementation, crystallization, or replacement.

Lithosphere The solid part of the Earth, consisting of the crust and the outermost portion of the mantle.

Lithostatic pressure The pressure exerted by the weight of overlying rock.

Loess (luhss) Nonstratified silt, commonly mixed with scattered particles of sand and clay, that has been deposited from suspension by the wind.

Longitude An angle measuring distance east or west of Greenwich, England; used in conjunction with *latitude* to determine locations on the Earth's surface or on maps.

Longitudinal dune A long, ridge-like sand dune that is aligned in the general direction of the wind that formed it.

Longitudinal profile The aggregate of all the local gradients of a stream.

Longshore current An ocean current parallel to the shore.

Longshore drift A general term for the movement of sediment parallel to the shore as a result of the activity of shoaling and breaking waves.

Luster The property of a mineral related to the amount and quality of light it reflects.

L wave (long wave) A seismic wave that travels along the surface of the Earth.

Mafic (MAY-fic) Igneous rock rich in iron and magnesium.

Magma Molten rock beneath the surface of the Earth.

Magnetism A force of attraction or repulsion exerted through space by certain substances, such as iron.

Mantle The layer of solid rock between the core and the crust of the Earth.

Mantle plumes Cylindrical parts of the mantle having high heat flow.

Mass number The number of neutrons plus the number of protons in the nucleus of an atom.

Map scale The ratio between a distance shown on a map to the actual corresponding distance on the ground.

Mass-wasting The displacement of a mass of bedrock or regolith adjoining a slope, in which the center of the mass advances outward and downward.

Matrix Fine particles that are cemented among coarser particles.

Matter Anything that has mass and volume.

Meander A great sweeping curve in the course of a stream.

Mechanical weathering See **physical weathering**.

Metallic bond A bond in which positively charged ions of an element are packed together, and detached electrons wander freely through the structure.

Metamorphic facies An assemblage of metamorphic minerals that formed under the same environmental conditions.

Metamorphic rock Rock created as a result of extreme modification of already formed rock by heat, pressure, or chemical action.

Metamorphism The process by which igneous or sedimentary rocks are converted into different kinds of rock under the influence of heat, pressure, and circulating fluids.

Meteoric water All water that precipitates from the atmosphere as rain or snow.

Mid-oceanic ridge A system of underwater mountainous ridges that extends along the floors of an ocean basin and displays relief of hundreds of meters.

Migmatite A mixed rock consisting of both granitic rock and schist.

Mineral A naturally occurring solid having a definite chemical composition and an ordered crystalline structure, resulting in a set of relatively uniform chemical and physical properties.

Modified Mercalli scale A scale used to express earthquake intensity that is based on phenomena experienced by people.

Mohorovicic discontinuity A seismic interface between the crust and the mantle.

Moraine A large body of drift that has been shaped into a rounded ridge.

Morphology The shape of the Earth's surface.

Mountain Any part of the Earth's land area having relief of more than 700 m.

Mudflow A debris flow containing an abundance of fine particles.

Natural levee A broad, low ridge of fine sand or silt that has been deposited along the banks of a channel by repeated overflow.

Neap tide A tide whose amplitude is less than normal. (See *tidal amplitude*.)

Nuclear energy Energy released when the atomic nuclei are split (fission) or joined together (fusion).

Nuée ardente (noo-ay ar-DAHNT) A glowing avalanche of white-hot fragments ejected from a volcano.

Obsidian Silicic volcanic glass.

Oöid (OH-oh-id) A small (less than 2 mm) spherical particle composed of calcium carbonate and having concentric or radial internal structure.

Orbicular granite A type of granite containing round or egg-shaped clusters of mineral grains.

Ore A natural deposit from which some material can be extracted at a profit.

Organic texture A texture of sedimentary rocks designating features secreted by organisms.

Orogeny (or-AH-juh-nee) A general term for the deformation of orogenic belts and origin of mountains by folding, faulting, regional metamorphism, and emplacement of granitic batholiths.

Outwash The sediments deposited by streams that issue from a glacier.

Ox-bow lake A crescent-shaped lake formed when a meander has been cut off.

Oxidation A chemical process in which oxygen combines with another element.

Oxide A mineral in which ions, usually of a metal, are combined with oxygen.

Pahoehoe (pah-hoey-hoey) The smooth, curved or ropy surface of fluid mafic lava.

Paleomagnetism The study of magnetic fields surrounding the magnetic particles of ancient rocks.

Parabolic dune A sand dune resembling an elongated crescent, with the horns extending upwind from the main body.

Parent rock The kind of pre-existing rock that has been changed into a metamorphic rock.

Peat Partially decomposed plant matter that can be burned as fuel.

Pedalfer (pe-DAL-fur) Soil of the humid-temperate regions, rich in aluminum and iron; divided into three varieties: conifer soils, leafy-tree soils, and prairie or grassland soils.

Pediment A gently sloping eroded surface between the foot of a mountain and a bajada.

Pedocal (PE-doh-cal) Soil of arid or semi-arid regions, in which calcium has been precipitated as calcite in the A zone; divided into grassland soils, forest soils, and desert soils.

Pegmatite Igneous rock with particles larger than 10 mm.

Peléean A term applied to a violent volcanic eruption in which all of the magma mixes with gas and is expelled in the form of nuées ardentes.

Peneplain (PEE-neh-plain) A land area eroded to an almost flat surface at or near sea level.

Permafrost Permanently frozen regolith.

Permeability The capacity of a porous material to transmit a fluid.

Petrified Turned to stone by a process in which organic molecules are replaced by inorganic minerals.

Petroleum Fluid hydrocarbons consisting of crude oil and/or natural gas.

Photic zone The upper, lighted portion of the ocean where sunlight supports plant growth.

Photosynthesis The process by which green plants manufacture plant tissue from water and carbon dioxide with the aid of energy from sunlight.

Phreatic A term applied to an eruption in which gas is released when magma comes into contact with groundwater or sea water.

Physical geology The study of the processes that shape the Earth.

Physical weathering The processes which break apart rocks without altering their chemical compositions. Also called *mechanical weathering* or *disintegration*.

Piedmont glacier A glacier formed by the flowing together of two or more valley glaciers.

Pillow structure A volcanic structure consisting of clusters of ellipsoidal pillow-like forms, created when lava erupts underwater.

Plagioclase A feldspar mineral, lacking potassium, in which sodium and calcium ions can substitute for one another and in which aluminum correspondingly substitutes for silicon in the tetrahedra.

Plate tectonics The concept that the Earth's lithosphere is constructed of a dozen or so moving plates.

Playa (PLY-ya) The flat floor of a desert basin.

Playa lake An ephemeral lake in a desert basin.

Playfair's law The principle that streams do not simply occupy valleys, but create them.

Plinian A term applied to a volcanic eruption consisting of the explosive emission of large amounts of lava, which causes the collapse of the upper mountain to form a caldera.

Plucking See **quarrying**.

Plug dome A volcanic dome created by the upthrusting of a plugged conduit by the pressure of magma pushing from below.

Pluton Any body of intrusive igneous rock; classified as concordant, discordant, tabular, lenticular, or massive.

Plutonic metamorphism The conversion of very deep sedimentary rocks into coarse-grained metamorphic rocks resembling igneous rocks.

Pluvial lake A desert lake whose former large size is evidence of past abundant rainfall.

Polar reversal The reversal of the Earth's magnetic field which occurs every half million years or so.

Porosity The proportion, in per cent, of the total volume of rock or sediment that is not occupied by solid particles.

Porphyritic An adjective applied to an igneous rock having nonuniform texture in which less than 25 per cent of the crystals are distinctly larger than the others.

Porphyry An igneous rock having a nonuniform texture in which at least 25 per cent of the crystals are distinctly larger than the others.

Potential energy Latent or stored energy.

Pothole A rounded depression caused by the erosion of bedrock by falling water.

Pre-geologic time The period consisting of a gap of 0.8 billion years between the age of the Earth (4.6 billion years) and the age of the oldest rocks (3.8 billion years).

Proglacial lake A lake that receives meltwater from a glacier.

Pumice A frothy, silicic volcanic rock formed by the solidification of a felsic lava froth.

P wave (compressional wave) A seismic wave which alternately pushes and pulls particles in a direction that parallels the direction of the material through which it travels.

Pyroclastic An adjective which collectively designates materials that have been ejected explosively from a volcano.

Quarrying The removal of large blocks of bedrock by an advancing glacier. Also called *plucking*.

Radiant energy Energy from the Sun transmitted to the Earth in the form of electromagnetic radiation. Also called *electromagnetic energy*.

Radioactivity The spontaneous decay of isotopes of certain elements that is accompanied by the emission of great heat and of various particles.

Recharge The addition of water to the zone of saturation.

Reef A wave-resistant ridge of rock in the ocean, especially one made of coral.

Regional metamorphism Widespread metamorphic changes affecting the rocks throughout a considerable area.

Regolith Loose, noncemented fragments of rocks underlying the surface of the Earth.

Relief The difference in altitude between any two points on the Earth's surface.

Remanent magnetism Evidence of ancient magnetic fields preserved in rock.

Replacement The exchange of one solid element for another, as when one material is dissolved and another is deposited in its place.

Reserves Known, identified deposits from which some material can be extracted profitably with existing technology and under present economic conditions.

Residual regolith Regolith formed by the weathering of the underlying bedrock.

Richter magnitude scale A scale that ranks the amount of ground motion created by an earthquake.

Rift zone A zone of fissures.

Rill A tiny channel in soil carved by sheet flow resulting from heavy rainfall.

Rip current A narrow, seaward-flowing current that cuts through the breaker zone.

Roches moutonnées (ROASH moo-tahn-AY) Asymmetrical round hillocks created by glacial abrasion and quarrying. The slope of a roche moutonné is smooth and gentle on the side from which the glacier flowed and steep and jagged on the side toward which the glacier flowed.

Rock avalanche Downslope movement involving large amounts of rock debris traveling at high speed.

Rock cycle A continuous cycle of change which includes the creation, weathering, transporting, deposition, and alteration of natural materials to create regolith, soils, sediments, and rocks.

Rock-dilatancy theory A theory explaining the relationship between the dilation of rocks and the changes in the speeds of P waves before an earthquake.

Rock fall Rapid downslope movement involving the relatively free falling of detached blocks of bedrock.

Rock flour Fine chips and powder created by the abrasion of rocks embedded in the basal and side layers of a glacier against one another and against solid bedrock.

Rock glacier A tongue-shaped body of large rocks, found in some alpine and arctic mountain regions.

Rock slide Rapid downslope movement of many newly detached segments of former bedrock.

Runoff The flow of water over the surface of the Earth.

Salt dome A dome formed above a salt plug.

Salt-dome trap Petroleum that has been trapped in a salt dome or near a salt plug.

Salt plug A cylindrical body of rock salt that has forced its way upward into the sedimentary strata which initially overlay the salt layer.

Saltation A process by which bed-load particles being driven by water or by wind move downcurrent in a series of short excursions or jumps.

Sandstorm A windstorm that hurls sheets of saltating sand just above ground level.

Scoriae (singular: scoria) Porous mafic volcanic rock formed by solidification of vesicular mafic lava.

Sea-floor spreading The outward movement of newly formed crust away from a mid-oceanic ridge.

Sediment Regolith that has been transported at the surface of the Earth and deposited as strata, usually in low places.

Sedimentary rock Rock formed by the accumulation and solidification of layers of sediment.

Sediment yield The amount of sediment supplied per unit of land area.

Seepage pressure The force exerted on regolith by downward-percolating water.

Seismic activity The passage of seismic waves through the Earth.

Seismograph An instrument for detecting and recording seismic waves.

Seismology The study of seismic waves, earthquakes, and the interior of the Earth.

Shearing The tendency of materials to slip along parallel planes.

Sheet flow A flow of surface water in a rather uniform, thin layer.

Sheeting joint A joint parallel to the ground surface, formed by the upward expansion of bedrock.

Sheetwash A broad surface runoff laden with suspended sediment.

Shoreline The place where the edge of the sea meets the land at any particular time.

Shore zone The region extending from the level of lowest tides to the highest point on land that is subjected to wave action.

Sial (SEE-al) A layer composing and underlying the continents, made primarily of rocks rich in silicon and aluminum.

Silicate A mineral composed primarily of silicon and oxygen. Silicates are either ferromagnesians or nonferromagnesians.

Sill A concordant tabular pluton formed when magma has spread laterally between layers of weak rock.

Sima (SEE-ma) A layer below the sial, composed mainly of rocks rich in silicon and magnesium.

Sinkhole A closed depression in the land surface, formed by the collapse of a cave roof.

Slaty cleavage The breaking of slates along smooth surfaces.

Sliderock The angular blocks composing a talus.

Slope-normal component The resisting force of gravity, which pushes material toward a slope and thus tends to keep it from moving.

Slope-parallel component The pulling force of gravity, which acts along and down a slope.

Slope situation The height of a slope, its shape, and its direction of inclination with respect to sunlight and to rain-bearing winds.

Slump The downward and outward movement of a mass of bedrock or regolith along a distinct surface of failure.

Snowfield A large area of snow that lasts from one winter to the next.

Snowline The altitude above which snow lasts throughout the year.

Soil The upper part of the regolith that is capable of supporting the growth of rooted plants.

Soil profile A vertical section of a soil from its surface down through all its zones to the parent material from which it was formed.

Solar system The Sun, and the group of nine major planets which orbit it.

Solar wind Streams of high-energy particles and energy waves that convey solar radiation outward from the Sun into interplanetary space.

Solifluction A type of creep that involves water-saturated soil, occurring in cold regions where the ground freezes deeply.

Solum The collective name for the two distinct upper zones, A and B, of the soil profile.

Sorting The degree of uniformity of sizes of aggregates of sedimentary particles. Sediments are said to be well sorted if the fine particles have been separated from the coarser material, and poorly sorted if they have not.

Spatter An accumulation of clots of liquid lava welded together.

Spatter cone A small volcanic cone produced by the eruption of small clots of liquid lava.

Specific gravity The numerical expression comparing the difference in density of a solid substance with that of water.

Speleothem A mineral deposit in the shape of an icicle, a slab, or a mound, found in caves.

Spheroidal joint A crack in rock occurring along concentrically curved shells so as to make a cube of rock more and more spherical.

Spontaneous liquefaction The transformation of a solid-like body of sediment into a liquid-like body by any process that causes the particles to move out of continuous contact with one another.

Spring A surface stream of flowing water that emerges from the ground.

Spring tide A tide whose amplitude is larger than normal. (See **tidal amplitude**.)

Stalactite A deposit of minerals that has grown vertically downward from the ceiling of a cave.

Stalagmite A deposit of minerals that has grown upward from a cave floor toward a drip source on the ceiling.

Steppe A semi-arid region surrounding a desert.

Stock A massive discordant pluton with an exposed area of less than 100 square kilometers.

Strata (singular: stratum) Layers of sediment, sedimentary rock, or volcanic rock that were spread out, one at a time, with the oldest at the base.

Stratification An attribute of sediments and sedimentary rocks that are arranged into definite layers named strata.

Stratified drift A term for all glacially derived sediments deposited by streams.

Stratigraphic trap A petroleum trap formed by variations in porosity and permeability resulting from depositional processes.

Streak The color of a finely powdered mineral.

Stream All water flowing in channels, ranging from creeks to rivers.

Striae (singular: stria) Tiny linear grooves scratched in rock by a glacier parallel to the direction of its flow. Also, the parallel lines on the surface of a mineral caused by the intersection with the surface of closely spaced parallel internal planes of the lattice.

Strombolian A term applied to relatively regular eruptions occurring as moderate explosions that throw out incandescent cinder, lapilli, and bombs.

Strong interaction force The force that keeps together the whirling arrangements of sub-atomic particles.

Structural geology The study and mapping of geologic structures and the analysis of crustal deformation.

Subaerial Existing or taking place on the land surface, in contact with the atmosphere.

Sub-atomic particle One of the small parts, such as protons, electrons, or neutrons, contained in an atom.

Subduction zone The convergent boundary between two plates, at which crustal material is destroyed.

Subsoil Zone C of the soil profile, consisting of residual or transported regolith; the parent material from which the soil forms. If regolith is residual, includes underlying bedrock.

Sulfide A mineral in which one or more sulfur ions are combined with a metallic ion.

Summit eruption An eruption in which volcanic material is ejected from the central vent at the mountain peak.

Surge A rapid but short-lived advance of a glacier.

Suspended load Particles supported by upward flowing fluid within turbulent eddies in a current of water or air.

S wave (shear wave) A seismic wave that tends to displace particles at right angles to the direction in which the wave travels.

Swell A water-surface wave that has traveled away from a storm center into a region of calm or of less-active winds.

Syncline A downward fold of rock strata.

Tabular Slab-like or table-shaped; usually used in relation to plutons.

Talus (TAY-lus) An accumulation of large, angular blocks that have fallen to the base of a slope.

Tectonic A term referring to features that have formed by large-scale movement of the Earth's crust.

Tephra (TEF-ra) A general term for various pyroclastic materials, including cinder, lapilli, ash, pumice, and bombs.

Terminal orogeny The deformation, metamorphism, and plutonism of orogenic belts.

Terrace A flat, bench-like area formed by various processes, such as the downcutting by a stream through its flat valley floor.

Thrust A great fault along which strata have been piled one on top of another.

Tidal amplitude The vertical difference in altitude of the water surface between the level of high water and that of low water.

Tide A rhythmic rise and fall of water level resulting from astronomical causes and from the ways in which extremely long waves move into and out of basins.

Till Sediment deposited directly by a glacier.

Tillite Sedimentary rock created by the cementation of till.

Tiltmeter An instrument that measures the slight tilting of the ground that occurs before a volcanic eruption or before an earthquake.

Topographic map Scale model of portions of the Earth's surface which show three dimensions of space in two dimensions. A map view of the ground is from vertically above.

Topography The study and mapping of morphologic features.

Transpiration The process by which water is given off by the leaves of plants and enters the atmosphere.

Transported regolith Regolith that has come from elsewhere and may be altogether different from the underlying bedrock.

Transverse dune A sand dune that is linear, usually short, and aligned at right angles to the prevailing wind.

Tsunami (tsoo-NAH-mee) A huge sea wave, not related to tide or wind, that is set off by the displacement of the sea floor by a volcanic eruption, earthquake, or undersea avalanche.

Tufa (TOO-fa) A siliceous product deposited by a hot spring around a vent.

Tuff Fine-grained sedimentary rock composed of volcanic materials.

Tundra soil Soil consisting of sandy clay and raw humus, characteristic of arctic regions.

Turbidity current A density current loaded with suspended sediment.

Turbulent flow A stream flow characterized by a variety of vortices and eddies that are continually forming and disappearing.

Ultramafic A term applied to igneous rocks that consist of only one or two minerals, chiefly olivine, pyroxene, or calcic plagioclase.

Uniformity of process See **doctrine of uniformitarianism.**

Unit cell A regular arrangement of ions that forms the basic building block of a mineral.

Upwelling The rise to the surface of cold subsurface water.

Vadose Water occupying the zone of aeration.

Valley glacier A glacier occupying a valley; usually found in mountain regions.

Varve A layer of sediment deposited during one year.

Ventifact A pebble or larger rock with smooth facets cut and polished by the abrasive action of windblown sand.

Vesicle A small cavity made in volcanic rock by expanding gas.

Vine-Matthews hypothesis A hypothesis explaining the relationships between sea-floor spreading, magnetic anomalies, and polar reversals.

Volcanic dust See **ash.**

Volcanic shield A broad, gently sloping conical mound of volcanic rock, created by repeated effusive eruptions of liquid mafic lava.

Volcano A vent through which hot molten material or gas passes from the Earth's interior onto its surface.

Vug An irregular cavity in rock, often lined with a mineral of a different composition from that of the surrounding rock.

Water cycle See **hydrologic cycle.**

Water table The upper boundary of the zone of saturation.

Wave height The vertical distance between the top of a crest and the bottom of a trough.

Wavelength The horizontal distance between two adjacent crests, or between two adjacent troughs.

Wave period The time required for a complete wave form to pass a given point.

Weathering A general term for all the changes in rock material that take place as a result of their exposure to the atmosphere.

Welded tuff Rock made of melted, welded ejecta.

Xenolith A block of country rock that has been broken loose and engulfed in magma so that it is now surrounded on all sides by igneous rock.

Yardang An outcrop of relatively soft rock, formed by abrasion, with a rounded upwind face and a slanting, elongated downwind face.

Zone A The top layer of the soil profile.

Zone B The second layer of the soil profile, below zone A.

Zone C The third zone of the soil profile; the subsoil or parent material from which the soil is formed.

Zone O An upper soil zone rich in decaying organic matter.

Zone of aeration The part of the ground in which the subsurface pore spaces are occupied partially by water and partially by air.

Zone of cementation The area in which groundwater reacts with rock material to precipitate minerals which become the cementing agents in the creation of sedimentary rock.

Zone of saturation The part of the ground in which all the subsurface pore spaces are filled with groundwater under hydrostatic pressure.

Zone R The zone of the soil profile consisting of bedrock.

Index

Aa lava, 87
Abrasion, 211; and glaciation, 344–345; of rock, 401; in stream, 278
Abyssal plain, 432
Aeration zone, 307
Aftershock, 441
Agassiz, Jean Louis Rodolphe, 331–332
Aggarwal, Yash, 456
Agglomerate, 213
Agriculture, water uses, 265
Air pollution and weathering, 180–182
Alpine glacier, 336
Alteration products, 205
Altitude: and climate, 365; and weathering, 166
Amorphous solids, 74
Angle of repose, 240
Angular momentum, 546
Annular pattern, running water, 286
Anorthosite, 148
Antarctic Ice Sheet, 338-339
Anthracite coal, 222
Anticline, 498
Aphanitic, texture, 138
Aphotic zone, 411
Aquifers, 310–312; recharge, artificial vs. induced, 315
Arcs and trenches, island, 16
Arête, 349
Arkose, 212
Armor, deflation, 392
Arroyo, 382
Ash flow, 87
Asthenosphere, 507; flow of, 509

Atmosphere, 30–31; human impact on, 43–45
Atmospheric conditions and climate, 364
Atom, 52
Atomic nucleus, 53
Atomic number, 53
Atomic size, 53
Aureole, contact-metamorphic, 224
Avalanche: glowing, 88, 106–108; rock, 242–243

Bacteria and soil, 186
Badlands, 283
Bagnold, Ralph A., 397
Bajada, 385
Banded iron formation, 534–535
Barchan dune, 400
Barrier, beach, 426–427
Barycenter, 419
Basalt, 8, 144
Basalt glass, 144
Base flow, stream, 271
Base level, stream, 273
Basement rocks, 493; and large domal uplifts, 499
Basin: deflation, 392; shape and tide, 420–421
Batholith, 136
Bauxite, 187
Beach, 422–428; kinds, 427–428; nourishment, 434; problems, 434; profiles, 426–427
Beach drifting, 427
Beach face, 427
Bearing, 561

Bed, 202; bottomset, 295; load, running water, 277–278
Bedrock, 8 [see also Metamorphic rock(s); Rock(s)]; erosion of, 275; glacial landforms eroded in, 346–351; glaciers in grinding into sediment, 334–335; pore spaces, and weathering, 167; and sedimentary particles, 197
Bedrock coast, 428
Belcher, Edward, 253
Beloussov, V. V., 485
Benioff zone, 481
Berm, 427
Bifirom, 469
Big Geysir, 118
Biosphere, 31–33; human impact on, 45; oceans as part of, 408–412; water and, 264–270
Bituminous coal, 222
Block, volcanic, 89
Block faulting, 506
Block lava, 87
Blowout, 392
Blume, John A., 462
Bog soil, 189
Bolson, 386
Bomb, volcanic, 89
Bonding: covalent, 59; ionic, 57; metallic, 59
Bonneville Dam, 266
Bosses, 135–136
Bottomset bed, delta, 295
Boulder fields, 173
Bowen, N. L., 150
Bowen's reaction principle, 150–151
Braided stream, 296–298

576

Breakers, 426
Breccia: collapse, 324; sedimentary, 212
Broken-glass rock, 116
Brunhes, Bernard, 476
Burke, Kevin, 479

Caldera, 93–95
Caliche, 189, 326, 385
Carbonate minerals, 77–78
Carbonate rock, 213
Carbonation, 175
Carson, Rachel, 184, 424
Cataclastic metamorphic rock, 220
Catastrophism, doctrine of, 5
Celerity, 416
Cellulose, 523
Cementation, 200–201; of sediments, 206–213, 324; zone, 324
Center for Short-Lived Phenomena, 121–122
Channel: flood plain river, 291; migration, 294; stream, 273
Chemical energy, 61, 62
Chemical weathering and decomposition, 173–175
Chert, 213
Cinder, volcanic, 89
Cinder cone, 91
Circulation and density currents, sea, 414–415
Cirque, 346–348
Civilization, and rivers, 266–270
Clastic rock, 554
Clastic sedimentary particles, 210
Clastic texture, 205–206
Clay minerals, changes in, 223
Cleavage: of minerals, 72–73; slaty, 218
Climate: and deep-sea sediment, 432; desert, and geologic cycle, 381–384; and desert lake, 389; and earth factors, 364–365; and external factors, 363–364; quaternary oscillation theories, 365–366; and weathering, 163–165; zones, and wind belts, 412–413
Coal, 222, 524–525
Coast, 421; bedrock, 428; classification and evolution, 429; tide-dominated, 421; wave-dominated, 422–428
Coastal Management Act (1972), 433
Coastal-ocean zone, 421
Coastal sediments, 430
Coastal zone, 421; management, 433–435

Coastline desert, 380–381
Col, 349–350
Cold-arid regions, and weathering, 165
Collapse breccia, 324
Collapse crater, 93
Color of minerals, 70–71
Competence of stream, 278
Complex mountains, 503
Complex weathering, 175–176
Components of gravity on slope, 234–235
Composite cones, 97
Composite volcanoes, 97
Compound, 56
Compressional wave, 449–450
Compromise boundary, 70
Conchoidal fracture, 143
Concordant plutons, 130
Concretion, 325–326
Cone: composite, 97; of depression, 314; volcanic, 91–92
Confined aquifer, 310
Conglomerate, 212
Conservation of Energy, Law of, 63
Constancy of Angles, Law of, 68
Contact-metamorphic aureoles, 224
Contact-metamorphic deposits, 534
Contact-metamorphic minerals, 224–226
Continent: altitude, and climate, 365; and orogenic belts, 516–518
Continental drift, 5, 17, 20, 471–472; and paleomagnetism, 472–474; and plate tectonics, 484–486; and sea floor, 473–474
Continental-interior desert, 378–379
Continental rise, 431
Continental shelf, 430
Continental shield, 516
Continental slope, 431
Continuous reaction series, 151
Contour maps, 562–563
Convergent boundary, 479
Coral dunes, 398–399
Core of Earth, 470
Coriolis effect, 413–414
Corrosion, and running water, 275
Cousteau, Jacques-Yves, 431
Covalent bonding, 59
Crater, volcanic, 92–95
Creep, 249, 252; talus and rock-glacier, 253–254
Crevasses, 342
Cross strata, 203
Crustal deformation and plate tectonics, 486
Crystal, 68; mineral form, 71–72

Crystalline textures, 206; metamorphic rock with, 217–220
Crystallization of salts in weathering, 175–177
Curie point, 66–67, 472–473
Current: rip, 426; tidal, 421
Cuvier, Baron Georges, 5–6

Daily tide, 420
Darcy's law, 312
Darwin, Charles, 6, 332
Davis, William Morris, 289
Debris flow, 244–245, 248
Debris slide, 244
De Buffon, Comte, 546
Decomposition and chemical weathering, 173–175
Deep-sea sediments, 431–433
Deep-water waves, 415–416
Deflation, 391–392
Deflation armor, 392
Deflation basin, 392
De Geer, Gerard, 356–357
Delta, stream, 294–295
Democritus, 52
Dendritic pattern, running water, 286
Density, minerals, 73
Density currents and circulation, sea, 414–415
Depression, cone of, 314
Depth zones and habitats, sea, 410–411
Desert: climate, and geologic cycle, 381–384; coastline, 380–381; continental-interior, 378–379; defined, 374–377; geologic history, 403; lakes, 389–391; landforms, 385–389; mass-wasting, 385; pavement, 392; rainshadow, 379–380; rock weathering in, 172–173; subtropical, 377–378; weathering, 384–385
Detrital sedimentary rock, 212
Detritus, 197–198
Dewey, John F., 479
Diatom, 199
Dietz, Robert S., 475
Differential erosion, 499
Differential weathering, 180
Dike, 130–131
Diorite, 143
Direction, expression of, 561
Discharge, from stream, 270–271
Discontinuous reaction series, 151
Discordant plutons, 130
Disintegration and physical weathering, 171–173

Disphotic zone, 411
Dissolution, 324
Dissolved load, running water, 275–276
Dissolved products, 205
Divergent boundary, 479
Dolerite, 144
Dolomite, 213–214
Domal uplifts, large, and basement rocks, 499
Dome, volcanic, 91
Downslope movement, 238; kinds, 246–247; and landscape, 232–238; rapid, 239–248; slow, 249–254
Drainage, interior, desert, 382–384
Drawdown, 314
Drift, 351 [see also Continental drift]; stratified, 355–356
Drumlin, 360
Dune: Coral, 398–399; formation, 397; kinds, 400–401; movement and stabilization, 400
Dunites, 145, 148
Dust, 43–44
Dust Bowl, 392, 395
Dust storm, 394–395
DuToit, Alexander, 472
Dynamic metamorphism, 497

Eades, James B., 294
Earth: cooling and shrinking, 510; core of, 470; crust, 8, 56–57, 67–74; crust-mantle boundary, 468–470; crust shifts, and climate, 365; forces, 63–67; interior of, 468–470; magnetic field variations and climate, 364; origin of, 545–547; rotation, and Coriolis effect, 413–414; structure of, 7; theories, 470–474
Earth flow, 253
Earthquake: defined, 441–443; effects, 459–463; epicenter, 442, 450–452; historic, 443–449; learning to live with, 457–463; prediction of, 452–457; and seismic waves, 441–443
Earth Resources Technology Satellite, 121
Ebb tide, 421
Economic geology, 532
Effusive eruption, 99
Einstein, Albert, 63
Ejecta, 89
Electrical charge, 53
Electrical conductivity of minerals, 73
Electrical energy, 61, 62

Electricity, and rivers, 266
Electron, 53
Elements(s), 54; in Earth's crust, 56–57
Energy, 59; forms, 60–62; geothermal, 123–125, 523; and mass, 62–63; nonrenewable sources, 523–530; relationships, stream, 270; renewable sources, 522–523; solar, 522; sources, 28
Energy-level shell, 57
Environment: and rivers, 268–270; subsurface, 220–223; of weathering, 161–179
Epeirogeny, and isostasy, 506–510
Epicenter, 442, 450–452
Equigranular texture, 138
Erg, 397
Erosion: of bedrock, 275; differential, 499; factors, 288; of glacial landforms, 346–351; and glaciation, 343-345; headward, 280–281; and lithosphere, 38–39, 509; measurement of, 288; rates, 286–288; by running water, 279–290; stages of cycle, 289–290; and weathering, 160; by wind, sediment and, 391–392
Erratics, glacial, 354
Eruption: effusive, 99; fissure, 98; flank, 98
Esker, 360
Estuary, 421
Evaporation and precipitation, desert, 375–376
Evaporites, 198, 213
Exfoliation, 172
Explosive eruption, 99
Extrusive igneous rock, 130

Fan: and braided stream, 296, 298; desert, 385–386
Fault, 167, 498–499
Fault-block mountain, 501–502
Feel of minerals, 73
Feldspars, 75
Felsic lava, 86
Felsic rock, 140, 141–143
Felsite, 143
Ferromagnesian silicates, 74
Filling, 434
Finch, R. H., 120
Firn, 333
Fissure eruption, 98
Fjord, 348–349
Flank eruption, 98

Flood, and flood control, 267–268
Flood bulge, stream, 271
Flood plain, 267
Flood-plain basin, 291, 292
Flood-plain river, 291–294
Flood tide, 421
Flow zone, glacial, 341–342
Fluid pressure, and water, 235–237
Fluorocarbon gas, 44–45
Focus of earthquake, 442
Fold, 498
Fold mountain, 502–503
Foliate, 217–219, 556
Foliation, 216–217
Forbes, J. D., 339–340
Foreset bed, delta, 295
Foreshock, 441
Formation water, 306
Fossil, 14
Fossil fuel: as energy source, 524–528; and Industrial Revolution, 35–36
Fracture: conchoidal, 143; and cleavage, 549
Fracture zone, 478; glacial, 341–342
Framework particles, 201
"Freon," and ozone layer, 44–45
Frost heaving, 173
Frost wedging, 173
Fuji-No-Yama, 146–147
Fumarole, 99, 117

Gabbro, 144
Gas, volcanic, 90
Gel, 74
Geode, 325, 326
Geologic cycle, 5; and desert climate, 381–384; human dependence on, 33–36; and igneous activity, 84; and life on Earth, 28–33; and orogenic belts, 517–518; and weathering, 160–161
Geologic dating of plutons, 137–138
Geologic structures, 497–498; mountains based on, 499–503
Geologic time, 4
Geologic time scale, 12, 13
Geology: beginnings, 5–7; current view, 23–24; defined, 4–5; economic, 532; physical, in perspective, 47–48; planetary, 11; and resources, 538–539; structural, 497
Geomorphology, 232
Geosyncline, 493, 503–504
Geothermal energy, 123–125, 523
Geothermal gradient, 221

Geyser(s), 110–111, 117; deposits around, 323; Old Faithful and Big Geysir, 118–119
Geyserite, 323
Glaciated valley, 348–349
Glaciation, 343; Pleistocene, 361–362
Glacier(s): background, 330–332; defined, 332; deposition, 351, 354–357; and depositional landforms, 357–361; geological significance, 334–335; as historical archives, 334; kinds, 335–338; landforms eroded in bedrock, 346–351; movements, 339–343; pre-Pleistocene, 362–363; rock, 254; world distribution, 338–339
Glacier ice, 333
Glass, 74; basalt, 144
Glassy rock, 138
Glomar Challenger, 478
Glowing avalanche, 88, 106–108
Gneiss, foliation of, 219
Gondwanaland, 472
Graben, 499
Graded layer, 202, 356, 415
Graded stream, 274
Grand Canyon, 284–285
Grand Coulee Dam, 266
Granite, 8, 141–142: weathering of, 177–178
Granitization, 227
Granular disintegration, 171
Granular texture, 138
Graphic granite, 142
Graphic scale, 560–561
Graphite, 222
Gravity, 63; components on slope, 234–235
Gravity sliding of strata, 511
Graywackes, 212
Great Lakes, 366–367; ancestral, 367–368; modern, 368–369
Great vertical uplift, 499
Greenhouse effect, 31
Greenland Ice Sheet, 339
Groins, 434
Grooves, 403; and glaciation, 345
Ground tilt and earthquake prediction, 457
Groundwater [see also Ocean(s); Running water; Sea water; Water]: consumption, 313; defined, 306; extracting problems, 314–315; and landscape, 318–323; mechanics of flow, 308–312; movement rate, 311–312; pollution of, 316–318; replenishment of, 308; in rock cycle, 324–326; source, 306; storage, 306–308; supply, 313–315; use problems, 316–318
Gullies and valley-side slopes, 283
Gyres, 414

Habitats and depth zones, sea, 410–411
Hall, Sir James, 149
Hanging valley, 348
Hardness of minerals, 73, 548
Hawaiian eruption, 108
Headward erosion, 280–281
Hearn, Lafcadio, 146
Heat, and pressure and generation of magma, 85–86
Heat energy, 60, 62
Helgafell, 122–123
Hess, Harry, and moving sea floor, 474–475
High Sierra, 512–513
Holmes, Arthur, 474–475
Hooke, Robert, 444
Hoover Dam, 266
Horizon vs. zone, 183
Horn, 350
Horst, 499
Hot springs, 117; deposits around, 323
Household water uses, 264
Hubbert, M. King, 528, 536
Humans: activities, and resource-use curves, 536–539; dependence on geologic cycle, 33–36; escape from nature, 46–47; as geologic force, 36–45; impact on atmosphere, 43–45; impact on biosphere, 45; impact on hydrosphere, 42–43; and nature, 40–41
Humid-temperate zones and weathering, 164
Humid tropics and weathering, 164
Humus, 183; and soil, 186
Hutton, James, 5, 6, 7, 160
Hydraulic gradient, and groundwater movement, 311–312
Hydrocarbons, 523
Hydroexplosions, 113
Hydrologic cycle, 30, 262–264
Hydrology, 262
Hydrolysis, 174–175
Hydrosphere, 29–30; human impact on, 42–43
Hydrostatic pressure, 221
Hydrothermal ore deposits, 534

Ice sheet, 337, 338–339
Igneous activity: causes, 84–86; forecasting, 121–122; and geologic activity, 84; living with, 120–121; and metallic ores, 534
Igneous rock(s), 9, 84, 552–553; extrusive vs. intrusive, 130; mafic, weathering of, 178; and magma, 149–151; mineral composition, 140–148; moon, 152–155; varieties, 149; textures, 138–140
Impact erosion, 401
Industrial Revolution and fossil fuel use, 35–36
Industry, water uses, 264–265, 269
Infiltration, 263
Inselberg, 387
"Instant" limestone, 213
Interior drainage, desert, 382–384
Intrusive igneous rock, 130
Ion(s), 53; and bonding, 57–59; sedimentary particles derived from, 198–199
Island, volcanic, 112–113
Island arcs, 99; and trenches, 16
Isostasy and epeirogeny, 506–510
Isotope, 54

Jet-age pollution, 43
Joint Oceanographic Institutions Deep Earth Sampling program, 478
Joint, 167
Jokulklaups, 116
Joyner, William B., 462
Juvenile water, 306

Kame, 361
Kame terrace, 361
Kant, Immanuel, 546
Karst morphology, 322–323
Kendall, George W., 250
Kerogen, 161, 222
Kettle, 361
Kinetic energy, 60, 235
Krakatoa, 102, 104–105
Krutch, Joseph Wood, 398

Laccolith, 134
Lake: desert, 389–391; playa, 387–388; pluvial, 390
Lake deposits, varved, 356–357
Lamina, 202
Laminar flow, stream, 274
Landform(s): depositional, and glaciers, 357–361; desert, 385–389;

glacial, eroded in bedrock, 346–351; volcano, 90–97
Landform barrier, 379
Landscape: and downslope movement, 232–238; and groundwater, 318–323
Land subsidence, 315
Land use, and mass-wasting, 254–257
Langford, Nathaniel Pitt, 110
Lapilli, 89
LaPlace, Pierre Simon de, 546
Large domal uplifts, and basement rocks, 499
Lateral moraine, 359
Laterite, 187
Latitude, 561
Latasols, 187
Lattice, 68
Laurasia, 472
Lava, 9; felsic, 86; kinds, 149; pahoehoe, 87; and underwater eruptions, 113, 116
Lava plain, 96–97
Law of the Sea Conference, 435–436
Leachate, 317
Leaching, 174
Lens-shaped pluton, 134
Lenticular plutons, 130
Lignite, 222
Limestone: "instant," 213; weathering of, 178
Liquefaction, spontaneous, 236–237
Lisbon earthquake, 444
Lithification, 200–201
Lithosphere, 29, 469; and erosion, 38–39, 509; human impact on, 37–39; and isostasy, 507–510
Lithostatic pressure, 221
Load, stream, 273–274
Location, expression of, 561
Lodestone, 75
Loess, 395–397
Longitude, 561
Longitudinal dunes, 400
Longitudinal profile, stream, 273
Longshore current, 426
Long wave, 450
Luster of mineral, 71, 548
Lyell, Charles, 6

Macdonald, Gordon A., 120
MacLeish, Archibald, 23
Mafic lava, 86
Mafic igneous rock, weathering of, 178
Mafic rock, 144

Magma, 9, 84; generation of, 85–86; and igneous rock, 149–151; and rock cycle, 11
Magnetic behavior of minerals, 73
Magnetic field, 66; and climate, 364
Magnetic segregation, 534
Magnetism, 15, 63, 66–67; remanent, 473
Magnetite, 75
Magnetometer, 73
Mammoth Cave, 319, 320–321
Managua earthquake, 448–449
Mantle, 468–470; hot spots, and volcanic islands, 481, 484; plumes, 479
Mantle-core boundary of earth, 470
Map, topographic, 560–563
Mapping of earthquake epicenter, 452
Marine sediments, 429–433
Marlow, Michael S., 485
Marsh, George P., 40
Mass, 53; and energy, 62–63
Massive pluton, 130
Mass number, 53
Mass-wasting: desert, 385; engineering aspects of, 255–256; importance of, 232–234; and land use, 254–257
materials, 28
Matrix, 201
Matter: atomic structure of, 52–57; states of, 56
Matthews, Drummond, 476
Mature stage, erosion, 289–290
Maximilian, Prince, 208
Mayer, Frank Blackwell, 320
Meander, flood-plain river, 293
Measurement of glacial movement, 339–341
Mechanical weathering, 171
Medial moraine, 359
Meinzer, O. L., 313
Mercalli, Giuseppe, 443
Metallic bonding, 59
Metallic ore, formation of, 533–535
Metallic resources, 532–533
Metamorphic activity, and metallic ores, 534
Metamorphic change, 215
Metamorphic facies, 224–226
Metamorphic granite facies, 226–228
Metamorphic rock(s) 10, 214–220, 497, 556 [see also Bedrock; Igneous rock(s); Rock(s); Sedimentary rock(s)]; cataclastic, 220; classification basis, 215–217; with crystalline texture, 217–220

Metamorphism, 214–215, 497
Meteoric water, 306
Metric conversion tables, 557–558
Meyerhoff, A. A., 485
Meyerhoff, Howard A., 485
Mid-oceanic ridge, 16
Migmatite, 227
Mineral(s), 67; carbonate and sulfate, 77–78; contact-metamorphic, 224–226; identification of, 548–552; oxide, 75, 77; properties, 70–74; reasons for studying, 76–77; silicate, 74–75; structure, 67–68; sulfide, 75
Mineral composition, 67; of metamorphic rock, 216; of sedimentary rock, 205
Mining and lithosphere, 37–38
Modified Mercalli Scale, 443
Mohorovicic, Andrija, 468
Mohorovicic discontinuity, 468
Mohs, Friedrich, 73
Mohs scale of relative hardness, 548
Molecule, 54, 56
Moon: composition, 153; igneous rocks, 152; origin, 154–155; volcanism, 154
Moraine ridge, 357, 359
Morphology, 166
Mountain(s) [see also Volcano(es)]: complex, 503; fault-block, 501–502; fold, 502–503; and geologic structures, 499–503; kinds, 494–495; and orogenic belts, 492–496; and plate tectonics, 486; unsolved problems about, 515–516
Movement [see also Downslope movement]: epeirogenic, 506; glacier, 339–343; groundwater, 311–312; sea water, 412–421
Mudflow, 244–245, 248, 382
Muir, John, 352, 512
Multiples and prefixes, 559
Mylonite, 220

National Environmental Policy Act (1969), 267
Natural levee, flood-plain river, 291
Nature, man and, 40–41
Neap tide, 418
Nearshore water circulation, 426
Near-surface fracture, 171
Negative anomalies, sea floor, 476
Neutron, 53
New Madrid earthquake, 445–446
Newton, Isaac, 63
Nonclastic rock, 555

Nonclastic texture, 206
Nonconfined aquifers, 310
Noncrystalline solids, 74
Nonferromagnesian silicates, 74
Nonfoliates, 219–220, 556
Nuclear energy, 61
Nuclear fission, 530
Nuée ardente, 88, 107–108

Oasis, 384
Obsidian, 143
Ocean(s) [see also Sea; Sea floor]: and biosphere, 408–412; circulation, and climate, 364–365; distribution and characteristics, 408–411; ownership of, 435; as possible source of future resources, 539
Ocean-dumping Act, 435
Oceanography, 15–17
Oil pool, 527
Old age, erosion, 290
Old Faithful, 118
Oöids, 199
Orbicular granite, 142
Ore: concept of, 531–532; metallic, formation of, 533–535; secondarily enriched, 535; sedimentary, 534–535
Organic acid, and soil, 186
Organic material, 199–200
Organic matter, change to petroleum to graphite, 222
Organic textures, 206
Organisms: and organic material, and weathering, 167, 169, 177
Orientation, and expression of direction, 561–562
The Origin of the Continents and Oceans (Wegener), 471
Orogenic belt: anatomy of, 496–499; ancient, 516–517; and continents, 516–518; dynamic history, 503–506; and geologic cycle, 517–518; geosynclinal stage, 503–504; and mountains, 492–496; and plate tectonics, 511–515; rocks of, 496–497; terminal-orogeny stage, 504–506
Orogeny, 492; and lithosphere, 510; mechanisms of, 510–516; terminal, 504–506
Oscillation ripples, 423
Oscillation wave, 426
Outwash, 355–356
Ox-bow lake, 294
Oxides, 75, 77
Oxydation, 174
Ozone layer, and "Freon," 44–45

Page, Robert A., 462
Pahoehoe lava, 87
Paleomagnetism, and continental drift, 472–474
Pan, 403
Pangaea, 471; break-up of, 484–485
Parabolic dune, 401
Parent rock, 215
Peat, change to coal to graphite, 222
Pedalfers, 187–189
Pediment, 386–387
Permeability, factors affecting, 309–310
Pedocals, 189
Pegmatite, 138
Pelée, Mt., 105–108
Peléean eruption, 108
Pele's hair, 89
Peneplain, 288
Peridotites, 145, 148
Permafrost, 189
Petrified, defined, 324
Petroleum, 222
Photic zone, 411
Photosynthesis, 32
Phreatic eruption, 89
Piedmont glacier, 336
Pillowed rock, 113
Planetary geology, 11
Plant products: as energy source, 523; as renewable resource, 531
Plastic flow, glacial, 341
Plate boundaries, 479
Plate tectonics, 5, 17–20, 101; and continental drift, 484–486; dissenting opinions on, 485–486; and orogenic belts, 511–515; predictions and tests, 479–484; in relation to crustal deformation and mountains, 486
Playa, 387–388
Playa lake, 387–388
Playfair, John, 279
Playfair's law, 279
Plinian eruption, 102
Pliny the Elder, 102
Pliny the Younger, 102
Plucking, and glaciation, 344
Plug dome, 91
Pluton: geologic dating, 137–138; lens-shaped, 134; massive, 135–136; tabular, 130–133
Plutonic metamorphism, 226
Pluvial lake, 390
Point bar, flood-plain river, 293
Polar reversals, 476
Polar wandering, 473
Pollution: air, and weathering, 180–182; of Great Lakes, 369; of groundwater, 316–318; jet-age, 43; of oceans, 435; of rivers, 266–267; water, 42–43
Population, and resources, 45–47
Pore, underground, 308–310
Pore pressure, 236
Pore spaces in bedrock and regolith, and weathering, 167
Porosity, 236, 309–309; factors affecting, 309–310
Porphyrins, 526
Porphyritic granite, 142
Porphyritic texture, 139
Porphyry, 139
Positive anomalies, sea floor, 476
Potential energy, 60, 234
Potholes, 275
Powell, John Wesley, 273, 284
Powers of ten, 559
Precipitation, 30; and evaporation, desert, 375–376
Prefixes and multiples, 559
Pre-geologic time, 545
Pressure: heat, generation of magma, and, 85–86; pore, 236; in subsurface environments, 221
Prince William Sound earthquake, 446–447
Process, uniformity of, 5
Proglacial lake, 356
Proton, 53
Pulling force, 234; increase in, 237–238
Pumice, 89
Pyroxenites, 145

Quarrying, and glaciation, 344
Quaternary climatic oscillations, 365–366

Radial pattern, running water, 286
Radiant energy, 60, 62
Radical, 77
Radioactivity, 54; and time, 15
Radiolaria, 199
Raindrops, and valley-side slopes, 282–283
Rainfall: desert, 381; and erosion, 288
Rainshadow desert, 379–380
Rainwater, chemical composition of, 163
Reaction series, continuous vs. discontinuous, 151
Recessional moraine, 359

Recrystallization, and replacement, 201
Reef, 428–429
Refraction, wave, 426
Regional metamorphism, 225, 497
Regolith, 8; pore spaces, 167; residual and transported, 183; and soil, 183–186
Relativity, Theory of, 63
Relief, 166
Remanent magnetism, 473
Replacement, 324; and recrystallization, 201
Reserves, and resources, 532
Residual ore deposits, 534
Residual regolith, 183
Resisting force, 234; reduction of, 235–237
Resources: future, 539; and geology, 538–539; metallic, 532–533; nonmetallic, 535–536; nonrenewable, 531–536; and population, 45–47; renewable, 530–531; and reserves, 532; use curves, and human activities, 536–539
Rhyolite, 142–143
Richter, Charles F., 443
Richter Magnitude Scale, 443
Rift zone, 96
Rills, and valley-side slopes, 283
Ring of Fire, 99
Rip current, 426
River(s) [see also Running water; Stream(s)]: and civilization, 266–270; environmental impact on, 268–270; flood-plain, 291–294
Roches moutonnées, 351
Rock(s) [see also Bedrock; Igneous rock(s); Metamorphic rock(s); Sedimentary rock(s)]: basement, 493, 499; broken-glass, 116; carbonate, 213; clastic, 554; felsic, 140, 141–143; glassy, 138; identification of, 552–556; intermediate, 143–144; mafic, 144; natural sandblasting, 401–403; of orogenic belts, 496–497; parent, 215; pillowed, 113; rotten, 161; thermal effects, frost wedging, heating and cooling in desert, 172–173; ultramafic, 141, 145, 148, 151–152; weathering of, 177–179
Rock avalanche, 242–243
Rock cycle, 10–11, 29; groundwater in, 324–326
Rock-dilatancy theory, 456
Rock fall, 239–240
Rock flour, 344

Rock glacier, 254
Rock slide, 240–241
Rotten rock, 161
Running water [see also River(s); Stream(s); Water]: desert, 382; drainage patterns, 286; erosion by, 279–290; geologic work of, 275–279
Runoff, 263

Salinity, 409; of subsurface environments, 221
Salt, crystallization, in weathering, 175–177
Saltation, 277, 393
Sand, slumping in, 244
Sandblasting, natural, 401–403
Sand dune. See Desert; Dune
Sandstone, 212; weathering 178–179
Sandstorm, 394
San Francisco earthquake, 454–455
Satellites, in forecasting igneous activity, 121
Saturation zone, 307
Schist, foliation of, 218–219
Scholl, David W., 485
Scholtz, Christopher, 456
Scoria, 89
Sea, 408 [see also Ocean(s); Sea floor; Sea-floor spreading]: circulation, and density currents, 414–415; farming of, 411–412; political and economic aspects, 435–436
Sea floor, 5; and continental drift, 473–474; crust and sediment, 477; magnetic characteristics, 476
Sea-floor spreading, 17, 20, 474–479; and magnetic tape recorder, 476
Sea-level changes, and deep-sea sediment, 432–433
Sea water: movements, 412–421; properties of, 409–410
Sea wave, 417
Sediment, 9 [see also Sedimentary rock(s)]: cementation of, 206–213, 324; coastal, 430; continental margin, 431; and crust, sea floor, 477; deep-sea, 431–433; deposited by wind, 395–397; glacial, 351, 354–357; glaciers in grinding bedrock into, 334–335; marine, 429–433; shelf, 430; and time, 12, 14; transported by wind, 393–394; valley-fill, 299–300; and wind erosion, 391–392
Sedimentary breccia, 212
Sedimentary ore, 534–535

Sedimentary particles, 196–200
Sedimentary rock(s), 9 [see also Bedrock; Metamorphic rock(s); Rock(s); Sediment]: classification of, 204–214; classification basis, 204–206; clastic, 554; detrital, 212; formed by cementation of sediments, 206–213; nonclastic, 555; not formed by cementation of sediments, 213–214
Sedimentary strata, 201–204
Sediment load, running water, 276
Sediment yield, 288
Seepage pressure, 237
Seif dune, 400
Seismic activity, and volcanoes, 119–120
Seismic waves [see also Wave(s)]: changes in speed, 456; and earthquakes, 441–443; scientific use of, 450–452; and seismology, 449–452; types, 449–450
Seismograph, 119, 443
Seismogram, 443
Seismology, 17, 443; and seismic waves, 449–452
Semiconductors, 73
Semi-daily tide, 419
Serpentinite, 148
Settling rate, sediment, 276
Shadow zone, 469–470
Shale, 212
Shallow-water wave, 423
Shearing, 235
Shear wave, 450
Sheet flows and valley-side slopes, 283
Sheeting joints, 172
Sheetwash, 382
Shelf sediments, 430
Shield: continental, 516; and metallic ore deposits, 535; volcanic, 95–96
Shoreline, 421; tropical, 428
Shore zone, 421
Sigurgeirsson, Thorbjorn, 123
Silicates, 74–75
Sill, 131–133
Sinkholes, 322
Skeletal particles, 198
Slate, foliation, 218
Slaty cleavage, 218
Sliderock, 240
Sliding, glacial, 341
Slope, 232; continental, 431; gravity components on, 234–235; questions about, 234; of stream, 273

Slope failure, prediction and prevention of, 256–257
Slope-normal component, 234
Slope-parallel component, 234
Slope situation, and weathering, 166–167
Slump, 244
Slumping in sand, 244
Snowfall, and glacial ice, 332–333
Snowfield, 333
Snowline, 333
Soil: ancient, 190; defined, 182–183; desert, 385; latosol, 187; management of, 189–190; modern, 186–189; pedalfer, 187–189; pedocal, 189; realms of, 184–185; and regolith, 183–186; and weathering, 160
Soil profile, 183, 186
Soil zone, 183, 186
Solar energy, 522
Solar system, origin of, 546–547
Solar wind, 546
Solids, noncrystalline, 74
Solifluction, 252–253
Sorting, 210
Spatter cone, 91–92
Specific gravity: of minerals, 73; and sedimentary particles, 210
Speed, stream, 272
Speleothem, 318, 319
Spheroidal joints, 177
Spontaneous liquefaction, 236–237
Spring, 313–314
Spring tide, 418
Stalactites, 318–319
Stalagmites, 318–319
Steppes, 375
Stocks, 135
Strata, 14–15; gravity sliding of, 511; origin, 202–203; sedimentary, 201–204; surface features, 203–204
Stratification, 202
Stratified drift, 355–356
Streak, of minerals, 71
Stream(s), 263–264 [see also River(s); Running water; Water]; base level, 273; braided, 296–298; channels, 273; delta, 294–295; discharge from, 270–271; energy relationships, 270; flows, 263; graded, 274; laminar flow, 274; load, 273–274; profile, 273; slope, 273; speed, 272; terrace, 299; turbulent flow, 274–275
Striae, 549; and glaciation, 345–346
Stromatolites, 411
Strombolian eruption, 108
Strong interaction force, 57

Structural geology, 497
Sub-atomic particle, 52
Subduction, and trenches, 480–481
Subduction zone, 480
Subglacial eruption, 116
Subsidence, land, 315
Subsurface environments: conditions of, 220–221; changes in organic materials, 222; changes in rock-forming silicate minerals, 223
Subtropical desert, 377–378
Suess, Eduard, 470–471
Sulfate minerals, 78
Sulfides, 77
Summit eruption, 98
Surf, 426
Surface currents, ocean, 414
Surface water, movement of, 412–414
Surf zone, 426
Surge, glacial, 340–341
Surtsey, 113, 114–115
Survival products, 197, 205
Suspended load, running water, 276
Swell, 417
Sykes, Lynn, 456
Syncline, 498

Tabular pluton, 130–133
Talus, 240
Talus creep, 253–254
Taste, of minerals, 73
Tectonics, 17 [see also Plate tectonics]
Temperature: desert, 376–377, 381; sea water, 410; subsurface environments, 221
Tephra, 89
Terminal moraine, 357, 359
Terminal orogeny, 504–506
Terrace, stream, 299
Tetrahedron, 74
Texture: crystalline, 206, 217–220; igneous rock, 138–140; metamorphic rock, 216; porphyritic, 139; sedimentary rock, 205–206
Thoreau, Henry David, 64
Thrust, 498
Tidal amplitude, 418
Tidal current, 421
Tide, 418–421; astronomical aspects of, 419
Tide-dominated coast, 421
Till, 355
Tillite, 363
Tiltmeter, 121

Time, 4; geologic, 11–12; pre-geologic, 545; and radioactivity, 15; and sediments, 12, 14; and soil, 186
Time scale, geologic, 12, 13
Topographic maps, 560–563
Topographic profile, constructing, 562–563
Topography, 166
Topset bed, delta, 295
Topset plain, delta, 295
Transcurrent boundary, 480
Transform fault, 478
Translation, wave of, 426
Transpiration, 262–263
Transported regolith, 183
Transverse dunes, 401
Trellis-rectangular pattern, running water, 286
Trenches: and arcs, island, 16; and subduction, 480–481
Tsunami, 105, 417–418
Tufa, 323
Tuff, 213; welded, 89
Tundra soil, 189
Turbidite, 432
Turbidity current, 414–415
Turbulent flow, stream, 274–275

Ultramafic rock, 141, 145, 148, 553; origins, 151–152
Underground activity, 117–119
Underwater volcanoes: and island formation, 112–113; vs. subaerial volcanoes, 112; underwater eruptions and lava, 113, 116
Uniformitarianism, doctrine of, 5
Unit cells, 67–68
Uranium, 528, 530

Vadose water, 307
Valley: glaciated, 348–349; hanging, 348; glacier, 336
Valley-fill sediment, 299–300
Valley-side slopes, 281–282; rills, gullies, and badlands, 283; and sheet flows, 283
Varve, 356–357
Vegetation, desert, 377
Ventifact, 402
Very-shallow-water wave, 426
Vesicle, 89
Vine, Frederick J., 476
Vine-Matthews hypothesis, 476
Volcanic block, 89
Volcanic bomb, 89

Volcanic crater, 92–95
Volcanic dome, 91
Volcanic dust, and climate, 364
Volcanic islands, 112–113; and mantle hot spots, 481, 484
Volcanic material, sedimentary rock composed of, 213
Volcanic particles, 200
Volcanic shield, 95–96
Volcanism: in continental United States, 108–109; on moon, 154
Volcano(es): classic eruptions, 102–109; distribution, 99–101; Hawaiian eruption, 108; Krakatoan eruption, 102; landforms, 90–97; liquid products of, 86–87; Peléean eruption, 108; reducing damage from, 122–123; and seismic activity, 119–120; solid-gaseous mixtures from, 87–89; solids from, 89–90; Strombolian eruption, 108; underwater, 109, 112–116
Von Helmholtz, Hermann, 546
Von Laue, Max, 68

Wadi, 382
Warm-arid regions, and weathering, 164–165

Water [see also Groundwater; Ocean(s); Running water; Sea water]: and biosphere, 264–270; chemical composition of rain, 163; consumption of, 264–265; desert lake, 389; distribution problems, 265–266; and fluid pressure, 235–237; molecular structure, 162–163; nearshore circulation, 426; as renewable resource, 530–531; from rivers, 266; states of, 262; and weathering, 162–163
Water spreading, 315
Water table, 307–308; and land subsidence, 315
Wave(s), 415–418 [see also Seismic waves]; compressional, 449–450; height, 416; of oscillation, 426; period, 416; refraction, 426; shear, 450; of translation, 426
Wave-dominated coast, 422–428
Wavelength, 416
Weathering, 9; and air pollution, 180–182; chemical, 173–175; and climate, 163–165; combined physical and chemical, 177; of common rocks, 177–179; complex, 175–177; desert, 384–385; differential, 180; environment of, 161–179; and erosion, 160; and geologic cycle, 160–161; of granite, 177–178; and organisms and organic materials, 167, 169; physical, 171–173; and pore spaces in bedrock and regolith, 165; rates, 179–180; and slope situation, 166–167; visible evidence of, 169–170; and water, 162–163
Weathering potential index (WPI), 179
Wegener, Alfred, and continental drift, 471–472
Welded tuff, 89
Well, 314
Wind: desert, 384; and erosion of loose sediment particles, 391–392; sediments deposited by, 395–397; sediment transport by, 393–394; solar, 546
Wind belts, and climate zones, 412–413
Wind-generated waves, 417
Wisconsin glacier, 368
World Wide Standardized Seismograph Network (WWSSN), 452
Wright, Frank Lloyd, 457

Yardang, 402–403
Yellowstone Park, 110–111, 117
Youthful stage of erosion, 289

Chapter-opening photo credits *Introduction*: Surtsey, December 1, 1963 (Sigurgeir Jonasson, Vestmannaeyjar, Iceland). *Chapter 1*: Lake Taisho-Ike, Japan (Japan National Tourist Organization). *Chapter 2*: Electron micrograph of kaolinite clay from Georgia (Kenneth M. Towe, Smithsonian Institution). *Chapter 3*: Mt. Fujiyama, Japan (Japan National Tourist Authority). *Chapter 4*: Giant's Causeway, North Ireland (British Tourist Authority). *Chapter 5*: Petra, Jordan (United Nations). *Chapter 6*: Canyonlands, Utah (American Airlines). *Chapter 7*: Titus Canyon, Grapevine Mountains, Death Valley, California (National Park Service). *Chapter 8*: Waterfall, Jasper National Park, Alberta, Canada (Information Canada Photothèque). *Chapter 9*: Glow worm threads, Waitomo Caves, Aukland Province, New Zealand (Consulate General of New Zealand, New York). *Chapter 10*: Merger of the Gorner and Grenz Glaciers in the Monte Rosa area, Switzerland (Swiss National Tourist Office). *Chapter 11*: Sand dunes near Stovepipe Wells, Cottonwood Mountains, Death Valley (George Grant, National Park Service). *Chapter 12*: Squall, Jones Beach, New York (Rhoda Galyn). *Chapter 13*: Openings in the Earth caused by earthquakes, Andes Mountains, near Cuzco, Peru. (Carl Frank, Photo Researchers, Inc.). *Chapter 14*: The Afar Triangle (NASA). *Chapter 15*: Aerial view of Rainbow Gardens on the flank of Frenchman Mountain, Lake Mead, Nevada, showing spectacular faulting, center, right (Bill Belknap, Photo Researchers, Inc.). *Chapter 16*: Canadian coal miners (National Film Board of Canada, D60384). **Field Trip photo credits** *Chapter 1* (United Nations). *Chapter 2* (Baxter State Park Authority, Maine). *Chapter 3* (U.S. Geological Survey). *Chapter 4* (Japan National Tourist Organization). *Chapter 5* (U.S. Forest Service). *Chapter 6* (Engraving after sketch by Karl Bodmer). *Chapter 7* (American Airlines). *Chapter 8* (American Airlines). *Chapter 9* (American Airlines). *Chapter 10* (Canadian Consulate General). *Chapter 11* (Danish Information Office). *Chapter 12* (United Nations). *Chapter 13* (UPI). *Chapter 15* (Rhoda Galyn).